D1247312

CENOZOIC TECTONICS AND REGIONAL GEOPHYSICS

OF THE WESTERN CORDILLERA

The Geological Society of America, Inc.
Memoir 152

Cenozoic Tectonics and Regional Geophysics of the Western Cordillera

Edited by
ROBERT B. SMITH
AND
GORDON P. EATON

1978

4810476

Copyright © 1978 by The Geological Society of America, Inc.
Copyright is not claimed on any material prepared by
U.S. Government employees within the scope of their employment.
Library of Congress Catalog Card Number 78-55296
ISBN 0-8137-1152-5

Reprinted 1980

Published by
THE GEOLOGICAL SOCIETY OF AMERICA, INC.
3300 Penrose Place
Boulder, Colorado 80301

Printed in the United States of America

Contents

v

This publication consists of hardbound text and eight maps and one table folded separately,
all in slipcase.

Preface

In September 1975, the Geological Society of America sponsored a Penrose Conference at Alta, Utah—its subject, "The Cenozoic Tectonics and Regional Geophysics of the Intermountain West." Its basic purpose was to bring together geologists and geophysicists working on tectonic problems in the Intermountain region to discuss available geophysical and geological data in the light of the general theory of plate tectonics. Emphasis was placed on establishing interpretive constraints. Conference goals included the evaluation of various theoretical mechanisms to explain geological and geophysical observations in the region and the postulation of resource exploration models. The conference sought a consensus, but ended without really achieving one. A brief report on the highlights of the conference was published in the Society's journal, *Geology* (v. 4, no. 7, p. 437–438).

Two months later the Geophysics Division of the Geological Society of America convened a special symposium at the annual meeting of the Society in Salt Lake City (Geol. Soc. America Abs. with Programs, v. 7, no. 7). The title of that symposium is the one carried by this volume and reflects a broader geographic scope than the Alta meeting. Having received a number of expressions of desire to have the symposium papers published in a single volume, the Geophysics Division laid the matter before the authors. A decision was made to proceed with publication of this volume. Of the 16 papers included in the symposium program, 13 are published here. One other has already been published in journal form. A direct comparison of titles and authors of the other papers is obscured by the merging of two of the Salt Lake City papers and the splitting of another. There have been changes in the authorship and titles of still others. Nevertheless, the coverage of the present volume is basically that of the Salt Lake City symposium, constructed on the foundation of the Alta Penrose Conference.

Because many of the authors of papers in this volume were exposed to the data, interpretations, and thinking of the others, either at the Salt Lake City or Alta meeting, or even both, the reader may feel reasonably justified in anticipating a certain unanimity of thought and viewpoint. Such is not the case. The jury is still out on many points of fundamental importance. There are, for example, almost as many opinions about the origin of Basin and Range faulting here as there are elsewhere in the published literature. Significant questions such as whether or not the normal faulting, high heat flow, thin lithosphere, and low P_n-velocities in the Great Basin are products of an *externally driven* rapid extension or whether they are, instead, accompaniments of a broad arching of the local continental lithosphere, the formation of a triple junction, or back-arc spreading are not convincingly answered in these pages. Many new factual data are offered, but as the reader will see, the authors disagree as to their ultimate interpretation. This is as it should be. No one of the data sets presented, either alone or in combination with others, is fully compatible with a single conceptual model. At best they frame interpretive constraints. Ultimately the reader must choose for himself. In doing so he will probably be no less prejudiced than were the authors of this volume themselves.

Several new maps are published here for the first time. These consist of age distributions of igneous rocks, regional fault maps, seismicity, heat flow, gravity, magnetics, fault-plane solutions, P_n-velocities, and crustal structure of major portions of the western Cordilleras—the Great Basin and high volcanic plateaus to the northwest, the Snake River Plain, the Pacific Coast States, and the continental margin off the Western United States. Part of the value of this volume surely rests here. Quite apart from their interpretation, the maps and tabular data are useful and valuable additions to our understanding of a broad and complex region, one rich in natural resources. The new data and new compilations should be of aid in planning resource exploration programs, as well as in the development of an understanding of resource occurrence.

The reader is warned to examine the context in which two geographic terms are used by different authors throughout the volume; they are "Cordillera" and "Basin and Range province." Some authors use the first term to define the entire mountain system of the Western United States, the full width of which stretches from Denver, Colorado, to Eureka, California. They include the Colorado Plateaus. Others use the term "western Cordilleras" as we have above, to exclude the Colorado Plateaus, the central and southern Rocky Mountains, and much of the Laramide fold and thrust belt.

Individual use of the term "Basin and Range province" poses a greater problem because several authors have used it interchangeably with the term "Great Basin." As noted in more than one paper, however, some of the attributes of Great Basin lithosphere are sufficiently different from those of the rest of the Basin and Range province that a clear distinction should probably be made. Fortunately, the geographic focus of most papers lies north of lat 37°N, the boundary between Utah and Arizona. Use of the term "Basin and Range" for this region necessarily implies interchangeability with the term "Great Basin." Despite the explicitness of the volume's title, it is worth emphasizing that the chronologic focus is the Cenozoic Era. Keeping this firmly in mind will aid the reader in understanding the geologic scope and significance of proposed conceptual models.

To end on an upbeat note, we found the process of assembling these papers both intellectually rewarding and pleasant. Even though the consensus sought by the Penrose Conference at which it all began continues to escape the collective grasp, we hope the readers of this volume will have as much fun wading through it as we did. To the critic, fire away. There are many broad-sided barns here and a wider variety of philosophic stripe than one can imagine. Read diligently and you may even find a point-of-view with which you can agree—at least until next year.

ROBERT B. SMITH
University of Utah

GORDON P. EATON
U.S. Geological Survey

Geological Society of America
Memoir 152

1

Basin-range structure in western North America: A review

JOHN H. STEWART
U.S. Geological Survey
345 Middlefield Road
Menlo Park, California 94025

ABSTRACT

For more than 1,500 km along the western Cordillera of North America, late Cenozoic extensional faulting has produced block-faulted basin-range structure characterized by alternating elongate mountain ranges and alluviated basins. The faulting follows older geologic patterns, particularly those of Mesozoic and early Tertiary deformation and of early and middle Tertiary igneous activity. Basin-range structure is commonly inferred to represent either (1) blocks tilted along downward-flattening (listric) faults in which the upslope part of an individual rotated block forms a mountain and the downslope part a valley or (2) alternating downdropped blocks (grabens) that form valleys and relatively upthrown blocks (horsts) that form mountains. Such structure has been produced by extension estimated to be from 10% to 35% of the original width of the province and as much as 100% in specific areas. The province is characterized by anomalous upper mantle, thin crust, high heat flow, and regional uplift.

Current theories on the origin of basin-range structure can be grouped loosely into four main categories. In the first, the structure is presumed to be related to oblique tensional fragmentation within a broad belt of right-lateral movement and distributed extension along the west side of the North American lithospheric plate. This motion was initiated by the collision of the East Pacific Rise with the North American plate, which brought together the North American and Pacific plates to form the right-lateral San Andreas transform fault system. The second theory relates extension to spreading caused by upwelling from the mantle behind an active subduction zone (back-arc spreading). The third theory relates the basin-range structure to spreading that resulted from presumed subduction of the East Pacific Rise beneath part of North America. The fourth theory relates the basin-range structure to plate motion caused by deep-mantle convection in the form of narrow mantle plumes. The combination of anomalous upper mantle, thin crust, high heat flow, regional uplift, and extension with a previous history of high heat generation can best be related to back-arc spreading. The spreading may have been accelerated by slackening of confining pressure after the destruction of the subduction system along western North America and may have been accompanied by right-lateral shear because of the development of the transform western margin of North America.

INTRODUCTION

Basin-range structure consists of block-faulted mountain ranges and intervening alluviated valleys. It is one of the most distinctive geologic features of western North America and is more extensively developed there than in any other part of the world. Knowledge of its origin is critical in understanding the structural development of western North America, as well as of other regions where similar structures occur.

In this report, I will outline the geologic setting and characteristics of basin-range structure and current theories of origin. The subject involves many facets of geology that I can only summarize here. A review of the historical development of ideas concerning basin-range structure has been given by Nolan (1943) and Roberts (1968).

DISTRIBUTION OF BASIN-RANGE STRUCTURE

High-angle extension faulting extends throughout much of the western Cordillera of North America from Canada to northern Mexico (Fig. 1-1). Basin-range structure consisting of alternating mountains and valleys is best developed in the Basin and Range physiographic province (Fig. 1-2) that extends from southern Oregon and Idaho through most of Nevada and parts of California, Utah, Arizona, and New Mexico to northern Mexico—a total distance of more than 1,500 km. In the United States, the province is about 500 to 800 km across in the Great Basin region of Nevada and western Utah. The elevations of the valleys in the Great Basin are generally 1,300 to 1,600 m; mountain crests are commonly 2,000 to 3,000 m and locally about 3,600 m. Elevations of valleys and mountains are commonly 500 to 1,000 m lower in most of southeastern California, southern Arizona, and southwestern New Mexico. In the United States, basin-range structure extends around much of the Colorado Plateau province, a tectonically stable region underlain by relatively undeformed Paleozoic, Mesozoic, and Cenozoic sedimentary rocks, with a general surface elevation of 1,200 to 1,800 m. Extensional faulting along the east margin of the Colorado Plateau and southward to near the international boundary forms the Rio Grande rift valley (Chapin and Seager, 1975). In Mexico, basin-range structure extends southward along either side of the Sierra Madre Occidental, a high plateau (1,800 to 3,000 m) of relatively undeformed Cenozoic volcanic rocks. Elevations west of the Sierra Madre Occidental range from sea level along the coast to 1,000 to 2,000 m in some of the higher mountains inland. East of the Sierra Madre Occidental, elevations are generally higher; they range from about 1,000 to 1,500 m in valleys to 2,000 to 3,000 m in mountain ranges. Basin-range structure also occurs in Mexico along the east side of Baja California, north of lat 28°N. Some extensional faulting and possible basin-range structure also occur in the southern part of the Sierra Madre Occidental and in the so-called Central Mesa (really a physiographic basin) to the west.

GEOLOGIC SETTING

Western North America has had a complex history traceable into Precambrian time. The relation, if any, of older structures to the development of late Cenozoic basin-range structure is of critical importance. For example, is the distribution of basin-range structure largely determined by structures developed in Precambrian, Paleozoic, or Mesozoic time, or is the

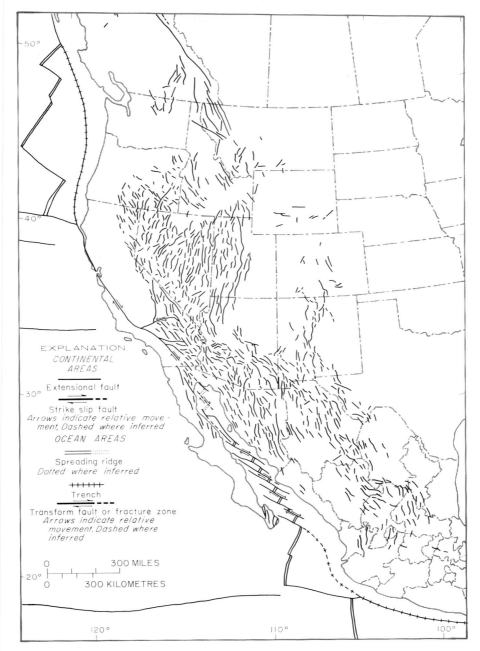

Figure 1-1. Distribution of late Cenozoic extensional faults and a few major strike-slip faults in western North America and present-day lithospheric plate boundaries. Faults are generalized and, in part, inferred. Based on various sources, including King (1969b).

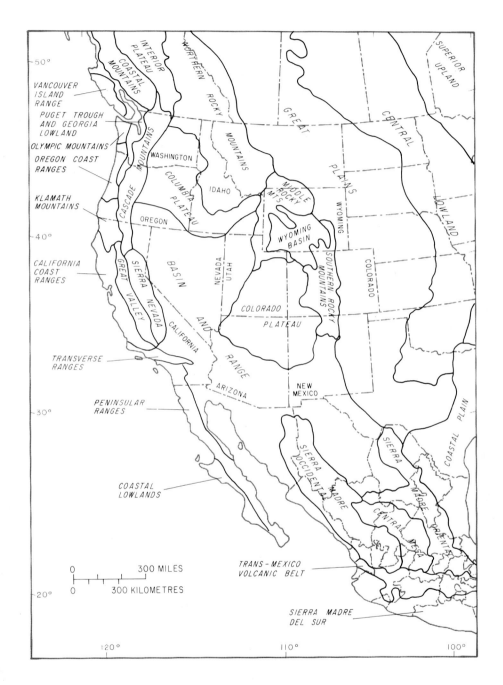

Figure 1-2. Physiographic provinces in western North America. Based on Canada Geological Survey (1970a), Fenneman (1946), Raisz (1959), Guzman and de Cserna (1963), and Gastil and others (1975).

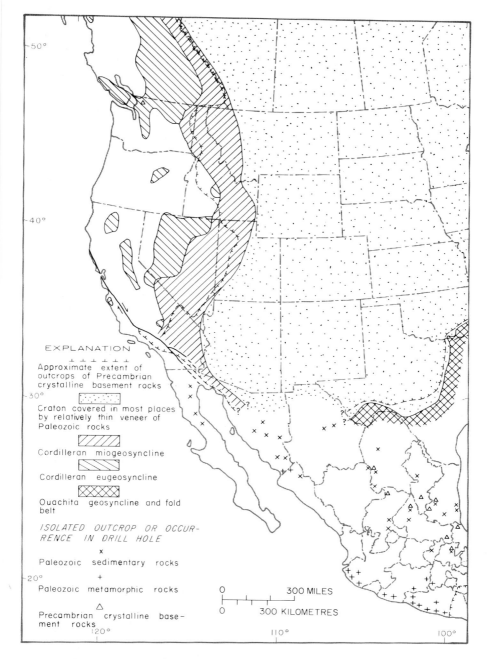

Figure 1-3. Pre-Mesozoic tectonic setting of western North America. Based on many sources, including Canada Geological Survey (1970b), Comite de la Carta Geologica de Mexico (1968), López-Ramos (1969), King and Beikman (1974), Gastil and others (1975), and Nicholas and Rozendal (1975).

distribution largely controlled by geologically short-lived and exceptional or unique Cenozoic events?

Precambrian crystalline rocks underlie the interior of the Western United States and Canada (King, 1969a) but do not crop out, and are presumed to be absent at depth, in the westernmost parts of these countries (Fig. 1-3). The original distribution of Precambrian basement rocks in Mexico is uncertain owing to the sparsity of outcrops. In most of the Western United States, Precambrian crystalline metamorphic rocks were formed in the Hudsonian orogeny (1,800 m.y. ago) although older metamorphic rocks occur locally. Along the south margin of the United States, within the region considered here, metamorphic and intrusive rocks involved in the Elsonian (1,300 to 1,500 m.y. B.P.) and Grenville (about 1,000 m.y. B.P.) events are also recognized.

In the Western United States and Canada, relatively unmetamorphosed Precambrian Y (800 to 1,600 m.y. B.P.) sedimentary and volcanic rocks occur in deep epicratonic troughs or as shelf deposits on Precambrian crystalline basement (Stewart, 1976). These rocks include such units as the Belt and Purcell Supergroups and the Uinta Mountains Group.

By latest Precambrian time, the Cordilleran geosyncline (Fig. 1-3) was well defined (Stewart and Poole, 1974). This feature dominated the tectonic setting of the Western United States and Canada during late Precambrian and Paleozoic time (Burchfiel and Davis, 1972, 1975). The geosyncline contains deposits formed along the western continental margin of North America and, farther west, in back-arc basins and on island arcs. During middle and late Paleozoic time the geosyncline was involved in orogenic events that resulted in the emplacement of eugeosynclinal rocks on the continent. The Cordilleran geosyncline can be traced into northern Mexico, but its trend from there is unknown. It may curve eastward and join the Ouachita geosyncline of Texas and Oklahoma, or extend farther southward in Mexico.

The possibility that the Paleozoic Cordilleran and Ouachita geosynclines join across northern Mexico leads to interesting speculations about the geologic evolution of Mexico. If the trace of such a geosynclinal belt was the edge of a continent during Paleozoic time, then Mexico must have been welded to North America after the Paleozoic Era. Morris (1974) has suggested such a history and indicated that Mexico was connected to the rest of North America during the Triassic Period. The presence of serpentinite near Ciudad Victoria in northeastern Mexico (Carrillo-Bravo, 1961) may indicate one of perhaps several suture lines related to such a join. The change in position and character of Mesozoic and earliest Tertiary deformational belts near the United States–Mexico border (Fig. 1-4) may indicate that Mesozoic and Tertiary deformation followed the trend of older structures related to the Paleozoic continental margin.

During the Mesozoic Era, the tectonics of the western margin of North America was dominated by plate subduction along an Andean-type continental margin (Burchfiel and Davis, 1975). Large plutonic masses were emplaced (Fig. 1-4). A complex history of sedimentation, deformation, and igneous activity is evident along the western coastal margin of North America during this time. Inland, deformation took place by folding and thrusting in a well-defined belt that extended from Canada to Mexico and beyond (Fig. 1-4). This deformation occurred primarily during the Sevier (latest Jurassic to latest Cretaceous) and Laramide orogenies (latest Cretaceous to middle or late Eocene). The Laramide orogeny included thrusting and folding along the same trend as the earlier Sevier orogeny as well as an eastern belt of basement uplifts.

During early and middle Cenozoic time, the tectonics of western North America was also dominated by events related to a subduction system along the margin of the continent (Lipman and others, 1972; Snyder and others, 1976). This time interval is notable for the widespread eruption of silicic volcanic rocks (Fig. 1-5) in the Great Basin region of Nevada and Utah,

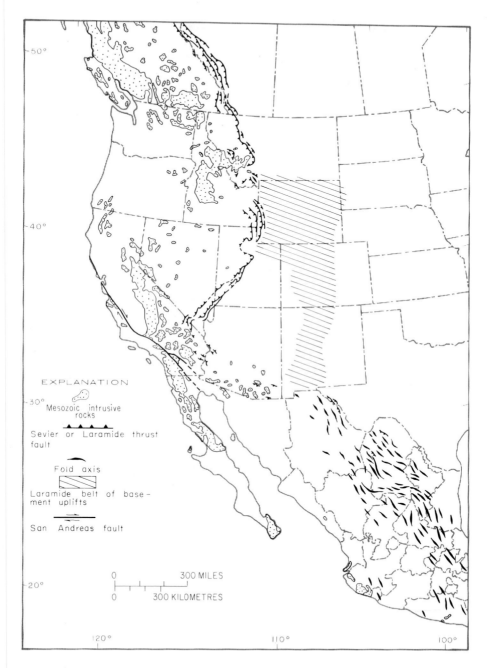

Figure 1-4. Mesozoic intrusive rocks and late Mesozoic and earliest Cenozoic deformational belts in western North America. Compiled from Canada Geological Survey (1970b), King (1969b), King and Beikman (1974), Comite de la Carta Geologica de Mexico (1968), Burchfiel and Davis (1975), and Guzman and de Cserna (1963).

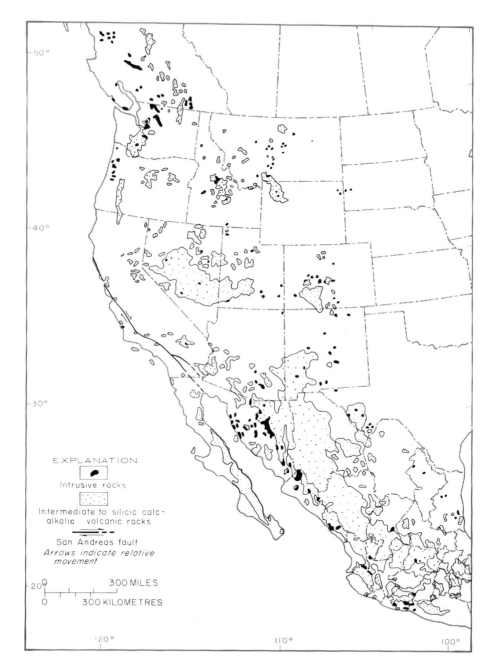

Figure 1-5. Early and middle Cenozoic (65 to 17 m.y.) igneous rocks in western North America. Compiled from Canada Geological Survey (1970b), King and Beikman (1974), Stewart and Carlson (this volume), Comite de la Carta Geologica de Mexico (1968), and Gastil and others (1975).

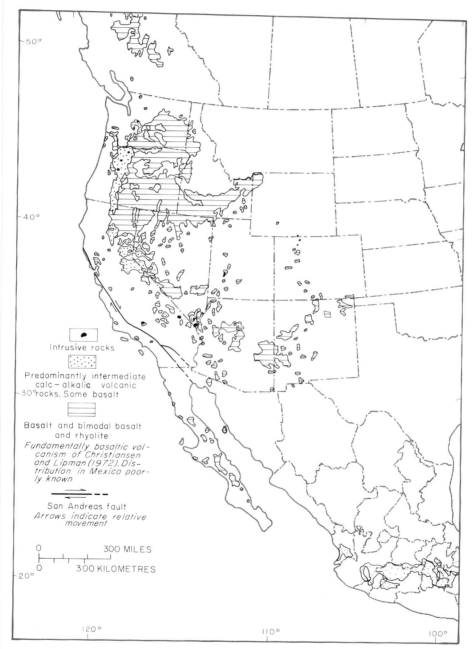

Intrusive rocks

Predominantly intermediate
calc-alkalic volcanic
rocks. Some basalt

Basalt and bimodal basalt
and rhyolite
*Fundamentally basaltic vol-
canism of Christiansen
and Lipman (1972). Dis-
tribution in Mexico poor-
ly known*

San Andreas fault
*Arrows indicate relative
movement*

0   300 MILES
0   300 KILOMETRES

Figure 1-6. Late Cenozoic (17 m.y. to present) igneous rocks in western North America. Distribution of rocks of this age is poorly known in Mexico, and scattered basaltic rocks of this age are known, but are not shown, in Mexico east of the Gulf of California. Compiled from Canada Geological Survey (1970b), King and Beikman (1974), Stewart and Carlson (this volume), Comite de la Carta Geologica de Mexico (1968), and Gastil and others (1975).

in Colorado, southern Arizona, southwestern New Mexico, and in Mexico. Elsewhere in western North America, intermediate calc-alkalic volcanic rocks are dominant.

Early and middle Cenozoic igneous activity generally moved southward across the Western United States (Lipman and others, 1972; Christiansen and Lipman, 1972; Armstrong and Higgins, 1973; Stewart and Carlson, 1976; Stewart and others, 1977), from Washington, about 45 to 57 m.y. ago, to central Idaho and northwestern Wyoming, about 38 to 49 m.y. ago, to northern Nevada, about 43 to 33 m.y. ago, to east-central Nevada and west-central Utah, about 20 to 34 m.y. ago (see Stewart and Carlson, this volume). Volcanic activity in Arizona, Colorado, New Mexico, and Mexico occurred at various times during the early and middle Cenozoic and does not fit into this picture of a general southward migration of igneous activity. The concept proposed by Armstrong and others (1969) and McKee (1971) that volcanic activity in the Great Basin region of Nevada and Utah migrated outward from a central core area does not appear to be valid for early and middle Cenozoic time, although the late Cenozoic shift of volcanic activity to the margins of the Basin and Range region is well defined.

During late Cenozoic time, the tectonic and igneous history of western North America changed markedly (McKee, 1971; Christiansen and Lipman, 1972; Noble, 1972). Basalt or bimodal basalt and rhyolite assemblages of volcanic rocks became dominant (Fig. 1-6), and the initiation of extensional tectonics led to the development of basin-range structure. Andesitic rocks related to a subduction system persisted in the Cascade magmatic arc parallel to the western margin of the United States, and tremendous volumes of basaltic lava were erupted at the same time east of this arc in the Columbia Plateau area of Washington, Idaho, and eastern Oregon (McBirney and others, 1974). Basaltic igneous activity was concentrated near the margins of the Basin and Range province (Best and Brimhall, 1974) and was relatively sparse within the province itself. The concentration of basalt near the margins is particularly marked along the mutual boundary of the Colorado Plateau and the Basin and Range province. Within the Basin and Range province, basalt was erupted from vents along basin-range faults in only a few areas.

The distribution and orientation of extensional faults in western North America (Fig. 1-1) clearly follows patterns established long before late Cenozoic time. Some similarity in the distribution of late Cenozoic extensional faulting and the position of the Paleozoic and latest Precambrian Cordilleran miogeosyncline can be noted, particularly the correspondence of the eastern margin of the Basin and Range province and the eastern margin of the miogeosyncline in western Utah (the Wasatch line). A close correlation can be seen in the patterns of Mesozoic and earliest Cenozoic tectonic deformation, early and middle Cenozoic igneous activity, and late Cenozoic extensional faulting.

PLATE–TECTONICS SETTING

The relative motion of major lithospheric plates (Fig. 1-7) in the vicinity of western North America during middle and late Cenozoic time is well known from studies of ocean-floor magnetic anomalies (Atwater, 1970; Atwater and Molnar, 1973). During the middle Cenozoic, two major lithospheric plates, the Pacific and Farallon plates, separated by the East Pacific Rise (a spreading ridge), were present west of North America. The Farallon plate was being consumed at a subduction zone along the western margin of North America. About 29 m.y. ago, a notable change in this system occurred when the spreading ridge intersected the subduction zone; this caused a change in plate geometry and the development of a transform fault system along parts of the western margin of North America. According to Atwater's (1970) model, the spreading ridges in this system are passive features related to the pulling apart of plates rather than to a deep-mantle convective system. Thus, Atwater indicated that mantle upwelling at the spreading ridges ceased in areas where plates were no longer separating, such as along the transform boundary between the Pacific and North American plates. If upwelling does cease, a "hole" forms in the descending Farallon plate (Fig. 1-7) where no new lithosphere is being created east of the transform fault. The size of this hole may depend on whether the descending plate breaks into pieces and whether some of the pieces remain in the developing hole. If such a hole develops, it presumably is filled by material rising from the asthenosphere.

Morgan (1971, 1972a, 1972b) and Wilson (1973) have suggested that deep-mantle convective systems may be in the form of narrow, rising plumes and that such plumes actively drive plates apart along spreading ridges. According to this view, spreading ridges where they are related to plumes are active features, and upwelling would not stop merely because of changing

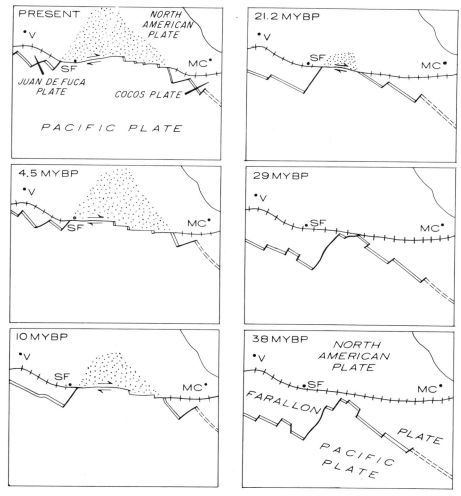

Figure 1-7. Plate tectonics in vicinity of western North America at various times in the Cenozoic Era. Double lines, spreading ridges; lines with cross bars, trenches; single lines, transform faults; dotted areas, "hole" in subducting Farallon plate; V, Vancouver; SF, San Francisco; MC, Mexico City. After Atwater and Molnar (1973).

plate geometry. Plume activity could persist along what was formerly a spreading ridge, even if the ridge and plume system were overridden by a continent. Thus, if a plume existed along the East Pacific Rise and was overridden by western North America, mantle upwelling would continue and could be an important tectonic factor. Wilson (1973) has suggested, for example, that such a plume is responsible for the uplift of the Colorado Plateau.

REGIONAL GEOPHYSICS

The regional geophysics of the Basin and Range province in the United States has recently been reviewed by Thompson and Burke (1974), and discussions of various aspects of the subject are included in the present volume. I will only briefly describe here those characteristics of the province that seem most pertinent to a review of the origin of basin-range structure.

The recent seismicity of the Western United States shows that earthquakes are relatively widespread, although unevenly spaced. Many earthquakes are confined to the Intermountain seismic belt (Smith and Sbar, 1974), extending from the Northern Rocky Mountains in western Montana through the Middle Rocky Mountains of western Wyoming, eastern Idaho, and northern Utah, into southern Utah. The belt dies out farther south in Arizona. A zone of high seismicity extends westward from the belt across southern Utah and Nevada. Another belt of high seismicity trends northward across western Nevada and part of eastern California and is coextensive with a zone of historic faulting (Ryall and others, 1966). Earthquake activity, of course, is also widespread in California, principally along the San Andreas and associated faults. The widespread distribution of earthquakes in the Western United States has been interpreted (Atwater, 1970) as indicating that the relative motion between the Pacific and North American lithospheric plates has been absorbed across a wide "soft" zone of western North America.

Focal depths of earthquakes in the Great Basin and Intermountain West are mostly less than 15 km (Smith and Sbar, 1974), and probably most are less than 12 to 14 km (R. B. Smith, 1976, oral commun.). This depth appears to be coincident with a low-velocity zone in the crust (Braile and others, 1974; Keller and others, 1975). This low-velocity zone may represent a layer of low rigidity that accommodated Cenozoic deformation (Smith and Sbar, 1974). First-motion studies indicate general northwest-southeast extension across the Great Basin, with local areas of strike-slip motion (Smith and Sbar, 1974). In general, the most reasonable interpretation of first-motion studies and the distribution and focal depths of small earthquakes, when considered in combination with surface geology, suggests steeply dipping faults (Ryall and Malone, 1971; Sbar and others, 1972; Smith and Sbar, 1974).

The Basin and Range province of the Western United States is characterized by thin crust, anomalous upper mantle, and high heat flow (Thompson and Burke, 1974; and various articles in this volume). The crust is generally 20 to 35 km thick as compared with 35 to 50 km in the Colorado Plateau and Rocky Mountains. Upper-mantle velocities are generally lower than 7.9 km/s compared with 7.9 to 8.1 km/s in the Colorado Plateau and Rocky Mountains and may indicate partial melting of the mantle (York and Helmberger, 1973). Heat-flow values greater than 2 μcal $\cdot$ cm^{-2} $\cdot$ s^{-1} characterize the Basin and Range province compared with average values of about 1.5.

The Basin and Range province is in nearly perfect isostatic balance. Regional isostatic anomalies average no more than 10 mgal (Thompson and Burke, 1974). As described by Thompson (1972) and Thompson and Burke (1974), this balance means that any loss of mass by near-surface crustal spreading is almost perfectly matched by lateral backflow in the mantle. They noted that if a 10-km-thick plate were merely attenuated by 10%, an isostatic anomaly of about 100 mgal would result. Because no such anomalies exist, flow of mantle material into the area at depth is required.

CHARACTER OF BASIN-RANGE STRUCTURE

Basin-range structure consists of a highly complex system of normal faults along which movement has resulted in the relative uplift of linear segments of the crust to form mountains and the relative sinking of adjacent segments to form valleys (Fig. 1-8). The mountains are usually about 15 to 20 km across and are separated by alluviated valleys of comparable width. In detail, the patterns of mountains and valleys are highly complex (Pl. 1-1, in pocket) and the ranges, although they are generally elongate, are locally equidimensional or elliptical in plan view.

Basin-range structure has generally been related to block faulting, whereby ranges are formed by vertical movements along profound faults on one or both sides of the mountain block. This theory accounts for the gross pattern of mountains and valleys, and for the uniform internal structure of some of the mountain blocks. The concept of block faulting, on the other hand, can imply that the structure is simple, whereas in reality, faulting is not merely confined to the sides of mountains but is distributed throughout the mountain areas and presumably in the suballuvial rocks of the valleys as well. Major structural blocks, therefore, should not be viewed as rigid coherent masses, but rather as aggregate structural units that generally move in a more or less uniform manner relative to adjacent structural units.

Basin-range structure is typically developed in the Great Basin region of Nevada, western Utah, and parts of Oregon, Idaho, California, and Arizona. Here the general distribution of major basin-range faults, the tilting of fault blocks, and the general attitudes of Tertiary rocks are well known (Pl. 1-1). A small-scale map, however, cannot adequately portray the complexities of the structure. Numerous known extension faults have been omitted because they are too small to show. Also omitted are faults that are inferred to bound major blocks but are covered by alluvium and poorly located.

In the Great Basin, basin-range structure is clearly the expression of a system of major blocks, many with significant tilts. In places, the mountains (horsts) and valleys (grabens) are bounded on both sides by normal faults. Tilts of the mountain ranges are generally from a few degrees to about 30°, although higher tilts are known and locally Tertiary units are overturned (Anderson, 1971; Proffett, 1972). Within the Great Basin as a whole, about two-thirds of the tilts are to the east and one-third to the west (Fig. 1-9). Patterns of east- or west-directed tilts are complex, but near the eastern margin of the province, most tilts are to the east; near the western margin, most are to the west (Proffett, 1972; Ekren and others, 1974), although less consistently so. The pattern of tilts near the margins of the Great Basin extends into adjacent provinces; to the east, adjacent parts of the Rocky Mountains and Colorado Plateau are tilted east, and to the west the Sierra Nevada is tilted west; thus, the Great Basin is a central, highly extended part of a giant uplift (Le Conte, 1889). The westward tilts in the western part of the province and eastward tilts in the eastern part contribute to an overall bilateral symmetry of the Great Basin that is also expressed in regional topography and Bouguer gravity, thickness of lithosphere, and pattern of inception of fundamentally basaltic volcanism (G. P. Eaton, 1976, written commun.; Best and Hamblin, this volume).

The amount of valley fill in major valleys in the Great Basin is from a few hundred metres to more than 3,000 m. The structural relief between the lowest bedrock areas under valleys

Figure 1-8 (facing page). Aerial photographs showing basin-range structure in central Great Basin. Upper photograph, low-altitude oblique photograph in central Nevada looking northwest across the Toquima Range at the Toiyabe Range in the background. Fault-bounded eastern margin of Toquima Range is well defined. Photograph by the author. Lower photograph, high-altitude oblique aerial photograph looking northeast at Snake Range in easternmost Nevada. Western Utah in background. Block-like character of range and fault-bounded western margin are clearly visible. U.S. Geological Survey–U.S. Air Force photograph.

to the highest adjacent mountains is generally from 2,000 to 5,000 m. Structural relief along the eastern escarpment of the Sierra Nevada in California and the western escarpment of the Wasatch Mountains in Utah is locally about 6,000 m (Christiansen, 1966).

Strike-slip faults are locally an important part of the tectonic framework of the Great Basin. The most conspicuous group of these faults occurs in a northwest-trending belt (the Walker lane) of right-lateral displacement and disrupted structure in the western part of the Great Basin (Albers, 1967). Total right-lateral displacement along this belt is probably 130 to 190 km and occurs in part as fault slip and in part as a more pervasive large-scale drag (Stewart

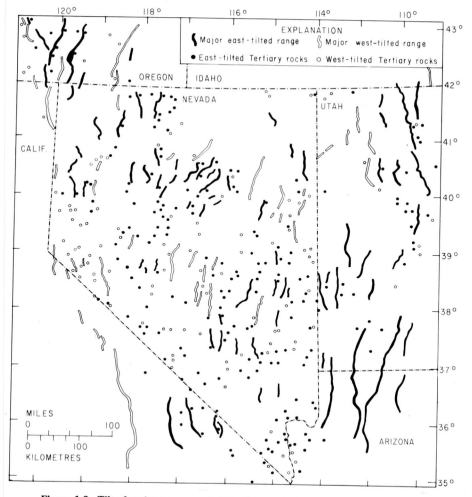

Figure 1-9. Tilt of major ranges and of Tertiary rocks in the Great Basin region.

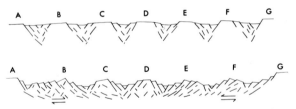

MODEL A. HORSTS AND GRABENS FORMED BY FRAGMENTATION
ABOVE PLASTICALLY EXTENDING SUBSTRATUM

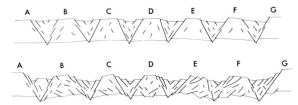

MODEL B. HORSTS AND GRABENS FORMED BY FRAGMENTATION
AND SEGMENTATION INTO BUOYANT BLOCKS

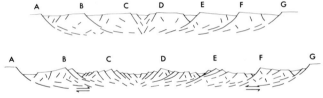

MODEL C. TILTED BLOCKS RELATED TO DOWNWARD-FLATTENING
(LISTRIC) FAULTS

Figure 1-10. Horst-and-graben (models A and B) and tilted-block (listric fault; model C) of basin-range structure showing early and late stages of development.

and others, 1968). Displacement on the Las Vegas shear zone, a part of the Walker lane in southern Nevada, occurred predominantly from 15 to 11 m.y. ago (Anderson and others, 1972), and much of the displacement elsewhere along the Walker lane could be of late Cenozoic age. Major zones of right-lateral strike-slip faulting have been described in southeastern Oregon (Lawrence, 1976). The Garlock fault in eastern California is a left-lateral fault, and left-lateral movement also has occurred on small faults in the western and southern part of the Great Basin. A rhomboid pattern of faulting is seen in some parts of the Great Basin, particularly in southeastern Oregon, and has been attributed at least in part to strike-slip movement (Donath, 1962; Sales, 1966). Strike-slip displacements have occurred on historic faults in Nevada and have been cited as evidence of the importance of strike-slip movement in developing basin-range structure (Shawe, 1965). Thompson and Burke (1973), on the other hand, after a careful study of historic faulting in the Dixie Valley area of western Nevada, observed that the faulting was related to a uniform direction of extension and that the strike-slip displacements are related to movement on fault traces oriented at an angle to the prevailing spreading direction.

Two general models of basin-range structure in the Great Basin have been proposed. One relates the structure to a system of structural blocks rotated along curving, downward-flattening normal faults, sometimes called listric faults (for example, Lowell and others, 1975). The uptilted part of an individual block forms a mountain, and the downtilted part a valley (Fig. 1-10, model C). The second model, the horst-and-graben model, relates the structure to a system of relatively downthrown blocks (grabens) that form valleys, and relatively upthrown blocks (horsts or tilted horsts) that form mountains (Fig. 1-10, models A and B). Evaluating the relative merits of these two models is not easy, and structure in the Basin and Range province probably involves elements of both models.

In a previous article (Stewart, 1971), I discussed the two models of basin-range structure and indicated a preference for the horst-and-graben model. In this model, basin-range structure is produced by the fragmentation of an upper crustal slab over a plastically extending substratum. Extension of the substratum causes the basal part of the slab to be pulled apart along narrow, systematically spaced zones; this, in turn, causes the downdropping of complex horizontal prisms (grabens) of the brittle upper crust. Tilting is produced either by (1) the development of asymmetric grabens in which the mountain block on one side of the graben slumps downward, (2) the rotation of large blocks along downward-flattening faults after the initial formation of the grabens, or (3) rotation of large buoyant blocks due to mass differences. In support of the horst-and-graben model, I outlined geologic and geophysical evidence indicating that many of the valleys in the Great Basin are underlain by grabens and that the alluvial fill in valleys is in many cases much too thick to be accounted for in a simple tilted-block (listric fault) model. In southeastern Oregon, the horst-and-graben structure is particularly well developed, and tilting relatively minor. Structure in this area may represent an initial stage in the development of basin-range structure, before widespread tilting occurs. Studies of first motions of earthquakes and of the distribution of microearthquakes generally indicate high-angle faults (Sbar and others, 1972; Smith and Sbar, 1974) compatible with the horst-and-graben model but not the tilted-block model, in which low-angle faults are predicted at depth.

The horst-and-graben model can be related to the fragmentation of an upper crustal slab above a plastically extending substratum (model A, Fig. 1-10) or to the fragmentation and segmentation of the upper crustal slab into buoyant blocks that float on the substratum (model B, Fig. 1-10; Fig. 1-11). The buoyant block model (Taber, 1927; de Sitter, 1959; Sales, 1976)

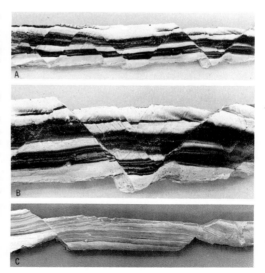

Figure 1-11. Plaster-barite mud models of basin-range structure made by John K. Sales (see Sales, 1976). The photographs were supplied by Sales and are reproduced here with his permission. The models were produced by pouring a layer of molding plaster over a layer of soupy barite drilling mud and then, after the plaster reached the proper consistency, allowing the plaster to extend by backing off a movable piston at one end of the box that holds the model. The plaster is less dense than the barite and floats on it. The photograph shows the plaster after it has hardened. The plaster slab is about 10 cm thick. The now-disrupted markings (representing layering) were originally horizontal.

Photograph A shows widespread high-angle normal faulting. Photograph B is a detail of A showing a keel caused by the sinking of the graben prism. Photograph C shows a graben with elevated marginal blocks that tilt outward. Such geometry is caused by the greater buoyancy of the marginal blocks adjacent to the graben (Taber, 1927; de Sitter, 1959; Sales, 1976). Several grabens in Oregon show this general structure (see Pl. 1-1).

is particularly attractive because it allows for the rotation of large blocks on a scale that could cause entire ranges to tilt. Because of the shape of the large blocks, excess buoyancy and deficient mass is created in blocks adjacent to grabens, thus accounting for elevated margins and the tilt of mountain blocks in opposite directions away from a central graben (Fig. 1-11). Such oppositely directed tilts away from a central graben are observed at several places in southeastern Oregon (Pl.1-1).

Advocates of the tilted-block (listric fault) model have noted that (1) major mountain blocks in the Great Basin are commonly tilted by as much as 10° to 30° or more and (2) the simplest way to explain this tilting is by rotation of large structural blocks along downward-flattening faults. This concept is supported by surface observations that some faults do decrease in dip downward (Longwell, 1945; Hamblin, 1965; Proffett, 1971, 1977) and that low-angle faults are numerous locally (Anderson, 1971; Wright and Troxel, 1973). In some areas, however, these low-angle faults may have been originally steep, and their present low dip is due to rotation of major structural blocks (Hunt and Mabey, 1966). Additional evidence for the rotation of large structural blocks above downward-flattening faults has been suggested by Moore (1960), who noted that the plan view of many fault traces are crescent-shaped in the Great Basin, indicative of spoon-shaped faults in the subsurface. Crescent-shaped traces are evident only locally on Plate 1-1, however.

The two models of basin-range structure are similar in some respects. Both probably involve the fragmentation of an upper-crustal slab over a plastically extending substratum. The thickness of this slab may be about 12 to 14 km, judged from the greatest depth of most earthquakes in the Great Basin. Both models require substantial internal readjustments along many small faults. These faults may tend to have mostly high angles; new faults may develop as older ones are rotated into low-angle attitudes (Proffett, 1972; Chamberlain, 1976). In both models, tilting of blocks has been related to (1) gravity sliding off regional highs (J. G. Moore, in Wallace, 1964), (2) lateral flow in the crust below the fragmenting slab (Proffett, 1972), or (3) mass differences that cause the rotation of individual blocks.

With present data, the character of basin-range structure at depth cannot be firmly established. A better understanding will come from integrated studies of surface geology, deep seismic profiles, first-motion studies of earthquakes, and the spatial distribution of individual earthquake swarms and aftershocks.

AMOUNT OF EXTENSION

Estimates of the amount of extension necessary to produce basin-range structure have varied considerably. Hamilton and Myers (1966) have suggested between 50 and 100 km of generally east-west extension across the Great Basin, an 8% to 18% increase in the width of the province, in late Cenozoic time on the basis of an average estimate for horizontal extension on each of 25 major range-bounding faults along lat 40°N. They and Elston (1976, 1977) have suggested that extension may have occurred in the early and middle Cenozoic Era as well, a concept not generally accepted, and that total Cenozoic extension in this case could have been as great as 100%. In 1971, I estimated total extension across the Great Basin of about 72 km (13%) in late Cenozoic time on the basis of the "graben rule" of Hansen (1965), in which the cross-sectional area of a graben can be related to the amount of lateral motion (Stewart, 1971). Also in 1971, Proffett, on the basis of detailed geologic work in the Yerington mining district in western Nevada, indicated local extension of more than 100%. He suggested from reconnaissance work that 50% extension was likely in the western and eastern parts of the

Great Basin and that less occurred in the central part. Wright and Troxel (1973) and Wright (1976) indicated 30% to 50% extension in part of the Death Valley area of eastern California, on the basis of detailed mapping. In 1974, Thompson and Burke estimated the total amount of extension across Dixie Valley to be 5 km on the basis of geophysical exploration of the valley and, using this as a typical amount of extension for major grabens, suggested about 100 km of total extension (18% increase in width) across the Great Basin. An extension of 30% to 50% (Thompson, 1972) is required if the thinness of the crust in the Basin and Range province is due entirely to lateral spreading, but as Thompson (1972) noted, this estimate may not be meaningful because of the possibility of phase conversions between crust and mantle.

Variations in estimates of extension are the result, in part, of uncertainties in interpreting basin-range structure at depth. If the structure is related to a system of downward-flattening faults, then the total extension necessary to produce the structure is probably much larger than if it is related to generally steep faults. Even extension related to steep faults, however, may lead to a large amount of spreading. This is illustrated by the Turnagain Heights translatory landslide produced by the 1964 Alaskan earthquake. During this slide, total extension was about 100%, and displacement apparently took place on generally steep faults (Hansen, 1965). The complex surface patterns of elongate ridges and valleys produced during the landslide are similar, except in scale, to those of the Basin and Range province.

The amount of extension can also be studied in relation to the thinning of an upper crustal slab and related regional uplift. These relations (shown in Figs. 1-12 and 1-13, Table 1-1) apply best to models A and C in Figure 1-10 where basin-range structure occurs above a specific level in the crust.

In Figure 1-12, the lower diagram represents in cross section a slab of the upper crust with dimensions a and b. In the upper diagram, the slab is shown with new dimensions c and d after extension and the development of basin-range structure. Some erosion of material from the uplifted blocks has taken place, and the material has been deposited in the basins. In the model, all of the material eroded is considered to be trapped in the basins, an assumption approximately true in the Great Basin where drainage is predominantly internal. Thus the total eroded area of the ranges is approximately equal in cross section to the total depositional area of the basins, not counting the relatively small increase in volume as a result of disaggregation

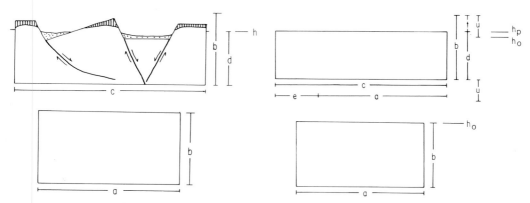

Figure 1-12. Diagram illustrating average elevation (h) in relation to basin-range structure. Vertically lined pattern, eroded areas; dotted pattern, depositional areas. Further explanation in text.

Figure 1-13. Diagram illustrating extension and uplift during development of basin-range structure. Further explanation in text.

TABLE 1-1. RELATION OF EXTENSION TO UPLIFT

Percent crustal extension $(e/a) \times 100\%$ (%)	Amount of crustal extension $e = c - a$ (km)	Thinning of slab $t = b - d$ (m)	Total uplift $u = h_p - h_o + t$ (m)		
			$h_o = 0$ m	$h_o = 700$ m	$h_o = 1{,}200$ m
5	30	639	2,539	1,839	1,339
10	58	1,296	3,196	2,496	1,996
15	83	1,937	3,837	3,137	2,637
20	107	2,610	4,510	3,810	3,310
25	128	3,250	5,150	4,450	3,950
30	148	3,910	5,810	5,110	4,610
35	166	4,553	6,453	5,753	5,253
40	183	5,206	7,106	6,406	5,906

Note: Geometric relations between variables are shown in Figures 1-12 and 1-13; description is in text.

and deposition. If these relations are approximately correct, then the average elevation h in the upper diagram (Fig. 1-12) can be used to calculate the total cross-sectional area of the upper diagram (cd). As no material has been gained or lost from the system, the cross-sectional area of the lower diagram (ab) is equal to that of the upper diagram (cd).

In Figure 1-13, the relations derived from Figure 1-12 are applied to the Great Basin along lat 40°N. Here, the average present elevation h_p is approximately 1,900 m, the width c of the Great Basin is 640 km, and the thickness d of the slab is approximately 13 km (based on average maximum focal depth of earthquakes). The average original elevation h_o of the Great Basin at lat 40°N before development of basin-range structure is poorly known, but probably was between sea level and about 1,200 m from the types of early Tertiary flora in California and Nevada (Axelrod, 1966, Fig. 4). Axelrod (1957) estimated that the average elevation of western Nevada was about 600 to 750 m in Miocene-Pliocene time, also on the basis of fossil flora. These estimates of elevations are uncertain, but nonetheless appear to indicate the extremes that should be considered. The relations between the amount of extension e, the thinning t of the slab, and the amount of uplift u, (see Fig. 1-13) are listed in Table 1-1.

Table 1-1 indicates that extension is accompanied by thinning of the upper crustal slab and that the larger the extension, the greater the thinning. As a consequence, areas of the Great Basin that have undergone the largest amount of extension should have the thinnest upper crust and thus be at the lowest elevations, unless regional uplift of the slab is greater in these same areas. If we assume that regional uplift is uniform throughout the Great Basin, differential extension perhaps explains why some areas of the Great Basin are significantly lower than others. For example, the Lahontan depression in the western part of the Great Basin and the Bonneville depression in the eastern part, both sites of Pleistocene lakes, have lower elevations than other areas in the northern part of the Great Basin. The lower elevations may be a result of greater extension at these places, which seems to be borne out by a more complex basin-range structure here (Stewart, 1971).

TIME OF DEVELOPMENT

Basin-range structure is clearly related predominantly to late Cenozoic tectonic events. In many areas of western North America, extensional faults cut widespread tuffs of early and middle Cenozoic age that formed as highly mobile ash flows, which tend to fill troughs much like water. The occurrence of individual ash-flow units at many different topographic and structural levels can only be explained by faulting.

Two types of evidence have been used to date precisely the development of basin-range structure. The first is indirect and is based on the assumption that the transition from calc-alkalic volcanic rocks in early and middle Cenozoic time to fundamentally basaltic volcanic rocks in late Cenozoic time marks the change from predominantly compressional tectonics (related to a subduction zone) to extensional tectonics (related to wrench faulting, back-arc spreading, or some other factor). The second type of evidence is direct and based on the first appearance of fault-controlled sedimentary basins and topographic forms approximately resembling those seen today.

The volcanic transition in most areas of the Great Basin was about 17 m.y. ago (McKee and others, 1970; McKee, 1971). It was much later than this, however, along the Cascade belt of calc-alkalic volcanism parallel to the western margin of North America in the westernmost Great Basin and the Sierra Nevada of California. Snyder and others (1976) have carefully documented that calc-alkalic volcanism along this belt ended perhaps as much as 20 m.y. ago at lat 35°N and progressively more recently northward to near lat 40°N, where it is still active. In southern Arizona and New Mexico, the transition may have been somewhat older than in the Great Basin. Christiansen and Lipman (1972) indicated that the transition date there was between 23 and 37 m.y. ago; Armstrong and Higgins (1973) suggested a general range of 25 to 30 m.y.; McKee and Noble (1974) suggested about 21 m.y. Several authors (Christiansen and Lipman, 1972; Proffett, 1972; Best and Hamblin, this volume) have suggested a general northward migration of the time of inception of basin-range faulting in the Western United States. Data are not complete enough to generalize about the time of transition from calc-alkalic to basaltic rocks in Mexico, although basaltic rocks in Baja California (Gastil and others, 1975) have the same spread in ages as those in most areas of the Western United States.

Direct evidence of the time of development of basin-range structure is sparse and more difficult to evaluate. In the Great Basin, large sedimentary basins, presumably formed by extensional faulting, were well defined about 11 to 13 m.y. ago (late Barstovian to Clarendonian) (Axelrod, 1957; Robinson and others, 1968; Gilbert and Reynolds, 1973). In a study in western Nevada, Gilbert and Reynolds (1973) described one such basin that was in existence from about 12.5 to 8 m.y. ago, but noted that it was much more extensive than present basins in the area. Sedimentary basins that have a distribution similar to those of today did not develop in that area until approximately 7.5 m.y. ago. In the Nevada Test Site area of southern Nevada, evidence of the late development of basin-range structure has also been described (Ekren and others, 1968). In this area, two systems of faulting have been recognized. The earlier system consists of two sets of faults—one striking northeast and the other northwest—and is developed only in rocks older than 17 m.y. The younger system, a single set, strikes north and cuts 14-m.y.-old tuff that is not cut by the older system of faults. Thus, the north-trending system started to develop sometime between 17 and 14 m.y. ago, but high relief related to this faulting did not develop rapidly. That the area had low relief about 11 m.y. ago is indicated by a widespread tuff of that age that does not vary significantly in thickness between present-day valleys and mountains, a situation that would be impossible if present-day topographic relief had developed by then. By the time another tuff had erupted about 7 m.y. ago, however, the topographic grain was much as it is today. This younger tuff lapped up against some of the ranges and in places flowed into valleys that are the sites of present-day streams.

In summary, the present basins and ranges clearly are a late Cenozoic structure. Extensional

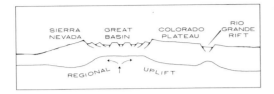

Figure 1-14. Schematic diagram illustrating regional uplift in western North America.

faulting probably started about 17 m.y. ago in much of the Great Basin, but may have started slightly earlier in southern Arizona and New Mexico. The grain of present-day topography, however, probably did not develop before 10 m.y. ago, at the earliest.

REGIONAL UPLIFT

The Basin and Range province lies within a broad region of late Cenozoic uplift (Fig. 1-14). This uplift is documented by mid-Cenozoic floras in the Cordilleran region that indicate elevations at least 1 km below those at present (see discussion and references in Proffett, 1972; Suppe and others, 1975). More precise data on uplift are available west of the Great Basin, where regional uplift of more than 1,800 m in the past 9 m.y. has been described by Christiansen (1966). The Colorado Plateau east of the Great Basin also was uplifted in late Cenozoic time, primarily between 5 and 10 m.y. ago, according to McKee and McKee (1972). Chapin and Seager (1975), however, indicated that the earliest uplift may be as old as 24 m.y. Within the Great Basin, an uplift of 2,000 to 3,000 m in late Cenozoic time seems likely (Table 1-1).

THEORIES OF ORIGIN OF BASIN–RANGE STRUCTURE

Current theories of the origin of basin-range structure can be loosely grouped into four main categories: wrench faulting, back-arc spreading, subduction of the East Pacific Rise, and mantle plumes.

Wrench Faulting

The wrench-faulting concept relates the development of basin-range structure to oblique tensional fragmentation within a broad belt of right-lateral movement along the western side of the North American lithospheric plate. This theory is based on concepts developed by Carey (1958), Wise (1963), Shawe (1965), Hamilton and Myers (1966), Sales (1966), and Slemmons (1967) and has been put in terms of plate-tectonics theory by Atwater (1970) and Christiansen and McKee (this volume). According to this view, western North America is within a broad belt of right-lateral movement related to differential motion between the North American and Pacific plates. Some of the right-lateral movement is taken up on the San Andreas fault and related zones of right-lateral shear, such as the Walker lane in the western Great Basin. The movement is also thought to produce distributed extension and tensional crustal fragmentation (including basin-range structure) along trends oriented obliquely to the trend of the San Andreas fault. This concept is illustrated in Figure 1-15, a map drawn on a Mercator projection about the pole of relative rotation between the Pacific and North American plates (Atwater, 1970). On the map, pure strike-slip motion occurs along horizontal lines, pure tension on vertical lines, and oblique tensional fragmentation on lines having other orientations. Many basin-range faults, as shown on the map, have an oblique orientation consistent with the concept of wrench faulting.

The concept of a broad zone of right-lateral displacement in the Western United States also is consistent with the concept that the broad dextral-curving pattern of Paleozoic eugeosynclinal rocks and of Mesozoic plutonic rocks in the Western United States is due to large-scale oroclinal folding (Mendocino and Idaho oroclines). The oroclinal folding from this point of view is part of the same tectonic pattern that developed basin-range structure (Hamilton and Myers, 1966).

The wrench-faulting concept is particularly attractive when accounting for major zones of strike-slip faulting in the Basin and Range province and for strike-slip first motions of earthquakes in some regions of western North America. The concept is also supported by the contemporaneous development of a transform boundary and of basin-range structure, and the similarity in latitudinal extent (Fig. 1-1) of both structures. The concept does not easily account for regional uplift that has affected the western Cordillera, nor does it entirely explain why shear stress was transmitted over such a broad zone of western North America rather than being confined to the edge of the plate, or why the Basin and Range province should extend (increase in area) instead of simply shearing without an increase in area. These and other factors of plate interactions possibly related to the development of basin-range structure are discussed by Christiansen and McKee (this volume).

Back-Arc Spreading

The back-arc–spreading theory relates basin-range structure to mantle upwelling and associated spreading above a subduction zone (Fig. 1-16). The concept was originally developed to explain the origin of island arcs in the western Pacific that, according to the theory, drifted out from the Asian continent by spreading (back-arc spreading) on the continent side of the arc at the same time that the arc was being underthrust by the Pacific plate (Karig, 1971, 1974; Matsuda and Uyeda, 1971). The theory relates spreading to high heat generated along a Benioff zone by friction and the upwelling of mantle material that causes spreading in the crust and, in the case of the western Pacific, leads to the development of oceanic crust in back-arc

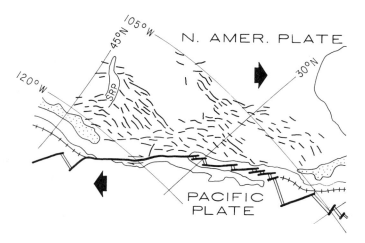

Figure 1-15. Map projected about the pole of relative rotation of Pacific and North American plates, showing relation of extensional faulting to shear in western North America. After Atwater (1970). Double lines, spreading ridges; lines with cross bars, trenches; broad lines, transform faults; narrow, short lines, extensional faults and a few strike-slip faults; dotted areas, late Cenozoic calc-alkalic volcanic rocks; SRP, Snake River Plain.

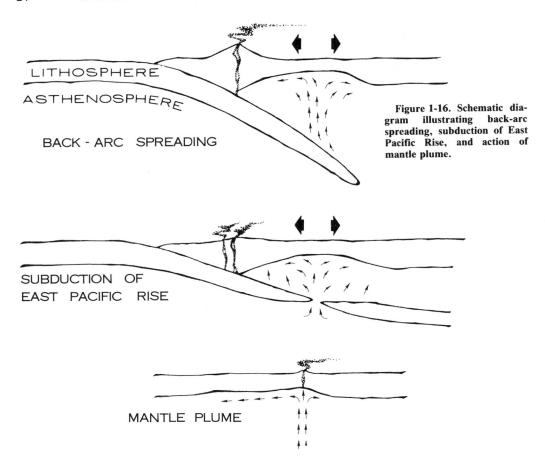

Figure 1-16. Schematic diagram illustrating back-arc spreading, subduction of East Pacific Rise, and action of mantle plume.

basins. The process is slow; calculation by Hasebe and others (1970) suggests that a subduction system would have to be active for 100 m.y. to reach the heat flow observed in the Japan arc system. The back-arc–spreading concept has been applied to western North America by Scholz and others (1971), Karig and Jensky (1972), Thompson (1972), and Thompson and Burke (1974).

The back-arc–spreading model is particularly appealing when explaining such characteristics of the Basin and Range province as high heat flow, thin crust, anomalous upper mantle, regional uplift, and the similarity in distribution of late Cenozoic extensional faulting to the distribution of late Mesozoic and early to middle Cenozoic igneous and tectonic features. High heat flow and anomalous upper mantle are characteristics that the Basin and Range province has in common with the back-arc basins of the western Pacific (Scholz and others, 1971). The regional uplift of the western Cordillera of North America can be explained as the result of high heat flow and thermal expansion or phase changes in the mantle or crust. The similarity of distribution of basin-range structure and Mesozoic and Cenozoic igneous and deformational events indicates a long history of heat generation in the Basin and Range province. The abundance of mid-Cenozoic silicic volcanism in the Basin and Range province is particularly significant. Karig and Jensky (1972) have noted the association of silicic volcanism with rifting in both marginal ocean basins and in volcano-tectonic rift zones. The broad zone of Mesozoic and Cenozoic igneous activity in the Western United States and northern Mexico is anomalous in comparison with other parts of western North and South America; this suggests that somehow the anomalously broad zone and spreading are related. Uyeda and Miyashiro (1974) also noted an association of spreading in the western Pacific with a broad zone of igneous activity.

Why did back-arc spreading, if this concept is correct, start in late Cenozoic time? Perhaps the generation of heat along a subduction system had reached some critical point where mantle upwelling and crustal spreading could start. The spreading could have been triggered by the factors, as yet unknown, that caused an episode of intense volcanic activity in much of the circum-Pacific area during middle Miocene time (McBirney and others, 1974). The voluminous basalts of the Columbia Plateau, themselves indicative of back-arc spreading, were erupted at virtually the same time as this mid-Miocene episode. Scholz and others (1971), on the other hand, have suggested that spreading was accelerated by the cessation of subduction and development of a transform boundary along the western margin of North America. They reasoned that subduction created a confining pressure on western North America and that once this pressure was relieved, the spreading rate would increase. The fact that extention is greatest in regions of North America inland from the transform boundary (Fig. 1-1) supports this concept.

Bott (1973) has outlined how oceanward spreading might occur along a stable Atlantic-type continental margin, and a similar process might take place after termination of compression. Terman (1975) has also suggested that extension is related to lack of confining pressure, although in a somewhat different context than here.

East Pacific Rise

Basin-range structure has been related to convection currents and lateral spreading on the flanks of the East Pacific Rise, which according to this theory, extends under western North America (Menard, 1960; Cook, 1969; McKee, 1971; Gough, 1974). In support of this view, Menard (1960) noted that the East Pacific Rise, if projected northward, would extend under the Great Basin area and that both the Great Basin and the East Pacific Rise are characterized by high heat flow. In addition, Menard pointed out that the ridges and troughs analogous to the basins and ranges also occur on the ocean bottoms on the flanks of the East Pacific Rise.

Present-day plate-tectonics theories discount the possibility that the East Pacific Rise extends under western North America. In Atwater's (1970) model, the upwelling of mantle material along the East Pacific Rise is considered to be a passive response to the separation of plates—not the driving mechanism of plate separation. Upwelling should cease, according to this view, once plate separation ceased because of the development of a transform boundary (Fig. 1-7). Therefore, upwelling related to the East Pacific Rise should not occur beneath western North America. An additional argument against any relation between the Basin and Range province and the East Pacific Rise is the widespread presence of basin-range structures in Mexico, in areas far east of where the ridge system still exists at the mouth of the Gulf of California and locally as spreading centers within the gulf.

A somewhat different concept can be considered if basin-range structure is not related directly to upwelling along the East Pacific Rise but to the presence of hot oceanic lithosphere, a remnant of the Farallon plate, beneath the Great Basin. This lithosphere would be relatively young and therefore hot, because the youngest parts would have been created from upwelling magma immediately prior to the collision of the ridge and trench. Its heat could help drive the spreading in the Great Basin. This concept apparently requires that part of the Farallon plate ceased its descent under North America and remained under the Great Basin. If, on the other hand, the plate did continue to descend, a hole would be left in the plate where no new lithosphere was being created inland of the transform system (Fig. 1-7). If this were true, material from the asthenosphere would presumably ascend into the hole as the plate descended, creating a convection system much like that along a ridge system. Both the "hot plate" and "hole filling" concepts are difficult to apply to Mexico, where basin-range structure lies east of areas where the ridge system still exists.

Mantle Plumes

Deep-mantle convection in the form of narrow rising plumes has been thought to play a major role in the late Cenozoic history of the Western United States (Matthews and Anderson, 1973; Wilson, 1973; Smith and Sbar, 1974; Suppe and others, 1975). According to Morgan (1971, 1972a, 1972b), such mantle plumes provide the driving mechanism for continental drift and cause plates to rift and to be driven apart from one another. The Snake River Plain–Yellowstone area in southern Idaho and northwestern Montana has been suggested (Suppe and others, 1975) as the trace of such a mantle plume. According to this theory, the plume lay under the western part of the Snake River Plain about 15 m.y. ago and has traced a path relatively eastward as the North American plate moved over it. Its present position is under the Yellowstone region of northwest Wyoming. Matthews and Anderson (1973), Smith and Sbar (1974), and Suppe and others (1975) have suggested that this plume may have caused the breakup of a large segment of the Western United States. Others (Eaton and others, 1975) related the volcanism in the Snake River Plain to a major crustal fracture that propagated northeastward, guided by structures in Precambrian rocks.

As of now, mantle plumes similar to those proposed for oceanic areas have not been proved to underlie western North America, nor is the effect of such plumes, if they exist, certain. Seemingly, several widely scattered plumes would be necessary to produce the extensional faulting observed in such a large segment of the Cordillera of western North America. As yet only one area (Yellowstone) can credibly be considered to be underlain by a plume, and even that is a matter of controversy.

CONCLUSIONS

Of the many characteristics of the Basin and Range province that should be considered in evaluating the origin of basin-range structure, the most important appear to be (1) low seismic velocity of the upper mantle, indicative of partial melting; (2) thin crust; (3) high heat flow; (4) regional uplift and extension; (5) previous history of deformation in Mesozoic and earliest Cenozoic time, and of widespread siliceous volcanism in middle Cenozoic time; and (6) position inland from a transform plate boundary.

Considered together, these characteristics generally fit well with the concept of back-arc spreading. In this system, heat generated by friction along a descending slab provides the energy for upwelling of mantle material and the resultant thinning of the crust, regional uplift, and near-surface spreading. High heat generation can be assumed because of long history of subduction along the western margin of North America that led to the widespread emplacement of plutonic rocks in the Mesozoic Era and to the eruption of voluminous volcanic rocks, particularly silicic types, in early and middle Cenozoic time. The back-arc setting is further indicated by the distribution of volcanic rocks in late Cenozoic time (Snyder and others, 1976; Stewart and Carlson, this volume) that clearly shows that an active andesitic magmatic arc existed along the western margin of the Great Basin and adjacent parts of the Sierra Nevada in California at the same time that extensional faulting was occurring farther east.

As proposed by Scholz and others (1971), back-arc spreading may have been accelerated by the slackening of confining pressure owing to the destruction of the subduction system along western North America and the development of a transform boundary. Such a concept explains why basin-range structure is mostly confined to areas inland from the transform boundary of the western margin of North America. In addition, a major component of right-lateral shear related to distributed stress from the transform system can be incorporated in the back-arc model to account for right-lateral motion along the Walker lane and other fault zones in the Western United States.

ACKNOWLEDGMENTS

I am indebted to many geologists for help during the preparation of this report. I am particularly indebted to John K. Sales, who kindly supplied photographs of his plaster-barite models of basin-range structure. My knowledge increased greatly from discussions at the Penrose Conference on Regional Geophysics and Tectonics of the Intermountain West convened by G. P. Eaton, D. R. Mabey, and R. B. Smith. The maps were improved considerably from help by M. G. Best, P. M. Blacet, M. D. Crittenden, Jr., W. A. Duffield, R. K. Hose, F. J. Kleinhampl, and W. J. Moore. The manuscript was improved by thoughtful reviews by D. R. Shawe and G. A. Thompson.

REFERENCES CITED

Albers, J. P., 1967, Belt of sigmoidal bending and right-lateral faulting in the western Great Basin: Geol. Soc. America Bull., v. 78, p. 143–156.

Anderson, R. E., 1971, Thin skin distension in Tertiary rocks of southeastern Nevada: Geol. Soc. America Bull., v. 82, p. 43–58.

Anderson, R. E., Longwell, C. R., Armstrong, R. L., and Marvin, R. F., 1972, Significance of K-Ar ages of Tertiary rocks from the Lake Mead region, Nevada-Arizona: Geol. Soc. America Bull., v. 83, p. 273–288.

Armstrong, R. L., and Higgins, R. E., 1973, K-Ar dating of the beginning of Tertiary volcanism in the Mojave Desert, California: Geol. Soc. America Bull., v. 84, p. 1095–1100.

Armstrong, R. L., Ekren, E. B., McKee, E. H., and Noble, D. C., 1969, Space-time relations of Cenozoic silicic volcanism in the Great Basin of the Western United States: Am. Jour. Sci., v. 267, p. 478–490.

Atwater, Tanya, 1970, Implications of plate tectonics for the Cenozoic tectonic evolution of western North America: Geol. Soc. America Bull., v. 81, p. 3513–3535.

Atwater, Tanya, and Molnar, Peter, 1973, Relative motion of the Pacific and North American plates deduced from sea-floor spreading in the Atlantic, Indian, and South Pacific Oceans, in Kovach, R. L., and Nur, Amos, eds., Tectonic problems of the San Andreas fault system, Conf., Proc.: Stanford Univ. Pubs. Geol. Sci., v. 13, p. 136–148.

Axelrod, D. I., 1957, Late Tertiary floras and the Sierra Nevadan uplift [California-Nevada]: Geol. Soc. America Bull., v. 68, p. 19–45.

——1966, The Eocene Copper Basin flora of northeastern Nevada: California Univ. Pubs. Geol. Sci., v. 59, 125 p.

Best, M. G., and Brimhall, W. H., 1974, Late Cenozoic alkalic basaltic magmas in the western Colorado Plateaus and the Basin and Range transition zone, U.S.A., and their bearing on mantle dynamics: Geol. Soc. America Bull., v. 85, p. 1677–1690.

Best, M. G., and Hamblin, W. K., 1978, Origin of the northern Basin and Range province: Implications from the geology of its eastern boundary, in Smith, R. B., and Eaton, G. P., eds.,

Cenozoic tectonics and regional geophysics of the western Cordillera: Geol. Soc. America Mem. 152 (this volume).

Bott, M.H.P., 1973, Shelf subsidence in relation to the evolution of young continental margins, *in* Tarling, D. H., and Runcorn, S. K., eds., Implications of continental drift to the earth sciences, Vol. 2: New York, Academic Press, p. 675–683.

Braile, L. W., Smith, R. B., Keller, G. R., Welch, R. M., and Meyer, R. P., 1974, Crustal structure across the Wasatch front from detailed seismic refraction studies: Jour. Geophys. Research, v. 79, p. 2669–2677.

Burchfiel, B. C., and Davis, G. A., 1972, Structural framework and evolution of the southern part of the Cordilleran orogen, Western United States: Am. Jour. Sci., v. 272, p. 97–118.

——1975, Nature and controls of Cordilleran orogenesis, Western United States—Extensions of an earlier synthesis: Am. Jour. Sci., v. 275-A, p. 363–396.

Canada Geological Survey, 1970a, Physiographic regions of Canada: Canada Geol. Survey Map 1254A, scale 1:5,000,000.

——1970b, Geologic map of Canada, *in* Douglas, R.J.W., Geology and economic minerals of Canada: Canada Geol. Survey Econ. Geology Rept. 1, scale 1:5,000,000.

Carey, S. W., 1958, The tectonic approach to continental drift, *in* Carey, S. W., ed., Continental drift, a symposium: Hobart, Tasmania, Tasmania Univ. Geology Dept., p. 177–355.

Carrillo-Bravo, José, 1961, Geología del Anticlinorio Huizachal-Peregrina al N-W de Ciudad Victoria Tamps: Asoc. Mexicana Geólogos Petroleros Bol., v. 13, nos. 1–2, p. 1–98.

Chamberlain, R. M., 1976, Rotated early-rift faults and fault blocks, Lemitar Mountains, Socorro County, New Mexico: Geol. Soc. America Abs. with Programs, v. 8, p. 807.

Chapin, C. E., and Seager, W. R., 1975, Evolution of the Rio Grande rift in the Socorro and Las Cruces areas, *in* Seager, W. R., Clemons, R. E., and Callender, J. F., eds., Guidebook of the Las Cruces country, 1975: New Mexico Geol. Soc. Guidebook 26, p. 297–321.

Christiansen, M. N., 1966, Late Cenozoic crustal movements in the Sierra Nevada of California: Geol. Soc. America Bull., v. 77, p. 163–181.

Christiansen, R. L., and Lipman, P. W., 1972, Cenozoic volcanism and plate-tectonic evolution of the Western United Sates. II, Late Cenozoic: Royal Soc. London Philos. Trans., v. 271, p. 249–284.

Christiansen, R. L., and McKee, E. H., 1978, Late Cenozoic volcanic and tectonic evolution of the Great Basin and Columbia intermontane region, *in* Smith, R. B., and Eaton, G. P., eds., Cenozoic tectonics and regional geophysics of the western Cordillera: Geol. Soc. America Mem. 152 (this volume).

Comite de la Carta Geologica de Mexico, 1968, Carta Geologica de la Republica Mexicana, scale 1:2,000,000.

Cook, K. L., 1969, Active rift system in the Basin and Range province, *in* The world rift system—Internat. Upper Mantle Comm., Upper Mantle Project sci. rept. 19: Tectonophysics, v. 8, p. 469–511.

Donath, F. A., 1962, Analysis of basin-range structure, south-central Oregon: Geol. Soc. America Bull., v. 73, p. 1–15.

Eaton, G. P., Christiansen, R. L., Iyer, H. M., Pitt, A. M., Mabey, D. R., Blank, H. R., Jr., Zietz, Isidore, and Gettings, M. E., 1975, Magma beneath Yellowstone National Park: Science, v. 188, p. 787–796.

Ekren, E. B., Rogers, C. L., Anderson, R. E., and Orkild, P. P., 1968, Age of basin and range normal faults in Nevada Test Site and Nellis Air Force Range, Nevada, *in* Nevada Test Site: Geol. Soc. America Mem. 110, p. 247–250.

Ekren, E. B., Bath, G. D., Dixon, G. L., Healey, D. L., and Quinlivan, W. D., 1974, Tertiary history of Little Fish Lake Valley, Nye County, Nevada, and implications as to the origin of the Great Basin: U.S. Geol. Survey Jour. Research, v. 2, p. 105–118.

Elston, W. E., 1976, Tectonic significance of mid-Tertiary volcanism in the Basin and Range province: A critical review with special reference to New Mexico, *in* Elston, W. E., and Northrop, S. A., eds., Cenozoic volcanism in southwestern New Mexico: New Mexico Geol. Soc. Spec. Pub. 5, p. 93–151.

——1977, Rifting and volcanism in the New Mexico segment of the Basin and Range province, southwestern USA, *in* Larson, B. T., and Neumann, E. R., eds., NATO Advanced Study Institute on Paleorift Systems, Oslo, Norway, Proc.: Dordrecht, Holland, D. Reidel (in press).

Fenneman, N. M., 1946, Physical divisions of the United States: U.S. Geol. Survey Map, scale 1:7,000,000.

Gastil, R. G., Phillips, R. P., and Allison, E. C., 1975, Reconnaissance geology of the state of Baja California: Geol. Soc. America Mem. 140, 170 p.

Gilbert, C. M., and Reynolds, M. W., 1973, Character and chronology of basin development, western margin of the Basin and Range province: Geol. Soc. America Bull., v. 84, p. 2489–2509.

Gough, D. I., 1974, Electrical conductivity under western North America in relation to heat flow, seismology, and structure: Jour. Geomagnetism and Geoelectricity, v. 26, p. 105–123.

Guzman, Eduardo J., and de Cserna, Zoltán, 1963, Tectonic history of Mexico, in Childs, O. E., and Beebe, B. W., eds., Backbone of the Americas: Am. Assoc. Petroleum Geologists Mem. 2, p. 113–129.

Hamblin, W. K., 1965, Origin of "reverse drag" on the downthrown side of normal faults: Geol. Soc. America Bull., v. 76, p. 1145–1164.

Hamilton, Warren, and Myers, W. B., 1966, Cenozoic tectonics of the Western United States: Rev. Geophysics, v. 5, p. 509–549.

Hansen, W. R., 1965, Effects of the earthquake of March 27, 1964, at Anchorage, Alaska: U.S. Geol. Survey Prof. Paper 542-A, p.A1–A68.

Hasebe, K., Fujii, N., and Uyeda, S., 1970, Thermal process under island arcs: Tectonophysics,v. 10, p. 335–355.

Hintze, L. F., 1975, Geological highway map of Utah: Brigham Young Univ. Geology Studies Spec. Pub. 3, scale 1:1,000,000.

Hunt, C. B., and Mabey, D. R., 1966, General geology of Death Valley, California—Stratigraphy and structure: U.S. Geol. Survey Prof. Paper 494-A, p. A1–A165.

Jennings, C. W., 1973, State of California preliminary fault and geologic map: California Div. Mines and Geology Prelim. Rept. 13, scale 1:750,000.

Karig, D. E., 1971, Origin and development of marginal basins in the western Pacific: Jour. Geophys. Research, v. 76, p. 2542–2561.

——1974, Evolution of arc systems in the western Pacific: Earth and Planetary Sci. Ann. Rev., v. 2, p. 51–75.

Karig, D. E., and Jensky, Wallace, 1972, The proto–Gulf of California: Earth and Planetary Sci. Letters, v. 17, p. 169–174.

Keller, G. R., Smith, R. B., and Braile, L. W., 1975, Crustal structure along the Great Basin–Colorado Plateau transition from seismic refraction studies: Journ. Geophys. Research, v. 80, p. 1093–1098.

King, P. B., 1969a, The tectonics of North America—A discussion to accompany the tectonic map of North America, scale 1:5,000,000: U.S. Geol. Survey Prof. Paper 628, 95 p.

King, P. B., compiler, 1969b, Tectonic map of North America: Washington, D.C., U.S. Geol. Survey Map, scale 1:5,000,000.

King, P. B., and Beikman, H. M., 1974, Geologic map of the United States: U.S. Geol. Survey Map, scale 1:2,500,000.

Lawrence, R. D., 1976, Strike slip faulting terminates the Basin and Range province in Oregon: Geol. Soc. America Bull., v. 87, p. 846–850.

Le Conte, Joseph, 1889, On the origin of normal faults and of the structure of the Basin region:

Am. Jour. Sci., 3rd ser., v. 38, p. 257–263.

Lipman, P. W., Prostka, H. J., and Christiansen, R. L., 1972, Cenozoic volcanism and plate-tectonic evolution of the Western United States. I, Early and middle Cenozoic: Royal Soc. London Philos. Trans., v. 271, p. 217–248.

Longwell, C. R., 1945, Low-angle normal faults in the Basin-and-Range province: Am. Geophys. Union Trans., v. 26, pt. 1, p. 107–118.

López-Ramos, Ernesto, 1969, Marine Paleozoic rocks of Mexico: Am. Assoc. Petroleum Geologists Bull., v. 53, p. 2399–2417.

Lowell, J. D., Genik, G. J., Nelson, T. H., and Tucker, P. M., 1975, Petroleum and plate tectonics of the southern Red Sea, in Fischer, A. G., and Judson, Sheldon, eds., Petroleum and global tectonics: Princeton, N.J., Princeton Univ. Press, p. 129–153.

Matthews, Vincent, III, and Anderson, C. E., 1973, Yellowstone convection plume and break-up of the Western United States: Nature, v. 243, p. 158–159.

Matsuda, T., and Uyeda, S., 1971, On the Pacific-type orogeny and its model-extension of the paired belts concept and possible origin of marginal seas: Tectonophysics, v. 11, p. 5–27.

McBirney, A. R., Sutter, J. F., Naslund, H. R., Sutton, K. G., and White, C. M., 1974, Episodic volcanism in the central Oregon Cascade Range: Geology, v. 2, p. 585–589.

McKee, E. D., and McKee, E. H., 1972, Pliocene uplift of the Grand Canyon region; Time of drainage adjustment: Geol. Soc. America Bull., v. 83, p. 1923–1931.

McKee, E. H., 1971, Tertiary igneous chronology of the Great Basin of Western United States—Implications for tectonic models: Geol. Soc. America Bull., v. 82, p. 3497–3502.

McKee, E. H., and Noble, D. C., 1974, Timing of late Cenozoic crustal extension in the Western United States: Geol. Soc. America Abs. with Programs, v. 6, p. 218.

McKee, E. H., Noble, D. C., and Silberman, M. L., 1970, Middle Miocene hiatus in volcanic activity in the Great Basin area of the Western United States: Earth and Planetary Sci. Letters, v. 8, p. 93–96.

Menard, H. W., Jr., 1960, The East Pacific Rise: Science, v. 132, p. 1737–1746.

Moore, J. G., 1960, Curvature of normal faults in the Basin and Range province of the Western United States, in Geological Survey research 1960: U.S. Geol. Survey Prof. Paper 400-B, p. B409–B411.

Morgan, W. J., 1971, Convective plumes in the lower mantle: Nature, v. 230, p. 42–43.

——1972a, Deep mantle convection plumes and plate motions: Am. Assoc. Petroleum Geologists

Bull., v. 56, p. 203–213.

——1972b, Plate motion and deep mantle convection, *in* Shagam, R., and others, eds., Studies in Earth and space science: Geol. Soc. America Mem. 132, p. 7–22.

Morris, R. C., 1974, Sedimentation and tectonic history of the Ouachita Mountains, *in* Dickinson, W. R., ed., Tectonics and sedimentation: Soc. Econ. Paleontologists and Mineralogists Spec. Pub. 22, p. 120–142.

Nicholas, R. L., and Rozendal, R. A., 1975, Subsurface positive elements within Ouachita foldbelt in Texas and their relation to Paleozoic cratonic margin: Am. Assoc. Petroleum Geologists, v. 59, p. 193–216.

Noble, D. C., 1972, Some observations on the Cenozoic volcano-tectonic evolution of the Great Basin, Western United States: Earth and Planetary Sci. Letters, v. 17, p. 142–150.

Nolan, T. B., 1943, The Basin and Range province in Utah, Nevada, and California: U.S. Geol. Survey Prof. Paper 197-D, p. 141–196.

Proffett, J. M., Jr., 1971, Late Cenozoic structure in the Yerington district, Nevada, and the origin of the Great Basin: Geol. Soc. America Abs. with Programs, v. 3, p. 181.

——1972, Nature, age, and origin of Cenozoic faulting and volcanism in the Basin and Range province (with special reference to the Yerington district, Nevada [Ph.D. thesis]: Berkeley, Univ. California, 77 p.

——1977, Cenozoic geology of the Yerington district, Nevada, and implications for the nature and origin of basin and range faulting: Geol. Soc. America Bull., v. 88, p. 247–266.

Raisz, E. J., 1959, Landforms of Mexico: Cambridge, Mass., Inst. Geog. Explor. Harvard Univ., scale 1:2,500,000.

Roberts, R. J., 1968, Tectonic framework of the Great Basin, *in* A coast to coast tectonic study of the United States: UMR Jour., no. 1 (V. H. McNutt-Geology Dept. Colloquim Ser. 1, Univ. Missouri, Rolla) p. 101–119.

Robinson, P. T., McKee, E. H., and Moiola, R. J., 1968, Cenozoic volcanism and sedimentation, Silver Peak region, western Nevada and adjacent California, *in* Coats, R. R., Hay, R. L., and Anderson, C. A., eds., Studies in volcanology—A memoir in honor of Howel Williams: Geol. Soc. America Mem. 116, p. 577–611.

Ryall, Alan, and Malone, S. D., 1971, Earthquake distribution and mechanism of faulting in the Rainbow Mountain–Dixie Valley–Fairview Peak area, central Nevada: Jour. Geophys. Research, v. 76, p. 7241–7248.

Ryall, Alan, Slemmons, D. B., and Gedney, L. D., 1966, Seismicity, tectonics, and surface faulting in the Western United States during historic time: Seismol. Soc. America Bull., v. 56, p. 1105–1135.

Sales, J. K., 1966, Structural analysis of the Basin and Range province in terms of wrench faulting [Ph.D. thesis]: Reno, Nevada Univ., 178 p.

——1976, Model studies of continental rifting: Geol. Soc. America Abs. with Programs, v. 8, p. 1083.

Sbar, M. L., Barazangi, M., Dorman, J., Scholz, C. H., and Smith, R. B., 1972, Tectonics of the Intermountain seismic belt, Western United States: Microearthquake seismicity and composite fault plane solutions: Geol. Soc. America Bull., v. 83, p. 13–20.

Scholz, C. H., Barazangi, M., and Sbar, M. L., 1971, Late Cenozoic evolution of the Great Basin, Western United States, as an ensialic interarc basin: Geol. Soc. America Bull., v. 82, p. 2979–2990.

Shawe, D. R., 1965, Strike-slip control of basin-range structure indicated by historical faults in western Nevada: Geol. Soc. America Bull., v. 76, p. 1361–1378.

Sitter, L. U. de, 1959, Structural geology: New York, McGraw-Hill Book Co., 552 p.

Slemmons, D. B., 1967, Pliocene and Quaternary crustal movements of the Basin and Range province, USA, *in* Sea level changes and crustal movements of the Pacific—Pacific Sci. Cong., 11th, Tokyo, 1966, Symposium 19: Osaka City Univ. Jour. Geosciences, v. 10, p. 91–103.

Smith, R. B., and Sbar, M. L., 1974, Contemporary tectonics and seismicity of the Western United States with emphasis on the Intermountain seismic belt: Geol. Soc. America Bull., v. 85, p. 1205–1218.

Snyder, W. S., Dickinson, W. R., and Silberman, M. L., 1976, Tectonic implications of space-time patterns of Cenozoic magmatism in the Western United States: Earth and Planetary Sci. Letters, v. 32, p. 91–106.

Stewart, J. H., 1971, Basin and Range structure: A system of horsts and grabens produced by deep-seated extension: Geol. Soc. America Bull., v. 82, p. 1019–1044.

——1976, Late Precambrian evolution of North America: Plate tectonics implications: Geology, v. 4, p. 11–15.

Stewart, J. H., and Carlson, J. E., 1974, Preliminary geologic map of Nevada: U.S. Geol. Survey Misc. Field Studies Map MF-609, scale 1:500,000.

——1976, Cenozoic rocks of Nevada—Four maps and a brief description of distribution, lithology, age, and centers of volcanism: Nevada Bur. Mines and Geology Map 52, scale 1:1,000,000.

——1978, Generalized maps showing distribution, lithology, and age of Cenozoic igneous rocks in the Western United States, *in* Smith, R. B., and Eaton, G. P., eds., Cenozoic tectonics and regional geophysics of the western Cordillera: Geol. Soc. America Mem. 152, scale 1:5,000,000 (this volume).

Stewart, J. H., and Poole, F. G., 1974, Lower Paleozoic and uppermost Precambrian Cordilleran miogeocline, Great Basin, Western United States, *in* Dickinson, W. R., ed., Tectonics and sedimentation: Soc. Econ. Paleontologists and Mineralogists Spec. Pub. no. 22, p. 28–57.

Stewart, J. H., Albers, J. P., and Poole, F. G., 1968, Summary of regional evidence for right-lateral displacement in the western Great Basin: Geol. Soc. America Bull., v. 79, p. 1407–1413.

Stewart, J. H., Moore, W. J., and Zietz, Isidore, 1977, East-west patterns of Cenozoic igneous rocks, aeromagnetic anomalies, and mineral deposits, Nevada and Utah: Geol. Soc. America Bull., v. 88, p. 67–77.

Suppe, John, Powell, Christine, and Berry, Robert, 1975, Regional topography, seismicity, Quaternary volcanism, and the present-day tectonics of the western United States: Am. Jour. Sci., v. 275A, p. 397–436.

Taber, Stephen, 1927, Fault troughs: Jour. Geology, v. 35, p. 577–606.

Terman, M. J., 1975, Plate-tectonic implications of crustal extension models in east Asia and North America: Geol. Soc. America Abs. with Programs, v. 7, p. 1295.

Thompson, G. A., 1972, Cenozoic basin range tectonism in relation to deep structure: Internat. Geol. Cong., 24th, Montreal 1972, Proc., p. 84–90.

Thompson, G. A., and Burke, D. B., 1973, Rate and direction of spreading in Dixie Valley, Basin and Range province, Nevada: Geol. Soc. America Bull., v. 84, p. 627–632.

——1974, Regional geophysics of the Basin and Range province: Earth and Planetary Sci. Ann. Rev., v. 2, p. 213–238.

Uyeda, Seiya, and Miyashiro, Akiho, 1974, Plate tectonics and the Japanese Islands: A synthesis: Geol. Soc. America Bull., v. 85, p. 1159–1170.

Walker, G. W., 1973, Preliminary geologic and tectonic maps of Oregon east of the 121st meridian: U.S. Geol. Survey Misc. Field Studies Map MF-495, scale 1:250,000.

Wallace, R. E., 1964, Structural evolution, *in* Mineral and water resources of Nevada: U.S. 88th Cong., 2nd sess., Senate Doc. 87, p. 32–39; Nevada Bur. Mines Bull. 65, 314 p.

Wilson, E. D., Moore, R. T., and Cooper, J. R., 1969, Geologic map of Arizona: Tucson, Arizona Bur. Mines, scale 1:500,000.

Wilson, J. T., 1973, Mantle plumes and plate motions: Tectonophysics, v. 19, p. 149–164.

Wise, D. U., 1963, An outrageous hypothesis for the tectonic pattern of the North American Cordillera: Geol. Soc. America Bull., v. 74, p. 357–362.

Wright, L. A., 1976, Late Cenozoic fault patterns and stress fields in the Great Basin and western displacement of the Sierra Nevada block: Geology, v. 4, p. 489–494.

Wright, L. A., and Troxel, B. W., 1973, Shallow-fault interpretation of Basin and Range structure, southwestern Great Basin, *in* de Jong, K. A., and Scholten, Robert, eds., Gravity and tectonics: New York, John Wiley and Sons, Inc., p. 397–407.

York, J. E., and Helmberger, D. V., 1973, Low-velocity zone variations in the southwestern United States: Jour. Geophys. Research, v. 78, p. 1883–1886.

MANUSCRIPT RECEIVED BY THE SOCIETY AUGUST 15, 1977

MANUSCRIPT ACCEPTED SEPTEMBER 2, 1977

Printed in U.S.A.

Geological Society of America
Memoir 152

2
Mesozoic-Cenozoic Cordilleran plate tectonics

Peter J. Coney
Department of Geosciences
University of Arizona
Tucson, Arizona 85721

ABSTRACT

Motion of the Africa plate with respect to Tristan da Cunha since 135 m.y. B.P. and motion of the Pacific plate with respect to Hawaii since 80 m.y. B.P. are combined with relative motion between the North America and Africa plates and between the Kula-Farallon and Pacific plates to compute plate reconstructions and relative motion between the North American Cordilleran margin and Farallon-Kula-Pacific plates during Mesozoic-Cenozoic time. Cordilleran tectonic timing and distribution of petro-tectonic assemblages appear grossly consistent with inferred plate reconstructions and interactions.

Major tectonic transition from Paleozoic modes characterized by rifting and by arcs colliding with the quiet continental margin to the Mesozoic-Cenozoic mode of active Andean-type continental margins coincides with initiation of disruption of Pangea and opening of the Atlantic Ocean–Gulf of Mexico in Late Triassic–Early Jurassic time. North America moved northwestward during Jurassic–Early Cretaceous time, apparently overriding the Farallon plate from Alaska southward. Transition from the Sevier-Columbian orogeny to the Laramide orogeny about 80 m.y. ago coincides with the end of a major magnetic quiet period, initiation of separation of North America from Eurasia, and a general reorganization of plates. The Laramide orogeny (70 to 45 m.y. B.P.) progressed during more rapid westward motion of North America, accompanied by accelerated northeast-southwest convergence between the North America and Farallon plates. The end of the Laramide orogeny is broadly synchronous throughout the Cordillera and coincides with the age of the Hawaii-Emperor "elbow" and a drop in the North America–Farallon convergent rates. From 40 to about 20 m.y. B.P., vast ignimbrite eruptions in the southern Cordillera correlate with final subduction of the Farallon plate. Widespread basalt eruption, block faulting, and collapse of the Basin and Range province occurred since 20 m.y. B.P. and coincided with cessation of subduction, growth of the San Andreas–Queen Charlotte transform faults, and interaction between the North America and Pacific plates.

INTRODUCTION

In order to understand the tectonic evolution of the North American Cordillera, it is necessary to precisely reconstruct plate interactions across and along the Pacific margin of the North American plate. This is difficult to do in that the margin was generally convergent; thus, there is no direct method to determine relative plate motions across the boundary. Furthermore, much necessary evidence, both submarine and on-land, that might be used in reconstructions has been destroyed by the interactions themselves. In the absence of direct means of calculation, one must resort to indirect means. Unfortunately these indirect means and reconstructions that derive from them are at present suspect, owing largely to incomplete submarine geophysical data and to lack of agreement regarding certain assumptions on which the reconstructions are based.

This overview of post-Paleozoic Cordilleran plate tectonics will examine the relationship between continental regional tectonics and plate reconstructions inferred from data in adjacent oceans. The reconstructions (Figs. 2-2 through 2-7) are tentative and represent efforts to portray gross tectonic evolution by summarizing plate regimes and continental tectonic response in the Cordillera from Permian-Triassic time to the present.

BASIS FOR THE RECONSTRUCTIONS

The North American Cordillera has evolved through most of Mesozoic-Cenozoic time as an Andean-type continental-margin mountain system on the western edge of the North America plate (Hamilton, 1969a, 1969b; Coney, 1972, 1973; Burchfiel and Davis, 1972, 1975; Dickinson, 1976). As a result, it would be helpful to know the relative motions between North America and oceanic "Pacific" plates to the west. It would also be helpful if motions of these plates with respect to the mantle could be approximated, because some have argued that these motions, as well as relative motions of plates to one another, influence tectonic response (Coney, 1971, 1973; Wilson and Burke, 1972).

Studies of relative motions between the North America plate and various "Pacific" plates derive from construction of relative-motion vector circuits by vector addition (Le Pichon and others, 1973) at a series of specific times across the boundaries of as many as six plates (Atwater and Molnar, 1973), extending from the Pacific plate through the Antarctica, Australia-India, Africa, and North America plates. Vector addition by computer yields "instantaneous" relative motion between North America and the Pacific plate at specific times. From these "instantaneous" rotations, extrapolated finite rotation reconstructions can be made. These reconstructions are considered very reliable between 40 m.y. B.P. and the present. Because of uncertainties in the Indian Ocean and within the Antarctic plate, they are less reliable from 40 to 80 m.y. B.P. (Molnar and Atwater, 1973). No reconstructions tying the "Pacific" plates to North America have been published for times prior to 80 m.y. B.P.

An alternative approach to the problem is to use the so-called "hot-spot" frame of reference (Morgan, 1971, 1972) to reconstruct past plate motions. This is based on the assumption that aseismic ridges and seamount chains on the floor of the world's oceans and "anorogenic" alkalic ring-dike complexes and plateau flood-basalt provinces on continents record, albeit imperfectly, motion of the plates over assumed semifixed "hot spots" in the mantle. There is currently much debate about the assumed fixed positions of the melting anomalies with respect to one another (Burke and others, 1973; Molnar and Atwater, 1973; Molnar and Francheteau, 1975) and also whether the seamount chains are tracks of past motion or simply leaky cracks in the lithosphere of one sort or another (Turcotte and Oxburgh, 1973).

In applying the "hot-spot" hypothesis to the North American Cordillera (Fig. 2-1), the procedure is first to estimate rotational poles and rates of the Pacific plate over Hawaii and other "hot spots" from trends and ages along lines of dead volcanoes on the floor of the Pacific Ocean. Then, using relative-motion data from magnetic anomalies and fracture zones formed at the boundaries of the Pacific, Farallon, and Kula plates, motions of the Farallon and Kula plates with respect to the "hot-spot" frame of reference can be derived. Next, poles and rates for motion of the Africa plate over "hot spots" in the Atlantic, such as Tristan da Cunha, are estimated. Then, using relative-motion data for North America with respect to Africa, motions of North America with respect to the "hot-spot" frame of reference are derived. Finally, working back in time with a computer, the position of North America and the "Pacific" plates can be reconstructed and, by vector subtraction, the relative motion between the North America plate and the various "Pacific" plates can be computed, all with respect to the "hot-spot" frame of reference (assumed fixed).

In the Pacific, such reconstructions are an artifact of the inferred motion of the Pacific plate during the past 80 m.y., based on the trend of the Hawaii-Emperor seamount chain. This trend is northwest from Hawaii to the prominent "elbow," then almost northward to the junction with the Aleutian Trench. Extensive dating along these trends, particularly from Hawaii to the elbow, places an age of just over 40 m.y. on the elbow and about 80 m.y. for the northern end of the chain at the Aleutian Trench (Morgan, 1972; Clague and Jarrard, 1973). This permits the suggestion that from 80 to 40 m.y. B.P., the Pacific plate rotated northward and from 40 m.y. B.P. to the present, its motion has been more northwestward. Before 80 m.y. B.P., the motion is very difficult to estimate, but it was probably more westward.

In the Atlantic, it is assumed the Walvis aseismic ridge was produced by motion of the Africa plate over a Tristan da Cunha hot spot, between 135 m.y. B.P. (start of the opening of the South Atlantic Ocean) and the present. Dating along the trend is very meager, but the data available are consistent with the model (Pastouret and Goslin, 1974; Perchnielsen and Supko, 1975). Before 135 m.y. B.P., the motion of Africa is very difficult to estimate.

Working back in time from the present to 40 m.y. B.P., the reconstructions that derive from these two indirect approaches are very similar and reasonably reliable. From 40 to 80 m.y. B.P., the reconstructions differ (Coney, 1976), and the reliability of both methods decreases dramatically. At ages older than 80 m.y., comparison is not possible. In the overview which follows, the "hot-spot" frame of reference is used. I am aware that the reconstructions may not be unique. Some will not accept the assumption on which they are based. In Figures 2-2 through 2-7 continental tectonic data are generalized into various petro-tectonic assemblages (Dickinson, 1972) that are assumed to have plate-tectonics significance. For an excellent recent summary of distribution of petro-tectonic assemblages in Western United States and Canada, see Dickinson (1976).

PERMIAN–TRIASSIC TIME

Figure 2-2 is a generalized abstraction of critical paleotectonic features that relate to the North American Cordillera during late Paleozoic and early Mesozoic time. It is presented as a point of departure, because the period represented was the most profound tectonic transition in the entire Phanerozoic history of not only the North American continent but also the Cordillera on the western margin of the continent.

A suture is shown marking a zone of oceanic closure and final collision between North

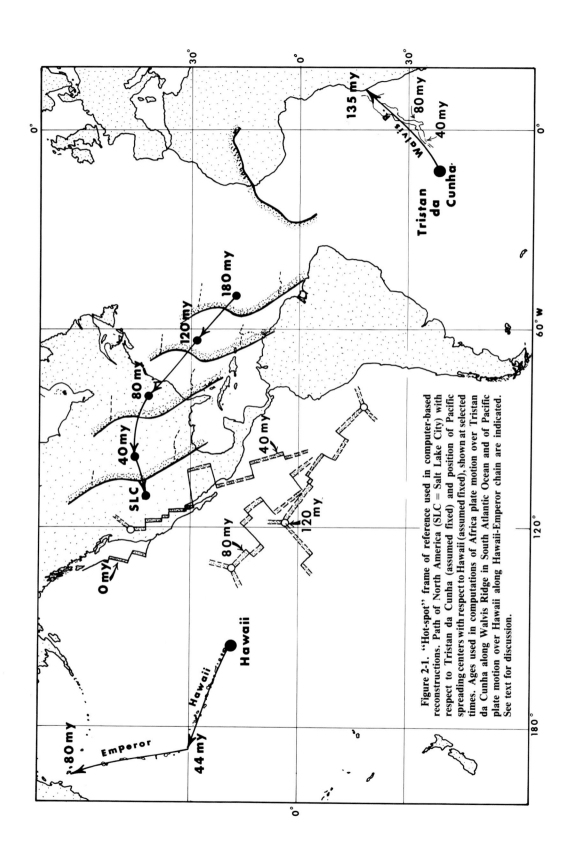

Figure 2-1. "Hot-spot" frame of reference used in computer-based reconstructions. Path of North America (SLC = Salt Lake City) with respect to Tristan da Cunha (assumed fixed) and position of Pacific spreading centers with respect to Hawaii (assumed fixed), shown at selected times. Ages used in computations of Africa plate motion over Tristan da Cunha along Walvis Ridge in South Atlantic Ocean and of Pacific plate motion over Hawaii along Hawaii-Emperor chain are indicated. See text for discussion.

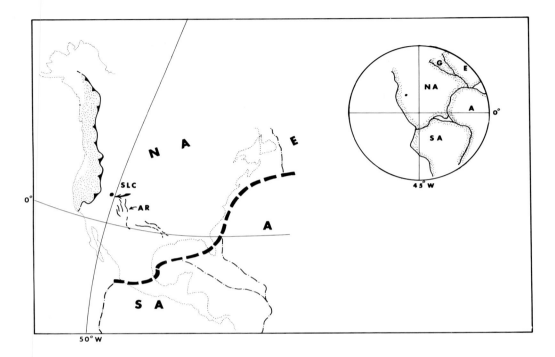

Figure 2-2. Permian-Triassic configuration. Major paleotectonic features are discussed in text. Heavy dashed line is late Paleozoic suture, lines labeled AR are Wichita–Ancestral Rockies trends, dotted area behind barbed line is the accreted "oceanic" terrane emplaced along Pacific margin of North America. Position of North America is in "hot-spot" frame of reference (present geographical coordinates). SLC is Salt Lake City; Pangea is depicted in global projection at right. On this and all succeeding maps NA = North America, SA = South America, A = Africa, E = Europe, G = Greenland, H = Hawaii, and TC = Tristan da Cunha.

America–Europe and Africa–South America. The collision is inferred to have occurred in late Paleozoic time. The only place this suture is probably exposed at the surface today is along a belt of subsequently much modified ultramafic rocks, known as the Motagua-Polychic fault zone, in southern Guatemala. Zones of late Paleozoic tectogenesis north of the suture in the Guatemala highlands and in the Marathon area and Ouachita Mountains of the Southern United States are presumed to have resulted from convergence and collision between South America and North America. The important point is, regardless of whether one accepts Paleozoic closure of a proto-Atlantic Ocean and Gulf of Mexico, most now accept some variation of the fit of Africa against North America and subsequent Mesozoic-Cenozoic opening of the central Atlantic Ocean (Pitman and Talwani, 1972). Recalling that what one does with Africa one must also do with South America at this time (they were joined until latest Jurassic time), one is left with a very embarrassing overlap of South America onto southern Mexico and Central America. The reader is welcome to choose a preferred fit, but overlap will be very difficult to eliminate. The reconstruction shown here, by no means unique, simply falls out of one fit of Africa against North America. The inference remaining, nearly a geometric necessity, is that all of Mexico and Central America within the overlap was either not in existence or not in its present position prior to the late Paleozoic–early Mesozoic time. (This will be dealt with in subsequent figures.)

Another feature shown in Figure 2-2, which probably relates to late Paleozoic convergence in the Gulf of Mexico, is the Wichita–Ancestral Rockies deformation (Ham and Wilson, 1967; Eardley, 1962). This enigmatic structural belt, curiously amagmatic, may have produced lines of weakness that were reactivated during Laramide time in the Rocky Mountains of the United States. The Wichita–Ancestral Rockies belt, in turn, may have been a reactivation of a later Precambrian aulacogen (Dickinson, 1974).

A vast terrane of puzzling lithology, extending from central California northward along the Pacific margin to Alaska, is portrayed in Figure 2-2. Represented are such terranes as the Calaveras Formation of the Sierra foothills (Schweickert and Cowen, 1975), the Golconda thrust allochthon of western Nevada (Silberling, 1975), the Cache Creek Group of western Canada (Monger and others, 1972; Monger, 1975), and what here is termed a "Klondike" terrane (Tempelman-Kluit and others, 1976) in the Yukon and central Alaska. These terranes are best described simply as of "oceanic affinity" and include cherts, argillites, variable volcanic rocks, and some minor limestone. They were emplaced above, beneath, or against terranes of shelf environment on the edge of the North American continent during early Mesozoic time. The event seems to have closely followed final closure of the Atlantic Ocean and seems to have been roughly synchronous along the entire Pacific margin.

This event, termed the Sonoma "orogeny" in the United States (Speed, 1971; Silberling, 1975), marks the end of a pattern of tectonic response in the Cordillera that prevailed through much of Paleozoic time; the event seems to have resembled current offshore island-arc tectonics in the western and South Pacific, off Australia and east Asia (Churkin, 1974). From this time forward the Cordillera shifted to its course of Andean-type continental-margin tectonics, which has characterized most of Mesozoic-Cenozoic time. The change in behavior coincided with a major global reorganization of plate regimes, namely the initiation of break-up of Pangea and opening of the Atlantic Ocean, and westward motion of the North America plate.

EARLY TO MIDDLE JURASSIC TIME (180 TO 155 M.Y. B.P.)

Late Triassic and Early to Middle Jurassic volcano-plutonic terranes (Fig. 2-3), assumed to be of arc affinity, are scattered along the Pacific margin from northwestern Mexico and southern Arizona northward into Canada (Armstrong and Suppe, 1973; Gabrielse and Reesor, 1974). It is not clear how far, or indeed if, the trend extends south across Mexico. These early Mesozoic marginal arcs stood on Precambrian crust in Arizona and Mexico, but northward they were apparently largely insular (or peninsular) in character, standing offshore from shelf seas (Dickinson, 1976) and built in part upon the "oceanic" terranes inherited from Paleozoic time. The inference made is that these rocks record east-dipping subduction of oceanic lithosphere from the west. Scattered back-arc thrusts are recorded and are best known in Nevada (Burchfiel and Davis, 1975), but the offshore character of the arc source areas was not conducive to widespread detrital-wedge development eastward onto the Cordilleran foreland.

Shown in Figure 2-3 are several terranes located off the margin of North America, principally in the region of the Sierra Nevada foothills (Schweickert and Cowen, 1975), Cascade Mountains–Vancouver Island (Dickinson, 1976), the Alexander terrane of southeastern Alaska, and most of southern Alaska south of the Denali fault. These are identified in part by the fact that ophiolite belts and possible lines of suture separate them from rocks closely linked to the North American continent. These terranes are inferred to have been accreted onto North America during Mesozoic time, possibly in some cases as late as Cretaceous time. They are best described as simply "suspect" terranes. In this way, the question of how far they have traveled is avoided. Paleomagnetic data (Jones and others, 1977; Irving and Yole, 1972)

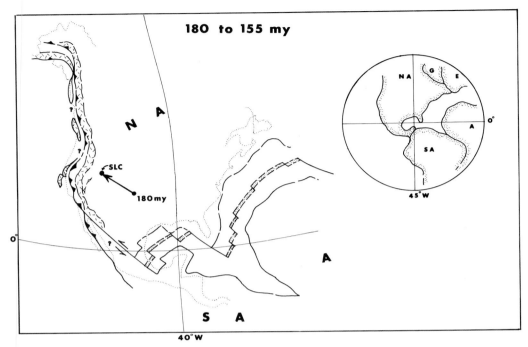

Figure 2-3. Early to Middle Jurassic configuration (180 to 155 m.y. B.P.). Position of North America at 155 m.y. B.P. after early opening of Atlantic Ocean and Gulf of Mexico. Geometry of spreading centers hypothetical. Arc terranes shown by dashed pattern, subduction zones by heavy barbed line, and thrust belts in light barbed lines. Heavy black arrow on Salt Lake City (SLC) shows motion of North America during this period.

on some, however, can be inferred to suggest large north-south distances of travel with respect to continental North America.

Shown in Figure 2-3 is the Silver and Anderson (1974) "mythical megashear," slicing across Mexico to the southwest corner of the Gulf of Mexico. The feature is shown with full awareness that even the creators of the concept are not exactly sure where on the ground it is and that some doubt its existence. The principal reason for showing it is to emphasize that, regardless of the eventual fate of the concept, it is the only really new and provocative idea advanced since the advent of plate tectonics that offers a solution to the infamous overlap of South America onto Mexico–Central America and the enigma of the origin of the Gulf of Mexico.

The feature is drawn as described by Silver and Anderson (1974) in Sonora and the southwestern United States, where left offset of Precambrian terranes suggested it. This general trend is extrapolated southeastward to the southwest corner of the Gulf of Mexico where it is inferred to link with spreading centers then in the opening Gulf of Mexico. This approximately N55°W extrapolation is reinforced by fractures and faults of Jurassic through Cretaceous age in this direction, slicing through southern Arizona (Titley, 1976), and by the fact that this direction lies close to a small circle about the pole of opening of the Atlantic Ocean and the separation of North America from Africa–South America (Pitman and Talwani, 1972). Thus, the inference is that the megashear was a transform fault. It is assumed the transform fault operated sometime between initiation of the opening of the central Atlantic (180 m.y. B.P.) and initial opening of the south Atlantic (135 m.y. B.P.). I infer that during this time, the Gulf of Mexico opened but was not open across the transform into the Pacific Ocean. As a result of the barriers, Tethys water and occasional Pacific spills (Burke, 1975) evaporated to form Gulf of Mexico salt deposits while widespread continental conditions prevailed on much of surrounding lands. Near Oxfordian time (±155 m.y. B.P.) there was widespread marine transgression and deposition of limestone in regions adjacent to the Gulf of Mexico. This event may mark the breaking away of South America from Mexico–Central America and the flooding of the Gulf and Caribbean region with Pacific water.

The "hot-spot" model yields a northwesterly movement of North America during this time. The geometry of Pacific spreading centers is unknown.

LATE JURASSIC TO LATE CRETACEOUS TIME (155 TO 80 M.Y. B.P.)

Late Jurassic to Late Cretaceous tectonic patterns in the Cordillera are clearer and more unified, at least from southern Mexico to Canada (Fig. 2-4). A well-developed volcano-plutonic arc terrane can be tracked from the central western Pacific margin of Mexico through northern Baja California into the Sierra Nevada and Idaho batholiths, and northward into Canada (Krummenacher and others, 1975; Armstrong and Suppe, 1973; Armstrong, 1974b; Gabrielse and Reesor, 1974; Lanphere and Reed, 1973). Accreted subduction assemblages, such as the Franciscan assemblage, and associated fore-arc basin deposits, such as the Great Valley assemblage, are well identified in the southern Cordillera, standing west of the arc terranes (Dickinson, 1976).

The trajectory through Canada and into Alaska is not so clear. During Late Jurassic–Early Cretaceous time a narrow, largely submarine, volcanic arc extended through southeastern Alaska (Berg and others, 1972) from at least Ketchikan northward into the eastern Alaska Range. The Alaska Range–Aleutian Peninsula arc, on the other hand, was quiet. Farther inboard Late Jurassic to Cretaceous plutons occur along the Omineca crystalline belt in Canada, across the Yukon-Tanama complex, and in the Kuyukuk basin of central Alaska (Gabrielse and Reesor, 1974; Patton, 1973). Between the two arc trends in Canada were the Bowser-Tyaughton-Mathow basins, which Dickinson (1976) interpreted as fore-arc basins linked to the eastern arc trend. Final suturing of the "exotic" terranes, including the Cascades–Vancouver Island, the Alexander terrane, and the Alaska Range–Aleutian Peninsula, may not have occurred until Early Cretaceous time (see discussion in Dickinson, 1976). In any event, by Late Cretaceous time the arc had migrated westward and was well established along most of the present trend of the Canadian coastal batholith.

North of the Kuyukuk Basin there was intense telescoping in the Brooks Range. The Brooks Range is shown in Figure 2-4 as having rifted away from Arctic Canada in early Mesozoic time, rotated through the Canada Basin, and collided with central Alaska in Late Jurassic–Early Cretaceous time (Tailleur, 1973).

On the Cordilleran foreland, east-verging thrust belts began to evolve near the Paleozoic shelf-slope break throughout most of the length of the Cordillera. They are best defined in the United States where they are related to the Sevier orogeny (Armstrong, 1968a, 1968b). These thrust belts stood east of another feature (shown separately in Fig. 2-4 for emphasis), the so-called Cordilleran metamorphic core complexes (Misch, 1960) or infrastructure (Armstrong and Hansen, 1966). These terranes are today characterized by widespread low-dipping foliation and lineation, cataclastic gneiss, migmatite, and gneiss dome culminations (Armstrong, 1968b;

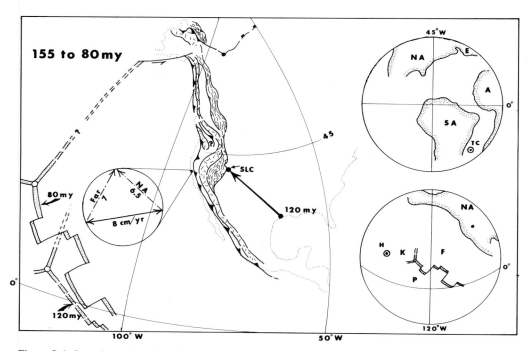

Figure 2-4. Late Jurassic to Late Cretaceous configuration (155 to 80 m.y. B.P.) of North America during Sevier orogeny. Symbols as on previous figures. Areas with fine, wavy lines contain metamorphic core complexes. Vector diagram shows inferred motion of Farallon (Far) and North America (NA) plates (dash-dot lines) with respect to hot-spot frame of reference and vector subtraction (solid line) for North America–Farallon relative motion. Velocities in centimetres per year.

Campbell, 1970; Reesor, 1970). Scattered granitic plutons, some of two-mica type, but mixed with more normal biotite-hornblende types, are typical (Gabrielse and Reesor, 1974; Miller and Engels, 1975). It is certain these belts have had complex histories, but it appears that they may have initiated vertical uplift in Late Jurassic time, as recorded in sedimentary petrology in basins on their flanks (Eisbacher, 1974). The complexes are probably related to thermal disturbances common to arc regions in a continental and compressional environment. Those with the most dramatic vertical uplift seem to lie just west of strongly telescoped foreland thrust belts and may have been influenced by massive westward cratonic underthrusting beneath the foreland thrust belts (Coney, 1973). In the Northern United States and Canada, the complexes appear to have remained active until early Tertiary time. South of the Snake River Plain they behaved differently and were either perpetuated through, reactivated, or initiated in Oligocene-Miocene time.

The geometry of sea-floor spreading centers shown in the Pacific is largely defenseless. The position of the spreading centers at 120 m.y. B.P. is simply extrapolated from the position at 80 m.y. B.P., but is similar to the position shown by Larson and Pitman (1972). The position at 80 m.y. B.P. results from rotations of the anomaly pattern about pole positions interpreted from the Hawaii-Emperor seamount trend. If one accepts these reconstructions, the entire Pacific margin of North America from at least the Gulf of Alaska southward interacted with the Farallon plate. North America–Europe is shown still moving northwestward away from Africa. After 135 m.y. B.P., South America separated from Africa as well.

LATE CRETACEOUS TO LATE EOCENE TIME (80 TO 40 M.Y. B.P.)

The period just discussed, from Late Jurassic to Late Cretaceous time, saw a maturation of the Cordillera into a definite continental-margin Andean-type orogen. One salient characteristic by the latter part of this period was linear continuity along the chain, manifested particularly in volcano-plutonic arc terranes. I infer that these arcs stood above east-dipping Benioff zones. Only in Alaska and the Caribbean does the continuity disintegrate. Here, paleogeographic complexities existed that are still not fully understood.

The simple pattern of a well-defined volcano-plutonic belt of arc affinity along the Pacific margin began to break up in Late Cretaceous time (Fig. 2-5), about the same time that the initial effects of the Laramide orogeny began in the Rocky Mountains foreland of the Western Interior of the United States (Armstrong, 1974a; Coney, 1976). In the south, the Cretaceous magmatic pattern spread northeastward out of Sonora into the Mexican Plateau and southern Arizona and New Mexico to produce a belt of occasionally copper-laden Laramide plutons (Livingston and others, 1968; Anderson and Silver, 1974). By the end of Cretaceous time, most of the Pacific Northwest was quiet, only to erupt violently in Eocene time (Armstrong, 1974b, and this volume). Meager data suggest that much of the Canadian coastal batholith cooled during Late Cretaceous to early Tertiary time (80 to 40 m.y. B.P.; Roddick and Hutchinson, 1974), and plutons of this age are well known in the Alaska Range and Aleutian Peninsula (Lanphere and Reed, 1973). Between these Laramide igneous belts, only scattered igneous centers are known, the infamous Laramide igneous "gap" in the central Western United States (Armstrong, 1974a; Coney, 1976). Much has been made of the fact that this igneous "gap" lay directly west of, and in the region of, the classic Laramide Rocky Mountain uplifts.

A distinct belt of generally asymmetric basement-cored uplifts, many bounded by thrust faults, extends from Montana and Idaho south into Arizona and New Mexico (Grose, 1972; Tweto, 1975), including the monoclines of the Colorado Plateau (Kelley, 1955). These structural uplifts and the basins between them were the result of Laramide orogeny, which progressed between Late Cretaceous and late Eocene time (Coney, 1972, 1976). This deformational phase is clearly separated from thin-skinned decollement-style folding and thrusting, found to the west in Utah and Nevada, which evolved during Late Jurassic to Late Cretaceous time and is termed the Sevier orogeny (Armstrong, 1968a). The boundary between the two belts appears to have been in part controlled by stratigraphy. Thin-skinned low-angle thrust-fault styles were favored in the Paleozoic miogeocline. Rocky Mountain basement uplifts were favored on the thin cratonic shelf (Burchfiel and Davis, 1975; Coney, 1973, 1976). Northward in Canada and southward in Mexico, distinction between Laramide and Sevier thrusting is less clear (Armstrong, 1974), but it can be argued that most of the thin-skinned deformation in both areas is Laramide in age (Coney, 1976).

One of the most astonishing facts about the Laramide orogeny is that it ceased virtually synchronously from Alaska to the Caribbean by middle to late Eocene time (50 to 40 m.y. B.P.; Coney, 1971, 1972, 1973, 1976). This very important tectonic event is clearly reflected in igneous and structural patterns throughout the entire North American Cordillera. For example, foreland thrusting in Canada, the United States, and Mexico ceased by this time, the Canadian coastal batholith was extinguished, and the Greater Antilles arc-trench system stopped activity. It seems obvious that only a fundamental reorganization of plate motions can explain this tectonic transition.

The positions of Pacific spreading centers at the beginning and end of the Laramide orogeny are shown in Figure 2-5. It is to be noted that at 80 m.y. B.P. the Farallon-Kula spreading center was pointed at southern Alaska; by 40 m.y. B.P. it had swept southward to Vancouver.

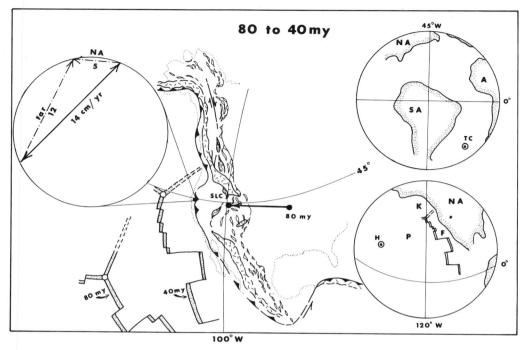

Figure 2-5. Late Cretaceous to late Eocene configuration (80 to 40 m.y. B.P.) showing plate regimes during Laramide orogeny. Symbols as on previous figures. See text for discussion. K = Kula plate, F = Farallon plate, and P = Pacific plate.

These relationships are an artifact of the inferred motion of the Pacific plate during the past 80 m.y., based on trends of the Hawaii-Emperor seamount chain. Recalling that the prominent elbow is dated at just older than 40 m.y., the major change in Pacific plate motion inferred at this time coincides with the end of the Laramide orogeny. This change in direction of motion from northward (between 80 and 40 m.y. B.P.) to northwestward (after 40 m.y. B.P.) reduced the rates of Farallon–North America plate convergence from their high during the Laramide orogeny. The convergence rates decreased by almost one-half after 40 m.y. B.P. As I have stated elsewhere, the initiation of the Laramide orogeny around 80 m.y. B.P. correlates with major changes in plate motions, also on a global scale. The Laramide orogeny thus appears bracketed by major changes in global patterns of sea-floor spreading and motions of plates (Coney, 1976). Finally, this orogeny, certainly the most widespread and profoundly compressional period of deformation in the entire Phanerozoic history of western North America, may coincide with high convergence rates between the North America plate and the Farallon plate to the west.

Most of the complex igneous and structural patterns of the southern Cordillera during Laramide time are probably explained by rapid northeast-southwest–directed convergence between the North America plate and a young, hot Farallon plate. The convergence produced elevated stress in North America lithosphere which took advantage of zones of weakness inherited from Precambrian time and particularly from the Paleozoic Ancestral Rockies. These zones of weakness allowed failure of the foreland as much as 1,500 km inboard of the trench. The rapid convergence and subduction of a young, hot Farallon plate may have caused pronounced flattening of dip of Farallon slabs being subducted beneath of southern Cordillera; this could result in gaps and rapid shifting of igneous activity over the same region.

LATE EOCENE TO EARLY MIOCENE TIME (40 TO 20 M.Y. B.P.)

At the end of the Laramide orogeny, tectonic and igneous patterns changed again, and the next 20 m.y. (Fig. 2-6), from late Eocene to early Miocene time, comprise one of the most puzzling times in Cordilleran history, particularly in the southern half.

Widespread erosion and beveling of the landscape in the southern Cordillera is well documented (Epis and Chapin, 1975). Soon after, an unusual igneous pattern evolved from Laramide trends and rather quickly spread across much of the Laramide igneous "gap" (Armstrong, 1974a, 1974b, and this volume) and the present Cordillera Occidental of Mexico. This is the widespread ignimbrite "flareup" that so typifies Oligocene–early Miocene time in the southern Cordillera (Lipman and others, 1971; Noble, 1972). Voluminous eruptions of ash-flow sheets, batholithic in proportions, from caldera complexes veneered much of the beveled Laramide landscape. The area of volcanism narrows abruptly northward along the Cascade arc trend. Northward through Canada, igneous activity is very sparse, but volcano-plutonic complexes reappear in southern Alaska (Lanphere and Reed, 1973).

The ignimbrite "flareup" of the southern Cordillera is a most fascinating phenomenon (Noble, 1972). Plate-tectonics reconstructions have always clearly linked the outburst to subduction of the Farallon plate (Lipman and others, 1971). This seems obvious since north of the Kula–Farallon–North America triple junction where the Kula or Pacific plate transformed northward past North America, little or no arc activity is recognized. In southern Alaska, subduction is indicated by normal arc activity. The main problem is to explain the voluminous calc-alkalic outbursts and the complex shifting patterns in the southern Cordillera extending as much as about 1,000 km inland. From reconstructions it can be argued that rather young, hot oceanic lithosphere was being subducted, perhaps at a shallow angle, beneath the "flareup." The low angle was perhaps inherited from Laramide time. In any event, high geothermal gradients may be inferred in the arc area where the "flareup" occurred.

In the same region as the ignimbrite eruptions and at the same time, Cordilleran metamorphic core complexes were either perpetrated, reactivated, or initiated from the Snake River Plain south into Mexico. Rapid uplift and cooling of these gneissic intrusive complexes with resulting cataclasis and gravitation-driven denudation faulting of unmetamorphosed cover into adjacent basins are well documented by a growing literature (Armstrong and Hills, 1967; Armstrong, 1972; Coney, 1974; Snoke, 1975; Mauger and others, 1968; Damon and others, 1963; Davis, 1975; Campton and others, 1977). The rising core complexes and the erupting calderas present a most unusual and enigmatic tectonic-igneous manifestation that must be related to profound thermal disturbance in an arc setting.

Changes in geometry, for the first time well controlled, evolved quickly. From its position at 40 m.y. B.P., most of the ridge system was progressively subducted by 20 m.y. B.P. from southern Mexico north (Atwater, 1970). The Kula-Pacific Ridge, if it was not extinguished about 50 m.y. B.P., must have been subducted by about 30 m.y. B.P. in the Aleutian–southern Alaska trench, but subduction of the Pacific plate continued (Grow and Atwater, 1970). The Pacific-Farallon Ridge, first at the Mendocino Fracture Zone, progressively evolved into two Pacific–North America transforms, the San Andreas and Queen Charlotte faults, halting subduction except for the remaining fragment of the Farallon plate off the Pacific Northwest. In the Caribbean–Central America region, major plate reorganization is inferred at 40 m.y.

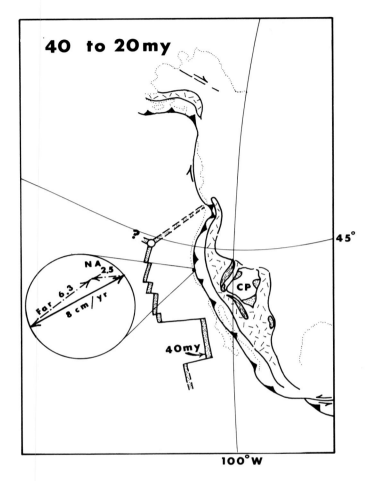

Figure 2-6. Late Eocene to early Miocene configuration (40 to 20 m.y. B.P.) showing plate tectonics of the ignimbrite "flareup." CP is the Colorado Plateau. Dotted areas within North America are large caldera complexes. Areas of fine, wavy lines are core complexes. North America and spreading centers shown in 40-m.y. B.P. positions.

B.P. (Coney, 1972). The Greater Antilles north-facing Laramide arc-subduction complex shut down and the Lesser Antilles east-facing system erupted. This change marked initiation of the Caribbean plate as we now know it. Figure 2-6 shows this plate, including Central America, starting to move eastward relative to North America. Nuclear Central America is inferred to have lain south of Mexico in Laramide time, and the inferred movement explains offset Laramide structural and igneous trends.

EARLY MIOCENE TIME TO THE PRESENT (20 M.Y. B.P. TO PRESENT)

In Miocene time (about 15 to 20 m.y. B.P.), tectonics in the Cordillera changed again (Fig. 2-7). The principal change was initiation of extensional faulting in the Basin and Range province (Hamilton and Myers, 1966) and sparse bimodal (mainly basaltic) volcanism (Lipman and others, 1971). This pattern extends from Canada south through Mexico, but is widest and best developed in the Western United States. This pattern of rifting coincides with growth of the Pacific–North America transform faults, particularly the San Andreas transform, and a genetic link is obvious (Atwater, 1970). Today, arc-trench terranes are limited to southern

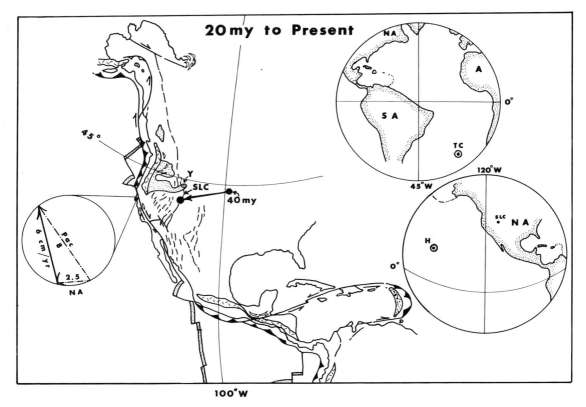

Figure 2-7. Early Miocene time (20 m.y. B.P.) to present plate tectonics of the Basin and Range province. Basin-range block faulting and San Andreas and other transforms generalized by black lines. Dotted areas are basalts of Columbia Plateau and Snake River Plain. Plate geometry is that of present.

Alaska—where the Pacific plate is being subducted—the Washington-Oregon Cascades, and southern Mexico–Central America—where remaining scraps of the Farallon plate converge on the North American margin. Outpourings of flood basalt occurred behind the Cascade arc, and a "hot spot" migrated up the Snake River Plain to its present position beneath the Yellowstone Plateau.

The principal enigma of this period, and much of the concern of the symposium and this book, is the Basin and Range province. It is to be remembered that the block faulting and basaltic volcanism extend far outside the Great Basin of western Utah and Nevada and cover much of the Cordillera from at least central British Columbia south to southern Mexico. Models for the evolution of the Basin and Range and related provinces are discussed elsewhere (Stewart, this volume; Hamilton, 1975; Smith, this volume; Scholz and others, 1971). I wish simply to make several observations that are emphasized by this overview.

I remain impressed by the major change in relative plate motions implied by growth of transform faults, such as the San Andreas, as North America began to interact with a northwest-moving Pacific plate rather than a northeast-moving Farallon plate. Growth of this change coincides with development of basin-range faulting and cessation of subduction. I am also impressed by the patterns suggested by examination of Figures 2-6 and 2-7. The basin-range faulting seems best developed within the region that had previously suffered widespread thermal disturbances, possible thinning and melting of continental lithosphere, and the vast ignimbrite eruptions of Oligocene time. As relative motion changed from convergence to transform, the thermally weakened ignimbrite belt may have failed as a wide transform and spreading boundary; this, of course, was essentially Atwater's original (1970) suggestion and in most ways remains the most satisfactory model of all.

ACKNOWLEDGMENTS

I am grateful to Gordon Eaton, Robert Smith, and Don R. Mabey, who convened the Alta Penrose Conference that led to the symposium. Bringing together geophysicists and geologists proved very rewarding.

An overview of plate tectonics and the North American Cordillera must draw heavily on the work and ideas of many scientists. Many people have shared views freely and contributed much to this paper. I am particularly grateful to W. R. Dickinson, D. L. Jones, W. Hamilton, and J. Morgan. Joan Grette and Kevin Furlong, former students, were a great help in developing computer programs. Research Corporation kindly helped fund Cordilleran investigations. Warren Hamilton and Jason Morgan reviewed this paper and provided many helpful suggestions to improve it.

REFERENCES CITED

Anderson, T. H., and Silver, L. T., 1974, Late Cretaceous plutonism in Sonora, Mexico and its relationship to circum-Pacific magmatism: Geol. Soc. America Abs. with Programs, v. 6, p. 484.

Armstrong, R. L., 1968a, Sevier orogenic belt in Nevada and Utah: Geol. Soc. America Bull., v. 79, p. 429–458.

——1968b, Mantled gneiss domes in the Albion Range, south Idaho: Geol. Soc. America Bull., v. 79, p. 1295–1314.

——1972, Low-angle (denudation) faults, hinterland of the Sevier orogenic belt, eastern Nevada and western Utah: Geol. Soc. America Bull., v. 83, p. 1729–1754.

——1974a, Magmatism, orogenic timing, and orogenic diachronism in the Cordillera from Mexico to Canada: Nature, v. 274, p. 348–351.

——1974b, Geochronometry of the Eocene volcanic-plutonic episode in Idaho: Northwest Geology, v. 3, p. 1–14.

——1978, Cenozoic igneous history of the U.S. Cordillera from lat 42° to 49°N, in Smith, R. B., and Eaton, G. P., eds, Cenozoic tectonics and regional geophysics of the western Cordillera: Geol. Soc. America Mem. 152 (this volume).

Armstrong, R. L., and Hansen, E. C., 1966, Metamorphic infrastructure in the eastern Great Basin: Am. Jour. Sci., v. 264, p. 112–127.

Armstrong, R. L., and Hills, F. A., 1967, Rb-Sr and K-Ar geochronologic studies of mantled gneiss domes, Albion Range, southern Idaho, USA: Earth and Planetary Sci. Letters, v. 3, p. 114–124.

Armstrong, R. L., and Suppe, J., 1973, Potassium-argon geochronometry of Mesozoic igneous rocks in Nevada, Utah, and southern California: Geol. Soc. America Bull., v. 84, p. 1375–1392.

Atwater, T., 1970, Implications of plate tectonics for the Cenozoic tectonic evolution of western North America: Geol. Soc. America Bull., v. 81, p. 3513–3536.

Atwater, T., and Molnar, P., 1973, Relative motion of the Pacific and North American plates deduced from seafloor spreading in the Atlantic, Indian, and South Pacific Oceans: Stanford Univ. Pubs. Geol. Sci., v. 13, p. 136–148.

Berg, H. C., Jones, D. L., and Richter, D. N., 1972, Gravina-Nutzotin belt—Tectonic significance of an upper Mesozoic sedimentary and volcanic sequence in southeastern Alaska: U.S. Geol. Survey Prof. Paper 800-D, p. 1–24.

Burchfiel, B. C., and Davis, G. A., 1972, Structural framework and evolution of the southern part of the Cordilleran orogen, western United States: Am. Jour. Sci., v. 272, p. 97–118.

——1975, Nature and controls of Cordilleran orogenesis, western United States: Extensions of an earlier synthesis: Am. Jour. Sci., v. 275-A,

p. 363–395.

Burke, K., 1975, Atlantic evaporites formed by evaporation of water spilled from Pacific, Tethyan, and Southern oceans: Geology, v. 3, p. 613–616.

Burke, K., Kidd, W.S.F., and Wilson, J. T., 1973, Relative and latitudinal motion of Atlantic hot spots: Nature, v. 245, p. 133–137.

Campbell, R. B., 1970, Structural and metamorphic transitions infrastructure to superstructure, Cariboo Mountains, B.C.: Geol. Assoc. Canada Spec. Paper no. 6, p. 67–71.

Churkin, M., 1974, Paleozoic marginal ocean basin-volcanic arc systems inf the Cordilleran fold belt: Soc. Econ. Paleontologists and Mineralogists Spec. Pub. no. 19, p. 174–192.

Clague, D. A., and Jarrard, R. D., 1973, Tertiary Pacific plate motion deduced from the Hawaiian-Emperor Chain: Geol. Soc. America Bull., v. 84, p. 1135–1154.

Compton, R. R., Todd, V. R., Zartman, R. E., and Nasser, C. W., 1977, Oligocene and Miocene metamorphism, folding, and low-angle faulting in northeastern Utah: Geol. Soc. America Bull., v. 88, p. 1237–1250.

Coney, P. J., 1971, Cordilleran tectonic transitions and motion of the North American plate: Nature, v. 233, p. 462–465.

——1972, Cordilleran tectonics and North American plate motion: Am. Jour. Sci., v. 272, p. 603–628.

——1973, Non-collision tectogenesis in western North America, in Tarling, D. H., and Runcarn, S. H., eds., Implications of continental drift to the earth sciences: New York, Academic Press, Inc., p. 713–727.

——1974, Structural analysis of the Snake Range "decollement," east-central Nevada: Geol. Soc. America Bull., v. 85, p. 973–978.

——1976, Plate tectonics and the Laramide orogeny: New Mexico Geol. Soc. Spec. Pub. no. 6, p. 5–10.

Damon, P. E., Erickson, R. C., and Livingston, D. E., 1963, K-Ar dating of Basin and Range uplift, Catalina Mountains, Arizona: Nuclear Geophysics–Nuclear Sci. Ser., no. 38, p. 113–121.

Davis, G. H., 1975, Gravity-induced folding off a gneiss dome complex, Rincon Mountains, Arizona: Geol. Soc. America Bull., v. 86, p. 979–990.

Dickinson, W. R., 1972, Evidence for plate-tectonic regimes in the rock record: Am. Jour. Sci., v. 272, p. 551–576.

——1974, Plate tectonics and sedimentation: Soc. Econ. Paleontologists and Mineralogists Spec. Pub. no. 22, p. 1–27.

——1976, Sedimentary basins developed during evolution of Mesozoic–Cenozoic arc-trench system in western North America: Canadian Jour. Earth Sci., v. 13, p. 1268–1287.

Eardley, A. J., 1962, Structural geology of North America (2nd ed.): New York, Harper and Row, Pubs., 743 p.

Eisbacher, G. H., 1974, Evolution of successor basins in the Canadian Cordillera: Soc. Econ. Paleontologists and Mineralogists Spec. Pub. no. 19, p. 274–291.

Epis, R. C., and Chapin, C. E., 1975, Geomorphic and tectonic implications of the post-Laramide, late Eocene erosion surfaces in the Southern Rocky Mountains: Geol. Soc. America Mem. 144, p. 1–40.

Gabrielse, H., and Reesor, J. E., 1974, The nature and setting of granitic plutons in the central and eastern parts of the Canadian Cordillera: Pacific Geology, v. 8, p. 109–138.

Grose, L. T., 1972, Tectonics, in Mallory W. W., ed., Geologic atlas of the Rocky Mountain region: Denver, Colo., Rocky Mtn Assoc. Geologists, p. 35–44.

Grow, J. A., and Atwater, T., 1970, Mid-Tertiary tectonic transitions in the Aleutian arc: Geol. Soc. America Bull., v. 81, p. 3715–3721.

Ham, W. E., and Wilson, J. L., 1967, Paleozoic epeirogeny and orogeny in central United States: Am. Jour. Sci., v. 256, no. 5, p. 332–408.

Hamilton, W., 1969a, The volcanic central Andes—A modern model for the Cretaceous batholiths and tectonics of western North America: Oregon Dept. Geology and Mineral Industries Bull. 65, p. 175–184.

——1969b, Mesozoic California and the underflow of Pacific mantle: Geol. Soc. America Bull., v. 80, p. 2409–2430.

——1975, Neogene extension of the western United States: Geol. Soc. America Abs. with Programs, v. 7, no. 7, p. 1098.

Hamilton, W., and Myers, W. B., 1966, Cenozoic tectonics of the western United States: Rev. Geophysics, v. 4, p. 509–549.

Irving, E., and Yole, R. W., 1972, Paleomagnetism and the kinematic history of mafic and ultramafic rocks in fold mountain belts: Canada Dept. Energy, Mines and Resources Earth Physics Br. Pub. 42, p. 87–96.

Jones, D. L., Silberling, N. J., and Hillhouse, John, 1977, Wrangellia—A displaced terrane in northwestern North America: Canadian Jour. Earth Sci., v. 14, p. 2565–2577.

Kelley, V. C., 1955, Regional tectonics of the Colorado Plateau and relationship to the origin and distribution of uranium: New Mexico Univ.

Pubs. Geology, no. 5, 120 p.

Krummenacher, D., Gastil, R. G., Bushee, J., and Doupont, J., 1975, K-Ar apparent ages, Peninsular Ranges batholith, southern California and Baja California: Geol. Soc. America Bull., v. 86, p. 760–768.

Lanphere, M. N., and Reed, B. L., 1973, Timing of Mesozoic and Cenozoic plutonic events in circum-Pacific North America: Geol. Soc. America Bull., v. 84, p. 3773–3782.

Larson, R. L., and Pitman, W. C., 1972, World-wide correlation of Mesozoic magnetic anomalies, and its implications: Geol. Soc. America Bull., v. 83, p. 3645–3662.

Le Pichon, X., Francheteau, J., and Bonnin, J., 1973, Plate tectonics: Elsevier Scientific Pub. Co., 300 p.

Lipman, P. W., Prostka, H. J., and Christiansen, R. L., 1971, Evolving subduction zones in the western United States, as interpreted from igneous rocks: Science, v. 174, p. 821–825.

Livingston, D. E., Mauger, R. L., and Damon, P. E., 1968, Geochronology of the emplacement, enrichment, and preservation of Arizona porphyry copper deposits: Econ. Geology, v. 63, p. 30–36.

Mauger, R. L., Damon, P. E., and Livingston, D. E., 1968, Cenozoic argon ages on metamorphic rocks from the Basin and Range province: Am. Jour. Sci., v. 266, p. 579–589.

Miller, F. K., and Engels, J. C., 1975, Distribution and trends of discordant ages of the plutonic rocks of northeastern Washington and northern Idaho: Geol. Soc. America Bull., v. 86, p. 517–528.

Misch, P., 1960, Regional structural reconnaissance in central-northeast Nevada and some adjacent areas—Observations and interpretations, in Geology of east central Nevada: Intermtn. Assoc. Petroleum Geologists, 11th Ann. Field Conf., 1960 Guidebook, p. 17–42.

Molnar, P., and Atwater, T., 1973, Relative motion of hot spots in the mantle: Nature, v. 246, p. 288–291.

Molnar, P., and Francheteau, J., 1975, The relative motion of "hot spots" in the Atlantic and Indian oceans during the Cenozoic: Royal Astron. Soc. Geophys. Jour., v. 43, p. 763–774.

Monger, J.W.H., 1975, Correlation of eogeosynclinal tectonostratigraphic belts in the North American Cordillera: Geosciences Canada, v. 2, p. 4–9.

Monger, J.W.H., Souther, J. G., and Gabrielse, H., 1972, Evolution of the Canadian Cordillera: A plate tectonic model: Am. Jour. Sci., v. 272, p. 575–602.

Morgan, W. J., 1971, Convection plumes in the lower mantle: Nature, v. 230, p. 42–43.

——1972, Plate motions and deep mantle convection, in Shagam, R., and others, eds., Studies in Earth and space sciences (Hess volume): Geol. Soc. America Mem. 132, p. 7–22.

Noble, D. C., 1972, Some observations on the Cenozoic volcano-tectonic evolution of the Great Basin, western United States: Earth and Planetary Sci. Letters, v. 17, p. 142–150.

Pastouret, L., and Goslin, J., 1974, Middle Cretaceous sediments from the eastern part of Walvis Ridge: Nature, v. 248, p. 495–496.

Patton, W. W., Jr., 1973, Reconnaissance geology of the northern Yukon-Koyukuk province, Alaska: U.S. Geol. Survey Prof. Paper 774-A, 17 p.

Perchnielsen, K., and Supko, P. R., 1975, JOIDES Leg 39: Geotimes, March, p. 26–28.

Pitman, W. C., III, and Talwani, M., 1972, Seafloor spreading in the North Atlantic: Geol. Soc. America Bull., v. 83, p. 619–646.

Reesor, J. E., 1970, Some aspects of structural evolution and regional setting in part of the Shuswap metamorphic complex: Geol. Assoc. Canada Spec. Paper no. 6, p. 73–86.

Roddick, J. A., and Hutchinson, W. W., 1974, Setting of the coast plutonic complex, British Columbia: Pacific Geology, v. 8, p. 91–108.

Scholz, C. H., Barazangi, M., and Spar, M. C., 1971, Late Cenozoic evolution of the Great Basin, western United States, as an ensialic interarc basin: Geol. Soc. America Bull., v. 82, p. 2979–2990.

Schweickert, R. A., and Cowan, D. S., 1975, Early Mesozoic tectonic evolution of the western Sierra Nevada, California: Geol. Soc. America Bull., v. 86, P. 1329–1336.

Silberling, N. J., 1975, Age relationships of the Golconda thrust fault, Sonoma Range, north-central Nevada: Geol. Soc. America Spec. Paper no. 163, 28 p.

Silver, L. T., and Anderson, T. J., 1974, Possible left-lateral early to middle Mesozoic disruption of the southwestern North America craton margins: Geol. Soc. America Abs. with Programs, v. 6, no. 7, p. 955–956.

Smith, R. B., 1978, Seismicity, crustal structure, and intraplate tectonics of the interior of the western Cordillera, in Smith, R. B., and Eaton, G. P., eds., Cenozoic tectonics and regional geophysics of the western Cordillera: Geol. Soc. America Mem. 152 (this volume).

Snoke, A. W., 1975, A structural and geochronological puzzle: Secret Creek Gorge area, northern Ruby Mountains, Nevada: Geol. Soc. America Abs. with Programs, v. 7, no. 7, p. 78.

Speed, R. C., 1971, Permo-triassic continental margin tectonics in western Nevada: Geol. Soc. America Abs. with Programs, v. 3, no. 2, p. 199.

Stewart, J. H., 1978, Basin and range structures in western North America: A review, *in* Smith, R. B., and Eaton, G. P., eds., Cenozoic tectonics and regional geophysics of the western Cordillera: Geol. Soc. America Mem. 152 (this volume).

Tailleur, I. L., 1973, Probable rift origin of Canada Basin, Arctic Ocean, *in* Am. Assoc. Petroleum Geologists Mem. 19, p. 526–535.

Tempelman-Kluit, D. J., Gordey, S. P., Read, B. C., 1976, Stratigraphic and structural studies in the Pelley Mountains, Yukon Territory: Geol. Soc. Canada Spec. Paper 76-1A, p. 97–106.

Titley, S. R., 1976, Evidence for a Mesozoic linear tectonic pattern in southeastern Arizona: Arizona Geol. Soc. Digest, v. 10, p. 71–101.

Turcotte, D. L., and Oxburgh, E. R., 1973, Mid-plate tectonics: Nature, v. 244, p. 337–339.

Tweto, O., 1975, Laramide (Late Cretaceous–early Tertiary) orogeny in the Southern Rocky Mountains: Geol. Soc. America Mem. 144, p. 1–44.

Wilson, J. T., and Burke, K., 1972, Two types of mountain building: Nature, v. 239, p. 448–449.

Manuscript Received by the Society August 15, 1977

Manuscript Accepted September 2, 1977

Geological Society of America
Memoir 152

3

Regional gravity and tectonic patterns: Their relation to late Cenozoic epeirogeny and lateral spreading in the western Cordillera

GORDON P. EATON
U.S. Geological Survey
National Center Mail Stop 911
Reston, Virginia 22092

RONALD R. WAHL
HAROLD J. PROSTKA
DON R. MABEY
M. DEAN KLEINKOPF
U.S. Geological Survey
Denver Federal Center
Denver, Colorado 80225

ABSTRACT

A new, simple Bouguer gravity map for that part of the United States west of long 109°W is examined in terms of its relationship to other geophysical and geological parameters. Unifying geophysical and tectonic characteristics define a large central province that has the Great Basin as its principal geographic element, but also includes parts of the Sierra Nevada, western Colorado Plateaus, Columbia Plateaus, and Middle and Northern Rocky Mountains. It is characterized by a high average elevation (>1.5 km), a low Bouguer gravity field (<-110 mgal) with a bilaterally symmetrical distribution of long–wavelength anomalies in its southern half (120°-opposed tracks of progressively outward–migrating silicic volcanism continue this symmetry in its northern half), high heat flow, the presence of two-thirds of all thermal springs in the conterminous United States, a concentration of Quaternary volcanic rocks at its east and west margins, and pervasive extensional faulting throughout. Much of the boundary of this geophysical province is moderately sharp, and parts of it cut across the interiors of classic physiographic provinces, as well as earlier geologic provinces. With the exception of the gravity and volcanic symmetry, none of these characteristics alone is unique to the province; uniqueness lies in their collective assemblage. The northern and southern ends of the province are not sharply defined; it narrows notably and some of its characteristics merge with those of adjoining regions both to the north and south.

Long-wavelength gravity-anomaly patterns within the province are interpreted as reflecting extensional, thermomagmatic episodes in the late Cenozoic history of the lithosphere. Kinematics and patterns of faults, dikes, and geophysical anomalies suggest that east-west spreading of late Cenozoic age was preceded by significant northeast-southwest spreading of greater latitudinal extent in Miocene time. The latter is interpreted as back-arc spreading. Northwest-southeast oblique spreading commonly attributed to the Great Basin as a whole appears to be restricted largely to its western one-third and to the northeast-trending Humboldt zone.

Most regional topographic features of the province are in approximate isostatic equilibrium. Compensating masses for some appear to be within the crust, which in the Great Basin is nearly coextensive with the lithosphere. Some are in the upper mantle. Load differences for individual basin and range structures are supported by the strength of the crust and lithosphere. Faults that block out basins and ranges do not penetrate deeply into the crust, but tend to dip less steeply with depth. This interpretation is supported both by maximum earthquake focal depths and by the observed local response to surface loading. Geothermal gradients and material properties suggest that lateral extension below depths of 15 to 20 km probably takes place by plastic flow aided perhaps by pervasive injection of basaltic magma.

Because the lithosphere of the Great Basin is thin, it probably does not greatly modify the temperature field of the upper asthenosphere. For this reason the origin of compensating masses and major regional gravity gradients is thought to be the complex sum of (1) lateral temperature distributions in the lithosphere and asthenosphere, (2) distribution of Cenozoic intrusive masses reflecting earlier thermal events, (3) high temperature metamorphism related to both injection and heating of the crust, and (4) variations in the degree of extension and resultant thickness of both lithosphere and crust.

Volcanic activity extended across the entire province in Cenozoic time, but is youngest at the east and west margins. In this aspect and in the details of heat flow, topography, broadly distributed extension, and the symmetry of its geophysical anomalies, the region contrasts sharply with both oceanic spreading ridges and major intracontinental graben systems. The difference may be attributed to active, as opposed to passive, spreading processes. Those features of the western Cordillera not related directly to subduction and regional dextral shear are interpreted as the products of lithospheric heating, injection, uplift, and basal traction resulting from the rise and divergent flow of hot, asthenospheric mantle. Associated phenomena include thermal tumescence of the lower lithosphere and brittle faulting of the upper lithosphere.

A "hot spot" of very large dimensions is thus identified near the western, transform boundary of the North American plate. Its proximity to this boundary has led to a complex, episodic and cyclic interaction of two profound stress fields, producing young structures on the west indicative of both thermal doming and dextral shear. To the east the principal deformation is simple east-west extension resulting from rapid spreading that accompanied major regional doming and collapse.

INTRODUCTION

Gravity anomalies reflect departures from radially symmetrical, mass homogeneity in a concentrically layered earth. For this reason their patterns and relation to other geophysical and geological parameters can be used to infer something of the constitution of the western Cordilleran lithosphere and underlying asthenosphere. A new gravity map was prepared on this premise. (Pl. 3-1, in pocket, and Fig. 3-1).

The Bouguer gravity field of the United States west of long 109°W is dominated by an extensive gravity low having well-defined boundaries and a near-closure of 90 mgal (Fig. 3-1). It is approximately 1,000 by 1,300 km in extent and includes parts, but not all, of several physiographic provinces. It coincides with a region of pronounced and broadly distributed crustal extension, high heat flow, repeated Cenozoic igneous activity, abundant hot springs, peripheral seismicity, peripheral Quaternary volcanism, and outward tilted margins. It stands a full kilometre above its surroundings on all but one side, and, in so doing, doubles the

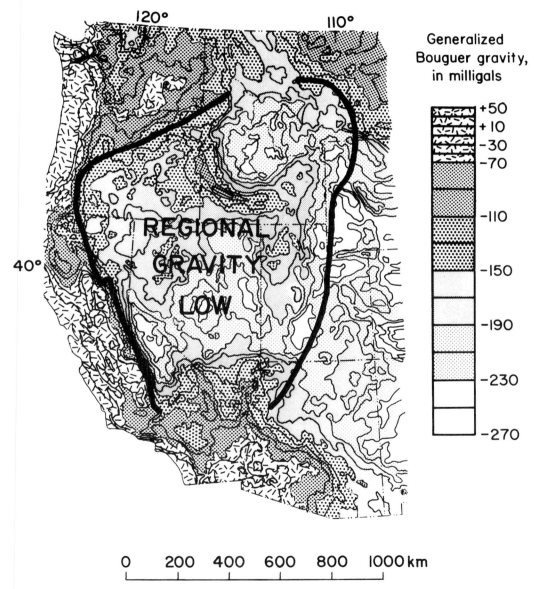

Figure 3-1. Boundaries of the regional gravity low which is the principal subject of this paper. Note that patterns of gravity intensity and contour interval both differ from those of Plate 3-1. Figures in text follow patterns of Figure 3-1.

width of the elevated Cordillera of Central and North America. Its north and south ends, which cut across physiographic provinces, seem unrelated in any simple way to the location of the evolving late Cenozoic trench and transform systems at the western edge of the North American plate (Atwater, 1970). For this reason, and because a well-developed bilateral symmetry is evident in the southern half of the regional gravity low, despite a long and complex sequence of anisotropy-producing and heterogenizing crustal deformations and magmatic events that began in mid-Paleozoic time, we believe that the features of the region are not readily explained in terms of simple plate interactions alone. They are more likely related to local, but profound, lithospheric thinning and heating caused by the localized upwelling of hot, asthenospheric mantle, and perhaps locally, some subduction derived melt. The process is an active one, and its product contrasts with oceanic spreading ridges, where both volcanism and highest heat flow are concentrated in a narrow zone at the axis of symmetry as a result of passive plate separation and the drawing-up of basaltic magmas from the asthenosphere. The region does *not* represent an onshore extension of the East Pacific Rise, with which it contrasts sharply.

CHOICE OF THE BOUGUER GRAVITY FIELD

Any gravity anomaly represents a difference between a measured value and a theoretical value predicted for the point of observation. Free-air anomalies incorporate the fewest assumptions about the nature of the Earth at and near the point of observation. They are poorly suited for investigation of the region under discussion, however, because no local density variations, *including those due to the contrast between rock and air at the surface*, are taken into account in computing anomaly values. They reflect *all* departures from mass homogeneity, including that of the topography, which in the western Cordillera is by far the largest effect. Not only do free-air anomalies reflect variations in local topography, but even those of regional topography in areas where isostatic equilibrium has been achieved, because compensating subsurface masses are usually so far removed vertically from the point of observation that their own gravity expression is spread over a broad region.

Because our objective was a study of the patterns of subsurface density variation, we removed the effect of topography by making the Bouguer correction. The Bouguer anomaly is the best presentation of the gravity effect of both local and regional mass anomalies over large regions with major surface relief. In preparing the Bouguer map we assumed a density of $2.67 \text{ g} \cdot \text{cm}^{-3}$ for all material above sea level, thus the anomalies reflect masses of any other densities above sea level and lateral density variations below sea level. The first of these is an error; fortunately it is small. For example, the artifact component produced by a laterally extensive, 1,000-m-high topographic edifice consisting entirely of rocks of density $2.40 \text{ g} \cdot \text{cm}^{-3}$ (instead of the $2.67 \text{ g} \cdot \text{cm}^{-3}$ value assumed) is only 10 mgal.

We faced a practical problem in opting for the computation of simple Bouguer values instead of more rigorous complete Bouguer values. It was that of having to compute terrain corrections for more than 112,000 points of observation (Pl. 3-2, in pocket). Many of the available data files were of simple Bouguer values only. We employed several hundred complete Bouguer values in making local comparisons of anomaly patterns displayed by both types of maps. Although anomaly amplitudes were seen to vary, patterns of distribution did not.

The absence of terrain corrections was ameliorated somewhat by gridding, which tended to suppress the influence of measurements on high peaks where terrain corrections may reach many tens of milligals. The term "gridding," as used here, refers to preparation of a contour map from a regular array of grid-point values calculated from nearby, irregularly spaced observational data. The gridding interval employed in preparing Plate 3-1 was different from that used in plotting Figures 3-7, 3-8 and 3-10; hence the difference in detail. The grid acts as a filter tending to lessen the impact of singular extreme values. The bulk of the observations incorporated in the map were made along roads, where near-in terrain effects tend to be minimized.

RELATION OF THE BOUGUER GRAVITY FIELD TO OTHER FEATURES OF THE WESTERN CORDILLERA

General Statement

Our approach in this study was an item-by-item examination of the relation of the regional gravity field to other geophysical fields. Neither the gravity field nor any of the others *alone* provides comprehensive insight into the nature of the crust, lithosphere, and asthenosphere of the western Cordillera. Similarly, the relation of these fields to one another varies over the region in complex ways. As examples, the interiors of the Great Basin and Colorado Plateaus have identical gravity values even though their crustal thicknesses differ by 15 km. The southern Great Basin and northern Sonoran Desert sections, on the other hand, have nearly identical crustal thicknesses and gravity values that differ by 100 mgal. Only from simultaneous consideration of many different parameters does this area become unique.

Individual long-wavelength anomalies with gentle flanking gradients, shown in Plate 3-1, generally originate at depths greater than those of short-wavelength and steeper gradient. Anomaly amplitudes of mass distributions whose horizontal dimensions are many times their vertical dimensions are relatively insensitive to source depth. Thus, anomalies with wavelengths, amplitudes, and gradients similar to those of deep sources conceivably could be produced by near-surface masses that are wide and thin, if they had very gently sloping flanks, but these are uncommon in the western Cordillera. The long-wavelength (>100 km) anomalies of Plate 3-1 reflect lateral density variations at levels *beneath* the observable surface geology.

Density contrasts reflected in Bouguer gravity anomalies may originate in contrasts in composition; for this reason simple variations in crustal thickness can produce variations in the gravity field, as can lateral variations in average crustal composition. Other contrasts may originate from differences in phase assemblage within rocks of similar bulk composition. Still others may reflect variations in physical state, such as temperature. Because the coefficient of thermal expansion of most rock-forming silicates is low, however, simple temperature differences of several hundreds of degrees Celsius will give rise to only a few percent change in the gravity field. The presence of partial melt produces larger differences. We suggest that in the Great Basin, and possibly also in the region immediately to the north, large-scale density contrasts giving rise to long-wavelength regional gravity anomalies have a partial dependence on temperature. The relationship is unquestionably blurred by compositional variations and possibly by wet thermal metamorphism associated with invasion of the crust by magmas produced during earlier thermal events. Continental lithosphere is far more complex than oceanic lithosphere and hence more difficult to understand.

General Relation of the Gravity Field to Topography

Long-wavelength gravity anomalies mirror the regional topography in this region (Fig. 3-2, Pl. 3-1). The topography is not the source of the anomalies. Deep mass anomalies produce both the regional surface topography and observed Bouguer anomalies. The quantitative relationship between topography and gravity can be used to deduce something of the degree of isostatic compensation.

The western Cordillera, excluding the Colorado Plateaus and Rocky Mountains, is dominated by a coherent region of high, but variable, topography extending from approximately lat 36° to 47°N and from long 110° to 122°W (Fig. 3-2). On the east it is adjoined by the Colorado Plateaus and Rocky Mountains, which are as high as, or higher than, itself. The eastern boundary of the region of interest is defined by a major discontinuity in gravity trends just east of the Wasatch front, and by heat flow, seismicity, and fault patterns, as will be shown in sequence. The region is surrounded by a crudely arcuate band of lower topography on the south, west, and north.

The 4,000-ft (~1,200-m) contour encloses the central topographic high everywhere except on the open eastern side. The region of high topography inside it includes, as a coherent whole, parts of several physiographic provinces, but its edge, as defined by this contour, cuts across the interior of some of them. The Great Basin section of the Basin and Range province stands mostly above 5,000 ft (~1,500 m), whereas the rest of the province to the south lies mostly below 3,000 ft (~900 m). Similarly, the division that Fenneman (1931) referred to as the Columbia Plateaus is crossed by a moderately steep, northeast-trending topographic gradient having a maximum step-amplitude of 3,000 ft (~900 m). Clearly, there are subsurface density distributions whose edges do not everywhere coincide with, nor parallel, the boundaries of the physiographic provinces.

This coherent region of high topography may represent a young episode of epeirogeny that has selectively uplifted parts, but not all, of both the Basin and Range and the Columbia Plateaus provinces, tilting and/or uplifting the adjoining Sierra Nevada and western Colorado Plateaus. Alternatively, it may be that much of the entire Western United States once stood this high but has since subsided epeirogenically or been deeply eroded. As we note below, a difference in age for both the beginning and ending of Basin and Range faulting in the northern and southern parts of the province has been documented. These sections may once have been continuous, having shared a common origin, but the Great Basin has experienced continuing extension not suffered by the rest of the Basin and Range province, and this later extension has been of differing orientation.

Figure 3-3 shows Cordilleran topography in a broader context. It reveals that the part of the uplift above 1 km is nearly twice as wide between lat 36° and 47°N as it is anywhere else in North and Central America. The region of specific interest constitutes that part of this broad tract west of long 110°W. Figure 3-3 also reveals that the location of the abrupt widening bears little geographic correspondence to the location and latitudinal extent of the transform fault (the San Andreas fault system) separating the offset ends of the East Pacific–Gorda–Juan de Fuca rise system. Although the measurements of width were made obliquely along standard circles of latitude, the preceding statements are equally true for measurements made along circles of longitude about the pole of relative motion (53°N, 53°W) between the North American and Pacific plates (Morgan, 1968).

An imaginary eastward projection of the Mendocino fracture zone (the northern end of the transform) bisects the elevated area at lat 40°N. The southern end of the transform, at the head of the Gulf of California, lies well south of the south edge of the elevated region. Atwater's (1970) reconstructions of the relative locations of rise, transform, and trench systems bounding (or lying west of) the western edge of the American plate at earlier times in the Cenozoic Era seem to provide no ready explanation for this feature either. We view it as an entity that developed partly independently of plate boundaries and plate interactions. Data and arguments to support this view follow.

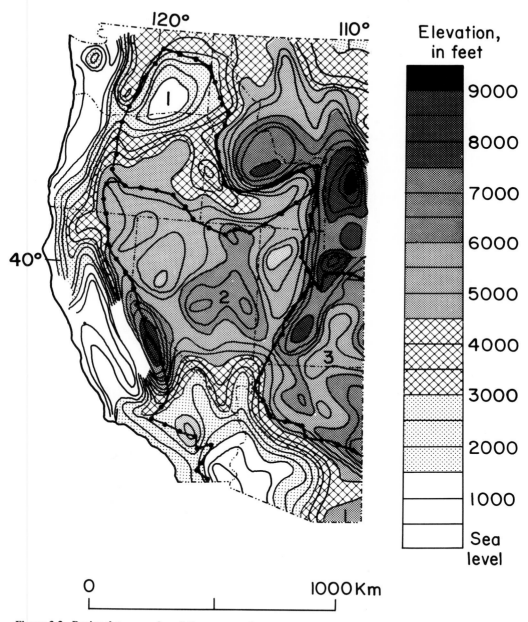

Figure 3-2. Regional topography of the western Cordilleras of the United States. Elevations averaged in 1° squares. Contour interval 500 ft. Modified from Strange and Woollard (1964). Beaded lines are selected physiographic province boundaries from Fenneman (1931). Shown are (1) Columbia Plateaus province of Washington, Oregon, and Idaho; (2) Basin and Range province of Oregon, Nevada, Utah, California, and Arizona; and (3) Colorado Plateaus province of Utah and Arizona.

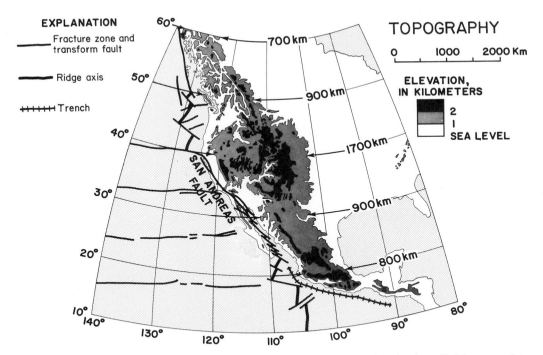

Figure 3-3. Regional topography of the Cordilleras of North and Central America (compiled from a variety of atlas sources). Arrows and kilometre figures indicate width of terrain more than 1 km above sea level. Note doubling of this width between lat 35° and 47°N and its lack of alignment with the latitudinal span of the San Andreas fault.

Quantitative Relationships between Bouguer Gravity and Regional Elevation

Correlations between regional topography and long-wavelength gravity anomalies suggest isostatic compensation. Woollard (1966, 1968, 1969, 1972) showed that large topographic features in the United States are mostly compensated and that there is notable interdependence of surface elevation, crustal thickness, and crustal and upper-mantle density values (as reflected in measured values of seismic compressional wave velocities).

In describing the absence of a consistent, systematic relationship between surface elevation and crustal thickness, Woollard (1968) noted that some major mountain ranges, such as the Sierra Nevada–White Mountains complex, appear to have roots, but that others, such as the Rocky Mountains, apparently do not. Something other than the Airy ("roots-of-mountains") mechanism of compensation is at play in some regions. According to Woollard (1972, p. 501):

There are areas of major changes in crustal and upper mantle parameter values where the underlying (seismic) velocity pattern suggests there is a density bias over a considerable depth range, and the overall isostatic compensation appears to conform to that of the Pratt concept of isostasy.

He went on to note that the interrelation of physical properties of the crust with seismic velocities of the mantle and that the dependence of the depth of seismic discontinuities in the mantle on associated velocities suggest the existence of a reversible physical process that affects velocity and density alike in both the upper mantle and crust. He attributed it to a thermal phenomenon. We concur. We now briefly examine the isostatic relations in and around the region of high topography between lat 35° and 47°N and long 110° and 123°W.

Colorado Plateaus and Snake River Plain. Computed regional gravity values derived from regional topography using the method of Mabey (1966) generally mimics measured Bouguer gravity across the boundary between the Great Basin and the Colorado Plateaus in Utah (Fig. 3-4). The largest discrepancies occur east of Salina, on the Colorado Plateaus. Isostatic compensation apparently is more regionalized under the plateaus and the lithosphere is stronger there, owing largely to its greater thickness.

Figure 3-5 shows a similar set of curves for the eastern Snake River Plain. Although the plain is only 90 km wide in this area, it appears to be in approximate isostatic equilibrium with the adjoining highlands. The curves of computed and observed Bouguer anomaly values are similar in configuration. Note, however, that a constant was added to the computed values to bring them into conformity with observed values. To the west, where the Snake River Plain narrows to about 60 km, it is still in approximate isostatic equilibrium (Mabey, 1976). The close relationship between topography and Bouguer gravity field for a feature the width of the Snake River Plain suggests that compensation is achieved at shallow depths. Over the western Snake River Plain, the compensation may occur in the upper 20 km of the crust, which suggests the possibility of plastic flow at levels this shallow.

Although the eastern Snake River Plain *appears* to be in approximate isostatic equilibrium with the adjoining highlands, a large region in eastern Idaho, southwest Montana, and western Wyoming, roughly centered around Yellowstone National Park, is characterized by less negative Bouguer anomaly values than occur at similar regional elevations in the central part of the Great Basin (Eaton and others, 1975). Figure 3-6 illustrates this difference. Note that the Nevada stations fall about 30 mgal below Idaho and Wyoming stations at the same regional elevation. The regression line for the Nevada data has a gravity intercept near zero. This anomaly is also seen on the profile across the eastern Snake River Plain (Fig. 3-5), where 30 mgal were added to the values computed from regional elevation to match measured Bouguer anomaly values. Average free-air anomaly values are also positive by approximately 30 mgal in the region around Yellowstone Park. Isostatic compensation apparently occurs at relatively shallow depths for topographic features within this region, whereas the entire region is held in an elevated position by a non-isostatic condition. Perhaps dynamic forces are at play, or possibly a regional variation in the mechanical strength of the lithosphere exists (it being lower in the Great Basin). Alternatively, there may have been a time lag in the establishment of compensation following a major change in mass caused by some as yet undefined, but probably thermal, mechanism. Other geophysical data for the Snake River Plain do not shed light on this problem (compare Hill and Pakiser, 1966).

The Snake River Plain is aseismic, which suggests that it may deform by creep. A shallow electrical conductor (Stanley and others, 1977) and high heat flow (D. D. Blackwell, personal commun.) in the eastern plain are permissive of subsurface temperatures high enough at depth to induce creep.

Great Basin. Because of the large local gravity lows produced by low-density alluvial and lacustrine fill in most of the basins of the Basin and Range province, compensation of individual ranges is difficult to evaluate. The very largest mountain masses, for example, the Spring Mountains of southern Nevada, have associated gravity lows that suggest they may be partly compensated. Kane and Carlson (1961) found Bouguer lows over the high parts of ranges in southern Nevada, which suggests that these ranges are the elevated ends of blocks tilted in part by isostatic forces, with the less dense ends of the blocks uplifted.

Our data suggest that the typical range is supported by the strength of the crust and lithosphere,

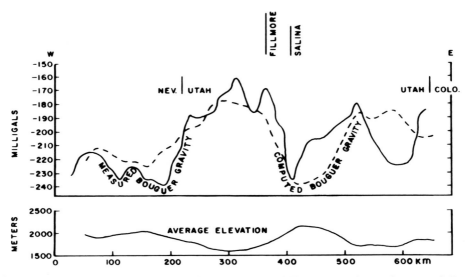

Figure 3-4. East-west profiles of average elevation, measured Bcuguer gravity, and computed Bouguer gravity across the eastern Great Basin and western Colorado Plateaus at lat 39°N. The town of Fillmore, Utah, marks the extreme eastern edge of the Great Basin. Average elevations (lower solid line) were computed for areas 64 km in radius about the points of gravity measurement. Derived values of computed Bouguer gravity (dashed line) are compared with measured Bouguer gravity (upper solid line).

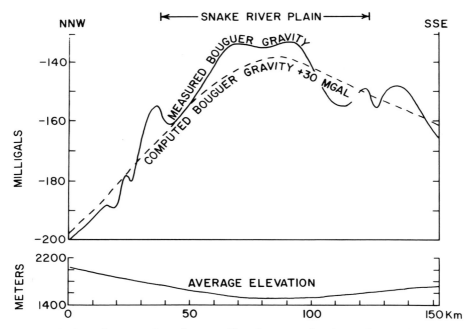

Figure 3-5. North-northwest–south-southeast profiles of average elevation and measured and computed Bouguer gravity across the eastern Snake River Plain. Average elevations (lower solid line) were computed for areas 64 km in radius about the points of gravity measurement. Derived values of computed Bouguer gravity (dashed line) are compared with measured Bouguer gravity (upper solid line). Thirty milligals have been added to values of computed Bouguer gravity.

even though the latter is thin (Walcott, 1970) and relatively weak compared to most continental lithosphere. Because of the probable shallow level of laminar flow in the crust of this region, as deduced in the section on seismicity which follows, most major Basin and Range faults probably become somewhat flatter at depth and do not penetrate deeply into the crust. They thus do not provide surfaces along which isostatic adjustments can occur. This conclusion is supported by the observation that isostatic response to relief from the water load of Lake Bonneville (Crittenden, 1963) was seemingly independent of, and in the opposite sense of vertical displacement from, active basin-range faulting along the margin of the lake. It is also supported by the model studies of Cloos (1968), by the geometric arguments of Moore (1960) and Hamilton and Myers (1966), and by the observations of Hamblin (1965).

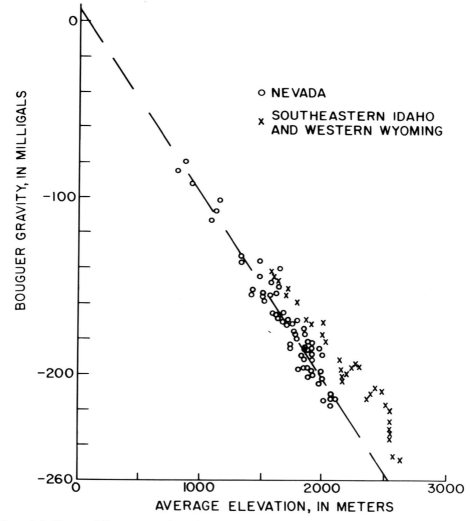

Figure 3-6. Measured Bouguer gravity values and average elevations for areas of 64-km radius about representative stations in Nevada and in southeastern Idaho and western Wyoming. Relationships suggest that topography is compensated, but note systematic discrepancy between the two regions, suggesting different average densities and different mechanisms of compensation.

Hamilton (1969, 1975) argued that the normal faults must decrease in dip downward in order to produce the rotations of layered rock sequences seen in the ranges, but rotation of the mountain blocks alone does not necessarily require curvature of the faults. The blocks may simply begin to tilt back with continued extension of the ductile substratum (see Proffett, 1977, for a description of the sequence of evolution). In his study of major normal faults in the westernmost Colorado Plateaus, however, Hamblin (1965) noted that most have dips that decrease perceptibly with depth. Movements on faults of this sort tend to pull blocks apart, as well as to displace them vertically. It is not implied that the toes of the faults are horizontal, only that the surfaces are concave upward.

It has been shown (Kanizay, 1962) that when rocks become ductile as a result of increasing mean stress or increasing temperature, the angle of internal friction of the rocks changes and slip surfaces may display systematic curvature. Lateral extension in regions of strong vertical temperature gradients like those of the Great Basin (Sass and others, 1976b) may thus produce normal faults that are concave upward. We think that no brittle fracturing is taking place in the lower crust of the region today, nor has it in the past few million years.

Relation of the Gravity Field to Crustal Thickness

Woollard (1966, 1968) noted that there is no systematic relationship between surface elevation and crustal thickness in the United States and that the relationship between regional gravity anomalies and crustal thickness is complex. Clearly no quantitative correlation between Bouguer anomaly values and crustal thickness exists in the western United States (compare Pl. 3-1 with Prodehl, 1970, Fig. 11; Warren and Healy, 1973; or Smith, this volume). The Sierra Nevada–White Mountains complex of eastern California has a maximum crustal thickness of between 36 and 45 km (Eaton, 1966; Prodehl, 1970; Carder and others, 1970), but has Bouguer gravity values similar to those of the Great Basin in central Nevada, where the crust is only 30 to 33 km thick. Similarly, the crust in the central Colorado Plateaus region is 42 km thick, yet its gravity values also match those of the central Great Basin. By contrast, a pronounced, east-trending, 90-mgal gradient crosses southern Nevada and Utah at lat 37°N and is not related to a change in crustal thickness. The gravity map is thus a poor guide to relative crustal thickness in the western Cordillera, and one must call upon other explanations to account for variations in the gravity field.

Variation in the thickness of the lithosphere may account for some gravity variation but cannot account for all of it, as we will demonstrate in the paragraphs on temperature and heat flow. The thickness of the mantle lid zone that must be present beneath the Colorado Plateaus to provide a match between the gravity field there and that of the Great Basin can be calculated easily using (1) averages of the many published values of compressional wave velocities of the crust and mantle in the Great Basin and Colorado Plateaus regions, (2) the empirical density-velocity relation of Nafe and Drake (1963), and (3) published values of crustal thickness. If one accepts the presence of asthenosphere immediately beneath the Great Basin crust, then average values of 2.90, 3.33, and 3.40 g·cm^{-3} for densities of the crust, asthenosphere, and lithosphere, respectively, require, under the Colorado Plateaus, a lid-zone thickness of nearly 80 km and a lithosphere thickness of 120 km. This figure lies in the uppermost part of the range of values derived by Biswas and Knopoff (1974) on different grounds and may be too high. The difference between it and that employed by Crough and Thompson (1976) stems from differences in the values of density and crustal thickness employed. Regardless of which figure is correct, both argue that the thicker crust of the Colorado Plateaus can be gravitationally balanced by differences in the thickness of the lithosphere so as to provide gravity values identical with those of the Great Basin. If thickness of the lithosphere is a function of the thermal structure of the mantle, as Crough and Thompson (1976) suggest, then the gravity field must also be a function of temperature, at least in part, just as it is a function of crustal thickness.

The difference in density contrast between crust and lithosphere on the one hand and asthenosphere and lithosphere on the other yields gravity effects that differ by nearly an order of magnitude. The velocity contrast between asthenosphere and normal lithospheric mantle in the western Cordillera suggests a density contrast of $0.08 \ \text{g} \cdot \text{cm}^{-3}$, whereas that between crust and lithosphere is $0.50 \ \text{g} \cdot \text{cm}^{-3}$. This is a significant factor in modeling the gravity field of continental regions.

Relation of the Gravity Field to Batholiths

Comparison of the gravity map (Pl. 3-1) with the geologic map of the United States (King and Beikman, 1974) gives the impression that the Sierra Nevada and Idaho batholiths are sources of pronounced negative gravity anomalies with minima of less than −200 mgal. The relatively light granitic rocks of those batholiths doubtless contribute to the negative anomalies, but close study reveals that the relationship is complex. Figure 3-7 shows the distribution of exposures of the Sierra and southern California (Peninsular Ranges) batholithic rocks in relation to the Bouguer gravity field. The gravity field increases by more than 150 mgal from the mid-latitude of the Sierra Nevada batholith to its southern end. Still farther south, in the southern California batholith, values rise by another 50 mgal. The abundance of large exposures of batholithic rocks in east-central and southeastern California suggests pervasive invasion of the shallow crust, yet within their area of occurrence the gravity field varies by more than 200 mgal.

The heavy dashed line in Figure 3-7 represents the eastern limit of outcrops of plutonic rocks of Mesozoic age. East of this line, all the way to the Great Plains of Colorado and New Mexico, almost all exposed plutonic rocks younger in age than Paleozoic are Cenozoic, and their areas of outcrop are notably smaller than those of the Mesozoic plutons.

The heavy solid line in Figure 3-7 marks the eastern escarpment of the Sierra Nevada. The mountain range coincides with a steep gravity gradient separating a regional gravity high on its west and a regional low on the east. This gravity low thus precisely corresponds neither with the uplifted mountain block nor with the batholith. A much closer relationship exists between Bouguer gravity and regional topography (compare Pl. 3-1 and Fig. 3-2), a relationship stemming in part from sources deeper than the surface geology. The asymmetrical regional gravity low whose axis lies at or very near the eastern escarpment of the Sierra Nevada · also corresponds to an area of locally elevated heat flow, as shown below. Both observations are relevant to the discussion of regional symmetry which follows.

The age of the Sierran topography is important to an understanding of the origin and the time of origin of the geophysical province under discussion. We return to this subject in a more general vein in later paragraphs but describe the age of the Sierran block here. According to Christensen (1966), the maximum age of the last major uplift of the Sierra Nevada is approximately 9 m.y. It was standing at its present elevation in late Pliocene time. Faulting of the eastern escarpment occurred less than 3 m.y. ago, and as much as 1.3 km of vertical faulting occurred following the earliest glacial stage. The relative youthfulness of much of this local topographic relief calls into question its direct genetic relation to the crustal root that extends to its east.

The *eastern* half of this gravity low, which is commonly attributed in the literature to the

Sierra Nevada, as well as the seismically mapped crustal "root" with which it is associated (Eaton, 1966; Carder and others, 1970), is actually an integral part of the western Great Basin. Sometime between 9 and 3 m.y. ago, the Great Basin apparently "grew" westward and consumed part of an originally broader, uplifted border, converting it to new structural basins and blocks. It left, as a high-standing remanent, the White Mountains near the California-Nevada border, well to the east. The gravity low is thus only partly related to today's surface geology.

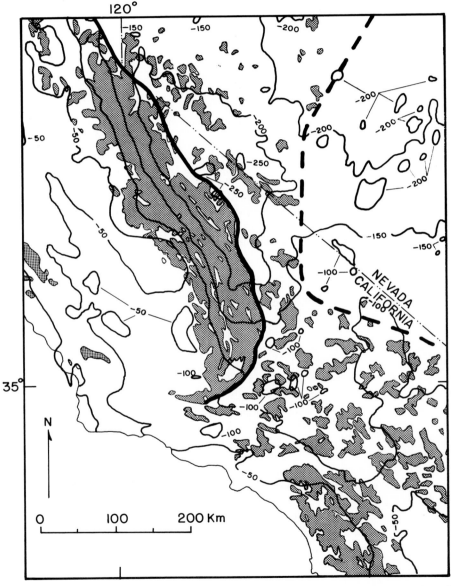

Figure 3-7. Observed Bouguer gravity field of the region of principal Mesozoic plutons in southern California and southwestern Nevada. Contour interval, 50 mgal. Mesozoic plutons shaded. Solid line marks eastern front of Sierra Nevada. Dashed line is eastern boundary of Mesozoic igneous rocks. Note lack of systematic relation of gravity field to either the Sierran block or the distribution of Mesozoic igneous rocks.

Symmetry of the Bouguer Gravity Field

General Statement. The gravity field is characterized by bilateral symmetry between lat 35° and 47°N. In Figure 3-8 we have isolated that part of the field over the Great Basin in order to emphasize this symmetry. On the left is the actual Bouguer gravity map and on the right the east half of the map and its mirror image folded about a general axis of symmetry. Bilateral symmetry in the distribution of long-wavelength anomalies is undeniable. Five anomaly pairs participate in this symmetry. Starting with the cardioid low on the south and progressing northward, their distances from the symmetry axis (their half separations) are 20 km, 100 km, 290 km, 100 km, and 230 km. The matching anomalies 290 km from the axis coincide with the uplifted west and east margins of the Great Basin. Those at 230 km overlie the Lahonton and Bonneville depressions. The high-frequency content of Figure 3-8 constitutes a gravity grain derived from near-surface geology. It crosses the patterns of long-wavelength anomalies and does not participate in the bilateral symmetry. The long-wavelength anomalies, which do, reflect deeper sources.

The north-trending axis of bilateral symmetry ends abruptly at the south edge of the Snake River Plain. The plain is identified on Plate 3-1 by the arcuate, high amplitude, positive gravity anomaly concave to the north at lat 43°N. Northeast of the plain, an oval gravity low trends northeast across central Idaho. This low consists of a northwestern half associated with the Idaho batholith and younger Eocene intrusive and volcanic rocks and a southeastern half associated with northwest-trending, Miocene basin-range structure involving Precambrian and younger Phanerozoic rocks, as well as the Boulder batholith of Montana. The long axis of this gravity low coincides with an axis of high (>2,100 m) topography.

The elevated topographic surface north of the central and eastern Snake River Plain (Fig. 3-2) is similar in location, configuration, and trend to a surface of Eocene topography reconstructed by Axelrod (1968), though the maximum elevation of the latter was approximately 1 km lower. The broadly domed Eocene surface coincides closely enough with the distribution of outcrops of Eocene volcanic and intrusive rocks (see King and Beikman, 1974) to suggest regional, thermomagmatic uplift. It is unlikely that a thermal tumescence of this magnitude, if that is what it was, would be preserved to the present without either a continuous sustaining or episodic renewal of its thermal energy. As will be seen below, a belt of hot springs with base temperatures exceeding 90°C trends northeast through this area, suggesting that the present uplift may indeed reflect a younger thermal event. Nevertheless, we tentatively interpret part of the amplitude of the gravity low north of the Snake River Plain as the "frozen" expression of Cretaceous to Eocene plutonism, the intrusive rocks of which are lighter than their crustal surroundings.

Origin of Symmetry in the Gravity Field. Before proceeding to a discussion of the origin of gravity symmetry in the Great Basin, it must be emphasized that the symmetry is not a symmetry of surface and near-surface geology; hence, the ages, compositions, and origins of rocks exposed at the surface in both elements of a given anomaly pair need not be similar. The surface geology is inherited from a long and complex crustal history and stands in marked contrast to that of most oceanic lithosphere, which is young and very simple. The gravity symmetry of the Great Basin is a symmetry of subsurface mass distributions and is reflected at the surface most obviously in topography and other geophysical fields.

One hypothesis offered to explain the symmetry relates the local gravity field directly to elevated temperatures. If one accepts the argument that the local lithosphere is thin and weak, geographic variations in the gravity field can be viewed as reflecting patterns of elevated temperature in the lower lithosphere and asthenosphere directly. Individual long-wavelength

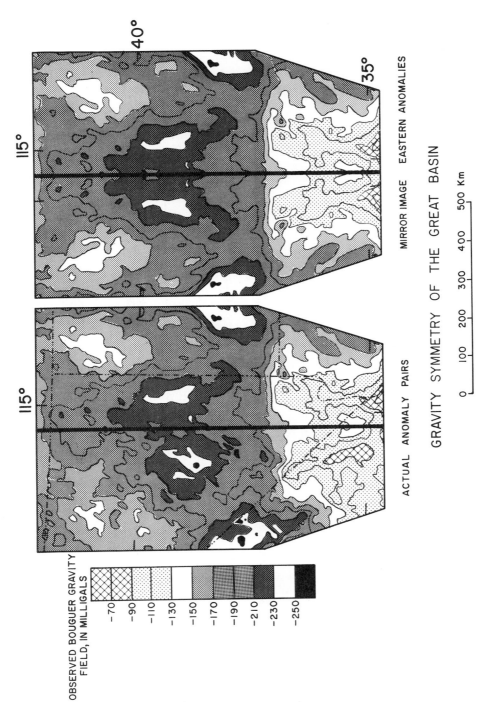

Figure 3-8. Bilateral symmetry of the observed Bouguer gravity field of the Great Basin and immediate surroundings. On the left is an isolated portion of the data from Plate 3-1, gridded at a larger interval. The heavy straight line marks the axis of *general* bilateral symmetry. On the right is the east half of this map and its mirror image folded about this symmetry axis.

anomalies would thus correspond to individual areas of differing subsurface temperature. Another hypothesis, indirectly related to the first, interprets the elements of symmetry as a "frozen" record of an earlier thermomagnetic event that resulted in symmetrical invasions by relatively light plutonic masses. Still another hypothesis suggests that paired anomalies represent paired thickenings and thinnings of the crust and lithosphere. We have already seen that the gravity field is an undependable reflection of variations in crustal thickness, yet the matching gravity highs in northwestern Nevada and northwestern Utah coincide with the Lahontan and Bonneville depressions, both of which mark areas of abnormal crustal thinning (Smith, this volume). The gravity data alone do not allow one to choose from among these hypotheses.

Whatever the origin of the symmetry, the pattern suggests a possibility of symmetry in other parameters and in the causal mechanism. Figure 3-9 shows a set of east-west geophysical profiles across the Great Basin, all within 100 km of the 38th parallel. All display a moderate degree of symmetry, as do the progressively migrating but opposed tracks of the foci of volcanism north of the Great Basin along the eastern Snake River Plain and Brothers fault zone (Christiansen and McKee, this volume). Quaternary volcanism is in general concentrated near the east and west margins of the province as far south as the Coso Mountains of southeastern California.

Figure 3-9 shows that the axis of symmetry is characterized by a low magnetic level. It

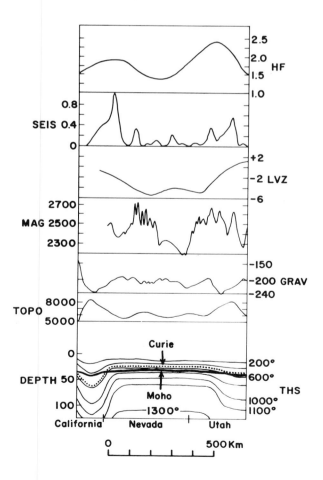

Figure 3-9. Geophysical symmetry across the southern Great Basin at lat 38°N. All profiles drawn within 100 km of 38th parallel. The heat-flow curve (HF) represents smoothed, observed heat-flow values in $\mu cal \cdot cm^{-2} \cdot s^{-1}$ and was prepared from Figure 3-12; the seismicity profile (SEIS) was prepared from data provided by C. W. Stover, U.S. Geological Survey, and represents the number of seismic events per 1,000 km^2/yr. in the period 1961 to 1973; the low-velocity-zone data (LVZ) represent time differences in seconds and were taken, with slight modification, from York and Helmberger (1973); the magnetic profile (MAG) is in gammas and was taken from Mabey and others (this volume) on a traverse somewhat oblique to the 38th parallel; the gravity profile (GRAV) is in milligals and was prepared from Plate 3-1 of this report; the smoothed topographic profile (TOPO) is in feet above sea level and was taken from Strange and Woollard (1964); the profile of the M-discontinuity (Moho) is from Prodehl (1970); and the cross sectional profiles of isothermal surfaces (THS), in degrees Celsius, calculated from an earlier compilation of heat-flow data, not from curve HF, above, are from Roy and others (1972). The calculated curve of maximum depth to the Curie isotherm is based on magnetite. Depth is expressed in kilometres.

represents a cross section through the "quiet zone" of Stewart and others (1977), the "quiet basement zone" of Mabey and others (this volume). A comparison of Figure 3-8 with Plate 4-1 of Mabey and others shows that the axis of gravity symmetry lies along the narrow, north-trending corridor of the quiet basement zone for the full length of the axis of symmetry. Intense hydrothermal alteration of ferrimagnetic minerals in the basement rocks and those that intruded them seems the simplest explanation for this magnetic feature, inasmuch as pre-Holocene volcanic rocks in the surface geology continue across it without modification (Stewart and others, 1977) and there is an absence of present-day volcanism at the axis.

A single narrow swath through the middle of the Great Basin, between long 115° and 116°W, includes the axis of regional topographic arching, the axis of symmetry of five pairs of matching, long-wavelength gravity anomalies, and a zone of low magnetic intensity within an otherwise magnetically busy terrain. The youngest volcanic rocks are found not along this axis but at the outer edges of the province, where they participate in the symmetry and provide constraint in the interpretation of its origin.

Artemjev and Artyushkov (1971) suggested that passive necking by simple, "pull-apart"

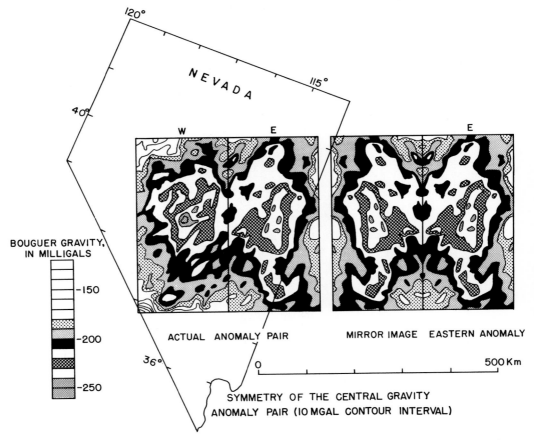

Figure 3-10. "Butterfly" Bouguer gravity anomaly pattern of central Nevada on the left. Line indicates axis of symmetry. Compare actual anomaly pair, on left, with image generated by folding eastern half of this pair about a best-fit axis of symmetry, on right. The axis of this anomaly pair is cocked counterclockwise to the general axis of symmetry in Figure 3-8. Note high-frequency signal grain rotated clockwise of symmetry axis in actual anomaly pair. Gridding interval same as in Figure 3-8.

extension best accounted for the bilateral symmetry of the Baikal Rift, another region of active continental extension. We regard this as an invalid model for the Great Basin. Given the following conditions: (1) the pronounced mechanical anisotropy and inhomogeneity of the local Great Basin crust (stemming from its long and complex pre-late Cenozoic history) and the relative weakness of its lithosphere; (2) the regular pattern of long-wavelength gravity anomaly pairs diverging northward from lat 35°N across a width of more than 800 km; (3) the marginal, as opposed to axial, concentration of young volcanism; and (4) the systematic and paired thickenings and thinnings of the crust, we think that an active extension driven by loose, traction-coupled, laterally divergent flow in the asthenosphere provides a better explanation.

Segmentation of the Axis of Symmetry. The pair of long-wavelength gravity lows closest to the axis of symmetry in central and eastern Nevada is cocked counterclockwise to the north-trending, general axis of regional symmetry. This pair is shown in isolation in Figure 3-10, where it is displayed next to a mirror image of its eastern element folded about a local axis of symmetry. Here, as in Figure 3-8, one sees a high-frequency grain that does not participate in the symmetry, but the longer wavelength features clearly are symmetrically disposed. This anomaly pair constitutes an isolated and rotated segment within the larger pattern of regional symmetry. Inspection of anomalies closest to and along the north-trending regional axis (Pl. 3-1) reveals that other such segments exist near the axis (Fig. 3-11A) and that some of them are also cocked at angles to the generalized, north-trending regional axis (Fig. 3-11B). We suggest that this set of anomalies may be related to an earlier episode of symmetry development now frozen in the crust. The axis of the gravity high along the western Snake River Plain may be one such segment. Supporting arguments are discussed in the section on regional fault patterns.

We turn now to an examination of relations between gravity, crustal temperature and heat flow, seismicity and faulting. The reader familiar with the geophysical literature of the region will recognize both the influence of and our departure from the views of Thompson (1959, 1972), Thompson and Burke (1974), Thompson and Talwani (1964), and Crough and Thompson (1976) in the paragraphs that follow.

Relation of the Gravity Field to Temperature and Heat Flow

General Statement. Blackwell (1969) identified a north-trending zone of high heat flow traversing the full extent of the western United States west of long 110°W. He named it the "Cordilleran thermal anomaly zone." Later compilations of observed heat flow (Roy and others, 1972; Diment and others, 1975; Sass and others, 1976b) all confirm the existence of this broad zone of elevated heat flow, higher than average for normal continental lithosphere and extending from Mexico to Canada. The region of low gravity and bilateral symmetry lies within this zone. Many of the observed heat-flow values greater than 2.5 HFU are found within it. Figure 3-12A shows that they cluster specifically within the regional gravity low.

Waring's (1965) map of thermal springs reveals that the region of low gravity and high elevation is uniformly endowed with hydrothermal circulation (Fig. 3-12B). A simple census of these springs reveals that fully two-thirds of all those in the conterminous United States occur within the province. They fall off in numbers sharply at the province borders, including the southern edge of the Great Basin.

Many of the hot springs of this region have geochemically estimated base temperatures exceeding 90°C (Renner and others, 1975), allowing inferences about elevated subsurface temperatures in the local absence of heat-flow measurements (Fig 3-12C). In particular, the distribution of hot springs having high base temperatures north of the Snake River Plain suggests the possibility of elevated subsurface temperatures in and around the Idaho batholith, where heat-flow data are generally sparse. Blackwell (this volume) has measured values in the Idaho batholith as high as 2 HFU. The local intrusive igneous rocks, including even the youngest Eocene rocks, are nevertheless too old to have retained enough heat to account for the abundant hot springs of high grade (Smith and Shaw, 1975).

A contrasting interpretation of these springs holds that they result from abnormally deep crustal circulation of meteoric water. According to this interpretation, thermal gradients and

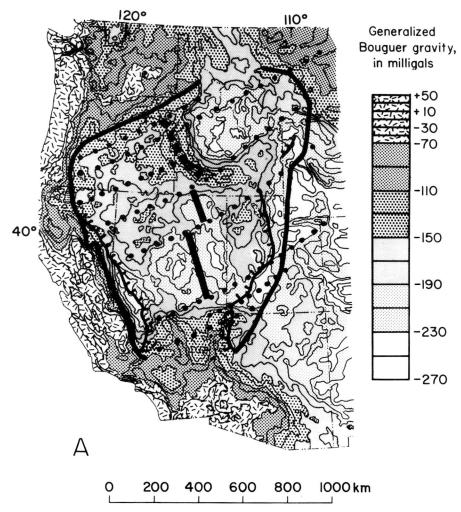

Figure 3-11. Generalized Bouguer gravity map of the western Cordilleras of the United States showing principal borders of the geophysical province identified by the large regional gravity low (curved solid lines) and the inward-facing mountain fronts at the margins of the Great Basin (lighter, hachured lines). (A) Individual segments of the axis of gravity symmetry (straight solid lines, dashed where less certain) and northeast-trending gravity lineaments (dotted lines). (B) Generalized axis of gravity symmetry (straight solid line) and regional topographic axis (straight solid line and lighter dashed lines), as well as generally west-trending gravity lineaments (dotted lines). See Plate 3-1 for full details of gravity field used in identifying lineaments. Figure 3-11 continued on facing page.

conductive heat flow in the region are more or less normal, but vertical permeability is high. Such permeability is ascribed to deeply penetrating faults and joints of steep dip and moderate to high areal density. Until definitive data are in hand, this interpretation can be regarded as tentative only. Fault densities are not higher north of the Snake River Plain than in the Great Basin and, in fact, are lower in the Idaho batholith.

If fracture-related vertical permeability is the correct explanation for the abundance of thermal springs, one must question why the southern and southeastern Basin and Range province in southeastern California and southern Arizona (a densely faulted region) is not the site of abundant hot springs. It is not a matter of mean annual precipitation and consequent ground-water recharge. Precipitation in southeastern Oregon, northern Nevada, and southwestern Idaho, the location of many springs with base temperatures exceeding 90°C, is identical to that of southern Arizona (U.S. Geological Survey, 1970, p.97). The range of values of average annual surface runoff is also similar (U.S. Geological Survey, 1970, p.188), and both regions have major rivers draining to the sea (the Snake on the north and the Colorado and Gila

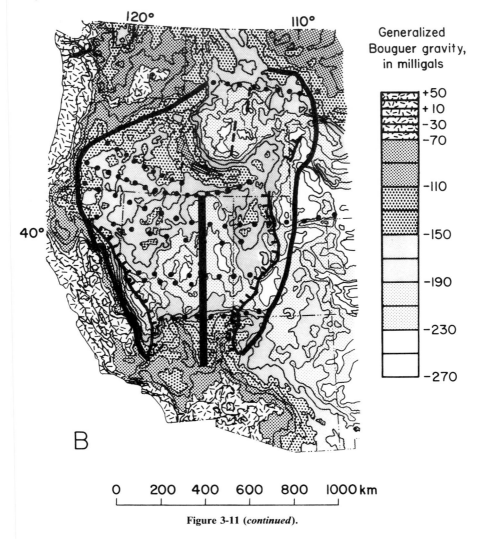

Figure 3-11 (*continued*).

on the south). We therefore view the striking contrast in numbers of high temperature hot springs in southern Idaho and Oregon as opposed to southern Arizona and California as a consequence of a difference in subsurface temperature, not just a difference in permeability and very deep convection.

Until definitive data prove otherwise, we view southern Oregon and Idaho as a region that is hotter than others of similar elevation in the Cordillera to the east and southeast (for example, Wyoming and eastern Utah) as a result of higher crustal temperatures. Only

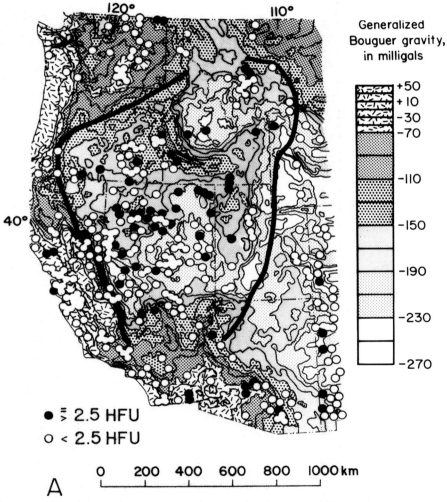

Figure 3-12. Observed heat-flow, hot-spring distribution, and generalized Bouguer gravity in the western Cordilleras. Heavy lines show edges of regional gravity low. Data from Plate 3-1 (this report), Sass and others (1976b, with additions from D. D. Blackwell, 1977, oral commun.), Waring (1965), and Renner and others (1975). (A) Heat-flow measuring sites. Those with values exceeding 2.5 heat-flow units (HFU) are concentrated largely in the region where gravity values are generally less than −110 mgal. (B) Thermal springs. All springs with temperatures more than 8°C above mean annual air temperature are shown. (C) Hot springs with base temperatures exceeding 90°C. Many of them cluster in that part of the regional gravity low west of the autochthonous Paleozoic carbonate sequence (King and Beikman, 1974). (D) Smoothed distribution of observed heat flow. Prepared from tabular data in Sass and others (1976b). Figure 3-12 continued on following three pages.

the southern Rocky Mountains are similar. Both regions bear important similarities to the Great Basin: a low Bouguer gravity field, high regional elevation, young extension, and probable high subsurface temperature. As will be shown in sequence, the region north and northwest of the Snake River Plain is characterized by extensional faulting originating in Miocene time and continuing only locally to the present. To the northwest, the evidence of brittle extension ends abruptly at the gravity gradient that trends northeast across Oregon and southeasternmost Washington.

Figure 3-12D shows a smoothed distribution of observed heat flow prepared by applying a low-pass filter to the data of Sass and others (1976b). The map, unfortunately, does not include many values recently measured by D. D. Blackwell in eastern Oregon, some of which are above that of normal continental lithosphere (compare Sass and others, 1976a). The filter wavelength employed was 100 km. This dimension and the technique used in smoothing place this heat-flow map on a basis comparable to the topography of Figure 3-2.

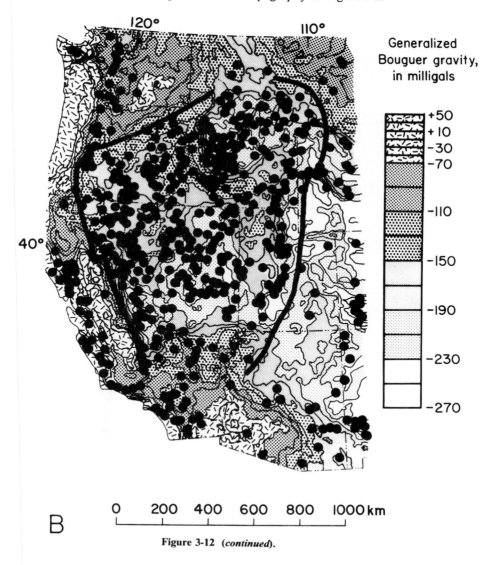

Figure 3-12 (*continued*).

The Great Basin is typified by variable, but generally high, heat flow. High values occur in longitudinal belts near its eastern and western margins. Heat flow is also high at the south edge of the Great Basin. At these locations, elevations *within* the province are lowest and presumably the crust has been extended most. A broad belt of very high heat flow, the Battle Mountain heat-flow high of Sass and others (1976b), trends northeastward across northern Nevada and southern Idaho. It coincides with the Humboldt zone of Mabey and others (this volume). It bears little relation to the regional topography, embracing areas that are both high and low in elevation. It is not reflected in the gravity field or in maps of crustal thickness (compare Fig. 3-12D with Fig. 8 of Warren and Healy, 1973, or Smith, this volume). The heat-flow field, like the gravity field, thus displays varying relations to other geophysical parameters in the region.

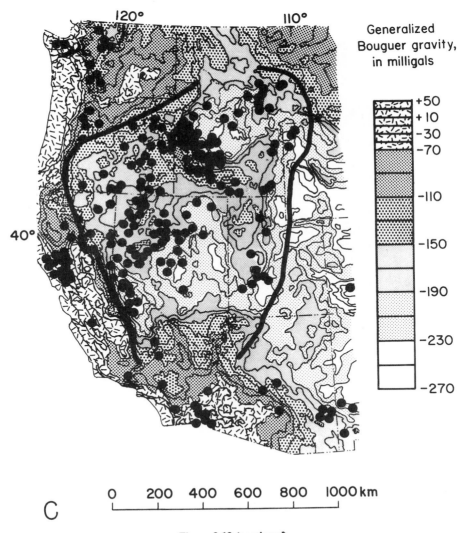

Figure 3-12 (continued).

Both heat-flow data and topography illustrate a fundamental difference between the Great Basin and a spreading oceanic ridge, both of which display bilateral symmetry. In the latter setting, the heat flow is highest along the axis of symmetry, as is the topography (excepting for the medial trough), and both decay systematically and progressively away from the axis. In the Great Basin, both heat flow and topography remain high across a region many hundreds of kilometres wide and roll off sharply only at the edges, well away from the axis of symmetry.

Crough and Thompson (1976) suggested that surface heat flow may be inversely related to thickness of the lithosphere. If this is so, the north and south boundaries of the area of low regional gravity might be regarded as attributable to contrasts in thickness of the lithosphere. Simple calculations indicate that this is not the case, however. To illustrate the point, we will examine relations near the southern end of the province in detail.

Origin of Gravity Gradients—Their Relation to Thermal Phenomena. Physiographers have long commented on contrasts between the Great Basin section of the Basin and Range province

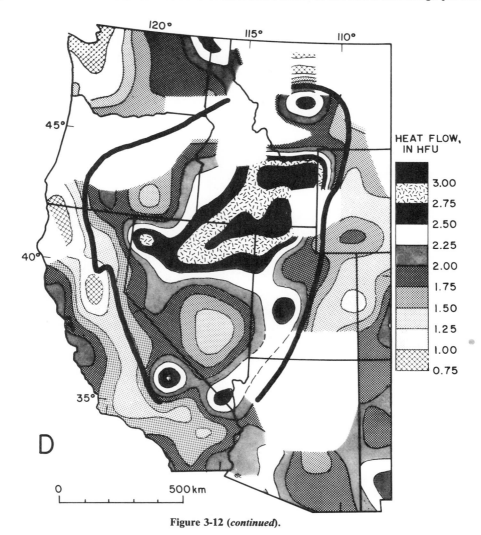

HEAT FLOW, IN HFU

3.00
2.75
2.50
2.25
2.00
1.75
1.50
1.25
1.00
0.75

Figure 3-12 (*continued*).

on the north and the Sonoran Desert section to the south. Fenneman (1931, p. 328) wrote that in the Great Basin:

Space taken by the mountains is about half the total. . . . South of this . . . is the Sonoran Desert section, much lower in altitude, in which mountain ranges are smaller and perhaps older, occupying perhaps one-fifth of the space. Moreover, large areas are without concave basins of internal drainage.

Lobeck (1939, p. 557) commented on the contrast in erosion:

In the southern part of the Basin and Range Province, notably in southern California (and) southern Arizona . . . the basin ranges have been almost annihilated by erosion. The presence and position of faults can only be inferred.

A quarter century later Thornbury (1965, p. 463–464) wrote:

An appreciable difference in appearance exists between the ranges in the northern and those in the southern parts of the province. Particularly in the Sonoran Desert section at the southwest the ranges are smaller, lower, comprised of older rocks, and fringed by more extensive piedmont slopes than in the Great Basin section.

These descriptions indicate that there are significant physiographic differences within the Basin and Range province between the Great Basin and Sonoran Desert sections. These differences reflect differences in the late Cenozoic histories of the two sections. Excluding the western Mojave Desert part, the Sonoran Desert, which stands lower in regional elevation than the Great Basin, is tectonically far less active. The broad boundary zone between them is the scalloped topographic rise that slopes irregularly upward from 35° to just north of lat 37°N, much of the slope occurring abruptly at 37°N.

Comparison of Plate 3-1 and Figure 3-2 shows that the drop in elevation between lat 37° and 35°N (from 1,700 m [5,500 ft] southward to 750 m [2,500 ft]) is accompanied by an increase in gravity of nearly 100 mgal. Quantitative analysis of this gradient on eight crossings at 37°N indicates that the maximum depth to the top of the causative geologic discontinuity lies in the range from 13.3 to 18.4 km. Its mean value is 15.2 km, with a standard deviation of 1.9 km. The method of analysis (Bancroft, 1960) is based on the relationship between gradient, maximum amplitude, and source depth, and makes no assumption concerning the density contrast involved. Its limitation lies in the fact that it yields a figure for the limiting maximum depth only. The top of the anomaly-causing body may be shallower but not deeper.

It is important to stress that this step does not originate at the base of the crust, which here lies at depths closer to 30 ± 3 km (Prodehl, 1970). As noted above, variations in crustal thickness in the vicinity of the step are gradual (Prodehl, 1970; Warren and Healy, 1973; Smith, this volume). Because the crust and lithosphere of the Great Basin are nearly coincident, the step likewise does not occur at the interface between lithosphere and asthenosphere. It is, *instead*, within the crust. What, then, is its origin?

Mabey and others (this volume) comment that Precambrian basement rocks exposed extensively south of lat 37°N are not observed to the north, either in outcrop or in terms of a magnetic expression. Perhaps they lie at very great depth. If they do, it would be tempting to ascribe the gravity gradient to a step on the surface of the Precambrian basement. Average bulk densities of our samples of crystalline basement and of those of the Paleozoic section are similar, however, suggesting that a simple step on the basement surface would not account for the steplike gravity anomaly. The explanation apparently lies elsewhere.

Lachenbruch and others (1976) cited basaltic melt in the Great Basin crust as a possible explanation for the strong, observed geothermal gradients and high heat flow. Is it possible that the presence of partial melt in the asthenosphere and its invasion of the lower lithosphere might alone account for the low gravity values one observes in the Great Basin? We think not. Heating of basalts at 1 atm through the temperature range of 800° to 1,400°C (including their melting interval) produces a change in bulk density of approximately 0.15 $g \cdot cm^{-3}$ (Murase and McBirney, 1973, Fig. 10). Correction for compressibility and thermal expansion appropriate for the pressure-temperature conditions existing in the lower lithosphere would alter this figure somewhat, but we are interested here only in the order of magnitude of the change produced by heating. Gravity calculations show that for a lithospheric basaltic melt with a density contrast of 0.15 $g \cdot cm^{-3}$ (with respect to its surroundings) to produce a 90 to 100 mgal negative anomaly, it would have to be present as a continuous layer 15 km thick. If its top were at a depth of 15 km, it would necessarily coincide with the entire lower half of the crust. Obviously this is not the correct interpretation either. Changes in compressional wave velocity with temperature are far more profound than those of density (Murase and McBirney, 1973, Fig. 13), and seismic refraction measurements of lower *crustal* velocities in the Great Basin do not reveal abnormally low values (Warren, 1968; Prodehl, 1970).

Abundant partial melt in that part of the asthenosphere below the province may be contributory, but only in part. Compilations of P_n-velocities (Herrin and Taggart, 1962; Herrin, 1969; Smith, this volume) indicate abnormally low values beneath the region of the province, and these are interpretable as evidence of the presence of partial melt. Smith's map shows a local increase in P_n at the south edge of the province, as one might anticipate, but values still farther south, where gravity values are much higher, are similar to those in the Great Basin. The width of the region having P_n-values lower than 7.8 $km \cdot s^{-1}$ is, however, narrower on the south.

Herrin's (1972) calculations showed that physical differences in the upper mantle beneath the Basin and Range province (specifically the Great Basin) and the Canadian Shield can be explained by lateral differences in temperature. The same is doubtless true of differences betweeen the Great Basin and the Colorado Plateaus. Subsurface temperature contrasts may thus account for differences across the east and west boundaries of the province, but not for those at the north and south.

The northeast-trending topographic gradient crossing the Columbia Plateaus province separates a region on the south with an average elevation exceeding 1,200 m (4,000 ft) from one on the north with an average elevation close to 300 m (1,000 ft). It is accompanied by a difference in Bouguer gravity level of approximately 80 mgal consisting of two parallel steps, centered at 100 and 140 mgal respectively, and separated by a terrace 80 to 100 km wide. A quantitative analysis of the two parts of this gradient on half a dozen crossings indicates that the maximum depth to the top of the causative geologic discontinuities lies in the range from 9.6 to 20.1 km. Their mean value is 14.3 km, with a standard deviation of 4.5 km. Crustal data are sparse in this region (Warren and Healy, 1973; Hill, this volume), but it appears from the gravity calculations that the top of the causative body here is also well within the crust. Hamilton and Myers (1966, Fig. 6) suggested that the boundary is one between normal and rifted continental crust to the south and younger Cenozoic volcanic crust (on former oceanic crust) to the north.

We have shown that computed maximum depths to the tops of the causative mass discontinuities at both the south and north ends of the province are approximately 15 km. Similar computations for the pronounced gradient that coincides with the Sierra Nevada yield values of an identical order of magnitude (13 to 23 km). All three gradients face into the province and have their sources within the crust. Properties of the crust therefore produce much of the 90-mgal regional gravity low that characterizes the interior of the province. It is possible that all three of these gradients owe their existence to very different causes, but we doubt it. Note, however, the northward continuation of the *upper part* of the western gradient beyond the latitude of central Oregon (Pl. 3-1), indicating that the regional gravity low reinforces another anomaly of different origin.

In order for a 25- to 30-km-thick crust to produce a steplike gravity anomaly of 90 mgal, a lateral density contrast of approximately 0.10 $g \cdot cm^{-3}$ extending through its entire thickness is required, or an even higher contrast if restricted to its upper part. This is roughly equivalent to the thermal expansion of a crust of average density 2.90 $g \cdot cm^{-3}$ resulting from an elevation of its normal temperature into the subsolidus temperature range for basalt. It is also equivalent to an abrupt contrast in total porosity of 3% to 5% (Clark, 1966). Neither of these seems geologically probable, although if the porosity were a fracture porosity it might be argued that it would support the vigorous hydrothermal circulation observed in the province. A sharp compositional contrast related to magmatic invasion and metamorphism seems geologically more plausible to us as an explanation for the crustal gravity low.

The southern gravity gradient corresponds in location to the abrupt southern edge of a vast field of Tertiary volcanic rocks (King and Beikman, 1974). According to Stewart and others (1977, Figs. 1,2,3), these rocks are predominantly rhyolitic in composition, with ages ranging from 34 to 6 m.y. The southern part of the field is younger than 17 m.y. Its southern edge migrated systematically southward, reaching its present position 6 to 8 m.y. ago. Similar data for Oregon (MacLeod and others, 1976, Fig. 2) indicate a systematic northwest migration of rhyolitic volcanism generally toward the northwest gravity gradient in the period from 10 to 0.6 m.y. ago. Both these volcanic fields represent invasions of the *shallow* crust by hot, silicic magmas of relatively young geologic age that migrated toward the edge of the province. The ages of initiation of rhyolitic volcanism in the eastern Snake River Plain show that a similar phenomena occurred there (Christiansen and McKee, this volume). We interpret the 90-mgal regional gravity low to be in part an expression of these shallow, crustal magmatic invasions. As with the alternative hypotheses discussed above, however, this cannot be the entire explanation. Rhyolitic lavas of Miocene and younger age are not seen everywhere throughout the province, nor even along all of its perimeter.

Gravity values and surface elevations in and adjoining the province are viewed, therefore, as complex functions of temperature, composition, and thickness of the crust and lithosphere. The Airy model of isostasy does not apply to much of the region, and the Pratt model, with a substantial dependence on temperature and/or composition, must be invoked.

Although pronounced gravity and topographic steps both mark the upper part of the broad, southern edge of the geophysical province under discussion, and although it is further delineated by the distribution of thermal springs and earthquakes (discussed in sequence), a key element—the axis of gravity symmetry—continues southward to 35°. Relatively high topographic relief between basins and ranges also continues south of the top of the topographic gradient to the Garlock fault, where Fenneman (1931) placed the boundary between the Great Basin and the Sonoran Desert. The high crest of the Sierra Nevada–Tehachapi Mountains continues this far south also, systematically decreasing in elevation from Mount Whitney, near 36°30′N, to 35°. If the initial south boundary of the province originally lay near the end of the axis of symmetry, high elevations that once may have extended this far south have been lost to subsidence and/or erosion following local cessation of tectonism. Best and Hamblin (this volume, Fig. 14-12) visualize a northward propagating front of basaltic volcanism and tectonism here. The southern edge of the Great Basin is clearly a highly complex tectonic area in spite of its apparent structural simplicity.

Relation of the Gravity Field to Regional Seismicity

The seismicity of the western Cordillera is the subject of two other papers in this volume (see Hill; Smith). It is discussed here primarily in relation to the geophysical province we have defined. The seismicity map of Barazangi and Dorman (1969) for shallow (<100 km) earthquakes shows a belt of seismicity that runs from northeastern California counterclockwise down the southwest side of the Great Basin, across Nevada at approximately lat 37°N, thence northeastward and northward across Utah, Idaho, and Wyoming, and, finally, northwestward into Montana. In terms of its intraplate extent, this belt of seismicity has no equal elsewhere in the Cordillera of the Americas. It is concentric with the west and south sides of the gravity gradient bordering the geophysical province under discussion. In its totality it swings around 200° to 240° of arc of the elevated region of low gravity. Inside this arc, a lower seismicity characterizes much of the interior of the Great Basin. As noted by Thompson and Burke (1974), however, evidence of recent faulting can be seen across the entire Great Basin, even though present-day seismicity is concentrated primarily near the margins.

The contrast in seismicity between the Great Basin section of the Basin and Range province and the Sonoran Desert and Mexican Highland sections is striking. In our view, the seismicity of the western Mojave Desert, in the *western* part of the Sonoran Desert section, has a different origin, as noted below.

Because heat flow in the province is high, we hypothesized earlier that stresses probably are relieved by ductile flow at relatively shallow levels. Turcotte (1974) suggested that 300°C is a limiting temperature for elastic brittle behavior on geological time scales. Calculations by Roy and others (1968a), Lachenbruch (1970), and Blackwell (1971) suggested that this temperature is achieved at levels shallower than 15 km in the Great Basin. If so, contemporary failure by fracture should be limited to depths shallower than 15 km, and one may postulate that only a loose mechanical coupling would exist between the brittle plate above and the zone of plastic flow below. This is in excellent agreement with the local data on earthquake focal depths (Ryall and Savage, 1969; Gumper and Scholz, 1971), which indicate a concentration in the depth range from 5 to 16 km and a general absence of earthquakes below depths of about 16 to 20 km. It is also in agreement with our own observation of the lack of isostatic compensation for individual basin-range blocks. Another view holds that the maximum depth for earthquakes is related to a change from stick-slip failure to stable sliding.

Several investigators have published fault-plane solutions for earthquakes bordering, or occurring within, the geophysical province under discussion (Smith and Lindh, this volume). Aside from a small area in western Montana, all those solutions indicative of right-lateral strike-slip faulting are restricted to a band in the southwest part of the Great Basin. The dotted line in Figure 3-13 is the northeastern limit of earthquakes displaying such mechanisms. It is in part coincident with the so-called Walker lane (Shawe, 1965, Fig. 5; Hamilton and Myers, 1966, p. 533), the northeastern limit of demonstrable large-displacement, right-lateral strike-slip faulting. This line also identifies the northeastern hinge line of large-scale sigmoidal bending (Albers, 1967; Stewart, 1967) and the trace of the eastern limit of the subduction-related, magmatic arc for the past 20 m.y. (Snyder and other, 1976). To the south, active faults in the Mojave Desert lie west of this line and, in our view, reflect dextral shearing associated with relative plate motion at the margin of the continent. The sparsity of earthquakes and active faults in southeasternmost California and most of Arizona suggests an absence of tectonism, in contrast to the Great Basin. Fault-plane solutions for events more than 150 km northeast of the California-Nevada border generally indicate simple extension only, with the exception of some in the Humboldt zone. Several solutions characterizing left-lateral, east-west strike-slip movement are seen at the latitude of southern Utah and southern Nevada.

They identify the southern edge of the region of active spreading.

Focal-plane mechanisms of two different types are recorded southwest of the dotted line of Figure 3-13: (1) right-lateral strike-slip motion in a northwest-southeast direction and (2) simple extensional motion in either of two different directions:—east-west (Pitt and Steeples, 1975) or northwest-southeast (Hamilton and Healy, 1969; Gumper and Scholz, 1971). Studies of microearthquakes reveal that these contrasting mechanisms are recorded in the same areas as discrete events, not in combination as oblique slip mechanisms. Although the tension axes are similarly oriented for both the strike-slip and northeast-trending normal fault mechanisms, the orientations of the maximum and intermediate principal stresses are reversed, indicating departure from simplicity. The stress patterns may be interpreted as originating in one or more of several ways: (1) from different stress fields acting on the same region in episodic and cyclic sequence, perhaps somewhat as implied by Gumper and Scholz (1971, p. 1427); (2) as a consequence of broad crustal arching in which steeply dipping fractures that develop with strikes oblique to the arch axis (Handin and others, 1972) are selectively enhanced by relative plate motion at the margin of the continent; or (3) as originating from a single stress field acting on inherited structures in a crust having residual stress of contrasting orientation. The concept of "oblique tensional fragmentation" (Hamilton and Myers, 1966) appears applicable only to that part of the Great Basin that includes the northern and central parts of eastern California and a belt approximately 150 km wide in western Nevada parallel to the California line. Latitudinal or near-latitudinal extension characterizes most of the remainder of the province and even part of the area of oblique extension itself (Pitt and Steeples, 1975). Normal fault orientations and kinematics support this argument (see, for example, Best and Hamblin, this volume). The Humboldt zone (Mabey and others, this volume) is another restricted area of exception.

Directions of extension based on fault-plane solutions for normal faults alone reveal a slight contrast between the east and west sides of the Great Basin (Fig. 3-14). Although there is pronounced scatter in both regions, the average direction of extension southwest of the Walker lane is 300° (120°). To the east, in the region characterized by extensional mechanisms only, the average direction is 263° (83°), or essentially latitudinal. Thus the two sides of the province not only display a contrast in terms of the kinds of focal-plane mechanisms observed but, to a lesser degree, in the average directions of extension revealed.

To summarize, the province is characterized by (1) a peripheral distribution of earthquake epicenters in the interior of the continental plate that has no analog elsewhere in the whole of the Cordillera; (2) northwest-trending dextral shear and west-northwest, as well as east-west, extension west of the Walker lane; (3) east-west extension alone over much of the rest of the province; and (4) marked east-west sinistral shear near its south end.

Relation of the Gravity Field to the Regional Fault System

General Statement. Descriptions and interpretations of the fault system of the Great Basin have been and continue to be the subject of many papers. Our primary purpose here is to demonstrate the geometric relationship of these faults to features of the gravity and other geophysical fields.

Atwater (1970), whose Cenozoic plate model for the western Cordillera is now widely accepted by many, fit the various tectonic elements of western North America into a single unifying and evolutionary scheme. Her explanation of Basin and Range faulting was kinematically similar to that of Hamilton and Myers (1966), but had the added impact of relating the time of initiation and growth of the transform boundary at the edge of the North American plate

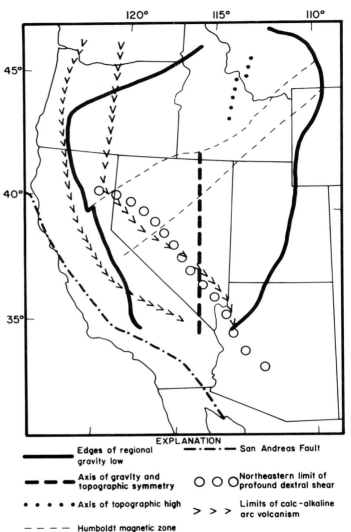

Figure 3-13. Sketch map of selected regional tectonic features. Outline and symmetry axis of gravity province under discussion are shown in relation to the San Andreas fault, the northeastern limit of profound dextral shear, the location of subduction-related arc volcanism in the past 20 m.y. (Snyder and others, 1976), and the Humboldt zone (modified from Mabey and others, this volume). Note general geographic correspondence of northeastern boundaries of dextral shear and arc volcanism in southwestern Nevada.

EXPLANATION

——— Edges of regional gravity low

— — — Axis of gravity and topographic symmetry

• • • • • Axis of topographic high

— — — — Humboldt magnetic zone

— · — · — San Andreas Fault

O O O Northeastern limit of profound dextral shear

> > > Limits of calc-alkaline arc volcanism

(the San Andreas fault system) to that of the onset and northward migration of Basin and Range faulting (earliest in the south and progressively younger northward). The Great Basin was viewed as part of a "very wide, soft boundary" between the rigid and moving Pacific and North American plates. Atwater (1970) did not say why it should be either wide or soft. The fact that the Great Basin is elevated a kilometre above the rest of the Basin and Range province and that it displays evidence of more than one direction of extension (see below) received no comment. The geophysical symmetry of the region was unrecognized.

Directions of Faulting. Strikes of steep faults of all ages in the western Cordillera box the compass (King and Beikman, 1974), but two trends dominate: (1) northwest and (2) north (in detail, north-northwest to north-northeast). The faults are especially densely concentrated in the region of low gravity. In order to study their relations to the gravity field, a set of directional (azimuthal) filters was applied to the faults shown by King and Beikman (1974)

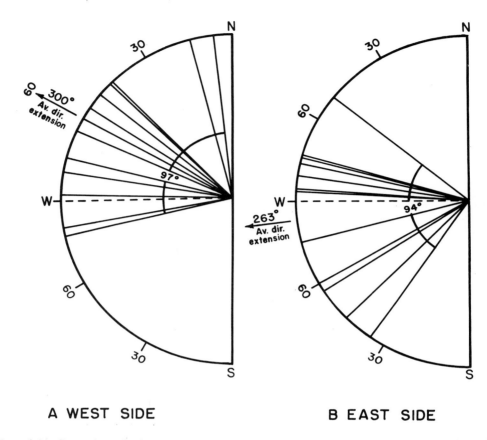

A WEST SIDE   **B EAST SIDE**

Figure 3-14. Comparison of azimuths of *T*-axes from focal-plane mechanisms of normal faulting on west and east sides of the Great Basin. Based on data in Smith and Lindh (this volume). Range of directions of extension for 13 solutions on west side is 97°, with an average azimuth of 300° (120°). Range of directions of extension for 12 solutions on east side is 94°, with average azimuth of 263° (83°).

and King (1969). The azimuthal limits of the filters were based on frequency of trends of gravity gradients in Plate 3-1. This approach rests on the decades-old observation that basement rock units of contrasting density, separated by steeply dipping surfaces, tend to influence later structures in overlying rocks and may be recognizable as gravity gradients.

The gentler gradients associated with long-wavelength gravity anomalies were emphasized in the compilation of strike frequencies because they are believed to represent deep structures of the region and therefore are less likely to have been influenced by inherited, near-surface structure. All gradients were length-weighted in the process of compilation (a gradient of 200-km length was assigned a weight twice that of one only 100 km long). This was done to assure emphasis for the features of larger scale.

The azimuthally filtered, northwest-trending faults are shown in Figure 3-15A. Their direction is that of the principal tectonic grain of the Cordillera. The great length of some of these faults is due to the fact that they are faults of profound, right-lateral strike-slip movement, which, since mid-Tertiary time, has been the direction of relative motion between the North American and Pacific plates (Atwater, 1970). The areal distribution of these northwest-trending faults is moderately uniform throughout the entire region, the only major gaps, or areas of sparsity, being along the northwest coast and in the central part of the map.

The north-trending faults (Figure 3-14B) second the northwest ones in total population and individual and aggregate length. Faults of this set are of dominantly dip-slip, normal displacement, and hence represent crustal extension. These faults display an areal concentration in the center of the map that is unmatched either by the northwest-trending set or by the north-trending set in any other location. This area of concentration is ringed on the west, south, and east by a region nearly devoid of faults having this orientation. Apart from the Rio Grande graben, much farther east, north-trending normal faults in the Cordillera appear to be concentrated in this central region, which extends from lat 35°N to approximately 47°N. The long axis of the central region of north-trending faults coincides with the axis of geophysical bilateral symmetry. The contrast in distribution of these faults between the Great Basin and the rest of the Basin and Range province is notable.

Although much of the Great Basin is characterized by north-trending ranges blocked out since early Miocene time, there is important but equivocal evidence of older block faulting of different trends (Loring, 1976a, 1976b). Older extensional faulting has been well documented for several areas in the Great Basin. Mid- to late-Tertiary block faulting there was followed

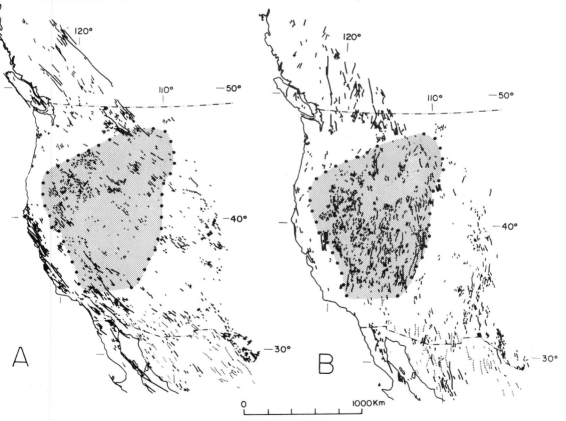

Figure 3-15. Azimuthally filtered, steep faults in the western Cordilleras. Prepared from fault distributions shown by King and Beikman (1974) and King (1969), with minor additions from Cohee (1961). Solid lines are mapped faults; dashed lines are axes of elongate block ranges whose bounding faults are obscured by Cenozoic alluvium. (A) Northwest-striking (297° to 337°) faults or principal Cordilleran grain system. Note general paucity in central region. (B) North-striking (337° to 33°) faults or principal Neogene extensional system. Note general concentration in central region. Heavy dots and shading identify borders and interior of the regional gravity low under discussion.

by younger block faulting of clearly differing orientation (James Gilluly, *in* Hunt and Mabey, 1966; Ekren and others, 1968; Gilbert and Reynolds, 1973; and Mabey and others, this volume). Geologic study of ranges in southern Nevada and Utah (R. E. Anderson and E. B Ekren, personal commun.) has revealed the presence of older normal faults with systematic northwest trends and stratal dips to the northeast or southwest, indicating northeast regional extension.

Orientations of dike swarms can also be used to infer directions of extension. Taubeneck (1969) summarized the extension implied by middle Miocene basaltic dikes of the Chief Joseph and Monument swarms in Oregon and Washington. The dikes trend north-northwest, indicating extension in an east-northeast direction. They reflect back-arc volcanism of Miocene age behind the andesitic Cascade Range arc. Mabey and others (this volume) and Christiansen and McKee (this volume) discuss a major, elongate magnetic anomaly in northern Nevada closely associated with a basaltic dike swarm of middle Miocene age. It trends north-northwest across the northern Great Basin and is likewise suggestive of northeast or east-northeast extension.

On the basis of all these observations, which span the region from southern Nevada to southeastern Washington, we suggest that a significant, early direction of Cenozoic extension in what was to become the geophysical province under discussion was northeast-southwest. It is seen in the strike direction of ranges in the tectonically inactive part of the Basin and Range province in southeastern California and southern Arizona, and in the Basin and Range structures north of the eastern Snake River Plain in Idaho, as well. It appears to be preserved in the gravity field as individual, northwest-trending segments of the axis of gravity symmetry. Inasmuch as this direction of extension would have been parallel to that of subduction of the Farallon plate, we interpret it as back-arc spreading. It spanned the full length of the Basin and Range province from central Mexico to at least as far north as northern Montana, and probably on into Canada.

The contrasting trends in block faulting (northwest versus north strikes) reflect a change in direction of extension relative to an absolute reference within the North American plate. It is significant that the earlier faults did not exert a mechanical influence on the younger ones. The younger faulting gave the Great Basin its present topography and north-south structural grain. This grain contrasts with that in the western Great Basin, west of the Walker lane (Albers, 1967) and also with that of the Sonoran Desert section. There is little or no evidence in the regional fault pattern or the seismicity to indicate "oblique extension" very far east of Fairview Peak, Nevada. Despite the abundant geologic evidence of shallow crustal anisotropy and inhomogeneity, the north-trending normal faults of much of the Great Basin are interpretable in terms of simple east-west extension that broke cleanly across earlier features of different trend. Movements resulting in the "oblique extension" of Hamilton and Myers (1966) can be viewed as a sum of incremental movements arising from separate stress fields or as the effect of young stresses acting on contrastingly prestressed rock in the region bounded by the Sierra Nevada and a line parallel to, but northeast of, the Walker lane.

The profound extension of the province as a whole raises a question as to the occurrence of transform structures parallel to the direction of spreading. Such structures are more certain guides to the movement direction than are the normal faults themselves (Morgan, 1968). Several gravity gradients on Plate 3-1 trend east-west; Figure 3-11B shows the location of these lineaments, but much of the supporting detail is visible only on Plate 3-1. The geophysical expression of east-trending structures in the western Cordillera has been described or illustrated by Fuller (1964), Mabey and Morris (1967), Herrin and Taggart (1962), Cook and Montgomery (1972), and Cook and others (1975). Stewart and others (1977), Ekren and others (1976), and Rowley and others (1978) described surface geologic features that trend east-west across the Great Basin, approximately normal to the basin ranges. Some of these, however, give evidence of existence prior to mid-Cenozoic time and may reflect renewed structures. They appear to have exerted structural control on the location of volcanic centers, ore deposits, and hot springs.

North of the Humboldt zone, the trend of most normal faults is northwest or north-northeast, a decided contrast with the Great Basin. Lawrence (1976) demonstrated the existence of relatively young transform faults trending west-northwest across Oregon. He argued against relating them to those of contrasting strike between the Walker lane and Sierra Nevada. The Oregon faults are perpendicular to a family of major normal faults of demonstrably young movement bounding Steens Mountain and related parallel ranges in eastern Oregon (G. W. Walker, 1977 oral commun.). This set of strike-slip faults is indicative of young west-northwest extension north of the Humboldt zone, and contrasts with the inferred direction of extension suggested by the older Miocene dikes both farther north and southeast.

As noted earlier, hot-spring data and sparse heat-flow measurements suggest a continuation of elevated temperatures in the region north of the Humboldt zone, and gravity and topographic data also indicate similarities between this northern region and the Great Basin. Although published geologic details are sparse, there is unequivocal evidence of Quaternary faulting well north and northwest of the Snake River Plain. B. F. Leonard (1977, written commun., and *in* Curtin and King, 1974) has mapped normal faults cutting Quaternary moraines and offsetting lichen-covered talus slopes in the area of the Eocene Challis Volcanics west of 45°N, 155°W. G. W. Walker's (1977, oral commun.) mapping in eastern Oregon has revealed normal faults that cut fans and pediments involving accumulations of Pliocene vertebrate fossils, at least as far north as lat 43°N. Ruppel (1964, 1968) mapped north-trending Quaternary faults near the Idaho-Montana line at lat 44° to 45°N. They cut very young surficial deposits, behead young alluvial fans, and have disrupted present drainage patterns. These examples all indicate minimal, but very young, steep faulting. The contrast with the Great Basin is apparently one of degree, not of kind. The area has experienced continuing extension, but its magnitude has been small and it apparently dies out northward. This extension, like that in the Great Basin, contrasts in direction with that of Miocene age, and was unaccompanied by Quaternary volcanism north of 44°N, at least on the west.

CONCLUSIONS

The gravity map of the Western United States identifies a unique province within the Cordillera. The gravity and topographic fields of the southern half of the province display a well-developed bilateral symmetry, as do opposed belts of progressively outward-migrating silicic volcanism in the northern half. The line of symmetry constitutes the axis of a broad, elongate, north-trending upwarp, which can be traced from southern California to southern Idaho. It terminates at the eastern Snake River Plain (part of the Humboldt magnetic zone). Quaternary extension continued north of the Snake River Plain, but in lesser degree, and it was not accompanied by Quaternary volcanism north of 44°N. Young extension of the entire region is believed to have resulted from the same stresses that produced the symmetrical distribution of deep mass anomalies. Spreading has been both rapid and complex and has varied in time and space. Earthquakes in the region are concentrated near the margins of the anomaly, where major changes in crustal structure exist. Crustal thinning and the youngest volcanic activity are pronounced near the edges of the province.

As a whole, these characteristics and the high regional heat flow define a feature closely akin to what elsewhere, and on a much smaller scale, has been called a "hot spot." This one differs in terms of (1) its lateral dimensions (900 by 1,200 km), (2) the profound nature of the extension, (3) its location near enough to a major plate boundary to have created interfering tectonic effects, and (4) its notable bilateral symmetry.

Uplift of the northwest edge of the province appears to have taken place sometime more recent than 10.5 m.y. This date is based on the age of the youngest middle and upper flows of the Tertiary Yakima Basalt affected by uplift there (D. A. Swanson, 1976, written commun.). The major episode of uplift and tilting of the Sierran block, approximately 9 m.y. ago (Christensen, 1966), may also define the age of uplift at the western edge of the province, because the western edge of the crustal mass-deficiency that underlies the province lies west of the eastern escarpment of the Sierra Nevada. According to Ekren and others (1968), present-day, north-trending fault-block topography in southern Nevada, near the southern edge of the province, developed between 11 and 7 m.y. ago. On the basis of this evidence, we suggest that the edges of the province, as seen today, developed sometime between 9 and a little less than 7 m.y. ago. It formed in a region that had already experienced northeast-directed back-arc spreading associated with subduction of the Farallon plate. Both the direction and magnitude of extension changed when subduction ceased. The new direction of extension was (and continues to be) east-west or nearly so, and the degree, and probably the rate, of extension increased markedly. However, the latitudinal extent of continuing extension was considerably more restricted than that of the back-arc spreading, which included the whole of the Basin and Range province. The broad southern boundary of today's active extension appears to have no explanation in the models of Atwater (1970) and Christiansen and Lipman (1972), just as the location of the north end, well beyond the contemporary Mendocino triple junction, likewise does not.

Given these various characteristics of the province and given also the fact that it had undergone repeated episodes of deformation and plutonism long before the Pliocene, all of which acted to produce crustal inhomogeneity and anisotropy, we think the province is most readily explained by the rise, and probably the lateral divergent flow of hot asthenospheric material accompanying major mantle upwelling. This model is similar in concept to that of Scholz and others (1971). It is supported by the argument of Thompson (1972), who noted that the absence of a large negative isostatic anomaly in the region requires mass flow in the mantle in order to balance the effect of thinning of the lithosphere.

We view the features of the region as the products of active, rather than passive, spreading, in sharp contrast with ocean ridges, which also display a bilateral symmetry. In the sea-floor spreading process the highest heat flow, seismicity, and active volcanism are concentrated at the ridge and decay rapidly outward. In the western Cordillera the seismicity and youngest volcanism are concentrated at the *outer* edges of the structure and high heat flow, distributed extension, and regionally high topography extend across its full width, all but the last one peaking near the margins. Although the topography of the Great Basin is symmetrical and slightly higher along its axis than just inside its margins, the margins themselves (the Sierra Nevada and Wasatch Mountains) are higher still.

Some regard the normal faulting, high heat flow, thin crust, and low P_n-velocities of the Great Basin as products of *externally driven*, rapid extension. Given the old, inhomogeneous and relatively weak nature of the lithosphere, we think this interpretation provides no explanation for the remarkable bilateral symmetry and young marginal volcanism of the province, for the continuing outward migration of its east and west margins, for the location of its north and south boundaries, for the absence of large, negative isostatic anomalies, nor for a driving mechanism. Postulation of an upwelling of specific location and dimension in the upper mantle, perhaps accompanied by lateral divergent flow, at least provides a rationale for these parameters, hence our inclination toward it at this time.

ACKNOWLEDGMENTS

The data and interpretations presented here constitute a synthesis of material offered piecemeal on several previous occasions (Eaton, 1975; Mabey and others, 1975; Eaton, 1976; Eaton and others, 1976; Smith and others, 1976). The reference cited last is an account of discussions held at the 1975 Penrose Conference in Alta, Utah. We are indebted to many people present at this conference as well as to others familiar with the geology and geophysics of the region. They willingly shared their detailed knowledge with us and provided many hours of stimulating discussion. Their guidance in interpretation was most helpful, even on those occasions when they disagreed with our conclusions. Those to whom we are especially indebted are M. G. Best, P. J. Coney, R. L. Christiansen. Lindrith Cordell, Mel Friedman, W. B. Hamilton, A. H. Lachenbruch, B. F. Leonard, P. W. Lipman, E. H. McKee, F. G. Poole, M. W. Reynolds, P. D. Rowley, E. T. Ruppel, D. R. Shawe, R. B. Smith, J. H. Stewart, D. A. Swanson, G. W. Walker, R. E. Wallace, and P. L. Williams; however, none of them should be assumed to endorse the views expressed here. Both our continuously evolving views and the final draft of the manuscript benefited considerably from critical and comprehensive reviews made of an embryonic first draft by W. B. Hamilton, P. W. Lipman, and G. A. Thompson.

The simple Bouger gravity map shown as Plate 3-1 was complied from gravity data in the files of the U.S. Geological Survey and the Defense Mapping Agency. We are indebted to M. F. Kane, U.S. Geological Survey, for support in this endeavor.

REFERENCES CITED

Albers, J. P., 1967, Belt of sigmoidal bending and right-lateral faulting in the western Great Basin: Geol. Soc. America Bull., v. 78, p. 143–156.

Artemjev, M. E., and Artyushkov, E. V., 1971, Structure and isostasy of the Baikal Rift and the mechanism of rifting: Jour. Geophys. Research, v. 76, P. 1197–1212.

Atwater, Tanya, 1970, Implications of plate tectonics for the Cenozoic tectonic evolution of western North America: Geol. Soc. America Bull., v. 81, p. 3513–3536.

Axelrod, D. I., 1968, Tertiary floras and topographic history of the Snake River basin, Idaho: Geol. Soc. America Bull., v. 79, p. 713–734.

Bancroft, A. M., 1960, Gravity anomalies over a buried step: Jour. Geophys. Research, v. 65, p. 1630–1631.

Barazangi, Muawia, and Dorman, James, 1969, World seismicity maps compiled from ESSA, Coast and Geodetic Survey, epicenter data, 1961–1967: Seismol. Soc. America Bull., v. 59, no. 1, p. 369–380.

Best, M. G., and Hamblin, W. K., 1978, origin of the Northern Basin and Range Province: Implications from the geology of its eastern boundary, in Smith, R. B., and Eaton, G. P., eds., Cenozoic tectonics and regional geophysics of the western Cordillera: Geol. Soc. America Mem. 152 (this volume).

Biswas, N. N., and Knopoff, Leon, 1974, The structure of the upper mantle under the United States from the dispersion of Rayleigh waves: Royal Astron. Soc. Geophys. Jour., v. 36, p. 515–539.

Blackwell, D. D., 1969, Heat-flow determinations in the northwestern United States: Jour. Geophys. Research, v. 74, p. 992–1007.

——1971, The thermal structure of the continental crust, in Heacock, J. G., ed., The structure and physical properties of the Earth's crust: Am. Geophys. Union Geophys. Mon. 14, p. 169–184.

——1978, Heat flow and energy loss in the Western United States, in Smith, R. B., and Eaton, G. P., eds., Cenozoic tectonics and regional geophysics of the western Cordillera: Geol. Soc. America Mem. 152 (this volume).

Carder, D. S., Quamar, Anthony, and McEvilly, T. V., 1970, Trans-California seismic profile—Pahute Mesa to San Francisco Bay: Seismol. Soc. America Bull., v. 60, p. 1829–1846.

Christensen, M. N., 1966, Late Cenozoic crustal movements in the Sierra Nevada of California: Geol. Soc. America Bull., v. 77, p. 163–182.

Christiansen, R. L., and Lipman, P. W., 1972, Cenozoic volcanism and plate-tectonic evolution of the Western United States, II, Late Cenozoic: Royal Soc. London Philos. Trans., v. 271, p. 249–284.

Christiansen, R. L., and McKee, E. H., 1978, Late Cenozoic volcanic and tectonic evolution of the Great Basin and Columbia intermontane region, *in* Smith, R. B., and Eaton, G. P., eds., Cenozoic tectonics and regional geophysics of the western Cordillera: Geol. Soc. America Mem. 152 (this volume).

Clark, S. P., Jr., ed., 1966, Handbook of physical constants: Geol. Soc. America Mem. 97, 587 p.

Cloos, Ernst, 1968, Experimental analysis of Gulf Coast fracture patterns: Am. Assoc. Petroleum Geologists Bull., v. 52, p. 420–444.

Cohee, G. V., 1961 (see entry under U.S. Geol. Survey and Am. Assoc. Petroleum Geologists).

Cook, K. L., and Montgomery, J. R., 1972, East-west transverse structural trends in the eastern Basin and Range province as indicated by gravity data [abs.]: Internat. Geol. Cong., 24th, Montreal, Canada, sec. 9, p. 175.

Cook, K. L., Montgomery, J. R., and Smith, J. T., 1975, Structural trends in Utah as indicated by gravity data: Geol. Soc. America Abs. with Programs, v. 7, no. 5, p. 598.

Crittenden, M. D., Jr., 1963, New data on the isostatic deformation of Lake Bonneville [Utah]: U.S. Geol. Survey Prof. Paper 454–E, p. E1–E31.

Crough, S. T., and Thompson, G. A., 1976, Thermal model of continental lithosphere: Jour. Geophys. Research, v. 81, no. 26, p. 4857–4862.

Curtin, G. C., and King, H. D., 1974, Antimony and other metal anomalies south of Stibnite, Valley County, Idaho with a section on Geology, by B. F. Leonard: U.S. Geol. Survey open-file rept. 74–111, 54 p.

Diment, W. H., Urban, T. C., Sass, J. H., Marshall, B. V., Munroe, R. J., and Lachenbruch, A. H., 1975, Temperatures and heat contents based on conductive transport of heat, *in* White, D. E., and Williams, D. L., eds., Assessment of geothermal resources of the United States—1975: U.S. Geol. Survey Circ. 726, p. 84–103.

Eaton, G. P., 1975, Characteristics of a transverse crustal boundary in the Basin and Range province of southern Nevada: Geol. Soc. America Abs. with Programs, v. 7, no. 7, p. 1062–1063.

——1976, Fundamental bilateral symmetry of the western Basin and Range province: Geol. Soc. America Abs. with Programs, v. 8, no. 5, p. 583–584.

Eaton, G. P., Christiansen, R. L., Pitt, A. M., Mabey, D. R., Blank, H. R., Zietz, Isidore, and Gettings, M. E., 1975, Magma beneath Yellowstone National Park: Science, v. 188, p. 787–796.

Eaton, G. P., Prostka, H. J., Oriel, S. S., and

Pierce, K. L., 1976, Cordilleran thermotectonic anomaly: I. Geophysical and geological evidence of coherent late Cenozoic intraplate magmatism and deformation: Geol. Soc. America Abs. with Programs, v. 8, no. 6, p. 850.

Eaton, J. P., 1966, Crustal structure in northern and central California from seismic evidence, *in* Bailey, E. H., ed., Geology of northern California: Calif. Div. Mines Geol. Bull. 190, p. 419–426.

Ekren, E. B., Rogers, C. L., Anderson, R. E., and Orkild, P. P., 1968, Age of Basin and Range normal faults in Nevada Test Site and Nellis Air Force Range, Nevada, *in* Eckel, E. B., ed., Nevada test site: Geol. Soc. America Mem. 110, p. 247–250.

Ekren, E. G., Bucknam, R. C., Carr, W. J., Dixon, G. L., and Quinlivan, W. D., 1976, East-trending structural lineaments in central Nevada: U.S. Geol. Survey Prof. Paper 986, 16 p.

Fenneman, N. M., 1931, Physiography of Western United States: New York and London, Mc-Graw-Hill Book Co., 534 p.

Fuller, M. D., 1964, Expression of E-W fractures in magnetic surveys in parts of the U.S.A.: Geophysics, v. 29, p. 602–622.

Gilbert, C. M., and Reynolds, M. W., 1973, Character and chronology of basin development, western margin of the Basin and Range province: Geol. Soc. America Bull., v. 84, p. 2489–2510.

Gumper, F. J., and Scholz, Christopher, 1971, Microseismicity and tectonics of the Nevada seismic zone: Seismol. Soc. America Bull., v. 61, no. 5, p. 1413–1432.

Hamblin, W. K., 1965, Origin of "reverse drag" on the downthrown side of normal faults: Geol. Soc. America Bull., v. 76, p. 1145–1164.

Hamilton, R. M., and Healy, J. H., 1969, Aftershocks of the Benham nuclear explosion: Seismol. Soc. America Bull., v. 59, no. 6, p. 2271–2281.

Hamilton, Warren, 1969, Mesozoic California and the underflow of Pacific mantle: Geol. Soc. America Bull., v. 80, p. 2409–2430.

——1975, Neogene extension of the Western United States: Geol. Soc. America Abs. with Programs, v. 7, no. 7, p. 1098.

Hamilton, Warren, and Myers, W. B., 1966, Cenozoic tectonics of the Western United States: Rev. Geophysics, v. 4, no. 4, p. 509–549.

Handin, J., Friedman, M., Logan, J. M., Pattison, L. J., and Swolfs, H. S., 1972, Experimental folding of rocks under confining pressure—Pt. I. Buckling of single-layer rock beams, *in* Flow and fracture of rocks (Griggs volume): Am. Geophys. Union Geophys. Mon., v. 16, 28 p.

Herrin, Eugene, 1969, Regional variations of P-wave velocity in the upper mantle beneath North

America, *in* Hart, P. J., ed., The Earth's crust and upper mantle: Am. Geophys. Union Geophys. Mon. 13, p. 242–246.

——1972, A comparative study of upper mantle models—Canadian Shield and Basin and Range provinces, *in* Robertson, E. C., ed., The nature of the solid earth: New York, McGraw-Hill Book Co., p. 506–543.

Herrin, Eugene, and Taggart, James, 1962, Regional variations in P_n velocity and their effect on the location of epicenters: Seismol. Soc. America Bull., v. 52, p. 1037–1046.

Hill, D. P., 1978, Seismic evidence for the structure and Cenozoic tectonics of the Pacific Coast States, *in* Smith, R. B., and Eaton, G. P., eds., Cenozoic tectonics and regional geophysics of the western Cordillera: Geol. Soc. America Mem. 152 (this volume).

Hill, D. P., and Pakiser, L. C., 1966, Crustal structure between the Nevada test site and Boise, Idaho, from seismic refraction measurements: Am. Geophys. Union Geophys. Mon. 11, p. 391–419.

Hunt, C. B., and Mabey, D. R., 1966, Stratigraphy and structure, Death Valley, California: U.S. Geol. Survey Prof. Paper 494-A, 162 p.

Kane, M. F., and Carlson, J. E., 1961, Gravity anomalies, isostasy, and geologic structure in Clark County, Nevada, *in* Short papers in the geologic and hydrologic sciences: U.S. Geol. Survey Prof. Paper 424-D, p. 274–279.

Kanizay, S. P., 1962, Mohr's theory of strength and Prandtl's compressed cell in relation to vertical tectonics: U.S. Geol. Survey Prof. Paper 414-B, 16 p.

King, P. B., compiler, 1969, Tectonic map of North America: U.S. Geol. Survey Map, scale 1:5,000,000.

King, P. B., and Beikman, H. M., 1974, Geologic map of the United States: U.S. Geol. Survey Map, scale 1:2,500,000.

Lachenbruch, A. H., 1970, Crustal temperature and heat production: Implications of the linear heat-flow relation: Jour. Geophys. Research, v. 75, p. 3291–3300.

Lachenbruch, A. H., Sass, J. H., Munroe, R. J., and Moses, T. H., Jr., 1976, Geothermal setting and simple heat conduction models for the Long Valley caldera: Jour. Geophys. Research, v. 81, p. 769–784.

Lawrence, R. D., 1976, Strike-slip faulting terminates the Basin and Range province in Oregon: Geol. Soc. America Bull., v. 87, p. 846–850.

Lobeck, A. K., 1939, Geomorphology, an introduction to the study of landscapes: New York, McGraw-Hill Book Co., 731 p.

Loring, A. K., 1976a, The age of Basin-Range faulting in Arizona: Arizona Geol. Soc. Digest, v. 10, p. 229–257.

——1976b, Distribution in time and space of late Phanerozoic normal faulting in Nevada and Utah: Utah Geology, v. 3, no. 2, p. 97–109.

Mabey, D. R., 1966, Relation between Bouguer gravity anomalies and regional topography in Nevada and the eastern Snake River Plain, Idaho: U.S. Geol. Survey Prof. Paper 550-B, p. 108–110.

——1976, Interpretation of a gravity profile across the western Snake River Plain, Idaho: Geology, v. 4, no. 1, p. 53–55.

Mabey, D. R., and Morris, H. T., 1967, Geologic interpretation of gravity and aeromagnetic maps of Tintic Valley and adjacent areas, Tooele and Juab Counties, Utah: U.S. Geol. Survey Prof. Paper 516-d, 10 p.

Mabey, D. R., Eaton, G. P., and Zietz, Isidore, 1975, Regional gravity and magnetic patterns of the Cordillera: Geol. Soc. America Abs. with Programs, v. 7, no. 7, p. 1181–1182.

Mabey, D. R., Zietz, I., Eaton, G. P., and Kleinkopf, M. D., 1978, Regional magnetic patterns in part of the Cordillera in the Western United States, *in* Smith, R. B., and Eaton, G. P., eds., Cenozoic tectonics and regional geophysics of the western Cordillera: Geol. Soc. America Mem. 152 (this volume).

MacLeod, N. S., Walker, G. W., and McKee, E. H., 1976, Geothermal significance of eastward increase in age of upper Cenozoic rhyolitic domes in southeastern Oregon, *in* Second United Nations Symposium on the Development and Use of Geothermal Resources, Proc., San Francisco, Calif., 20–29 May, 1975, p. 465–474.

Moore, J. G., 1960, Curvature of normal faults in the Basin and Range province of the Western United States, *in* Short papers in the geological sciences: U.S. Geol. Survey Prof. Paper 400-B, p. 409–411.

Morgan, W. J., 1968, Rises, trenches, great faults, and crustal blocks: Jour. Geophys. Research, v. 73, p. 1959–1982.

Murase, Tsutomu, and McBirney, A. R., 1973, Properties of some common igneous rocks and their melts at high temperatures: Geol. Soc. America Bull., v. 84, p. 3563–3592.

Nafe, J. W., and Drake, C. L., 1963, Physical properties of marine sediments, *in* Hill, M. N., ed., The sea, Vol. 3: New York, Interscience, p. 794–815.

Pitt, A. M., and Steeples, D. W., 1975, Microearthquakes in the Mono Lake–northern Owens Valley, California, region from September 28 to October 18, 1970: Seismol. Soc. America Bull., v. 65, no. 4, p. 835–844.

Prodehl, Claus, 1970, Seismic refraction study of crustal structure in the Western United States: Geol. Soc. America Bull., v. 81, p. 2629–2646.

Proffett, J. M., Jr., 1977, Cenozoic geology of the Yerington district, Nevada, and implications for the nature and origin of Basin and Range faulting: Geol. Soc. America Bull., v. 88, p. 247–266.

Renner, J. L., White, D. E., and Williams, D. L., 1975, Hydrothermal convections systems, in White, D. E., and Williams, D. L., eds., Assessment of geothermal resources of the United States—1975: U.S. Geol. Survey Circ. 726, p. 5–57.

Rowley, P. D., Lipman, P. W., Mehnert, H. H., Lindsey, D. A., and Anderson, J. J., 1978, Blue Ribbon Lineament, an east-trending structural zone within the Pioche mineral belt of southwestern Utah and eastern Nevada: U.S. Geol. Survey Jour. Research, v. 6, no. 2 (in press).

Roy, R. F., Blackwell, D. D., and Birch, F., 1968a, Heat flow in the United States: Jour. Geophys. Research, v. 72, p. 5207–5221.

——1968b, Heat generation of plutonic rocks and continental heat flow provinces: Earth and Planetary Sci. Letters, v. 5, 12 p.

Roy, R. F., Blackwell, D. D., and Decker, E. R., 1972, Continental heat flow, in Robertson, E. C., ed., The nature of the solid earth: New York, McGraw-Hill Book Co., p. 506–543.

Ruppel, E. T., 1964, Strike-slip faulting and broken basin-ranges in east-central Idaho and adjacent Montana, in Geological Survey research 1964: U.S. Geol. Survey Prof. Paper 501-C, p. C14–18.

——1968, Geologic map of the Leadore quadrangle, Lemhi County, Idaho: U.S. Geol. Survey Geol. Quad. Map GQ-733, scale 1:62,500.

Ryall, Alan, and Savage, W. U., 1969, A comparison of seismological effects for the Nevada underground test BOXCAR with natural earthquakes in the Nevada region: Jour. Geophys. Research, v. 74, no. 17, p. 4281–4289.

Sass, J. H., Galanis, S. P., Jr., Munroe, R. J., Urban, T. C., 1976a, Heat-flow data from southeastern Oregon: U.S. Geol. Survey open-file rept. 76-217, 52 p.

Sass, J. H., Diment, W. H., Lachenbruch, A. H., Marshall, B. V., Munroe, R. J., Moses, T. H., Jr., and Urban, T. C., 1976b, A new heat-flow contour map of the conterminous United States: U.S. Geol. Survey open-file rept. 76-756, 24 p.

Scholz, C. H., Barazangi, Muawia, and Sbar, M. L., 1971, Late Cenozoic evolution of the Great Basin, Western United States, as an ensialic interarc basin: Geol. Soc. America Bull., v. 82, p. 2979–2990.

Shawe, D. R., 1965, Strike-slip control of Basin-Range structure indicated by historical faults in western Nevada: Geol. Soc. America Bull., v. 76, p. 1360–1378.

Smith, R. B., 1978, Seismicity, crustal structure, and intraplate tectonics of the interior of the western Cordillera, in Smith, R. B., and Eaton, G. P., eds., Cenozoic tectonics and regional geophysics of the western Cordillera: Geol. Soc. America Mem. 152 (this volume).

Smith, R. B., Mabey, D. R., and Eaton, G. P., 1976, Regional geophysics and tectonics of the Intermountain West: Geology, v. 4, p. 437–438.

Smith, R. B., and Lindh, A. G., 1978, Fault plane solutions of the Western United States: A compilation, in Smith, R. B., and Eaton, G. P., eds., Cenozoic tectonics and regional geophysics of the western Cordillera: Geol. Soc. America Mem. 152 (this volume).

Smith, R. L., and Shaw, H. R., 1975, Igneous-related geothermal systems, in White, D. E., and Williams, D. L., eds., Assessment of geothermal resources of the United States—1975: U.S. Geol. Survey Circ. 726, p. 58–83.

Snyder, W. S., Dickinson, W. R., and Silberman, 1976, Tectonic implications of space-time patterns of Cenozoic magmatism in the Western United States: Earth and Planetary Sci. Letters, v. 32, p. 91–106.

Stanley, W. D., Boehl, J. R., Bostick, F. X., and Smith, H. W., 1977, Geothermal significance of magnetotelluric soundings in the eastern Snake River Plain–Yellowstone region: Jour. Geophys. Research, v. 82, p. 2501–2514.

Stewart, J. H., 1967, Possible large right-displacement along fault and shear zones in the Death Valley–Las Vegas area, California and Nevada: Geol. Soc. America Bull., v. 78, p. 131–142.

Stewart, J. H., Moore, W. J., and Zietz, Isidore, 1977, East-west patterns of Cenozoic igneous rocks, aeromagnetic anomalies and mineral deposits, Nevada and Utah: Geol. Soc. America Bull., v. 87, p. 67–77.

Strange, W. E., and Woollard, G. P., 1964, The use of geologic and geophysical parameters in the evaluation, interpolation, and prediction of gravity: Aeronautical Chart and Information Center, USAF, St. Louis, Missouri, Contract AF23(601)-3879 (Phase 1), 225 p.

Taubeneck, W. H., 1969, Dikes of Columbia River basalt in northeastern Oregon, western Idaho, and southeastern Washington: Second Columbia River Basalt Symposium, Proc., March 21–23, 1969: Cheney, Washington, Eastern Washington State College Press, p. 73–96.

Thompson, G. A., 1959, Gravity measurements between Hazen and Austin, Nevada—A study of Basin-Range structure: Jour. Geophys. Research, v. 64, p. 217–229.

——1972, Cenozoic Basin-Range tectonism in relation to deep structure: Internat. Geol. Cong., 24th, Montreal, Canada, proc., sec. 3, p. 84–90.

Thompson, G. A., and Burke, D. B., 1974, Regional geophysics of the Basin and Range province: Ann. Review Earth and Planetary Sci., v. 2, p. 213–238.

Thompson, G. A., and Talwani, M., 1964, Crustal structure from Pacific Basin to central Nevada: Jour. Geophys. Research, v. 69, p. 4813–4837.

Thornbury, W. D., 1965, Regional geomorphology of the United States: New York, John Wiley & Sons, Inc., 609 p.

Turcotte, D. L., 1974, Are transform faults thermal contraction cracks?: Jour. Geophys. Research, v. 79, no. 17, p. 2573–2577.

U.S. Geological Survey, 1970, National atlas of the United States of America: Washington, 417 p.

U.S. Geological Survey and American Association of Petroleum Geologists, G. V. Cohee, committee chm., 1961, Tectonic map of the United States, exclusive of Alaska and Hawaii: Washington, D.C., U.S. Geol. Survey.

Walcott, R. I., 1970, Flexural rigidity, thickness and viscosity of the lithosphere: Jour. Geophys. Research, v. 75, p. 3941–3954.

Waring, G. A., 1965, Thermal springs of the United States and other countries of the World—A summary: U.S. Geol. Survey Prof. Paper 492, 383 p.

Warren, D. H., 1968, Transcontinental geophysical survey (35°–39°N.)—Seismic refraction profiles of the crust and upper mantle from 100° to 112°W. longitude: Washington, D.C., U.S. Geol. Survey Map I-533-D.

Warren, D. H., and Healy, J. H., 1973, Structure of the crust in the conterminous United States: Tectonophysics, v. 20, p. 203–213.

Woollard, G. P., 1966, Regional isostatic relations in the United States, in Steinhart, J. S., and Smith, T. J., eds., The Earth beneath the continents: Am. Geophys. Union Geophys. Mon. 10, p. 557–594.

——1968, The interrelationship of the crust, the upper mantle, and isostatic gravity anomalies in the United States, in The crust and upper mantle of the Pacific area—Internat. Upper Mantle Proj., Sci. Rept. 15: Am. Geophys. Union Geophys. Mon. 12, p. 312–341.

——1969, Regional variation in gravity, in Hart, P. J., ed., The Earth's crust and upper mantle: Am. Geophys. Union Geophys. Mon. 13, p. 320–340.

——1972, Regional variations in gravity, in Robertson, E. C., ed., The nature of the solid earth: New York, McGraw-Hill Book Co., p. 463–505.

York, J. E., and Helmberger, D. V., 1973, Low-velocity zone variations in the southwestern United States: Jour. Geophys. Research, v. 78, no. 11, p. 1883–1886.

MANUSCRIPT RECEIVED BY THE SOCIETY AUGUST 15, 1977

MANUSCRIPT ACCEPTED SEPTEMBER 2, 1977

Printed in U.S.A.

Geological Society of America
Memoir 152

4

Regional magnetic patterns in part of the Cordillera in the Western United States

DON R. MABEY
U.S. Geological Survey
Box 25046, Denver Federal Center
Denver, Colorado 80225

ISIDORE ZIETZ
U.S. Geological Survey
National Center Mail Stop 927
Reston, Virginia 22092

GORDON P. EATON
U.S. Geological Survey
National Center Mail Stop 911
Reston, Virginia 22092

M. DEAN KLEINKOPF
U.S. Geological Survey
Box 25046, Denver Federal Center
Denver, Colorado 80225

ABSTRACT

A residual aeromagnetic map of Idaho, western Montana, western Wyoming, southwestern Oregon, Nevada, Utah, western Colorado, and northern Arizona illustrates magnetic patterns that are related to regional geology. The magnetic map provides useful information on the development of the crust in this region of the western Cordillera since Precambrian time. A major feature of the map is a broad zone extending from southern Nevada to northern Idaho, where magnetic anomalies from basement rock are not apparent. This feature is here named the "quiet basement zone." To the west of this zone abundant magnetic anomalies are produced by Phanerozoic intrusive and extrusive igneous rocks. West-trending zones of magnetic anomalies in western Utah and Eastern Nevada contain abundant igneous rocks and most of the known mineral resources of this region. The major magnetic anomalies in the area east of the northern part of the Basin and Range province and east of the overthrust belts in southeastern Idaho and western Montana reflect lithologic contrasts in the Precambrian basement with prominent northeast and northwest trends.

A magnetic high in north-central Nevada and another over the western Snake River Plain suggest north-northwest–trending Miocene rifts. The eastern Snake River Plain and the Yellowstone caldera are part of a much more extensive northeast-trending feature here called the Humboldt zone. In east-central Idaho, Tertiary intrusive and extrusive igneous rocks produce large magnetic highs. The Idaho batholith has little magnetic expression, but the Boulder batholith and related volcanic rock of western Montana produce a major magnetic high.

INTRODUCTION

Airborne magnetometer data have been obtained over large areas of the Western United States since World War II. Early surveys were designed to aid in the study of local geologic problems, but in recent years surveys have been made to complete coverage over large areas. Aeromagnetic data from Idaho, western Montana, southeastern Oregon, western Wyoming, Nevada, Utah, western Colorado, and northern Arizona have been compiled as a residual total magnetic intensity map (Pl. 4-1, in pocket). This map illustrates the regional magnetic anomalies over parts of several physiographic provinces in the Western United States and provides the basis for an examination of the regional magnetic patterns. The map is based on numerous aeromagnetic surveys with varying specifications (Fig. 4-1). Flight elevations ranged from 150 to over 2,000 m above the surface, and flight-line spacing ranged from 0.8 to 8 km. Despite these wide differences in elevation and spacing of flight lines and the problems relating to the establishment of a common datum, we believe this map is a valid presentation of the pattern of broad regional magnetic anomalies in this part of the Cordillera. Although numerous interpretations of parts of the magnetic data from the region under discussion here have been published, only three reports (Zietz and others, 1969, 1971; Stewart and others, 1977) treated the regional features that are the subject of this report.

To a greater degree than other kinds of geophysical anomalies (gravity, electrical conductivity, heat flow, or seismic-wave velocities, for example), magnetic anomalies can reveal the products of both local and regional events that have been preserved from early geologic time. Magnetic anomalies reflecting events in Precambrian time are still apparent over large areas. In addition, aeromagnetic maps commonly contain much more finely detailed information than can be obtained practically with any other geophysical technique. A regional magnetic anomaly map may be composed of overprinted patterns that are the products of geologic events covering a wide span of geologic time, and pattern recognition (or isolation) may, on the average, prove more difficult with data collected over old, inhomogeneous continental crust than it is with data reflecting younger, more homogeneous crust.

Magnetic anomalies reflect lateral variations in magnetic properties at any level in the upper lithosphere. The strength of a magnetic field of a simple source configuration falls off approximately as the cube of distance from the source, so an aeromagnetic map tends to emphasize shallow features and to obscure deep ones. A static magnetic field, unlike a gravitational field, cannot originate below a maximum depth that is controlled by temperature. As a ferromagnetic mineral is heated, it eventually passes through a fundamental transition at its Curie temperature, above which it behaves as a paramagnetic substance and effectively loses its ability to produce measurable magnetic anomalies. The Curie temperatures of magnetic minerals vary with their composition and structure. Because, for a given mineral, this temperature may be realized at different depths in different areas, depending on the depth of the isogeotherms, the total thickness of the magnetic part of the lithosphere will vary from place to place, thus producing long-wavelength anomalies (for an example, see Shuey and others, 1973).

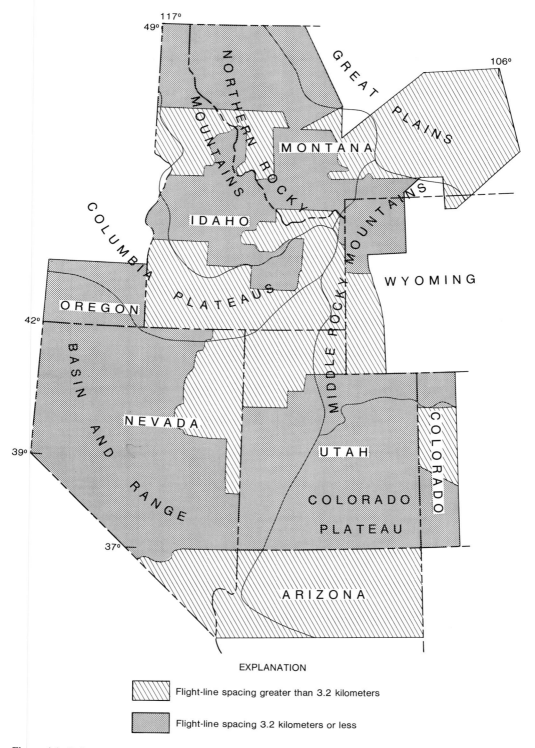

Figure 4-1. Index map showing flight-line spacing for aeromagnetic surveys and physiographic provinces from Fenneman and Johnson (1946).

Long-wavelength anomalies also can be associated with variations in depth to the top of the magnetic basement under large sedimentary basins. Quantitative analyses of individual anomalies can usually discriminate between these two long-wavelength effects, one caused by the top and the other caused by the bottom of the magnetic section.

Superimposed on these long-wavelength anomalies are anomalies of generally shorter wavelength produced by lithologic units of contrasting magnetization. Most of these anomalies in the Cordillera are associated with igneous or metamorphic rocks, and some, such as the anomaly over the western Snake River Plain, have considerable lateral extent.

In addition to defining specific areas or regions of distinctive magnetic "signature" or grain (textural patterns), Plate I also displays magnetic lineaments. As the term is used in this report, a magnetic lineament is any alignment of magnetic features that suggests a continuous, or nearly continuous, linear geologic structure. Some of these lineaments, such as the elongate north-northwest–trending magnetic high in north-central Nevada, are obvious and can be correlated with known geologic features. Others are subtle and have only a very tenuous correlation with known surface geology.

A broad, systematic variation in magnetic intensity is apparent across the map, with lower intensity in southwestern Utah and southeastern Nevada, and higher intensity in Idaho, western Montana, and northwestern Nevada. A zone of intermediate intensity occurs across northern Utah and Nevada. Part of this variation in residual intensity may be an artifact of the removal of the International Geomagnetic Reference Field (Fabiano and Peddie, 1969). Certainly another residual could be prepared that would minimize the differences in average magnetic intensity across the map. Some of the regional intensity changes, however, occur at major geologic boundaries; this suggests that the changes are also related in part to the regional geology. For example, the magnetic intensity north of the south edge of the Snake River Plain is higher than to the south, and the intensity over the Basin and Range area of southern Idaho is lower than over the Middle Rocky Mountains province to the east.

As noted above, zones of differing magnetic patterns or signatures are superimposed on the very broad regional variations in average magnetic intensity. These magnetic patterns reflect areal variations in the abundance, size, shape, trend, depth of burial, and magnetization of anomaly-producing masses. Examined qualitatively (and admittedly somewhat subjectively) along with other geophysical and geological data, these magnetic zones can be used to help define and interpret discrete, major, crustal tracts.

The heavy solid lines on Plate I represent boundaries between regions of different magnetic character. The lighter solid lines are perhaps somewhat less impressive as magnetic boundaries but not necessarily less significant geologically. Some are based entirely on the magnetic map but others on surface geology. The light dashed lines are individual lineaments discussed in the text. Most represent the axes of chain highs and lows, but several represent the aligned terminations of anomalies. The determinations of lineaments are obviously subjective, and we have not attempted to indicate all of the numerous lineaments that exist. We are not in complete agreement on which boundaries and lineaments are significant, but on Plate I we show features that one or more of us considers to be particularly important. We concerned ourselves primarily with variations in trend, wavelength, and level of intensity in drawing the line, but we have often been aided by more detailed magnetic maps. Because the magnetic expressions of younger geologic events have been superimposed across those of older ones (which may have different wavelengths and magnetic trends in different directions) and because the magnetic map is a reflection of the total of all events, magnetic field boundaries and lineaments may cross one another with propriety.

REGIONAL FEATURES

Colorado Plateau

The Colorado Plateau province provides a good opportunity to examine the characteristics of anomalies originating in the Precambrian crystalline basement (Case and Joesting, 1972). Sedimentary rocks, mostly flat lying but weakly magnetic, overlie a magnetic basement of highly varied lithology. Geologic mapping of basement exposures around the perimeter of the province, as well as in the Grand Canyon and the Uncompahgre Plateau, provides a notion of some of the physical characteristics, dimensions, and trends of lithologic units within the basement. Geologic and stratigraphic studies, in addition to abundant drill data in the interior of the plateau, offer knowledge of the depth and configuration of the basement surface. It is thus relatively simple to connect the major anomalies with variations in basement lithology. Local magnetic highs are produced by Tertiary laccolith groups in the La Sal (LS), Abajo (AB), and Henry Mountains (HM).

Three directions of magnetic lineaments dominate the Colorado Plateau. A northeast trend is more prominent in Arizona and southern Utah. Shoemaker and others (1974) have correlated northeast-trending magnetic features in northern Arizona with major fault systems. A set of northwest-trending lineaments, approximately parallel to a belt of salt anticlines and the Uncompahgre Uplift (UU) near the Utah-Colorado border, is apparent in southeastern Utah and the adjacent area of Colorado. West-trending lineaments, subparallel to but south of the Uinta Mountains (UM) and their related trough of Precambrian sedimentary rocks, are well developed in northeastern Utah and northwestern Colorado.

Basin and Range Province

We have been unable to identify magnetic anomalies from the Precambrian basement over most of the Basin and Range province in eastern Nevada and west-central Utah. Most of the major anomalies in this area appear to be explainable in terms of Phanerozoic rocks. In northwestern Utah, however, two zones of magnetic anomalies extend north and northwest from exposed Precambrian basement. The northwest-trending zone (D in Pl. 4-1) includes exposures of Precambrian basement in the Raft River (RR) and Albion Mountains (AM) at its northwest end. The north-trending zone (E in Pl. 4-1) in northwestern Utah also includes areas of exposed Precambrian rocks. We conclude that both of these zones are produced by magnetic units of Precambrian basement. South and west from these two zones, prominent magnetic anomalies occur in two major west-trending belts in central and southern Utah (those including lineaments A, B, and C in Pl. 4-1) that coincide with belts of stocks intruded into Phanerozoic sedimentary rocks. The larger anomalies are produced by exposed stocks, but the magnetic data suggest that the intrusive rock is more abundant in the subsurface (Mabey and others, 1964; Mabey and Morris, 1967). Thus the absence of magnetic expression of the Precambrian basement over most of this part of the Basin and Range province stands in marked contrast to the numerous basement anomalies that are apparent to the east in the Colorado Plateau.

In that part of eastern Nevada west of the west-trending zones just described, there is a north-trending zone of low magnetic relief. Stewart and others (1977) referred to this as a "quiet zone." It is characterized by an absence of high-frequency and high-amplitude anomalies. Actually, most of western Utah and eastern Nevada, exclusive of the area of west-trending zones mentioned above, is characterized by low magnetic relief and can be considered as a quiet magnetic province in its entirety. We use the term "quiet basement zone" here to refer to a more extensive area where basement magnetic anomalies are largely absent.

The western part of the quiet basement zone in eastern Nevada coincides with the axis of gravity and topographic symmetry described elsewhere in this volume (Eaton and others, this volume); however, the regional magnetic map (Pl. 4-1) gives no indication of any regional magnetic symmetry. The gravity symmetry is not caused by a symmetry of the surface and near-surface geology; instead, it reflects a symmetry of mass distributions in the deeper subsurface that has given rise to a symmetry in the regional topography. The aeromagnetic anomalies tend to emphasize the expression of rock masses in the shallow lithosphere; thus, no symmetry in the magnetic pattern comparable to that of the gravity pattern is to be expected.

Several reasons have been advanced for the apparent absence of magnetic anomalies originating in basement in this part of the Basin and Range province. Throughout most of the area, the thrust-thickened Phanerozoic sedimentary pile buries the crystalline basement to a great depth. Although the magnetic anomalies from such a deeply buried basement would not be nearly as strong as those seen over the Colorado Plateau, they should still be apparent in muted form with any depth of burial that is geologically reasonable. Thus the change in pattern and intensity of magnetic anomalies at the eastern edge of the Basin and Range province is not entirely the effect of depth of burial, but appears to reflect a major change in the magnetic properties of the basement rocks or the thickness of the magnetic basement. We know that the northern Basin and Range province is characterized by high heat flow (Sass and others, 1971; Diment and others, 1975); therefore, the Curie isotherm is elevated under the province, and the depth to which magnetic contrasts can extend is relatively shallow. Zietz and others (1970) proposed a shallow depth to the Curie isotherm as an explanation for the anomalous magnetic character of the Basin and Range province. Shuey and others (1973) examined the magnetic field at the eastern edge of the Basin and Range province and concluded that the magnetic change at this boundary could be due either to a variation in depth of the Curie isotherm or to a change in average susceptibility of the crust.

The three generally west-trending magnetic highs (A, B, and C in Pl. 4-1) that cross western Utah into eastern Nevada reflect the presence of Tertiary intrusive and extrusive igneous rocks. The northern lineament (A) runs through the mining districts at Bingham (BC) and Park City (PC), the middle lineament (B) crosses the Tintic area (TA), and the southern one (C) runs through Milford (MU) and Pioche (PN). These zones are similar to, but not identical with, west-trending mineral belts identified by other investigators (for example, Hilpert and Roberts, 1964; Shawe and Stewart, 1976). Because these zones are significantly different, we do not apply the mineral belt names to them. These magnetic zones, which encompass the areas that have produced most of the copper, lead, silver, gold, and zinc mined in Utah, have several characteristics in common: (1) they are confined to the Basin and Range province if the eastern boundary of that province is drawn east of the so-called "back valleys" in the Wasatch Range, as proposed by Shuey and others (1973) on the basis of geophysical evidence; (2) the western parts of the zones are poorly defined and open to differing geographic definitions; (3) they coincide geographically with areas having abundant volcanic rocks and exposed stocks, but the magnetic anomalies suggest a greater abundance of intrusive rock in the subsurface; and (4) a few major structures in the surface rocks parallel the trend of these zones (Ekren and others 1976; P. D. Rowley and others, 1976, written commun.).

Mabey and Morris (1967) concluded that the west-trending belts of igneous rocks in western Utah were controlled by deep-seated structures that influence either the formation or the upward movement of magma and that these structures were largely decoupled from the upper crust. A possibility worthy of further investigation is that they represent alignments of intrusions along profound, crust-penetrating, intracontinental transform faults associated with the regional east-west spreading. This is the interpretive position taken in a companion paper (Eaton and others, this volume). Stewart and others (1977) have discussed the west-trending magnetic lineaments in western Utah with particular regard to the distribution of Cenozoic volcanic rocks and mineral deposits. They suggest that these zones may be concentrations of igneous activity "localized along a southward-propagating transverse break or structural warp in a subducting plate."

West-trending magnetic features with a different character occur over central and western Nevada. Fuller (1964) described these features as "interruptions," because their most common expression is the interruption of north-trending magnetic anomalies. These interruptions are best defined by a much more detailed contour interval than that of Plate 4-1, and their existence is not well demonstrated on this map. An example of one of the more continuous of these features is an east-trending interruption in north-central Nevada (N) indicated in Plate 4-1. Some of these features can be traced for hundreds of kilometres; others are shorter and form en echelon patterns in long zones. Magnetic anomalies arising from rocks as young as Miocene are interrupted. Fuller believed as we do that these magnetic features reflect deep fractures in a layer that was partly decoupled from the upper crust. He also predicted that near-surface features might show left-lateral displacements on the order of 50 km along one of these east-trending features in Nevada. We do not agree with Fuller that the magnetic or geologic evidence suggests large lateral offset along any of these interruptions in Nevada. Where these east-trending interruptions intersect the Miocene Cortez rift in north-central Nevada (to be defined later), the magnetic anomaly is partly interrupted but not noticeably offset. Ekren and others (1976) described four of these lineaments in west-central Nevada and concluded that they are probably expressions of pre-Oligocene structural trends.

Eaton and others (this volume) defined a zone trending west to west-southwest near lat 37°N that divides the Basin and Range province into areas of contrasting crustal structure. On the magnetic map, the area to the north of lat 37°N has magnetic anomalies reflecting Phanerozoic intrusive and extrusive rocks, whereas the area to the south has anomalies that primarily reflect Precambrian basement. The magnetic data suggest that the magnetic character of the crust in the Basin and Range province south of lat 37°N more closely resembles that of the Colorado Plateau than that of the northern Basin and Range province.

The western two-thirds of Nevada displays high-amplitude magnetic anomalies produced by Phanerozoic igneous rocks. Two trends of individual anomalies are dominant: north and west-northwest. The west-northwest trend is dominant in the southwest, whereas the north trend, which parallels the Basin and Range structures, is dominant farther north. In the northwestern part of Nevada, and to a lesser extent in the southwestern part of the state, the magnetic anomalies in part reflect surface topography; Basin and Range structures displace magnetic units. In the eastern part of the Basin and Range province, there is generally little correlation between the magnetic anomalies and topography. In the east the Basin and Range faults do not appear to displace a magnetic basement, either because it does not exist or because the faults flatten at depth.

Southeastern Oregon and an adjacent area of Nevada is a region of relatively complex high-amplitude magnetic anomalies. Although abundant Tertiary volcanic rocks are present in this region, the magnetic anomalies do not correlate well with the volcanic rocks and probably are due to older intrusive rocks and feeder dikes. North and northeast trends dominate the magnetic pattern.

A prominent magnetic high trending slightly west of north across north-central Nevada reflects a structure we have named the Cortez rift. This high is produced in part by a partially exposed swarm of Miocene basalt dikes and related flows and in part by an associated deeper, presumably related, dikelike mass (Mabey, 1966; Robinson, 1970). Toward the south the anomaly narrows before terminating, and in the north it broadens to merge with an extensive area of magnetic anomalies produced by surface volcanic rocks of about the same age. These dikes and related structures appear to reflect an extensional rift that may have been the first major spreading event in the development of the region-wide Basin and Range structure. Normal faults as old as Oligocene in the Basin and Range province have been interpreted as Basin and Range structures (Loring, 1972). We believe the evidence available suggests that these pre-Miocene structures are local features and are not part of the regional extension that characterizes the whole of the Basin and Range province. The anomaly produced by the Cortez rift is broken by northeast-trending Basin and Range faults, and there is a suggestion of slight left-lateral offset of the dike swarm along these faults (Mabey, 1966). The dikes are more abundant near these northeast-trending faults than elsewhere; this suggests that the northeast-trending faults were controlled by older structures (Muffler, 1964). Stewart and others (1975) have inferred a connection between this rift in Nevada and a younger fault zone in Oregon and have named the combined feature the ''Oregon-Nevada lineament.'' We do not believe these two features are similar either in age or structure and see no evidence in the geophysical data that the primary features are connected. For this reason we prefer to label the structure in north-central Nevada the ''Cortez rift,'' for the area where it was first identified (Pl. 4-1).

The magnetic patterns over the western two-thirds of Nevada record a complex history of the devlopment of magnetic masses extending from Paleozoic thru Cenozoic time. Numerous igneous and tectonic events occurring in this area produced a magnetically heterogeneous crust. This heterogeneity is in pronounced contrast with the eastern part of Nevada, where the magnetic anomalies suggest a relatively uncomplicated distribution of magnetic rocks in the upper crust. The reader is referred to the summaries of late Precambrian and Phanerozoic history of this region by Burchfiel and Davis (1972, 1975). Westward accretionary growth of the North American continent, caused mainly by closing of marginal seas separating the continent and offshore arcs and by igneous processes related to subduction of oceanic lithosphere beneath the western part of the Cordillera, led to the geologic complexities reflected in the magnetic data.

Snake River Plain

The dominant magnetic feature of southern Idaho is a regional magnetic high over the Snake River Plain. Over the western Snake River Plain the magnetic field is relatively simple, with a positive anomaly (F in Pl. 4-1) along the southern edge of the plain and a negative one along the northern edge. Such a relationship is typical for a tabular plate of normally magnetized material. This anomaly could be produced by a simple slab of basalt underlying the plain. To the west along the Idaho-Oregon border, the high extends over Miocene basalts of the Columbia River Group. One interpretation of the magnetic anomaly is that a thick slab of these Miocene Columbia River basalts underlies the western Snake River Plain (Mabey, 1976). The trend of this anomaly is subparallel to the magnetic anomaly in north-central Nevada related to the Miocene Cortez rift. If the western Snake River Plain anomaly reflects a large rift containing Miocene basalts, these two features are probably related. Such a relationship would suggest that crustal extension occurred over a large area in Miocene time.

The magnetic expression of the eastern Snake River Plain is more complex, with a crude network of short-wavelength highs and lows. Two orientations of individual elongate anomalies are prominent: normal and parallel to the axis of the plain. Some of the individual anomalies can be correlated with source zones for the basalt flows; one such anomaly follows the northwest-trending Craters of the Moon rift (P in Pl. 4-1). A resistivity profile across the plain (Zohdy and Stanley, 1973) indicates that the basalt of the eastern Snake River Plain is one to two kilometres thick under its central part. Using the magnetic properties measured for samples of the basalt, E. S. Robinson (1971, written commun.) determined that the magnetic anomalies in the eastern plain could be produced by this thickness of basalt.

In the area of the Yellowstone caldera in northwest Wyoming (YC in Pl. 4-1), the magnetic intensity is relatively low; this reflects an area of high heat flow with resulting alteration of near-surface rocks and a relatively shallow Curie isotherm (Bhattacharyya and Leu, 1975). The high magnetic intensities northeast of the Yellowstone caldera are produced by both early Tertiary andesitic volcanic rocks and Precambrian crystalline basement rocks (Eaton and others, 1975).

Eaton and others (1975) have described a northeast-trending zone of magnetic anomalies that includes the eastern Snake River Plain, the Yellowstone Plateau, and a zone continuing northeastward across Montana. The magnetic anomalies in Montana are produced by lithologic units within the Precambrian basement. The zone is inferred to be an inhomogeneity that has controlled the northeastward migration of Cenozoic volcanism of the eastern Snake River Plain and the Yellowstone caldera. We here suggest that this zone extends to the southwest across northeastern and west-central Nevada between the two long dashed lines shown in Plate 4-1. In Nevada this zone contains many northeast-trending normal faults (King and Beikman, 1974), an area of high measured heat flow (Sass and others, 1976), and many major geothermal systems. Because several segments of the Humboldt River parallel this trend, we here name the entire feature, which extends from the Mono Lake area of eastern California to a part of northeast Montana near the Canadian border, "the Humboldt zone."

Although we believe strong justification exists for projecting the Humboldt zone southwest across Nevada, we recognize that the magnetic data alone provide only limited definition of the feature. Thus we have drawn on other data to support and define the zone in Nevada. We have defined the southeast edge of the zone to include the major northeast-trending faults that dominate much of the Basin and Range structure in this part of the region. As stated earlier, these faults are now active but appear to predate the Miocene Cortez rift. A major lineament identified from ERTS imagery within this zone has been named the Midas lineament by Rowan and Wetlaufer (1973). A major anomaly in P-wave arrival times has been observed parallel to and within the zone (Koizumi and others, 1973). As previously stated, existing data suggest this zone has an abnormally high heat flow. For all of these reasons, we believe the Humboldt zone is one of the major structures of the Cordillera, and is one deserving further specific study.

Central and Northern Idaho and Western Montana

Along the west border of central Idaho, large magnetic highs occur over basalts of the Columbia River Group, which are at least in part responsible for the highs. Just north of lat 45°N is a zone of more complex anomalies over Paleozoic and Mesozoic igneous rocks in the Hells Canyon area (HC).

The Idaho batholith, probably mostly of Cretaceous age, occupies an extensive area east and north of the western Snake River Plain. Over the southern part of the batholith the magnetic relief is generally low, with a north-trending grain on the west that appears to reflect the local topography. The western border of the Idaho batholith appears to be north trending and on strike with the west edge of the quiet basement zone to the north and south. The magnetic expression of the southern margin of the batholith is obscured by anomalies produced by younger volcanic rocks. The major local highs within the area of the batholith are produced by Tertiary intrusive bodies. The absence of major magnetic contrasts within the batholith suggests that it is more homogeneous than other batholiths in the Western United States, notably the Sierra Nevada and southern California (Peninsular Ranges) batholiths. Two extensive magnetic highs (G and H in Pl. 4-1) occur over the northern part of the Idaho batholith. In the area of the southern high (G), the batholith contains extensive roof pendants of metamorphosed rocks, and several Tertiary intrusive bodies have been mapped. However, the mapped surface distribution of these rocks does not explain the magnetic highs. Zietz and others (1971) pointed out that the northwest end of this magnetic high coincides with a quartz diorite pluton at the border of the batholith, and they speculated that the magnetic high may reflect a relatively mafic septum between large silicic plutons. They also recognized that the magnetic high occurs where Yates (1968) speculated that a zone of major left-lateral movement crosses Idaho. The northern high (H) might reflect a part of the batholith that contains more magnetite. The Idaho batholith appears to terminate abruptly on the north along an east-trending magnetic lineament (M in Pl. 4-1; also line A–A^1 of Zietz and others, 1971) that crosses but does not coincide with the Lewis and Clark Line (LC in Pl. 4-1).

In central Idaho, immediately east of the southern part of the Idaho batholith, the major magnetic anomalies occur over early Tertiary volcanic and intrusive rocks. These anomalies obscure any magnetic expression of the east edge of the batholith.

In southwestern Montana and the adjoining area of Idaho, large complex magnetic highs are produced by several batholiths of Cretaceous age and by Cretaceous and younger volcanic rocks. The Cretaceous intrusions in this region appear to be more magnetic than most of the Idaho batholith. The magnetic high in the area of the Boulder batholith (J in Pl. 4-1) is particularly prominent, as magnetic phases of the batholith combine their influence with that of the adjacent Cretaceous volcanic rocks to produce the anomaly.

The magnetic data suggest that several intrusive bodies are concealed in this region and that several others are larger than their exposures indicate. A few anomalies apparently produced by the Precambrian basement are also evident here. Northeast and northwest magnetic trends are apparent in this region, just as they are in the Great Plains section of Montana. Just southeast of the Boulder batholith magnetic high (J) is a parallel string of highs related to Cretaceous intrusions of the Tobacco Root Range and Precambrian gneisses of the Ruby Range. Another striking magnetic anomaly (K), in south-central Montana, is a low produced by the west-northwest-trending Stillwater Complex of Precambrian age. This anomaly reflects the remanent magnetism of the complex (Bergh, 1970).

The anomalies of the northwestern Montana plains display higher magnetic intensities than do those to the west. The two magnetic anomalies of highest amplitude on the entire map (P) and (Q) are in this area and are attributed to dioritic rocks of the Precambrian crystalline basement (M. R. Mudge, 1976, oral commun.). Both anomalies lie along the eastern edge of the Montana disturbed belt, but the nature of the genetic relationship between the disturbed belt and the anomalies is not apparent.

Another northeast-trending set of magnetic highs (O) in northwest Montana crosses the highly developed northwestward strike of the thrust sheets of the Montana thrust belt. The anomalies mark the southwestward projection of the Scapegoat-Banatyne trend, which is a transverse fault zone of Precambrian origin that extends across the Sweetgrass arch (SA; Kleinkopf and Mudge, 1972). The magnetic intensity decreases westward across the disturbed belt into the Belt basin. Over most of the Belt basin the regional magnetic field is relatively subdued. Individual magnetic highs reflect Cretaceous intrusive stocks within the Belt rocks. Elongate anomalies reflect relief and structure of magnetite-rich metasedimentary rocks in the basin. No magnetic expression of the deeply buried crystalline basement beneath the Belt basin is apparent, and we suggest that this region may be part of the same quiet basement zone described to the south.

The west-northwest–trending Osburn fault zone, a part of the Lewis and Clark Line (LC; Billingsley and Locke, 1939), is expressed both as an alignment of the termini of a set of anomalies of similar wavelengths and as a subtle magnetic low that is not well expressed by the contour interval of Plate 4-1. Other workers (Talbot and Hyndman, 1972) have considered the Osburn fault to be part of the "Montana lineament" (ML)—a more eastward-trending structural zone postulated to pass along the northern end of the Boulder batholith and then eastward to tie into the Lake Basin fault zone (LB) in south-central Montana. The magnetic patterns along this lineament are characterized by low intensities and by anomaly terminations. Along the western margin of the Belt basin in northern Idaho, a large magnetic high occurs over highly metamorphosed Belt rocks.

CONCLUSIONS

The regional magnetic anomalies in Plate 4-1 reflect some of the history of the shallow crust in this region. The history recorded in the magnetic anomalies is incomplete because some geologic events do not result in distribution of rocks that produce magnetic anomalies and because some geologic events change the magnetic properties of existing rocks or result in new anomalies that obscure pre-existing magnetic anomalies. Despite these limitations the magnetic-anomaly pattern provides some of the best and most detailed geophysical data available on the development of the crust. In this paper we have made only a cursory examination of the regional magnetic patterns.

The major local magnetic anomalies over the Colorado Plateau, the southern Nevada part of the Basin and Range province, western Wyoming, and the Great Plains of Montana reflect variations in the lithology of the Precambrian basement, with a few local anomalies produced by younger intrusive rocks within the section overlying the basement. The anomalies associated with these older rocks primarily record geologic events in the formation and modification of the North American Continent in Precambrian time. Magnetic trends of northwest, northeast, and east are dominant in the Precambrian rocks, with the northeast trend being the most prominent.

Extending from northwestern Montana and northern Idaho southward through eastern Nevada and western Utah is a quiet basement zone where no large magnetic anomalies associated with Precambrian basement are apparent, except for two local areas in southern Idaho and northern Utah. Little is known about the Precambrian basement in this zone, and the significance of the absence of magnetic anomalies is uncertain. High heat flow coupled with the thick cover of Phanerozoic rocks may be the primary cause. The crust, however, may also be more uniform, perhaps because it has experienced fewer geologic events. In general, the crust appears to be relatively thin in the magnetically quiet basement zone, and the velocity of the upper mantle appears to be relatively low (Herrin and Taggart, 1962). Phanerozoic sedimentary rocks are very thick, and major zones of overthrust faulting have developed in this region since Devonian time. Faults of this sort are uncommon to the east. Basin and Range structures occur throughout most of the quiet zone. Apparently the crust in this region has been relatively mobile throughout much of Phanerozoic time.

In central and western Nevada and extreme western Idaho is a segment of crust built by a series of igneous and structural events extending back at least to Paleozoic time. The magnetic anomalies here record a complex history related to the westward growth of the North American Continent.

In Mesozoic and early Tertiary time large volumes of igneous rock intruded into the shallow lithosphere and were extruded to form thick piles of volcanic rock in the magnetically quiet zone. The largest of the intrusive units is the Idaho batholith of Cretaceous age. In western Utah and eastern Nevada, west-trending zones of abundant intrusive and related extrusive rock developed. The rocks in these zones become younger southward. In east-central Idaho, southwestern Montana, and northwest Wyoming, abundant volcanic rocks of Eocene age were extruded, and related intrusive masses developed. Northwest- and west-northwest–trending structures appear to have controlled the igneous activity in this region.

In Miocene time, north-northwest–trending crustal rifts developed, and large volumes of basalt were extruded in southwestern Idaho and in north-central Nevada. The rift in Idaho developed into the western Snake River Plain, but at least the southern part of the Cortez rift in Nevada did not develop into a large structure. These old rifts may mark the beginning of the extensional tectonics that produced the Basin and Range province. Subsequent extension produced a more complicated pattern of block faulting with only local extrusions of basalt. Post-Miocene normal faults of the region apparently do not often displace a magnetic basement and therefore are not reflected in the magnetic data.

In Pliocene time an extensive area of silicic volcanism developed in southern Idaho and northwestern Nevada. The southeastern edge of this zone of volcanism appears to be controlled by the southeast edge of the Humboldt zone. The initial volcanism was widespread over northwestern Nevada and southwestern Idaho. Activity in this area declined as the volcanism migrated northeastward along the Humboldt zone toward the present center of activity in Yellowstone National Park. In Quaternary time magnetic patterns have continued to evolve in active volcanic areas such as the Yellowstone–Snake River Plain region and the Blackfoot lava field (BF) in southeastern Idaho.

REFERENCES CITED

Bergh, H. W., 1970, Paleomagnetism of the Stillwater Complex, Montana, *in* Runcorn, S. K., ed., Palaeogeophysics: London, Academic Press, p. 143–158.

Bhattacharyya, B. K., and Leu, Lei-kuang, 1975, Analysis of magnetic anomalies over Yellowstone National Park: Mapping of Curie point isothermal surface for geothermal reconnaissance: Jour. Geophys. Research, v. 80, p. 4461–4465.

Billingsley, Paul, and Locke, Augustus, 1939, Structure of ore districts in the continental framework: New York, Am. Inst. Mining, Metall. Engineers, 51 p.

Burchfiel, B. C., and Davis, G. A., 1972, Structural framework and evolution of the southern part of the Cordilleran orogen, Western United States: Am. Jour. Sci., v. 272, p. 97–118.

——1975, Nature and controls of Cordilleran orogenesis, Western United States; extensions of an earlier synthesis: Am. Jour. Sci., v. 275-A, p. 363–396.

Case, J. E., and Joesting, H. R., 1972, Regional geophysical investigations in the central Colorado Plateau: U.S. Geol. Survey Prof. Paper 736, 31 p.

Diment, W. H., Urban, T. C., Sass, J. H., Marshall, B. V., Munroe, J. R., and Lachenbruch, A. H., 1975, Temperatures and heat contents based on conductive transport of heat, *in* White, D. E., and Williams, D. L., eds., Assessment of geothermal resources of the United States—1975: U.S. Geol. Survey Circ. 726, p. 84–103.

Eaton, G. P., Christiansen, R. L., Pitt, A. M., Mabey, D. R., Blank, H. R., Zietz, Isidore, and

Gettings, M. E., 1975, Magma beneath Yellowstone National Park: Science, v. 188, p. 787–796.

Eaton, G. P., Wahl, R. R., Prostka, H. J., Mabey, D. R., and Kleinkopf, M. D., 1978, Regional gravity and tectonic patterns: Their relation to late Cenozoic epeirogeny and lateral spreading of the western Cordillera, in Smith, R. B., and Eaton, G. P., eds., Cenozoic tectonics and regional geophysics of the western Cordillera: Geol. Soc. America Mem 152 (this volume).

Ekren, E. B., Bucknam, R. C., Carr, W. J., Dixon, G. L., and Quinlivan, W. D., 1976, East-trending structural lineaments in central Nevada: U.S. Geol. Survey Prof. Paper 986, 16 p.

Fabiano, E. B., and Peddie, N. W., 1969, Grid values of total magnetic intensity IGRF-1965: U.S. ESSA Tech. Rept. C & GS 38, 55 p.

Fenneman, N. M., and Johnson, D. W., 1946, Physical divisions of the United States: U.S. Geol. Survey Map, scale 1:7,000,000.

Fuller, M. D., 1964, Expression of E-W fractures in magnetic surveys in parts of the U.S.A.: Geophysics, v. 29, p. 602–622.

Herrin, E., and Taggart, J., 1962, Regional variations in Pn velocity and their effect on the location of epicenters: Seismol. Soc. America Bull., v. 52, p. 1037–1046.

Hilpert, L. S., and Roberts, R. J., 1964, Metallic mineral resources—Uranium, in Mineral and water resource of Utah: U.S. 88th Cong., 2nd sess., Comm. Print, p. 28–34.

King, P. B., and Beikman, H. M., 1974, Geologic map of the United States: U.S. Geol. Survey, scale 1:2,500,000.

Kleinkopf, M. D., and Mudge, M. R., 1972, Aeromagnetic Bouguer gravity, and generalized geologic studies of the Great Falls–Mission Range area, northwestern Montana: U.S. Geol. Survey Prof. Paper 726-A, 19 p.

Koizumi, C. J., Ryall, Alan, and Priestley, K. F., 1973, Evidence for a high-velocity lithospheric plate under northern Nevada: Seismol. Soc. America Bull. v. 63, p. 2135–2144.

Loring, A. K., 1976, Distribution of time and space of late and latest Phanerozoic normal faulting in Nevada and Utah: Utah Geology, v. 3, no. 2, p. 97–109.

Mabey, D. R., 1966, Regional gravity and magnetic anomalies in part of Eureka County, Nevada: Soc. Exploration Geophysicists Mining Case Histories, v. 1, p. 77–83.

——1976, Interpretation of a gravity profile across the western Snake River Plain, Idaho: Geology, v. 4, p. 53–55.

Mabey, D. R., and Morris, H. T., 1967, Geologic interpretation of gravity and aeromagnetic maps of Tintic Valley and adjacent areas, Tooele and Juab Counties, Utah: U.S. Geol. Survey Prof. Paper 516-D, 10 p.

Mabey, D. R., Crittenden, M. D., Jr., Morris, H. T., Roberts, R. J., and Tooker, E. W., 1964, Aeromagnetic and generalized geologic map of part of north-central Utah: U.S. Geol. Survey Geophys. Inv. Map GP-422, scale 1:250,000.

Muffler, L.J.P., 1964, Geology of the Frenchie Creek quadrangle, north-central Nevada: U.S. Geol. Survey Bull. 1179, 99 p.

Robinson, E. S., 1970, Relation between geological structure and aeromagnetic anomalies in central Nevada: Geol. Soc. America Bull., v. 81, p. 2045–2060.

Rowan, L. C., and Wetlaufer, P. H., 1973, Structural geologic analysis of Nevada using ERTS-1 images, a preliminary report, in Symposium on significant results obtained from the Earth Resources Technology Satellite 1, Vol. I, Technical Presentations, Section A: U.S. Natl. Aeronautics and Space Adm. Spec. Pub. 327, p. 413–423.

Sass, J. H., Lachenbruch, A. H., Munroe, R. J., Greene, G. W., and Moses, T. H., 1971, Heat flow in the Western United States: Jour. Geophys. Research, v. 76, p. 6376–6413.

Sass, J. H., Diment, W. H., Lachenbruch, A. H., Marshall, B. V., Munroe, R. J., Moses, T. H., Jr., and Urban, T. C., 1976, A new heat-flow contour map of the conterminous United States: U.S. Geol. Survey Open-File Rept. 76–756.

Shawe, D. R., and Stewart, J. H., 1976, Ore deposits as related to tectonics and magmatism, Nevada and Utah: Soc. Mining Engineers Trans., v. 260, p. 225–230.

Shoemaker, E. M., Squires, R. L., and Abrams, M. J., 1974, The Bright Angel and Mesa Butte fault systems, in Karlstrom, T.N.V., Swann, G. A., and Eastwood, R. L., eds., Geology of northern Arizona with notes on archaeology and paleoclimate, Regional studies, Pt. 1: Flagstaff, Ariz., Northern Arizona Univ., p. 355–391.

Shuey, R. T., Schellinger, D. K., Johnson, E. H., and Alley, L. B., 1973, Aeromagnetics and the transition between the Colorado Plateau and Basin Range provinces: Geology, v. 1, p. 107–110.

Stewart, J. H., Walker, G. W., and Kleinhampl, F. J., 1975, Oregon-Nevada lineament: Geology, v. 3, p. 265–268.

Stewart, J. H., Moore, W. J., Zietz, Isidore, 1977, East-west patterns of Cenozoic igneous rocks, aeromagnetic anomalies, and mineral deposits, Nevada and Utah: Geol. Soc. America Bull., v. 81, p. 67–77.

Talbot, James, and Hyndman, Donald, 1972, Relationship of the Idaho batholith structures to Montana lineament: Northwest Geology, v. 2,

p. 48–52.

Yates, R. G., 1968, The trans-Idaho discontinuity, *in* Upper mantle (geological processes): Internat. Geol. Cong., 23rd, Prague 1968, Proc., sec. 1, p. 117–123.

Zietz, Isidore, Bateman, P. C., Case, J. E., Crittenden, M. D., Jr., Griscom, Andrew, King, E. R., Roberts, R. J., and Lorentzen, G. R., 1969, Aeromagnetic investigation of crustal structure for a strip across the Western United States: Geol. Soc. America Bull., v. 80, p. 1703–1714.

Zietz, Isidore, Andreasen, G. E., and Cain, J. C., 1970, Magnetic anomalies from satellite magnetometer: Jour. Geophys. Research, v. 75, p. 4007–4015.

Zeitz, Isidore, Hearn, B. C., Jr., Higgins, M. W., Robinson, G. D., and Swanson, D. A., 1971, Interpretation of an aeromagnetic strip across the Northwestern United States: Geol. Soc. America Bull., v. 82, p. 3347–3372.

Zohdy, A.A.R., and Stanley, W. D., 1973, Preliminary interpretation of electrical sounding curves obtained across the Snake River Plain from Blackfoot to Arco, Idaho: U.S. Geol. Survey Open-File Rept. 1829, 3 p.

MANUSCRIPT RECEIVED BY THE SOCIETY AUGUST 15, 1977

MANUSCRIPT ACCEPTED SEPTEMBER 2, 1977

Geological Society of America
Memoir 152

5

Fault-plane solutions
of the Western United States: A compilation

R. B. SMITH
Department of Geology and Geophysics
University of Utah
Salt Lake City, Utah 84112

A. G. LINDH
U.S. Geological Survey
345 Middlefield Road
Menlo Park, California 94025

The possibility of relating seismicity to Cenozoic tectonics in the Western United States has led to the need for a compilation and a map of fault-plane solutions for this area. For several years the U.S. Geological Survey has determined fault-plane solutions from their seismograph arrays, principally in California and Nevada. These data were summarized by Lindh and others (1973). Fault-plane solutions for the Intermountain region, including Utah, Idaho, Nevada, Montana, and Wyoming, were compiled by Smith and Sbar (1974). Following these compilations, many new fault-plane solutions were published as the result of detailed analyses of earthquakes recorded by portable and new permanent seismograph arrays.

Our intent here is to provide a tabulation (Table 5-1, in pocket) and map of the principal fault-plane solutions (Fig. 5-1, Pl. 5-1, in pocket) for the Western United States through 1975 and part of 1976. For each fault-plane solution, the tabulations list the type of solution (single event, S; composite, C), date, time (GMT), location, attitudes of *P*- and *T*-axes, nodal-plane attitudes, and the reference.

Because of the wide variety of types and sizes of seismograph arrays and the differences in data analyses, we have not attempted to analyze the quality of individual solutions; however, we have differentiated between single-event solutions (large symbols) and composite solutions (small symbols) on the map (Fig. 5-1). Generally, a single-event solution was determined from an earthquake of magnitude (*M*) 5.0 or larger, unless it was recorded within a dense array such as in the U.S. Geological Survey Nevada test-site array. Both long-period and short-period *P*-phases were generally used for the larger magnitude, single-event solutions, and they are considered more reliable. Composite solutions were usually from several small earthquakes recorded across detailed arrays. In most cases composite solutions were remarkably similar to single-event solutions in the same area, for example, Wasatch Front, Yellowstone, and

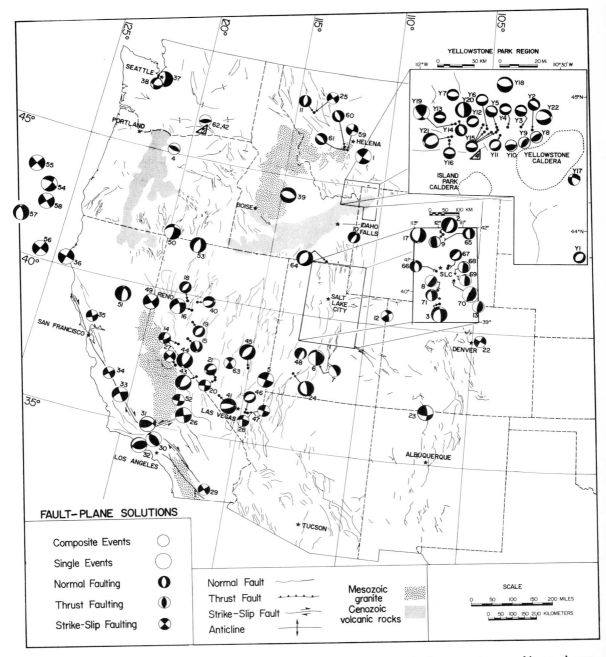

Figure 5-1. Map of fault-plane solutions of the Western United States. Projections are stereographic equal area, lower-hemisphere. Dark quadrants = compressional motion; light quadrants = dilatational motion. Numbers refer to tabulated entries in Table 5-1. Numbers *4* and *5* in triangles refer to number of similar solutions. Dots = epicenter or area from which the fault-plane solutions were determined.

southern Nevada (Fig. 5-1). However, we caution the user of our compilation to evaluate the original data from which the solutions were referenced.

A comparison of the fault-plane solution map (Fig. 5-1, Pl. 5-1) with the regional seismicity maps of the Western United States (Hill, this volume; Smith, this volume) shows that the number of reported earthquakes along the San Andreas fault system was about ten times greater than in other areas of the western interior, whereas only seven fault-plane solutions on the San Andreas fault and its branches were plotted in Figure 5-1. Incorporation of all published solutions from the San Andreas fault zone, which show primarily right-lateral solutions, would make a redundant and lengthy compilation which we did not consider feasible here. Our compilation for the San Andreas fault system was cut off at 1973, and the cut-off date for the rest of the compilation was 1975, except for a few events.

We have not interpreted the data presented in this report. Papers by Lindh and others (1973), Smith and Sbar (1974), Raleigh (1974), and Smith (1977 and this volume) have used parts or all of these data in generalized interpretations of the contemporary tectonics and directions of the regional stress fields.

REFERENCES CITED

Hill, D. P., 1978, Seismic evidence for the structure and Cenozoic tectonics of the Pacific Coast States, *in* Smith, R. B., and Eaton, G. P., eds., Cenozoic tectonics and regional geophysics of the western Cordillera: Geol. Soc. America Mem. 152 (this volume).

Lindh, A. G., Fischer, F. G., and Pitt, A. M., 1973, Nevada focal mechanisms and regional stress fields: EOS (Am. Geophys. Union Trans.) v. 54, p. 1133.

Raleigh, C. B., 1974, Crustal stresses and global tectonics, *in* Advances in rock mechanics: Proc. Third Cong. Internat. Soc. Rock Mechanics, v. 1, pt. A, p. 593–597.

Smith, R. B., 1977, Intraplate tectonics of the western North American plate: Tectonophysics, v. 37, p. 323–336.

——1978, Seismicity, crustal structure, and intraplate tectonics of the interior of the western Cordillera, *in* Smith, R. B., and Eaton, G. P., eds., Cenozoic tectonics and regional geophysics of the western Cordillera: Geol. Soc. America Mem. 152 (this volume).

Smith, R. B., and Sbar, M., 1974, Contemporary tectonics and seismicity of the Western United States with emphasis on the Intermountain seismic belt: Geol. Soc. America Bull., v. 85, p. 1205–1218.

MANUSCRIPT RECEIVED BY THE SOCIETY AUGUST 15, 1977

MANUSCRIPT ACCEPTED SEPTEMBER 2, 1977

Printed in U.S.A.

Geological Society of America
Memoir 152

6

Seismicity, crustal structure, and intraplate tectonics of the interior of the western Cordillera

ROBERT B. SMITH
Department of Geology and Geophysics
University of Utah
Salt Lake City, Utah 84112

ABSTRACT

Seismicity, fault-plane solutions, and Cenozoic geology are used to infer contemporary west to northwest extension and components of lateral slip between subplates of the western North American plate. Seismicity of the western interior of the Cordillera is characterized by earthquakes that occur in broad zones, up to 150 km wide, in the Nevada and the Intermountain seismic zones. Epicenters are scattered and when accurately located many do not coincide with mapped faults. Focal depths are shallow and seldom exceed 20 km. Secondary seismicity occurs around the margins of the Colorado Plateau, in central Idaho, and in eastern Oregon and Washington. The contemporary strain pattern of the Western United States, as interpreted from fault-plane solutions, suggests that three intraplate lithospheric blocks—the Sierra Nevada, the Great Basin–High Lava Plains, and the Northern Rocky Mountains–Columbia Plateau are moving west to northwest as "slivers" between the obliquely converging Pacific and North American plates. Intraplate extension within the Great Basin is primarily accommodated by north-south normal faulting in north-central Nevada and along the Wasatch Front, but significant components of strike-slip faulting occur along active seismic zones in southwestern Nevada and along the northern Intermountain seismic belt in Montana.

A thin crust, ~25 km, characterizes the east margin of the Great Basin, with average P_n-velocities of ~7.5 km/s. A crustal low-velocity layer, at 5 to 15 km depth, coincides with the eastern margin of the Great Basin. A thin crust, with an apparent P_n-velocity of ~7.7 km/s, occurs on the northwest margin of the Great Basin and beneath parts of the Oregon–High Lava Plains. The central part of the Great Basin has a thicker crust, ~30 km, and higher P_n-velocities, 7.7 to 7.9 km/s. The Colorado Plateau and the Rocky Mountains have thicker crusts, 40 to 50 km, and P_n-velocities of ~7.8 km/s.

An upper-mantle diapir or an upwelling thermal mechanism, facilitated by stress relaxation above the now-truncated Farallon subducting plate, is postulated to have uplifted, extended, and heated the crust of the Great Basin beginning at ~20 m.y. Late Cenozoic centers of volcanism appear to have progressed outward from a northern Great Basin thermal center in two divergent directions—northwesterly along a combined extensional–strike-slip zone of deformation in the High Lava Plains of southeastern Oregon and northeasterly along an extensional zone marked by the Snake River Plain. Lithospheric heating is thought to have produced a broad uplift of the Great Basin with concomitant crustal thinning and diminishing of upper-mantle P_n-velocities. The thermal mechanisms are hypothesized to have produced laterally divergent mantle flow to form symmetric zones of crustal thinning and low P_n-velocities that presently mark the east and west margins of the Great Basin.

INTRODUCTION

The prevailing plate-tectonics model for Cenozoic evolution of the Western United States requires that the principal displacement between the obliquely converging North American and Pacific plates occurs along the San Andreas right-lateral transform fault. Atwater (1970) suggested that the interior of the Western United States has accommodated some of the lateral motion between the plates as a broad transform fault. This zone of deformation extends eastward from the San Andreas fault up to 1,500 km inland to include the Sierra Nevada, the southern Columbia Plateau, the High Lava Plains of southern Oregon, and the Great Basin portion of the Basin and Range province (Fig. 6-1).

Crustal uplift and north-south block faulting constitute the principal modes of late Cenozoic tectonism of the Great Basin (Fig. 6-1). These features have been described by many authors. Reviews of structure and tectonic evolution of the Basin and Range province are found in Stewart (this volume), Christiansen and McKee (this volume), and Best and Hamblin (this volume).

Observations of the systematic Cenozoic deformation of the Western United States and the oroclinal offset of the Mesozoic granite batholiths of the western Cordillera (that is, the Idaho batholith and secondary granitic bodies are located northeast of the Sierra Nevada batholith) were noted by Carey (1958) in his explanation of continental drift. Hamilton and Myers (1966) elaborated on the tectonic evolution of the Western United States and also described the oroclinal offset of the Mesozoic plutons. Most recently, Atwater (1970) brought the interpretation of these tectonic features into the perspective of plate tectonics.

A contemporary tectonic model of the Western United States, proposed by Suppe and others (1975), suggested that the seismicity and Quaternary volcanism reflect microplates where westward-moving aseismic topographic highs connect stationary mantle plumes. All the plate-tectonics models published to date postulated eastward subduction of the Farallon plate beneath the Western United States that ceased as the East Pacific Rise was annihilated by overriding of the North American plate (Atwater, 1970; Scholz and others, 1971; Christiansen and Lipman, 1972).

On the assumption of (1) the oblique plate convergence, (2) a thinned and thermally weakened continental lithosphere, and (3) the buoyancy of the continental portion of the North American plate, I suggest a tectonic model in which late Cenozoic intraplate deformation has spread eastward into the Western United States to form the 1,500-km-wide zone of intraplate extension and oblique shear. This deformation is reflected by contemporary seismicity in broad zones of earthquakes, up to 150 km wide, with accompanying normal and oblique strike-slip faulting. The patterns of earthquakes, fault-plane solutions, Cenozoic volcanism, and faulting have been interpreted to outline active subplates, each with its own motion.

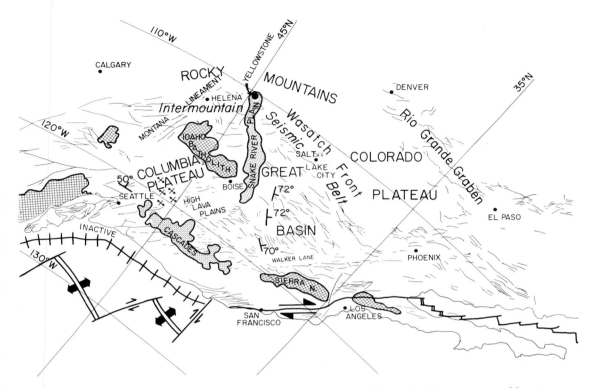

Figure 6-1. Generalized Cenozoic geology map of the Western United States plotted on a transverse-Mercator projection about the pole of rotation (lat 53°N, long 53°W) between the North American and Pacific plates (map modified from Atwater, 1970). Fine lines indicate faults. Shaded zones indicate important rock outcrops: cross pattern, Mesozoic batholiths; dash pattern, Quaternary rhyolite-basalt flows; generalized anticline symbols, folds in Columbia Plateau basalts; strike-dip symbols, attitudes of upper-mantle layers interpreted as paleosubduction zones (McKenzie and Julian, 1971; Ryall and others, 1973).

In this paper, seismicity, fault-plane solutions, crustal structure, and Cenozoic geology are used to infer the contemporary tectonic regime and to postulate a kinematic mechanism for the Cenozoic tectonics of the western North American plate. A condensed version of this paper was published by Smith (1977).

Important problems addressed here are the spatial relationships of seismicity to Quaternary faulting, temporal relationships of volcanism, relationships of crustal structure to surface tectonic features, and rates of contemporary uplift. Some difficulties persist because seismicity reflects only present-day or "instantaneous" deformation. Likewise, fragmentary late Cenozoic geology is difficult to interpret because of the limited extent of terrestrial deposits.

Within the framework of plate tectonics, the data and interpretation presented here are consistent with Cenozoic eastward subduction of the Farallon plate beneath the Western United States. Subsequent mantle diapirism or thermal upwelling, crustal uplift, and spreading are invoked in this model. The regime of eastward subduction was followed by west to northwest extension and uplift of the western interior of the Cordillera, which I postulate was accommodated by some intraplate right-lateral–oblique shear displacements subparallel to the San Andreas fault. Cumulative horizontal offset and drag along the intraplate shear zones may be a few hundreds of kilometres. The resultant effects of both subduction and oblique plate convergence have increased the area of the late Cenozoic tectonism and account for the widely scattered zones of seismicity in the western North American plate.

Continental convergence of up to 1,000 km along major east-west intraplate strike-slip faults in Asia has been described for the Indian-Eurasian continental collision (Molnar and Tapponier, 1975). The shear displacements reported for Asia are much larger than those postulated here for the Western United States, but an analogous mode of deformation may reflect similar mechanisms.

CENOZOIC TECTONIC PATTERNS

To emphasize the possible relationships between the motions of the North American and the Pacific plates, Figure 6-1 shows a generalized Cenozoic tectonic map (modified after Atwater, 1970) on a transverse-Mercator projection about the pole of rotation, lat 53°N, long 53°W, between the North American and the Pacific plates. If intraplate motions are analogous to those in sea-floor spreading, then lateral (transform-type) motion should occur along parallels, and extension (spreading-type motion) should be centered on meridians on this type of projection. Plotted in Figure 6-1 are the primary tectonic features that form the basis of my discussion: normal faults, Mesozoic batholiths, late Cenozoic volcanic rocks of the Snake River Plain and the High Lava Plains, and Cenozoic anticlinal folds of the Columbia Plateau.

Stewart (this volume), Eaton and others (this volume), Christiansen and McKee (this volume), and Best and Hamblin (this volume) present detailed maps of Cenozoic tectonic and volcanic features in conventional projections. I will describe briefly those tectonic features that are relevant to the seismicity, crustal structure, and contemporary stress distribution.

Cenozoic normal faults associated with central and northern Great Basin horsts and grabens generally trend north. Pronounced earthquake activity is coincident with zones of normal faulting along the east margin of the Great Basin (Wasatch Front) and in the west-central and southern Great Basin.

Only limited Cenozoic faulting and earthquake activity occurs within the Columbia Plateau. However, extensive west- to northwest-trending anticlinal folds in mid-Tertiary basalt flows are located north of the High Lava Plains (Fig.6-1) and are thought to reflect regional north-south compression.

A less conspicuous, but nevertheless important tectonic parameter of the Great Basin and possibly the eastern part of the Northern Rocky Mountains is Cenozoic strike-slip faulting. Stewart (this volume) has summarized the evidence for right-lateral displacements on northwest-trending strike-slip faults (Fig.6-1) along the Walker lane in the southern Great Basin. Stewart and others (1968) estimated up to 190 km of right-lateral offset along this fault zone. The lateral displacement includes both fault-slip and large-scale drag. In the central Great Basin, Thompson and Burke (1974) showed that the pervasive model of deformation was uniform extension, and they also described components of strike-slip displacement on fault traces oblique to the extension direction. Right-lateral strike-slip displacements along northwest-trending faults have also been recognized in the northwestern Great Basin and in the High Lava Plains south of the Columbia Plateau (Lawrence, 1976; Stewart, this volume).

The role of Cenozoic strike-slip faulting in the Northern Rocky Mountains has not been assessed in detail; however, right-lateral strike-slip faulting occurs along the northwest-trending Montana lineament, with up to 50 km of pre-Cenozoic lateral offset, in the same area as that of the northern Intermountain seismic belt. Strike-slip displacements are common components of the fault-plane solutions for this area and suggest some contemporaneous horizontal slip perhaps on pre-existing faults. Tectonism of the northern Intermountain region (northeastern Idaho and western Montana) appears similar to that produced in wrench-fault tectonics (Freidline and others, 1976) where lithospheric shearing produces both lateral slip and extension in the overlying and passive crustal layer.

The most direct evidence of subduction beneath the Western United States is the location and composition of the Cascade stratovolcanic cones that fit the pattern of andesitic volcanism above active island arcs. Indirect evidence of Tertiary subduction of the Farallon plate shows an eastward increase in alkaline content of igneous rocks toward the plate interior, similar to that above active Benioff zones (Lipman and others, 1971; Christiansen and Lipman, 1972). Teleseismic *P*-wave delays (McKenzie and Julian, 1971; Ryall and others, 1973) also suggest the presence of east-to southeast-dipping, high-velocity mantle layers interpreted to be remnant paleosubduction zones beneath the Columbia Plateau and the Great Basin (Fig. 6-1).

CENOZOIC VOLCANISM AND MANTLE DIAPIRISM

Cenozoic volcanism of the Western United States has been the subject of much discussion, and I refer to the following papers for details: Snyder and others (1976), Stewart and Carlson (this volume), White and McBirney (this volume), Armstrong (this volume), Christiansen and McKee (this volume), and Best and Hamblin (this volume). It is, however, important to describe some of the characteristics of the volcanism that relate to the regional geophysical data.

The initial stages of the widespread Cenozoic volcanism of the Great Basin began at ~43 m.y. B.P. where foci of volcanic activity marked by rhyolite flows, ash-flow tuffs, and andesites spread south to southwestward from the northeast corner of Nevada. This stage of volcanism corresponds to the ignimbrite "flareup" of Noble (1972). Volcanism apparently ceased at ~19 m.y. B.P., but began again at ~15 m.y. B.P. with a distinct change to bimodal basalt-rhyolite compositions (McKee, 1971).

At 15 m.y. B.P., rhyolites were first erupted at the west end of the Snake River Plain as part of a progression of silicic volcanic centers that migrated northeastward to the Yellowstone Plateau where the latest rhyolites were erupted 0.6 m.y. B.P. (Armstrong and others, 1975). The silicic rocks of the Snake River Plain were later covered by olivine basalts in the second stage of the bimodal volcanism. Foci of the systematic silicic volcanic activity progressed at rates of about 3 cm/yr (Armstrong and others, 1975) and was followed by the development of a low-lying topographic depression.

At about the same time that volcanism began in the Snake River Plain, another sequence of rhyolitic volcanism, in the form of rhyolite domes and ash-flow tuffs, began to progress northwesterly along the High Lava Plains. This volcanic activity, not as intense as that of the Snake River Plain, spread northwestward to form a progression of rhyolitic volcanism 1 to 3 cm/yr (MacLeod and others, 1975).

Contemporaneous with the inception of volcanism on the northern margin of the Great Basin, silicic and basaltic volcanism spread southward across the central Great Basin to complete a broader zone of volcanic deposition. Note that the Great Basin volcanism was initiated much earlier, ~43 m.y. B.P., and spread south to southwestward, whereas volcanism in the Snake River Plain and the High Lava Plain was initiated as a new stage of volcanic tectonism at ~18 m.y. B.P. Earlier volcanism in the Columbia Plateaus and the Northern Rocky Mountains, ~55 m.y. B.P., swept south to the Great Basin and stopped at 43 m.y. B.P. before renewed eruption of the Columbia River basalts at 18 m.y. B.P. (Armstrong, this volume).

The temporal and spatial relationships of the Cenozoic andesites and rhyolites of the western Cordillera are interpreted to have formed in a tectonic regime associated with the subduction of oceanic lithosphere beneath the Western United States. But the southward sweep of alkalic volcanism is difficult to reconcile with the eastward-dipping Benioff zone. The subduction regime persisted until 18 m.y. B.P., when the West Coast Trench was annihilated (Scholz and others, 1971; Christiansen and Lipman, 1972). Bimodal basalt-rhyolite volcanism followed with crustal extension beginning at about 15 m.y. B.P. and continues to the present. The most recent activity is dominantly basaltic.

Intermediate and silicic volcanism of early to mid-Cenozoic time is hypothesized to have developed by forceful injection into a compressed lithosphere at high magma-fluid pressures. This stress regime is thought to result from southwest-northeast horizontal convergence of 8 to 14 cm/yr of the North American plate with the Farallon plate (Coney, this volume), produced by active northeastward subduction. Bimodal basaltic-rhyolitic volcanism, in a relaxed extensional stress regime, has lead to the hypothesis that the mid to late Cenozoic Great Basin igneous activity resulted from a rising mantle diapir (Scholz and others, 1971). Horizontal stresses in this regime were reoriented at ~20 m.y. B.P. to produce oblique convergence parallel to the San Andreas fault and relaxed the horizontal stress, allowing crustal extension (Coney, this volume). Partial melting of the subducted slab represented by the ignimbrite flareup is hypothesized to have released material that rose through the mantle to form a laterally spreading diapir beneath the Great Basin. During the ignimbrite flareup ~43 m.y. B.P., high temperatures and partial melting are postulated to have begun uplift, thinning, and distension of the lithosphere beneath the Great Basin. This process was probably accelerated by the change in regional stress from compression, produced by northeast subduction at ~20 m.y. B.P., to east-west extension produced by oblique plate convergence. A thin crust, low upper-mantle velocities, high heat flow, and relatively high topography of the Great Basin probably reflect this major thermal event. This stage of tectonism has also been interpreted as the result of the development of an interarc basin with extensional tectonics and basaltic volcanism (Scholz and others, 1971) similar to that of western Pacific back-arc basins.

CRUSTAL UPLIFT AND EXTENSION

Evidence of Cenozoic uplift in the Great Basin is given by Eardley (1962) who showed that the average elevations of Tertiary deposits in the mountain and valley blocks were as much as 1.5 km above deposition datum. This uplift is also corroborated by elevation of Cenozoic floras of 1 km in the Cordillera (Axelrod, 1966). Stewart (this volume) suggests as much as 3 km of uplift of the Great Basin since late Cenozoic time, with as much as 10% to 35% extension of the original width.

Recent measurements of crustal deformation have been made using first order geodetic releveling across the Great Basin and the Snake River Plain. These new data described by Reilinger (1977) suggested contemporary asymmetric up-arching of the northern Great Basin, relative to the Wasatch Mountains, with surprisingly high values of 40 ± 8 cm. Contemporary uplift from 1976 releveling across the Yellowstone Plateau (Smith and Pelton, 1977) showed unusually large uplifts of 70 cm in 53 yr, with maximum uplift rates of 14 mm/yr centered near Quaternary resurgent domes. An average relative uplift of 15 cm in 53 yr occurs outside the Yellowstone caldera. The Snake River Plain, on the other hand, exhibited relative sinking and at lower rates (Reilinger and others, 1977). These large rates of contemporary deformation must be interpreted with caution since they suggest uplift of tenths of metres per century and must reflect some ephemeral variations and systematic errors. Yet the broad-scale uplifts must also reflect ongoing tectonism that originates in the mantle.

The area of present-day uplift in the Great Basin also correlates with an area of elevated

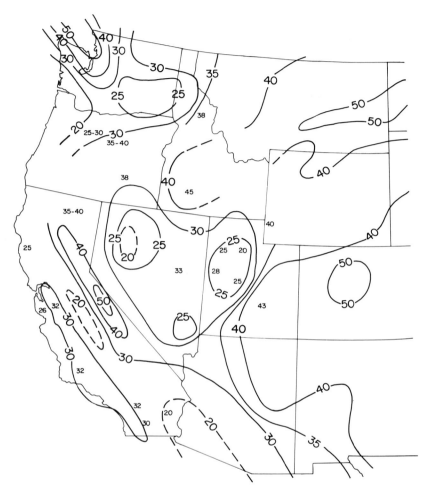

Figure 6-2. Map of crustal thicknesses of Western United States from compilation of Prodehl (1970), Warren and Healy (1974), Smith and others (1975), Hill (this volume), D. Boore (1976, written commun.), and Keith Priestley (1977, oral commun.). Contours in kilometres below surface.

topography that is highest, about 2.2 km, in the west-central Great Basin (Suppe and others, 1975; Eaton and others, this volume). Elevations decrease to less than 1.5 km on both the east and west sides of the uplift in areas of the pre-existing Pleistocene lakes Lahonton on the west and Bonneville on the east. Thus, direct evidence of crustal uplift centered near the zone of the early to mid Cenozoic volcanism is provided. These data give an additional constraint that must be matched by a tectonic kinematic model.

CRUSTAL STRUCTURE

The crustal structure of the Western United States, studied in more detail than many areas of the world, has proved to be complex. Early seismic refraction profiling by the U.S. Geological

118   R. B. SMITH

Survey and several universities (Pakiser and Steinhart, 1964) demonstrated the ability to determine crustal thickness and velocity structure. Since then, many papers have been written on individual surveys, and the reader is referred to Prodehl (1970) and Warren and Healy (1974) for bibliographies. Hill (this volume) discusses the crustal structure of the Pacific Coast areas including the Sierra Nevada, Columbia Plateau, and the western Great Basin. In this paper I will give an overview of the crustal structure of the Western United States and will elaborate on the details of the Great Basin and the transition to the Colorado Plateau.

Maps of crustal thickness (Fig. 6-2) and P_n-velocity (Fig. 6-3) have been prepared from a compilation of published data (Warren and Healy, 1974; Smith and others, 1975; Keller and others, 1975). The P_n-velocity map was based on explosion seismic refraction data and

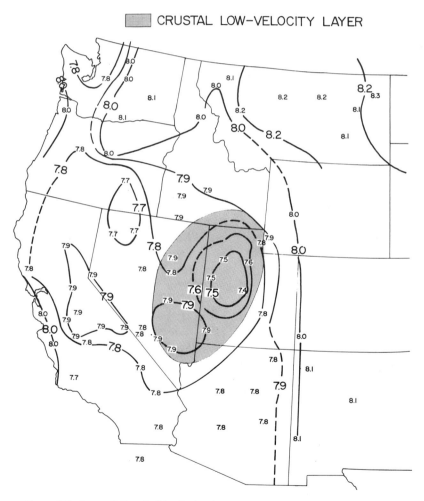

Figure 6-3. Map of P_n-velocities in km/s. Compilation from same references as in Figure 6-2.

does not include information determined using earthquake sources that were considered less reliable. It should be noted that much of the P_n-velocity data from the Western United States were determined from unreversed refraction profiles. This means that the measured velocities are apparent rather than true and may vary by ±0.2 km/s for Moho dips of ±10°. Thus, inferences of detailed variations of P_n-velocity must be carefully considered and should include an examination of the original data.

Great Basin

The crustal structure of the Great Basin demonstrates the diverse properties of a heterogeneous crust. This region is characterized by a 25-km crust along the east and an 18-km crust on the west side (Keith Priestley, 1977, personal commun.) of the Great Basin, with a 30-km-thick central zone. Likewise, P_n-velocities vary from 7.8 km/s in the central portion to a low of 7.7 km/s at the northwest margin (David Boore, 1976, personal commun.; Keith Priestley, 1977, personal commun.) and in the southern High Lava Plains. An area of low P_n-velocity, ~7.5 km/s, occurs along the eastern margin of the Great Basin (Braile and others, 1974; Keller and others, 1975). The southern part of the Great Basin is marked by a 25-km-thick crust where the P_n-velocity increases to ~7.9 km/s.

There is also some evidence that an upper crustal, low-velocity layer exists beneath the eastern Great Basin coincident with the area of low P_n-velocity and the thin crust. The low-velocity layer represents a decrease of about 0.3 km/s in a layer that extends from 5 km to 15 km in depth (Smith and others, 1975). Explanations of the low-velocity layer include crustal intrusion by granitic bodies (Mueller and Landisman, 1966), or an increase in pore pressure accompanied by high temperatures that also attenuates seismic velocity (Smith and others, 1975).

Colorado Plateau

The tectonically stable Colorado Plateau exhibits a thick crust, up to 45 km, and P_n-velocities of ~7.8 km/s. Because of limited refraction data in this area, details of the crustal structure are not known.

Rocky Mountains

Although little detailed refraction profiling has been conducted in the Middle and Southern Rocky Mountains (Colorado, New Mexico, Utah, Wyoming), the area seems to exhibit a 45-km to 50-km-thick crust with P_n-velocities of 7.8 km/s to 8.1 km/s.

The Northern Rocky Mountains (Montana, Idaho) show a range of crustal thickness from 35 km in northern Idaho to 40 km in Montana. P_n-velocities vary from 7.9 km/s to 8.2 km/s (Gary Crosby, 1976, personal commun.). Note that areas of the Northern Rocky Mountains exhibit relatively high heat flow (Blackwell, 1969) and thinner crust—different from that of the other Rocky Mountain areas.

Great Plains

An increase in crustal thickness to 50 km occurs beneath the northern Great Plains (Colorado, Montana, Wyoming). Apparent low heat flow coincides with high P_n-velocities of 8.2 km/s. in this area.

Sierra Nevada

The Sierra Nevada (Hill, this volume) exhibit an average crustal thickness of ~50 km with P_n-velocities from 7.7 km/s to 7.9 km/s. Of significance is the low heat flow in the Sierra Nevada (Lachenbruch and Sass, this volume) that is suggestive of a "cool" root.

Columbia Plateau–Cascade Mountains

The crustal thickness of the Northwestern United States varies from 25 km in eastern Washington (Hill, 1972) to 30 km in the southern Columbia Plateau in an area characterized tectonically much like that of the Great Basin. P_n-velocities here are 7.9 to 8.0 km/s, except for the western High Lava Plains in southeastern Oregon with 7.7 km/s (David Boore, 1976, personal commun.; Keith Priestley, 1977, personal commun.).

GREAT BASIN–COLORADO PLATEAU TRANSITION

The crustal structure of the Colorado Plateau–Great Basin transition is characterized by a thin crust similar to that of the Great Basin but which extends eastward at least 50 km beneath the Colorado Plateau (Braile and others, 1974; Smith and others, 1975; Keller and others, 1975). Thus, the Wasatch Mountains and the High Plateaus of Utah appear to lack deep roots. The easternmost extent of major Cenozoic faulting, seismicity, high heat flow, and thin magnetic crust (Fig. 6-4) also occurs at approximately the same distance east of the province boundary (Shuey and others, 1973; Best and Hamblin, this volume).

An east-west cross section at lat 38°N shown in Figure 6-4 illustrates the crustal structure of the Great Basin–Colorado transition. A thin crust is evident beneath the Wasatch Front with a corresponding low P_n-velocity of ~7.4 km/s and the presence of a crustal low-velocity layer. A shallow zone of high electrical conductivity also underlies the same area (Gough, 1974) in a zone of high regional heat flow.

Complementary data on crustal structure have been compiled from P_n-residuals for over 200 earthquakes on the Utah-Nevada border as recorded in southern Utah (Fig. 6-5). These data show a consistent negative 1.8-s P_n-delay interpreted as the result of a 9-km thinning of the crust. In the Wasatch Front area, two refraction profiles originating from the Bingham copper mine and oriented at 130° show low P_n-velocities of ~7.5 km/s. Although the profiles were not reversed, the close proximity of the profiles suggests that the low P_n-velocities are reasonably correct. This model was corroborated by a +60 mgal regional Bouguer anomaly over the P_n-residual area that was fit to the 9-km mantle upwarp model (Smith and others, 1975).

A possible correlation between the zone of crustal thinning and anomalous upper-mantle structure has been outlined by Jacob (1972) using teleseismic P-wave residuals. These results indicate up to a positive 2-s residual for arrivals along the southern Wasatch Front (Fig. 6-5), the largest in the conterminous United States. Jacob (1972) interpreted these residuals as the result of low P-wave velocities in the upper 200 to 300 km of the mantle.

Further corroboration of low upper-mantle velocities was demonstrated by large P-PL time differences (York and Helmberger, 1973). A maximum delay of 12 s was noted near the southern Wasatch Front above the inferred mantle upwarp (Fig. 6-5). These large delays were interpreted by York and Helmberger (1973) as the result of a pronounced upper-mantle, low-velocity zone suggested as a partial melt.

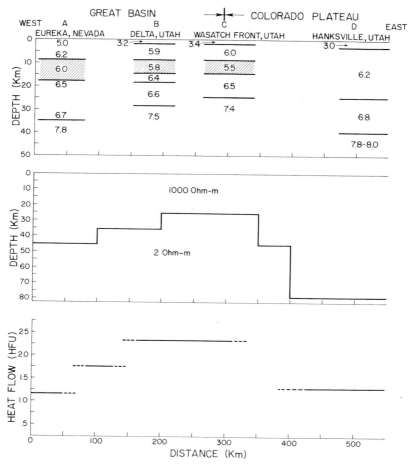

Figure 6-4. Cross section of crustal model along lat 38°N at Great Basin–Colorado Plateau transition zone. Top, P-velocities in km/s; pattern, crustal low-velocity layer. Middle, electrical resistivity. Bottom, heat flow in $\mu cal^{-2} s^{-1}$. Taken from Smith and others (1975).

Together, these data demonstrate that the transition between the Great Basin and the Colorado Plateau represents a low upper-mantle velocity, a distinct mantle upwarp (or crustal thinning), high heat flow, and an upper-crustal, low-velocity layer that may extend up to 50 km eastward beneath the Colorado Plateau. The anomalous crust and mantle structure underlies an area of intense late Cenozoic block faulting and basaltic volcanism that suggests a genetic relationship, perhaps one in which a zone of crustal ductility accommodates crustal extension. Pervasive listric faulting postulated by Hamilton (1976) and Eaton and others (this volume) to occur beneath the Great Basin may bottom in a ductile layer represented by the upper-crustal, low-velocity layer.

Blackwell (this volume) has mapped a linear zone of high thermal energy that parallels the north-south zones of seismicity, thin crust, low P_n-velocity, and crustal low velocity layer along the eastern Great Basin–Colorado Plateau transition. This suggests that the primary mechanism for the anomalous crustal properties is related to abnormally high heat flow rather than simple mechanic deformation.

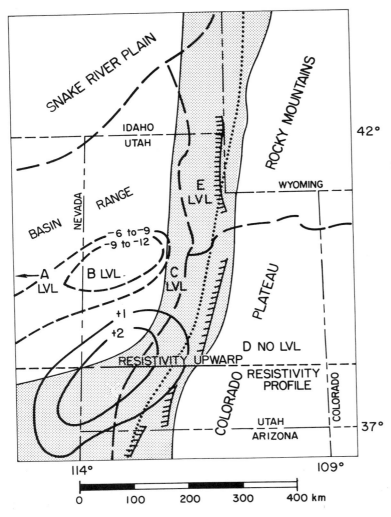

Figure 6-5. Generalized map of Great Basin-Colorado Plateau transition zone showing the following distribution: A to E = locations of crustal refraction data points in Figure 6-4; LVL = presence of crustal low-velocity layer. Dotted pattern = Intermountain seismic belt. Dotted line = heat-flow boundary (>1.5 HFU to west). Dashed line = contour of P-PL residual. Hachured line = eastern extent of Great Basin, Curie depth thickening. Solid line = contour of teleseismic *P*-wave residuals in seconds. Heavy dashed line = province boundary. Taken from Smith and others (1975).

On the basis of heat-flow data, Henyey and Lee (1976) demonstrated that the heat-flow transition between the Sierra Nevada and the western Great Basin occurs west of the physiographic boundary. Thus, it appears that in some areas the boundary between lateral variations in physical properties of the lithosphere of the Great Basin is displaced outward beyond the physiographic boundary. This finding requires that models include a component of lateral time transgression for the kinematic processes.

SEISMICITY

The seismicity of the western interior of the United States has been investigated in detail only during the last two decades (Ryall and others, 1966; Smith and Sbar, 1974). I will review these data particularly as they pertain to tectonic patterns and kinematics of subplate motion.

To illustrate the seismic patterns, an epicenter map of the western interior of the United States was plotted from the National Earthquake Information Service (NOAA) catalog of earthquakes (Pl. 6-1, in pocket; Fig. 6-6). These data pertain principally to earthquakes instrumentally located from about 1950 through 1976, but they contain some older historical data where locations were given on the basis of personal reports. It is important to note that the epicenter maps (Fig. 6-6; Pl. 6-1) contain data from both the California Institute of Technology and the University of California, Berkeley, epicenter catalogs. These additional data bias the map for earthquakes in the California region, because the data include many earthquakes of magnitude (M) less than 3.0. The user is thus cautioned not to use the map for relative seismicity inferences between California and the rest of the Western United States. Hill's (1977) map of the Western United States shows the relative distribution of earthquakes larger than M 4.0 and is a better representation of the relative seismicity of the entire region. Thus, the map (Fig. 6-6; Pl. 6-1) gives a general overview and guide to the seismicity of the Western United States. I will restrict my discussion to the interior of the western Cordillera.

The lower magnitude limit of locatable earthquakes decreased with the implementation of newer and more sophisticated seismographs principally after 1960 when the World Wide Standard Seismograph Network stations were installed. Before 1960 the minimum magnitude of locatable earthquakes on a regional scale was about M 3.5, and it decreased to about M 2.5 by 1970. In some cases where local earthquake networks provided additional data, such as in Yellowstone, the Nevada seismic zones, and the Wasatch Front, earthquakes of minimum magnitudes of M 1.5 were locatable from about 1970.

Some general observations from Figure 6-6 suggest that epicenters are scattered over broad zones up to 150 km wide, suggesting a complexly fractured crust. In most cases accurately located epicenters, locations accurate to ±1 km, do not coincide with mapped Cenozoic faults (for example, see Smith, 1974; Ryall and Priestley, 1975). Temporal variations of earthquake occurrence in the Great Basin suggest return intervals of several hundreds to thousands of years for M-7 earthquakes (Ryall, 1977) as opposed to only a few hundred years for earthquakes along the San Andreas fault.

The historic earthquake pattern of the Western United States is marked by episodic occurrences where an area may have been entirely inactive for tens, hundreds, or thousands of years, then exhibit a short period of intense sometimes swarm activity, and then return to a period of quiescence. This demonstrates that seismic zoning based solely upon epicenter maps of historic earthquakes or solely on evidence of Holocene faulting should not be done.

As a rule, focal depths are not accurately determined unless at least one station is within a focal depth in hypocentral distance. Thus, regional earthquake data with wide station spacings give poor focal-depth information. It has been only with detailed surveys of about six or more stations and with accurate velocity models that reliable focal depths have been calculated.

In Figure 6-7, a histogram of numbers of earthquakes versus focal depths is plotted for various areas throughout the Intermountain seismic belt. All of these data, ~1,200 focal depths, were taken from detailed earthquake surveys using up to 15 stations with station spacings from 3 to 15 km. The primary observation from these data is that 96% of the focal depths are less than 15 km. This observation puts a restriction on the depth limit of brittle deformation of the lithosphere to ~15 km for this area. Another aspect of these data shows a peak in the frequency of occurrence of focal depths between 5 and 10 km at Flathead Lake, Montana; Marysville-Helena, Montana; and Pocatello Valley, Idaho-Utah border area, where we will show in the next section that both components of normal faulting and strike-slip faulting occur; whereas the other areas with only normal faulting show the principal occurrence of earthquakes in the near-surface to 10-km focal-depth zone. This suggests that the strike-slip

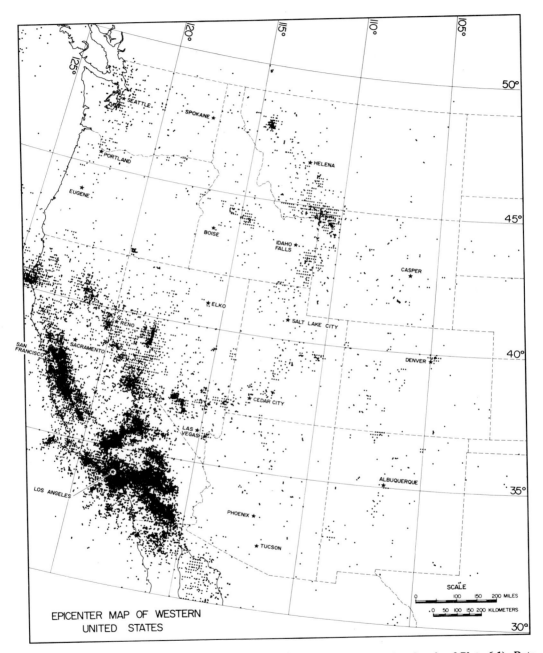

Figure 6-6. Epicenter map of the western interior of the United States (reduced scale of Plate 6-1). Data taken from NOAA, National Earthquake Information Service catalog. Instrumentally located earthquakes principally from 1950 through 1976, but some older earthquakes data are incorporated. Epicenters plotted from NOAA Hypocenter Data file by Paul Grim and are courtesy of Wilbur A. Rinehart, both of the National Geophysical Solar-Terrestrial Data Center of NOAA, Boulder, Colorado. It is important to note that this map contains data from both the California Institute of Technology and the University of California, Berkeley, epicenter catalogs. Magnitudes have not been differentiated. For California, the minimum-magnitude earthquakes plotted were of $M \sim 1$, and for the rest of the Western United States, they were of $M \sim 3$.

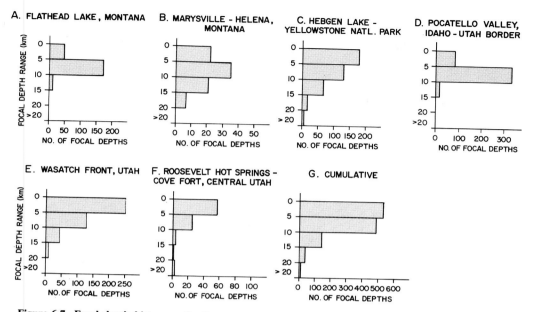

Figure 6-7. Focal depth histogram for Intermountain seismic belt. Data from the following sources: (A) Stevenson (1976); (B) Freidline and others (1976); (C) Smith and others (1977); (D) University of Utah (unpub. data); (E) Smith (1974) and University of Utah (unpub. data); and (F) Olsen and Smith (1976).

mechanisms for which the maximum and minimum stress directions are horizontal are constrained to depths deeper than 5 km in these areas.

Epicenter accuracies for regional data are generally no better than about ±5 km, and epicenter accuracies using older historical data are about ±15 km. Epicenters determined from detailed surveys can be as accurate as ±½ km. For the regional data in Figure 6-6, the most reliable parameter is the relative pattern of epicenters that best represents the overall earthquake distribution.

Intermountain Seismic Belt

The Intermountain seismic belt, is defined as the zone of seismicity (Fig. 6-6) that extends northward through Utah, Idaho, and Wyoming, then turns northwestward at Yellowstone National Park, and then continues through western Montana (Smith and Sbar, 1974). For this paper I have subdivided this zone into the northern, Yellowstone-Hebgen, and southern parts because they differ in trend and style of focal mechanisms.

Northern Intermountain Seismic Belt

Western Montana exhibits a northwest-trending seismic zone with diffuse and episodic earthquake activity. Three large earthquakes (M 6¾, 1925; M 6¼, and M 6¾, 1935) occurred in the vicinity of Helena, but none produced surface faulting. This area has been relatively aseismic since 1935, which may lead to an incorrect inference of long-term low seismicity in the Helena region.

During the 1935 Helena earthquake sequence, swarm-type activity occurred for 6 months (Freidline and others, 1976). During the sequence, earthquakes appear to have migrated from the Helena area 25 km toward the northwest. Temporal progression during the sequence, of at least 60 felt shocks, included a M-4½ shock that preceded the M-6¼ earthquake by 15 days.

Recent detailed monitoring in the Helena area—two 2-month surveys in 1973 and 1974, showed well-defined clusters of earthquakes extending from about 5 km northwest of Helena, to 20 km northwestward toward Marysville, Montana (Freidline and others, 1976).

Regional activity extends northwest from Helena along a diffuse zone to Flathead Lake where a series of episodic earthquakes began in 1964 (Stevenson, 1976). There is little coincidence of these epicenters with mapped faulting, but a possible relationship to reservoir loading has been suggested at Flathead Lake (Dunphy, 1972).

Yellowstone Park–Hebgen Lake Area

The area of most intense historic earthquake activity within the Intermountain region is that of the western Yellowstone Park–Hebgen Lake area (Fig. 6-8). Here at least four M-6 and one M-7.1 earthquakes have occurred in historic time (Trimble and Smith, 1975). The trend of this activity, unlike that of the southern and northern parts of the Intermountain seismic belt, is east-west. A 6.7-m scarp on an east-west normal fault was produced during the M-7.1 earthquake of 17 August, 1959, that occurred in the west Yellowstone Basin about 5 km west of the Yellowstone National Park boundary.

Results of detailed earthquake surveys by the University of Utah in 1972, 1973, 1975, and 1976 (Smith and others, 1974b; Trimble and Smith, 1975; Smith and others, 1977) are shown in Figure 6-8. From five to fifteen portable high-gain stations were used in these surveys that covered Yellowstone National Park and the west, north, and south periphery of the park.

Earthquakes extend 120 km from north of the Centennial Valley at the west side of the park eastward along a N80°W trend across the Hebgen Lake zone, but decreased near the boundary of the Yellowstone caldera. Epicenters in the Hebgen Lake zone occurred near and south of the scarps produced in the 1959 earthquake along a zone up to 15 km wide. Within Yellowstone the epicenter trend begins to change toward a northwesterly trend near the caldera. In some cases epicenters occurred on or near faults in the southern Gallatin Range, but in no case did significant activity occur within any geyser basin inside the caldera. Only scattered events occurred within the Yellowstone caldera, but activity commenced again near the northeast and southeast caldera boundary.

Maximum focal depths of ~17 km occurred in the Centennial Valley area and continued to the northwest part of the Yellowstone caldera (Fig. 6-9). Here they decreased to ~5 km across the caldera, but increased again at the southeast caldera boundary. Inversion of P_g-velocities and results from refraction profiling in Yellowstone Lake showed an eastward decrease from 6.1 km/s at the Madison Valley, to 6.0 km/s at Hebgen Lake, to 5.8 km/s 15 km northwest of the caldera, and 5.0 km/s beneath Yellowstone Lake. The decrease in frequency of occurrence, decrease in focal depths, and decrease in P_g-velocity beneath the Yellowstone caldera are thought to represent effects of high temperatures and anomalous pore pressure above a crustal magma body (Smith and others, 1977). Perhaps the increased temperature reduces the brittle fracture to stable sliding at depths greater than ~5 to 7 km (Brace, 1972).

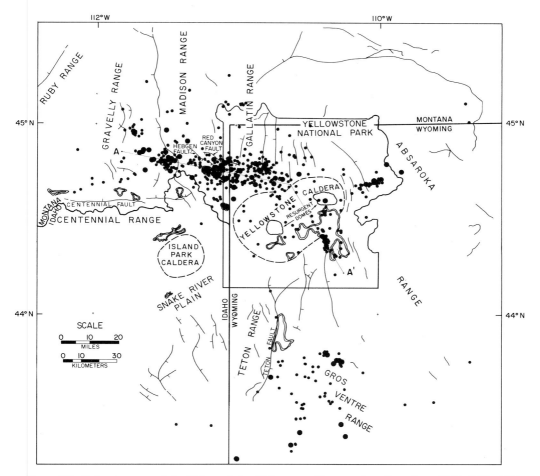

Figure 6-8. Epicenter map of the Hebgen Lake–Yellowstone–Teton region. Data from 1972, 1973, 1974, 1975, and 1976 University of Utah surveys. Small dots = focal depths <10 km; large dots = focal depths >10 km. Taken from Smith and others (1977).

Southern Intermountain Seismic Belt

Yellowstone appears to be a vertex of seismicity in the northern Intermountain region. A north-south trend begins at the southeast edge of the Yellowstone caldera (Fig. 6-8) and continues to the Wasatch Front.

Detailed surveys in the Teton and Gros Ventre Ranges were conducted in 1974 and 1975 (Smith and others, 1976). Along the base of the Teton Range, the Teton fault with 9 km of Cenozoic, vertical displacement and 15 m of Quaternary displacement marks one of the most spectacular fault scarps of the United States. However, little earthquake activity was observed on this fault in three months of observations.

To the east in the tectonically older Gros Ventre Range, many earthquakes occurred in a rather diffuse pattern. There is little evidence of late Cenozoic faulting in this area, but significant earthquake activity was coincident with areas of pronounced Quaternary landslide activity. That the prehistoric landslides were directly related to earthquakes cannot be

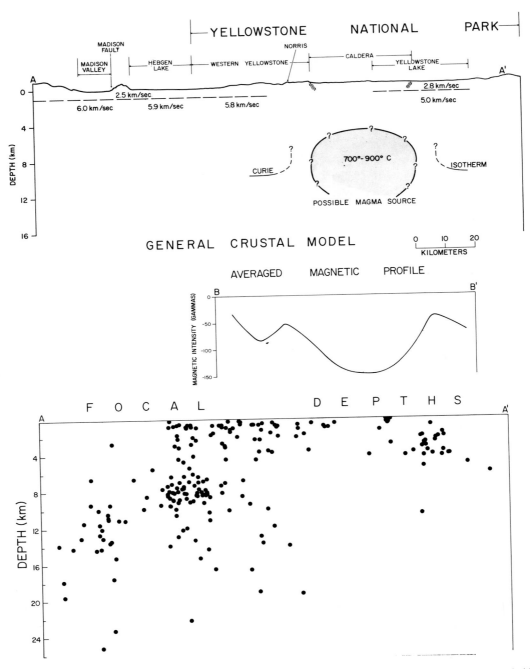

Figure 6-9. Focal-depth cross section for Hebgen Lake–Yellowstone earthquakes shown on profile A-A′ in Figure 6-8. Taken from Smith and others (1977). Vertical exaggeration 2.9:1.

documented, but the 1925 Gros Ventre landslide appears to have occurred about 20 hours following an earthquake of estimated M-4, suggesting a possible relationship (Smith and others, 1976).

Seismicity in southeastern Idaho is poorly known except that it has swarmlike characteristics (Smith and Sbar, 1974) and may reflect some earthquake triggering by crustal loading at the Palisades reservoir.

The Wasatch Front in northern Utah marks the next pronounced area of earthquake activity (Smith, 1974). The Wasatch fault with a total vertical displacement of ~4.5 km during Cenozoic time (Eardley, 1962) extends about 400 km north-south (Fig. 6-10). Recent movement is documented by Holocene fault scarps of up to 15 m near Salt Lake City (Fig. 6-11).

Historical seismicity of Utah, which has been documented since 1850, indicates sporadic activity concentrated on the southern and northern segments of the Wasatch Front. Six extensively damaging earthquakes (one of intensity IX, two of intensity VIII, and three of intensity VII) and at least ten with minor damage have been recorded.

The only historical earthquake in Utah that produced ground displacement was located at the north end of the Great Salt Lake. This M-6.6 earthquake, known as the Hansel Valley earthquake, occurred on March 12, 1934, at the edge of an alluvial-filled basin and produced up to 50 cm of vertical displacement on normal faults, extensive disruption of the land surfaces due to liquefaction, and subsidence of the alluvial basin.

The distribution of earthquakes along the Wasatch Front from July 1962 through December 1976 (Fig. 6-10) shows several features (from north to south): (1) a well-defined, north-south trend of epicenters along the East Cache fault zone near Logan (including a M-5.7 earthquake on August 30, 1962); (2) diffuse activity 15 to 25 km east of Ogden, but little activity along this segment of the Wasatch fault; (3) marked activity of low-magnitude earthquakes extending north and east of Salt Lake City; (4) diffuse activity along the west side of Salt Lake Valley; (5) a lack of significant activity along the segment of the Wasatch fault southeast of Salt Lake City including the area of young faulting shown in Figure 6-11; (6) a north-south zone of activity 25 km northeast of Provo; and (7) a well-defined zone of north-trending activity along the southern Wasatch Front. These seismic patterns including the areas of low seismicity are generally consistent with those of the earlier historic earthquake patterns.

The north-south distribution of earthquakes along the Wasatch Front coincides with that of the Intermountain seismic belt, but many earthquakes are located east of the Front, well east of the physiographic boundary between the Great Basin and the Middle Rocky Mountains.

Figure 6-10 shows areas of low seismic activity from Brigham City, 70 km south to Salt Lake City, and from Salt Lake City south 70 km to Provo. To explain the areas of low seismicity, we may make an analogy from plate tectonics where active plate boundaries are known to be in continuous motion. Averaged over a century or more, movement can be expected at all points along the plate boundaries. Gaps in the seismic activity could be developed along a boundary as a result of the past occurrence of a large earthquake, but eventually these gaps will be filled in by future earthquakes. Using this model, we may regard areas of unusually low seismicity and areas of previous faulting, such as those along parts of the central Wasatch Front, as having a reasonable probability for future large earthquakes.

Other possible explanations of the low seismicity along the Wasatch Front may be (1) the release of strain by seismic creep or by crustal rebound of Lake Bonneville, the predecessor to the Great Salt Lake; (2) seismicity of the Intermountain seismic belt may be moving slowly eastward, leaving aseismic zones in the wake of migrating tectonism; (3) our recording period may be too short to sample the long-term temporal patterns; or (4) the Wasatch fault is now inactive and not storing elastic energy.

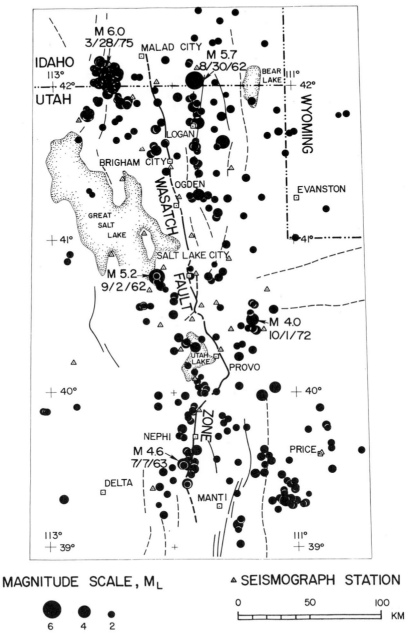

Figure 6-10. Epicenter map of Wasatch Front from 1962 through 1976. Faults shown by heavy solid lines. Taken from Smith (1974) and University of Utah (unpub. data).

LITTLE COTTONWOOD CANYON

FAULT SCARPS

MORAINE

Figure 6-11. Low-sun-angle photograph of Little Cottonwood Canyon, Wasatch Front, 15 km southeast of Salt Lake City, Utah, showing 15 m of Quaternary offset in glacial till and alluvium. Photograph courtesy of Lloyd Cluff and George Brogan.

Southern Utah shows a prominent change in the seismic pattern as the general belt of epicenters turns southwest across Nevada. A north-south zone of minor earthquakes extends southward into Arizona and was originally included in the Intermountain seismic belt (Smith and Sbar, 1974). This zone may be more genetically related to the mechanism of strain release around the Colorado Plateau than to the Intermountain seismic belt.

Southern Nevada

Seismicity continues east-west across southern Nevada at lat 37°N to lat 38°N in a trend that transects that of the north-south late Cenozoic structural grain. Notably, the seismicity is in a region where north-south normal fault blocks are also transected by east- to southeast-trending zones of lateral faulting (Stewart, this volume). Note that the dense cluster of epicenters at lat 37°N, long 116°W are principally from nuclear blasts and related aftershocks at the Nevada test site and do not reflect the true seismicity.

Central Idaho

Earthquakes in central Idaho occur principally along the east margin of the Idaho batholith (Pennington and others, 1974). Note, however, the suggestion of a northwest trend of earthquakes that extends across western Idaho to the Blue Mountains of Oregon and may extend all the way to the Cascade Mountains. Smith and Sbar (1974) postulated that the central Idaho zone of earthquakes was continuous with that of the Hebgen Lake trend. With newer data it is not clear if this zone of activity is related to the Intermountain seismic belt. There is, however, a definite correlation of the earthquakes in central Idaho with an east-west zone of hot springs.

Snake River Plain

The Snake River Plain during historic time has been aseismic. Detailed surveys in the western (Pennington and others, 1974) and the eastern Snake River Plain (Adrian Dahl, 1976, personal commun.) have not disclosed any significant earthquake activity. The lack of earthquakes in this large Cenozoic basalt-filled downwarp may be related to high temperatures documented by high heat flow (Brott and others, in prep.) that restricts brittle fracture but may permit aseismic creep. Possibly the crust beneath the Snake River Plain is simply not storing sufficient elastic strain energy for earthquake generation.

Colorado Plateau

A distinct feature of the regional epicenter map (Fig. 6-6) is the zone of seismicity around the Colorado Plateau. This pattern suggests that the interior of the Colorado Plateau is a rather rigid block that does not permit strain release in the form of earthquakes, but may allow uplift or rotation at its edges.

Epicenter locations in the northern Colorado Plateau (lat 39°30'N, long 110°30'W) represent earthquakes related to underground coal mining (Smith and others, 1974a). Induced earthquakes related to fluid injection in wells have occurred at the Rangely oil field, Colorado-Utah border (lat 40°N) and at the Denver arsenal disposal wells (lat 39°45'N, long 105°W).

Figure 6-12. Map of *P*- and *T*-axes and results of in-situ stress measurements for the western North American plate and eastern Pacific plate. Generalized compilation of data from selected fault-plane solutions of Smith and Lindh (this volume). Outward-directed solid arrows indicate directions of *T*-axes on normal faults, and inward-directed solid arrows denote direction of *P*-axes on thrust faults. Directions of inferred faulting on mechanisms of principal strike-slip motion are shown by small arrows; large arrows show direction of *T*-axes for the selected fault plane. Choices of strike-slip nodal planes were based upon correlations with either surface geology or directions of inferred extensions. Both single-event and composite fault-plane solutions were used. In-situ stress measurements from the United States Bureau of Mines (Aggson, J. R., 1978, in prep.). Horizontal components of principal stress axes from in-situ measurements (open arrows) indicate direction of minimum compressive stress; inward arrows indicate direction of maximum compressive stress.

Aseismic Zones and Cenozoic Faulting

A perplexing problem particularly evident from the seismicity map (Fig. 6-6) is the lack of earthquakes in areas of late Cenozoic faulting, such as in the east-central Great Basin and western Snake River Plain. To assume that the earthquake distribution depicted in Figure 6-6 reflects the long-term seismicity would be inaccurate. However, there is no kinematic model that allows a temporal variation or migration of seismicity back and forth across the Great Basin. I do not have an adequate answer to this problem except to say that a century of seismicity measurements is probably too short to sample the long-term seismicity, especially where return rates of large earthquakes are hundreds to thousands of years (Slemmons, 1967; Wallace, 1977; Ryall, 1977).

FAULT–PLANE SOLUTIONS

Mapping P-axes, T-axes, and slip vectors from fault-plane solutions of earthquakes has been used to deduce patterns of instantaneous plate motions (Minster and others, 1974) and to infer general directions of stress components. In-situ stress monitoring gives direct measurements of the stress components, but the measurements are more difficult and expensive to make.

A compilation of 92 fault-plane solutions of the Western United States in a standard projection, a table of solution parameters, and a reference list is included in the companion paper by Smith and Lindh (this volume). The fault-plane solutions were compiled from data of Smith and Sbar (1974), Lindh and others (1973), and Smith (1977, unpub. data). In-situ stress data are from the United States Bureau of Mines (Aggson, J. R., 1978, in prep.).

In Figure 6-12, 79 representative fault-plane solutions from Smith and Lindh (this volume) and 8 in-situ stress measurements are used to infer the present-day directions of subplate motions and directions of the regional stress components in the Western United States. The map projection in Figure 6-12 is the same as for Figure 6-1 and was chosen to emphasize the possible relationships and interactions between the North American and the Pacific plates.

It is important to point out some inherent problems in interpreting regional patterns of fault-plane solutions. For example, components of both strike-slip and normal faulting may occur in the same stress regime, because faulting on pre-existing surfaces with oblique strike and varying attitudes can produce oblique slip. Likewise, a locally heterogeneous deviatoric stress field could produce up to 90° rotations of the principal stress axes.

McKenzie (1969) discussed variations in the stress tensor for earthquakes and concluded that small earthquakes can occur on weak planes rather than by brittle failure of new rock. This is likely the case for small-magnitude earthquakes, $M \leqslant 2$, especially in a highly fractured crust such as that in an intraplate zone of deformation. McKenzie (1969) suggested in this case that the principal direction of stress, that is, the directions of the P-axes determined from the fault-plane solutions, could deviate up to ±45° from the true directions of maximum compressive stress. However, it is suggested from the data shown here (Fig. 6-12) that the directions of the P- or T-axes for smaller, $M \leqslant 5$, earthquakes are remarkably similar to those determined from the smaller shocks in the same area (see south-central Nevada, Hebgen Lake, Montana, Fig. 6-12). Usually one of the principal stress axes is consistent in direction for most solutions in a specific tectonic area, whereas the other axes may vary up to 90°.

Fault-plane solutions on the San Andreas fault (Fig. 6-12) show principally right-lateral, strike-slip faulting with north-south shortening and westward extension. Thrust mechanisms on secondary faults, north of Los Angeles where the fault bends eastward, are thought to reflect northward underthrusting of the North American plate by the Pacific plate.

Within the western North American plate, a zone of both northwest-trending strike-slip and north-south normal faulting occurs in the southern Great Basin. This zone of complex deformation is located north of the Sierra Nevada and in the Walker lane, but does not extend northward into central Nevada (Fig. 6-12). The motion implied by the fault-plane solutions reflects pervasive west to northwest extension on north-south normal faults with some components of right-lateral strike-slip motion on faults oblique to the main north-south tectonic grain. Note that the strike-slip solutions along the east side of the Sierra Nevada are similar in orientation to those of the San Andreas fault, whereas solutions farther east show an ~45° clockwise rotation of the T-axes.

The fault-plane solution (Fig. 6-12, no. 51) west of the Sierra Nevada reflects east-west extension probably on a normal fault. The direction of the T-axis is the same as that of the San Andreas fault solutions, but the P-axis is rotated to vertical. The consistency of the T-axes in the San Andreas–Sierra Nevada block appears to reflect the pervasive east-west extension. North of the Walker lane the solutions principally reflect northwest extension on normal faults. The north margin of the Walker lane thus appears to mark a boundary between fairly homogeneous extensional strain within the central Great Basin to that of extensional-lateral slip in a zone of pre-existing faulting to the south.

The continuation of the Intermountain seismic belt into southeastern Nevada shows left-lateral fault-plane solutions. These could be interpreted to reflect westward extension between the relatively stable portion of the Great Basin and a wide zone of active deformation in southern Nevada including the Walker lane.

Only a few fault-plane solutions are available for southern Oregon (Couch and MacFarlane, 1971). These solutions reflect oblique right-lateral, west to northwest motion with components of high-angle normal and reverse faulting. Thus, the High Lava Plains appear to represent a transition zone between normal faulting to the south in the Great Basin to oblique right-lateral strike-slip–normal faulting similar to that of the Walker lane. At the northern boundary of the High Lava Plains, a distinct change in the stress field is manifested by the change to north-south compression on thrust faults. This east-west stress boundary appears to coincide with the northern edge of the Great Basin zone of high heat flow and the northern extent of late Cenozoic normal faulting.

Along the eastern margin of the Great Basin, east-west extension occurs on north-trending normal faults (Fig. 6-12). Dips of north-south nodal planes decrease from high-angle reverse and steep normal faults at the south, opposite the northern Colorado Plateau, to moderate ~50° dips at the north, reflecting a change in the attitude of T-axes possibly by horizontal stress relaxation from south to north (Smith and Sbar, 1974).

An abrupt change in the stress field occurs at the intersection of the Snake River Plain with the Intermountain seismic belt at Yellowstone National Park. The T-axes rotate to a northerly direction along an east-west seismic zone north of the Snake River Plain. Nine single-event and composite fault-plane solutions along the Hebgen Lake–western Yellowstone trend show nearly identical solutions reflecting north-south extension on normal faults (Smith and others, 1977). These solutions were from earthquakes that ranged in magnitude from 7.1 to ~1. The similarity of solutions over the wide magnitude range is interpreted to reflect a homogeneous stress field.

The Yellowstone caldera marks the location where the stress field begins to reorient to that of east-west extension southward along the Intermountain seismic belt. Fault-plane solutions by Pitt and Weaver (1975) and Smith and others (1977) show that the trend of T-axes rotates from north-south, west of the Yellowstone caldera to generally east-west, south of the caldera.

Fault-plane solutions along the northern Intermountain seismic belt show a rotation in *T*-axes from north-south at Yellowstone to east-west at the termination of seismicity near lat 48°N (Freidline and others, 1976). The Helena, Montana, area exhibits both oblique strike-slip and normal faulting with consistent east-west *T*-axes. Stevenson (1976) computed several single-event fault-plane solutions for Flathead Lake, Montana, 200 km northwest of Helena, which showed east-west extension on dip-slip faults and some strike-slip solutions. But the consistent focal parameter was that of north-south *P*-axes for well-defined solutions.

Note that the northern Intermountain seismic belt parallels the San Andreas fault, and the orientations of the *T*-axes suggest westward extension oblique to the trend of the seismic belt. Intraplate motion could be accommodated here on pre-existing fractures of the northwest-trending Montana lineament.

Near Seattle, a *M*-6.5 earthquake at a focal depth of 60 km exhibited a normal faulting mechanism, perhaps in an extensional stress field near a plate boundary. Whereas to the east, fault-plane solutions from shallow earthquakes within the Columbia Plateau imply north-south thrusting. The seismicity in the Columbia Plateau is in a zone of late Cenozoic east-west anticlinal folding of basalts that also suggests past north-south crustal shortening within a compressing subplate.

Surrounding the northern Colorado Plateau, scattered strike-slip solutions with one thrust mechanism are difficult to interpret except to note that the area is not in primary horizontal extension. These solutions are consistent with the hypothesis that the Colorado Plateau is a stable uplifted block that may be sufficiently rigid to sustain rotation at its periphery.

Results from the in-situ stress measurements show fair agreement with the stress directions derived from the fault-plane solutions. In the southern Great Basin, the northwest-trending axis of minimum compression from the in-situ stress measurement is nearly the same as the direction of the *T*-axes from nearby fault-plane solutions. At the northern Colorado Plateau and near Denver, the minimum compression directions are within 30° of those determined from fault-plane solutions. In the aseismic Rocky Mountains, east of Salt Lake City, the axis of minimum compression from the in-situ stress measurement is within 30° of the *T*-axes determined from earthquakes 250 km west.

TECTONIC INTERPRETATION

Figure 6-13 summarizes the pattern of late Cenozoic deformation of the western North American plate interpreted primarily from seismicity, the mode of the contemporary stress fields deduced from the fault-plane solutions, and geologic evidence. Relative motions between the lithospheric blocks are generalized into four categories: (1) extension on normal faults, (2) lateral slip on strike-slip faults, (3) compression on high-angle reverse and oblique thrusts, and (4) zones of both oblique extensional- and lateral-slip components on normal faults.

Calculated motion between the North American and the Pacific plates, based upon instantaneous plate motions, is 5.5 cm/yr along the San Andreas fault and its main branches, whereas geologic and geodetic data suggest ~3.0 cm/yr (Minster and others, 1974). The discrepancy of ~2.5 cm/yr has been attributed to northwest intraplate extension within the Great Basin. That the present-day tectonic pattern is more complicated than pure extension, however, is suggested from Figures 6-12 and 6-13. Based on the locations of the active seismic zones (Fig. 6-6), four subplates (expanded from Smith and Sbar, 1974), each with its own characteristic motion, are defined: (1) the Great Basin–High Lava Plains, (2) the Colorado Plateau, (3)

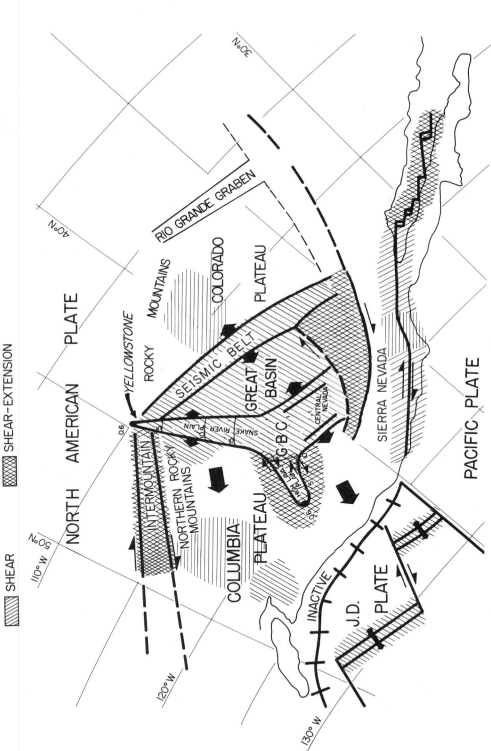

Figure 6-13. Summary of inferred contemporary deformation of the western North American plate. GBC = Great Basin Center of lithospheric uplift and divergence; J. D. = Juan de Fuca plate. Large arrows infer directions of subplate motions; small arrows indicate relative direction of motion on strike-slip fault zones. Fine lines across Snake River Plain-High Lava Plain indicate contours in m.y. B.P. of initiation of volcanism from Armstrong and others (1975) and MacLeod and others (1975).

the Northern Rocky Mountains–Columbia Plateau, and (4) the Sierra Nevada. These lithospheric blocks are separated by wide zones of deformation where the return rates of large earthquakes are lower than along plate boundaries.

Northwest extension is the principal mode of deformation of the western Great Basin, with westward extension on the east side. Primary extension occurs along the central Nevada seismic zone north of the Walker lane and along the southern Intermountain seismic belt, that is, the Wasatch Front in Utah and Idaho. However, the T-axes change direction from northwest-trending along the west and central Nevada seismic zones to east-trending on the Wasatch Front. This change in stress direction across the Great Basin suggests that some of the intraplate strain could be accommodated by aseismic deformation, perhaps plastic deformation of a heated lithospheric block.

The directions of the Great Basin stress field described here compare favorably with those determined by Wright (1976) from late Cenozoic fault patterns. Also, a generalized overview of the Western United States stress field by Raleigh (1974), as inferred from an earlier compilation of 24 fault-plane solutions, shows a similar east-west extension pattern in the Great Basin.

In the Western United States, three zones of combined extensional and shear strain are subparallel to the San Andreas fault: (1) the southern boundary of the Great Basin along the Walker lane, (2) the northern Intermountain seismic belt, and (3) a less conspicuous zone in the High Lava Plains of southeastern Oregon. I hypothesize that these intraplate zones accommodate some of the lateral convergence between the North American and the Pacific plates, perhaps like that of the Indian-Eurasian plate convergence.

In the southern Great Basin, contemporary extension and lateral slip is superimposed on a tectonic grain of north-south faulting. The normal faults accommodate part of the present extension, but lateral components of slip occur on pre-existing faults oblique to the spreading direction. Both contemporary right- and left-lateral slip, seen from the fault-plane solutions, occur in the southern Great Basin. If the pattern of oblique slip, inferred from the contemporary seismicity, reflects the late Cenozoic stress regime, the large-scale cummulative slip of 190 km along the Walker lane interpreted by Stewart and others (1968) is reasonable.

Seismicity and focal mechanisms in the northern Intermountain seismic belt of Montana are similar to those in the southern Great Basin with north-south extension at Yellowstone rotating to east-west extension at the north. The deformation occurs both on normal and strike-slip faults that are located, in part, along the Montana lineament. This left-lateral shear zone with evidence of pre-Laramide motion accommodates northwest extension by normal faulting on planes perpendicular to the extension direction and by lateral-slip oblique to the extension direction.

A less conspicuous zone of combined shear-extensional deformation occurs along the High Lava Plains of southern Oregon. Several right-lateral, strike-slip fault zones have been identified by Stewart and others (1975) and Lawrence (1976), including the northwest-trending Oregon-Nevada lineament and the Brothers, the Vale, and other fault zones. Focal mechanisms (Fig. 6-12) show east-west extension near the Nevada-Oregon border, but change to oblique slip with a component of thrusting in central Oregon. Couch and MacFarlane's (1971) interpretations of first motions from small earthquakes in the High Lava Plains indicated components of right-lateral slip on northwest trending faults; whereas, farther north at the edge of the Columbia Plateau, six thrust mechanisms demonstrate north-south crustal shortening.

Thus, at about the latitude of the Blue Mountains in Oregon, a boundary occurs between crustal compression, to the north and northwest-extension, to the south, with components of strike-slip in the High Lava Plains. This east-west zone separates the general extensional tectonism of the Great Basin from north-south compression of the Columbia Plateau. The zone represents a relatively aseismic intraplate boundary that may extend eastward to either the edge of the aseismic Snake River Plain or to the northwest-trending zone of earthquakes through Idaho (Fig. 6-6). Blackwell (this volume) has shown an east-west thermal-energy boundary that coincides with this stress boundary.

North-south crustal shortening is evident in the southern Columbia Plateau, possibly because of compression between the Northern Rocky Mountain subplate and the rigid North American plate to the north. Lithospheric shortening also occurs within the northern Colorado Plateau, perhaps because of the buttress effect between the western deforming zones and the eastward stable interior.

An important implication of the late Cenozoic volcanism of the Western United States is the possible development of continental zones of lithospheric fracturing with time-transgressive volcanism emanating outward from a central zone (Smith, 1977). These zones appear to be facilitated by continental spreading along the Snake River Plain and by combined extension and lateral slip in the High Lava Plains.

I suggest that an asthenospheric melting anomaly, perhaps manifested by a mantle diapir (Scholz and other, 1971), initially formed a lithospheric uplift coincident with the present areas of highest topography and contemporary uplift in the central and northern Great Basin. At cessation of Farallon plate subduction, ~20 m.y. B.P., plate compression was released sufficiently to allow the formation of the two continental zones of volcanism originating at the north end of the lithospheric uplift. These arms are oriented at 120° to the hypothesized uplift of the Great Basin—one, the Snake River Plain, the other along the High Lava Plains.

In this framework, volcanism of the Snake River Plain is then considered to be related to a northeast-propogating lithospheric zone of fracture or a transgressing melting anomaly that is now at Yellowstone. It was enhanced in width and intensity of volcanism over that in the High Lava Plains because it was in a direction parallel to that of lithospheric spreading or because it followed a pre-existing zone of weakness. The lack of extensive late Cenozoic basaltic volcanism in the High Lava Plains may have been produced by a sufficiently compressed lithosphere above the Juan de Fuca plate north of the Mendocino triple junction to restrict basic volcanism and pervasive extensional tectonism.

Recent studies by Smith and others (1974), Eaton and others (1975), and Smith and others (1977) described the Yellowstone Plateau as an area of very high heat flow, >3.0 HFU, intense seismicity around a relatively aseismic caldera, and low upper-mantle velocities. These characteristics suggest that Yellowstone, as a lithospheric thermal anomaly, has deep roots well within the mantle. Morgan (1972) and Suppe and others (1975) suggested that a mantle plume lies beneath Yellowstone. My model, while not invoking a plume, requires a deep-seated mechanism such as the tip of a lithospheric fracture or a thermal instability at the end of the Snake River Plain, that is, the Yellowstone Plateau.

The main uplift in the north and central Great Basin is hypothesized to have developed into a broader feature with accompanying crustal and lithospheric thinning and extension by uplift and lateral spreading. If the late Cenozoic Great Basin center of uplift and high heat flow has thinned the lithosphere through heating and partial melting, then the time-transgressive Quaternary volcanism along the Snake River Plain–High Lava Plains may reflect areas where the lithosphere is now the thinnest. This concept is corroborated by Brott and others (in prep.) who described a heat-flow model for the Snake River Plain where the heat flow increases from the western Snake River Plain with decreasing age to the area of the highest heat flow at Yellowstone. This model could be tested by combined reverse-refraction profiles, intermediate-period surface-wave dispersion, and *P*-wave delay monitoring. Low-velocity layers that may be definitive of partial melt would require a careful experiment to analyze wide-angle reflections and accurate amplitude characteristics.

Regional heat-flow data (Diment and others, 1975; Blackwell, this volume; Lachenbruch and Sass, this volume) show that the center of the postulated Great Basin uplift is located near an area of high heat flow, is underlain by a thin crust ~25 km, and exhibits low P_n-velocities, ~7.5 to 7.8 km/s probably reduced by partial melting. In this model (Smith, 1977) a lithospheric diapir or melting instability, similar to mechanisms postulated by Scholz and others (1971), Christiansen and McKee (this volume), and Eaton and others (this volume), is hypothesized to have uplifted and spread laterally from the northern Great Basin center in mid-Cenozoic time. As the lithospheric material spread laterally by ductile flow, it further thinned and heated the upper lithosphere. The current locations of the lithospheric heating, crustal thinning, and low P_n-velocities are located in a symmetric pattern on the east and west edges of the Great Basin. Thermal manifestations of the mechanisms are demonstrated in Blackwell's (this volume) discussion, which shows north-south zones of high heat flow, >3.0 HFU, coincident with the margins of the Great Basin (Scholz and others, 1971). This symmetry is similar to that of the gravity field of the Great Basin described by Eaton and others (this volume). The contemporary uplift of the Great Basin documented by the releveling suggests that the mechanism is still active.

Intensive seismic activity occurs along the western and eastern margins of the lithospheric anomaly in the Great Basin, whereas the Snake River Plain and the High Lava Plains and the Columbia Plateau exhibit relatively low seismicity. In this context the northern High Lava Plain and the Snake River Plain may be considered a diffuse subplate boundary. Seismicity in central Idaho and in eastern Oregon and Washington may reflect minor brittle strain relief along this diffuse zone.

Thompson (1977) made an analysis of the Cenozoic volcanism of the Western United States and concluded that post-Eocene tectonic and magmatic features east of the Sierra Nevada and Cascade Mountains are products of a shallow mantle diapir extracted from the Farallon plate at 40 m.y. B.P. He invoked the diapir as the source of heat and crustal attenuation, much in the same way that I envision this process.

I thus conclude that the contemporary tectonics of the western North American plate are the result of both convergent plate interaction, along large-scale tectonic zones of combined shear-extensional strain, and west to northwest extension above a lithospheric zone of uplift. Subplates are bounded by the active tectonic zones, where each subplate appears to be moving or "slivering" northwestward somewhat independently alongside one another to accommodate the lateral convergence. This model does not rely upon the necessity of an extension of the East Pacific Rise beneath the Great Basin, such as suggested by Menard (1964) and Cook (1969), but manifests some of the characteristics that these earlier studies were based upon.

In conclusion, a major lithospheric uplift centered over the north-central Great Basin appears to have formed as a result of a mid-Cenozoic thermal upwelling of the asthenosphere. Secondary lithospheric fracturing that formed the Snake River Plain was enhanced by stress relaxation following cessation of subduction. The zone of rhyolite domes of the High Lava Plains reflects a third but less intense volcanic arm in a zone of right-lateral slip. Axial extension and cooling of the central Great Basin since late Cenozoic time has resulted in a thicker and cooler crust with higher P_n-velocities. The margins of the Great Basin are now most affected by the asthenospheric lateral divergence that has resulted in a thin crust, ~25 km; low P_n-velocities, ~7.5 km/s; high heat flow, ~3.0 HFU; crustal low-velocity layers; and active seismicity.

ACKNOWLEDGMENTS

I thank D. P. Hill, L. Braile, W. J. Arabasz, W. T. Nash, and W. T. Parry for their review and criticism of this manuscript. J. Pelton, R. Nye, J. Bailey, T. Williams, and A. Kastrinsky assisted in preparation and compilation of these data. Valuable discussions with participants at the 1975 Alta Penrose Conference are greatly appreciated. In particular, G. P. Eaton, D. R. Mabey, R. P. Christiansen, D. D. Blackwell, J. Stewart, E. H. McKee, M. Best, and W. Hamilton offered important observations. This work was supported in part by the National Science Foundation grants DES74-13848 A01 and GA-43424.

REFERENCES CITED

Armstrong, R. L., 1978, Cenozoic igneous history of the U.S. Cordillera from latitude 42° to 49°N, *in* Smith, R. B., and Eaton, G. P., eds., Cenozoic tectonics and regional geophysics of the western Cordillera: Geol. Soc. America Memoir 152 (this volume).

Armstrong, R. L., Leeman, W. P., and Malde, H. E., 1975, K-Ar dating, Quaternary and Neogene volcanic rocks of the Snake River Plain, Idaho: Am. Jour. Sci., v. 275, p. 225–251.

Atwater, T., 1970, Implications of plate tectonics for the Cenozoic tectonic evolution of western North America: Geol. Soc. America Bull., v. 81, p. 3513–3536.

Axelrod, D. E., 1966, The Eocene Copper Basin flora of northeastern Nevada: Calif. Univ. Pubs. Geol. Sci., v. 59, p. 125.

Best, M. G., and Hamblin, W. K., 1978, Origin of the northern Basin and Range province: Implications from the geology of its eastern boundary, *in* Smith, R. B., and Eaton, G. P., eds., Cenozoic tectonics and regional geophysics of the western Cordillera: Geol. Soc. America Mem. 152 (this volume).

Blackwell, D. D., 1969, Heat flow in the Northwestern United States: Jour. Geophys. Research, v. 74, p. 992–1007.

——1978, Heat flow and energy loss in the Western United States, *in* Smith, R. B., and Eaton, G. P., eds., Cenozoic tectonics and regional geophysics of the western Cordillera: Geol. Soc. America Mem. 152 (this volume).

Brace, W., 1972, Laboratory studies of stick-slip and their application to earthquakes: Tectonophysics, v. 14, p. 189–200.

Braile, L., Smith, R. B., Keller, G. R., Welch, R., and Meyer, R. P., 1974, Crustal structure across the Wasatch Front from detailed seismic refraction studies: Jour. Geophys. Research, v. 79,

p. 1295–1317.

Carey, S. W., ed., 1958, The tectonic approach to continental drifting: Continental drift, a symposium: Hobart Tasmania Univ. Geology Dept., p. 177–355.

Christiansen, R. L., and Lipman, P. W., 1972, Pt. 2, Late Cenozoic, Cenozoic volcanism and plate-tectonic evolution of the Western United States: Royal Soc. London Philos. Trans., 271A, p. 249–284.

Christiansen, R. L., and McKee, E. H., 1978, Late Cenozoic volcanic and tectonic evolution of the Great Basin and Columbia intermontane region, *in* Smith, R. B., and Eaton, G. P., eds., Cenozoic tectonics and regional geophysics of the western Cordillera: Geol. Soc. America Mem. 152 (this volume).

Coney, P. J., 1978, Mesozoic-Cenozoic Cordilleran plate tectonics, *in* Smith, R. B., and Eaton, G. P., eds., Cenozoic tectonics and regional geophysics of the western Cordillera: Geol. Soc. America Mem. 152 (this volume).

Cook, K. L., 1969, Active rift system in the Basin and Range province: Tectonophysics, v. 8, p. 469–511.

Couch, R. W., and MacFarlane, W. T., 1971, A fault plane solution of the October, 1969 Mt. Rainer earthquake and tectonic movements of the Pacific Northwest derived from fault plane solutions and first motion studies: EOS (Am. Geophys. Union, Trans., v. 52, p. 428.

Diment, W. H., and others, 1975, Temperatures and heat contents based on conductive transport of heat, *in* White, D. E., and Williams, D. L., eds., Assessment of geothermal resources of the United States—1975: U.S. Geol. Survey Circ., v. 726, p. 94–95.

Dunphy, G. J., 1972, Seismic activity of the Kerr dam—S.W. Flathead Lake area, Montana, *in*

Earthquake research in NOAA: U.S. Dept. Commerce, NOAA Tech. Rept. ERL236-ESL121, p. 59–61.

Eardley, A. J., 1962, Structural geology of North America (2nd ed.): New York, Harper and Row, Pubs.

Eaton, G. P., Christiansen, R. L., Iyer, H. M., Pitt, A. M., Mabey, D. R., Blank, H. R., Jr., Zietz, I., and Gettings, M. E., 1975, Magma beneath Yellowstone National Park: Science, v. 188, p. 787–796.

Eaton, G. P., Mabey, D. R., Wall, R. B., and Kleinkopf, M. D., 1978, Regional gravity and tectonic patterns: Their relation to late Cenozoic epeirogeny and lateral spreading of the western Cordillera: in Smith, R. B., and Eaton, G. P., eds., Cenozoic tectonics and regional geophysics of the western Cordillera: Geol. Soc. America Mem. 152 (this volume).

Freidline, R. O., Smith, R. B., and Blackwell, D. D., 1976, Seismicity and contemporary tectonics of the Helena, Montana area: Seismol. Soc. America Bull., v. 60, p. 81–96.

Gough, D. E., 1974, Electrical conductivity under western North America in relation to heat flow, seismology and structure: Jour. Geomagnetism and Geoelectricity, v. 26, p. 105–123.

Hamilton, W., 1976, Neogene extension of the Western United States: Geol. Soc. America, Abs. with Programs (Ann. Mtg.), v. 7, p. 1098.

Hamilton, W., and Myers, W. B., 1966, Cenozoic tectonics of the Western United States: Rev. Geophysics, v. 4, p. 509–549.

Henyey, T. L., and Lee, T. C., 1976, Heat flow in Lake Tahoe, California-Nevada, and the Sierra Nevada–Basin and Range transition: Geol. Soc. America Bull., v. 87, p. 1179–1187.

Hill, D. P., 1972, Crustal and upper-mantle structure of the Columbia Plateau from long-range seismic refraction measurements: Geol. Soc. America Bull., v. 83, p. 1639–1648.

——1978, Seismic evidence for the structure and Cenozoic tectonics of the Pacific Coast States, in Smith, R. B., and Eaton, G. P., eds., Cenozoic tectonics and regional geophysics of the western Cordillera: Geol. Soc. America Mem. 152 (this volume).

Jacob, K. H., 1972, Global tectonic implications of anomalous seismic P travel times from the nuclear explosion Longshot: Jour. Geophys. Research, v. 77, p. 2556–2573.

Keller, G. R., Smith, R. B., and Braile, L., 1975, Crustal structure along the eastern margin of the Great Basin from detailed refraction measurements: Jour. Geophys. Research, v. 80, p. 1093–1099.

Lachenbruch, A. H., and Sass, J. H., 1978, Models of an extending lithosphere and heat flow in the Basin and Range province, in Smith, R. B., and Eaton, G. P., eds., Cenozoic tectonics and regional geophysics of the western Cordillera: Geol. Soc. America Mem. 152 (this volume).

Lawrence, R. D., 1976, Strike-slip faulting terminates the Basin and Range province in Oregon: Geol. Soc. America Bull., v. 87, p. 846–850.

Lindh, A. G., Fisher, F. G., and Pitt, A. M., 1973, Nevada focal mechanisms and regional stress fields: EOS (Am. Geophys. Union Trans.), v. 54, p. 1133.

Lipman, P. W., Prostka, H. J., and Christiansen, R. L., 1971, Evolving subduction zones in the Western United States, as interpreted from igneous rocks: Science, v. 174, p. 821–823.

Menard, H. W., 1964, The East Pacific Rise: Science, v. 132, p. 1737.

MacLeod, N. S., Walker, G. W., and McKee, E. H., 1975, Geothermal significance of eastward increase in age of upper Cenozoic rhyolitic domes in southeastern Oregon: Menlo Park, Calif., U.S. Geol. Survey Open-file rept. 75–348.

McKee, E. H., 1971, Tertiary igneous chronology of the Great Basin of Western United States: Implications for tectonic models: Geol. Soc. America Bull., v. 82, p. 3497–3502.

McKenzie, D. P., 1969, The relationship between fault plane solutions for earthquakes and the directions of the principal stresses: Seismol. Soc. America Bull., v. 59, p. 591–601.

McKenzie, D. P., and Julian, B., 1971, Puget Sound, Washington, earthquake and the mantle structure beneath the Northwestern United States: Geol. Soc. America Bull., v. 82, p. 3519–3524.

Minster, J. B., Jordan, T. H., Molnar, P., and Haines, E., 1974, Numerical modeling of instantaneous plate tectonics: Royal Astron. Soc. Geophys. Jour., v. 36, p. 541–576.

Molnar, P., and Tapponier, P., 1975, Cenozoic tectonics of Asia: Effects of a continental collision: Science, v. 189, p. 419–426.

Morgan, W. J., 1972, Plate motions and deep mantle convection, in Shagam, R., ed., Studies in earth and space sciences: Geol. Soc. America Mem. 132, p. 7–22.

Mueller, S., and Landisman, M., 1966, Seismic studies of the earth's crust in continents, 1, Evidence for a low-velocity zone in the upper part of the lithosphere: Royal Astron. Soc. Geophys. Jour., v. 10, p. 525–538.

Noble, D. C., 1972, Some observations on the Cenozoic volcano-tectonic evolution of the Great Basin, Western United States: Earth and Planetary Sci. Letters, v. 17, p. 142–150.

Olson, T. L., and Smith, R. B., 1976, Earthquake surveys of the Roosevelt Hot Springs and the Cove Fort areas, Utah: Final rept., no. 4, submitted to the Natl. Sci. Foundation, contract GI-43741.

Pakiser, L. C., and Steinhart, J. S., 1964, Explosion seismology in the Western Hemisphere, in Odishaw, H., ed., Research in geophysics: Cambridge, Mass., MIT Press, p. 123–147.

Pennington, W., Trimble, A., and Smith, R. B., 1974, A microearthquake survey of parts of the Snake River Plain and central Idaho: Seismol. Soc. America Bull., v. 64, p. 307–312.

Pitt, A. M., and Weaver, C., 1975, Aftershock studies of the June 30, 1975 Yellowstone earthquake [abs.]: Program, 1976 Joint Meeting, Seismol. Soc. America and Canadian Geophys. Union, p. 18.

Prodehl, C., 1970, Seismic refraction study of the Western United States: Geol. Soc. America Bull., v. 81, p. 2629–2646.

Raleigh, C. B., 1974, Crustal stresses and global tectonics, in Advances in rock mechanics: Proc. Third Cong. Internat. Soc. for Rock Mechanics, v. 1, pt. A, p. 593–597.

Reilinger, R. E., 1977, Vertical crustal movements from repeated leveling data in the Great Basin of Nevada and western Utah: EOS (Am. Geophys. Union Trans.), v. 58, p. 1238.

Reilinger, R. E., Citron, G. P., and Brown, L. D., 1977, Recent vertical movements from precise leveling data in southwestern Montana, western Yellowstone National Park and the Snake River Plain: EOS (Am. Geophys. Union Trans.), v. 58, p. 495.

Ryall, A., 1977, Earthquake hazard in the Nevada region: Seismol. Soc. America Bull., v. 67, p. 517–532.

Ryall, A., and Priestley, K. F., 1975, Seismicity, secular strain, and maximum magnitude in the Excelsior Mountains area, western Nevada and eastern California: Geol. Soc. America Bull., v. 86, p. 1585–1592.

Ryall, A., Slemmons, D. B., and Gedney, L. D., 1966, Seismicity, tectonism, and surface faulting in the Western United States during historic time: Seismol. Soc. America. Bull., v. 56, p. 1105–1135.

Ryall, A., Koizumi, C. J., and Priestley, K. F., 1973, Evidence for a high-velocity lithospheric plate under northern Nevada. Seismol. Soc. America Bull., v. 63, p. 2135–2144.

Scholz, C., Baranzangi, M., and Sbar, M., 1971, Late Cenozoic evolution of the Great Basin, Western United States, as an ensialic interarc basin: Geol. Soc. America Bull., v. 82, p. 2979–

2990.

Shuey, R. T., Shellinger, D. K., Johnson, E. G., and Alley, L. B., 1973, Aeromagnetics and the transition between the Colorado Plateau and the Basin Range provinces: Geology, v. 1, p. 107–111.

Slemmons, D. B., 1967, Pliocene and Quaternary crustal movements of the Basin and Range province, U.S.A.: Osaka City Univ. Jour. Geosciences, v. 10, p. 91–103.

Smith, R. B., 1974, Seismicity and earthquake hazards of the Wasatch Front, Utah: Earthquake Inf. Bull., v. 6, p. 12–17.

——1977, Intraplate tectonics of the western North American plate: Tectonophysics, v. 37, p. 323–336.

Smith, R. B., Lindh, A., 1978, A compilation of fault-plane solutions of the Western United States, in Smith, R. B., and Eaton, G. P., eds., Cenozoic tectonics and regional geophysics of the western Cordillera: Geol. Soc. America Mem. 152 (this volume).

Smith R. B., and Pelton, J. R., 1977, Crustal uplift and its relationship to seismicity and heat flow at Yellowstone: EOS (Am. Geophys. Union Trans.), v. 58, p. 495.

Smith, R. B., and Sbar, M., 1974, Contemporary tectonics and seismicity of the Western United States with emphasis on the Intermountain seismic belt: Geol. Soc. America Bull., v. 85, p. 1205–1218.

Smith, R. B., Shuey, R. T., Freidline, R. O., Otis, R. M., and Alley, L. B., 1974a, Yellowstone hot spot: New magnetic and seismic evidence: Geology, v. 2, p. 451–455.

Smith, R. B., Winkler, P. L., Anderson, J. G., and Scholz, C. H., 1974b, Source mechanisms of microearthquakes associated with underground mines in eastern Utah: Seismol. Soc. America Bull., v. 64, p. 1295–1317.

Smith, R. B., Braile, L., and Keller, G. R., 1975, Crustal low velocity layers: Possible implications of high temperatures at the Basin Range–Colorado Plateau transition: Earth and Planetary Sci. Letters, v. 28, p. 197–204.

Smith, R. B., Pelton, J. R., and Love, D. L., 1976, Seismicity and the possibility of earthquake related landslides in the Teton-Gros Ventre-Jackson Hole area, Wyoming: Wyoming Univ. Contr. Geology, v. 2, p. 57–64.

Smith, R. B., Shuey, R. T., Pelton, J. R., and Bailey, J. T., 1977, Yellowstone hot spot: Contemporary tectonics and crustal properties from new earthquake and aeromagnetic data: Jour. Geophys. Research, v. 82, p. 3665–3676.

Snyder, W. S., Dickinson, W. R., and Silberman,

M. L., 1976, Tectonic implications of space-time patterns of Cenozoic magmatism in the Western United States: Earth and Planetary Sci. Letters, v. 32, p. 91–106.

Stevenson, P. R., 1976, Microearthquakes at Flathead Lake, Montana: A study using automatic earthquake processing: Seismol. Soc. America Bull., v. 66, p. 61–80.

Stewart, J. H., 1978, Basin and Range structure in western North America: A review, in Smith, R. B., and Eaton, G. P., eds., Cenozoic tectonics and regional geophysics of the western Cordillera: Geol. Soc. America Mem. 152 (this volume).

Stewart, J. H., and Carlson, J. E., 1978, Generalized maps showing distribution, lithology, and age of Cenozoic igneous rocks in the Western United States, in Smith, R. B., and Eaton, G. P., eds., Cenozoic tectonics and regional geophysics of the western Cordillera: Geol. Soc. America Mem. 152 (this volume).

Stewart, J. H., Albers, J. P., and Poole, F. G., 1968, Summary of regional evidence for right-lateral displacement in the western Great Basin: Geol. Soc. America Bull., v. 79, p. 1407–1413.

Stewart, J. H., Walker, G. W., and Kleinhampl, F. J., 1975, Oregon-Nevada lineament: Geology, v. 3, p. 265–268.

Suppe, J., Powell, C., and Berry, R., 1975, Regional topography, seismicity, Quaternary volcanism, and the present-day tectonics of the Western United States: Am. Jour. Sci., v. 275-A, p. 397–436.

Thompson, G. A., and Burke, D. B., 1974, Regional geophysics of the Basin and Range province: Earth and Planetary Sci. Ann. Rev., v. 2, p. 213–238.

Thompson, R. N., 1977, Columbia/Snake River-Yellowstone magmatism in the context of Western U.S.A. Cenozoic geodynamics: Tectonophysics, v. 39, p. 621–636.

Trimble, A., and Smith, R. B., 1975, Seismicity and contemporary tectonics of the Hebgen Lake-Yellowstone Park region: Jour. Geophys. Research, v. 80, p. 733–741.

Wallace, R. E., 1977, Profiles and ages of young fault scarps, north-central Nevada: Geol. Soc. America Bull., v. 88, p. 1267–1281.

Warren, D. H., and Healy, J. H., 1974, The crust in the conterminous United States, in the structure of the earth's crust, in Mueller, S., ed., The structure of the earth's crust based on seismic data: Elsevier Scientific Pub. Co., p. 203–213.

White, C. M., and McBirney, A. R., 1978, Some quantitative aspects of orogenic volcanism in the Oregon Cascades, in Smith, R. B., and Eaton, G. P., eds., Cenozoic tectonics and regional geophysics of the western Cordillera: Geol. Soc. America Mem. 152 (this volume).

Wright, L., 1976, Late Cenozoic fault patterns and stress fields in the Great Basin and westward displacement of the Sierra Nevada block: Geology, v. 4, p. 489–494.

York, J. E., and Helmberger, D. V., 1973, Low velocity zone variations in the southwestern United States: Jour. Geophys. Research, v. 78, p. 1883.

Manuscript Received by the Society August 15, 1977

Manuscript Accepted September 2, 1977

Geological Society of America
Memoir 152

7

Seismic evidence for the structure and Cenozoic tectonics of the Pacific Coast States

DAVID P. HILL
U.S. Geological Survey
345 Middlefield Road
Menlo Park, California 94025

ABSTRACT

Processes involving interactions at the local boundaries between lithosphere plates have played a dominant role in shaping the structure and Cenozoic tectonics of the Pacific Coast States. The distribution of earthquake hypocenters indicates that contemporary deformation between the Pacific and North American plates includes a zone several hundred kilometres wide about the San Andreas fault zone between the Salton Trough and Cape Mendocino. Brittle deformation of the lithosphere in this zone, however, is confined to the upper 15 to 20 km of the crust as indicated by maximum earthquake focal depths. Low seismic activity north of Cape Mendocino suggests that subduction of the Kula plate beneath the Cascade andecitic volcanoes has either ceased or is proceeding at a very low rate. Moderate earthquake activity with focal depths as great as 40 to 80 km beneath the Cape Mendocino and Puget Sound regions at either end of the Cascade Range may, however, be related to contemporary local subduction processes.

The gross crustal and upper-mantle structure based on seismic body wave data for the major tectonic provinces within the Pacific Coast States is characterized by diversity. Total crustal thickness ranges from 60 km beneath the Sierra Nevada to 25 km beneath the Columbia Plateau and 20 km beneath the Salton Trough. In general, however, crustal thickness correlates poorly with average topographic elevation, and the depth to and thickness of the intermediate layer (the lower portion of the crust with P-wave velocities between about 6.5 to 7.0 km/s) varies widely from one province to another. Yet, the gross crustal structure determined for most of the provinces satisfies an approximate condition of mass balance (isostacy) consistent with average gravity anomalies measured over the respective tectonic provinces. The exceptions, which include structures determined for the western Snake River Plain, Idaho, and the Salinian block of the central California Coast Ranges, point to the possibility of tectonically significant anomalies in simple velocity-density relations or complications in the velocity structure not resolved by available seismic data.

INTRODUCTION

Current concepts in plate tectonics picture the development of the western margin of North America as being dominated by interactions at the boundaries between the North American, Farallon, Pacific, and Kula plates (Atwater, 1970; Coney, this volume). Seismology provides critical evidence on the evolution of these interactions from crustal and upper-mantle structure and on contemporary tectonic processes along existing plate boundaries from the distribution of earthquake activity. This paper reviews both lines of seismological evidence in the Pacific Coast States region, which encompasses about one-sixth of the present boundary between the Pacific and North American plates.

Major advances in seismic studies of crustal and upper-mantle structure have been realized over the past 15 yr (Pakiser and Steinhart, 1964; Warren and Healy, 1973), although knowledge of the crustal structure within the Pacific Coast States is still far from adequate for resolving most of the fundamental tectonic problems. Reasonable models exist for the gross structure in central California from the western Basin and Range province to the Pacific coast and in parts of southern California. Elsewhere in the Pacific Coast States evidence for even the gross structure is at best sketchy. Similarly, although considerable progress has been made in defining the average upper-mantle structure beneath the Western United States, existing data are sparse regarding regional variations of such tectonically critical properties as the depth to and thickness of the asthenosphere.

Resolution of both spatial and temporal patterns of earthquake activity (seismicity) in the Pacific Coast States has continued to improve following the installation of regional seismograph networks by the University of California (Berkeley) in central California and the California Institute of Technology in southern California during the 1930s. Installation of the Long-Range Seismic Measurement (LRSM) stations and World-Wide Standard Stations in the early 1960s served to establish regional coverage for earthquakes in Washington and Oregon. More recently, the series of telemetered seismograph arrays, with station spacings between 10 to 20 km, installed along sections of the San Andreas fault zone in central and southern California and the Puget Sound region in Washington permits the systematic location of earthquake hypocenters to within 5 km or better for events with magnitudes as low as 2 occurring within the arrays.

Our present understanding of the seismicity in the Pacific Coast States, however, remains uneven. Seismograph arrays with dense station spacing tend to be concentrated in regions of high seismic activity, such as along the San Andreas fault system. The distribution of small earthquakes and their focal mechanism in regions of low seismic activity, such as the Cascades or volcanic plateaus in Oregon, remain largely unknown.

EARTHQUAKES AND ACTIVE PLATE BOUNDARIES

The active boundary between the Pacific and North American plates is defined by a persistent band of earthquake activity along the axis of the Gulf of California and the San Andreas transform fault system from the Salton Trough to Cape Mendocino. Beyond Cape Mendocino the seismic activity follows a sawtooth pattern along the Mendocino Fracture Zone, the Gorda Ridge, the Blanco Fracture Zone, and the Juan de Fuca Ridge (Atwater, 1970). These patterns are clearly evident in the location of earthquake epicenters for the years 1970 through 1974 plotted in Figure 7-1.

Details in the seismicity in Figure 7-1 are strongly influenced by the particular time interval chosen. Aftershocks of the 1971 magnitude 6.5 San Fernando earthquake, for example, form

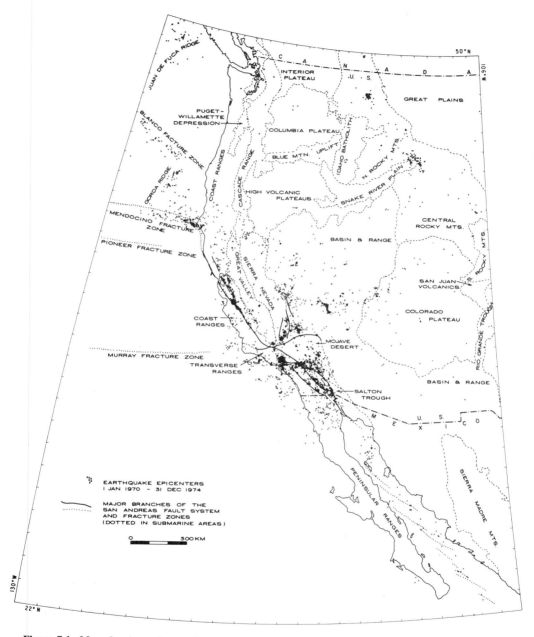

Figure 7-1. Map showing epicenter locations for earthquakes occurring between 1 January 1970 and 31 December 1974 in Western United States and northwestern Mexico. Major tectonic units are outlined by dashed lines; major branches of the San Andreas fault system are indicated by solid lines; submarine fracture zones and faults by dotted lines. Earthquake data are from National Earthquake Information Service tapes, which for this period include the catalogues from the Caltech network in southern California and the U.S. Geological Survey network in central California. Minimum magnitude for plotted earthquakes in these two regions is approximately 2; elsewhere minimum magnitudes range between 3 and 4. Compare with Figure 3 in Smith (this volume), which shows activity in the Intermountain seismic belt over an extended time period.

a conspicuous patch in the Transverse Ranges in southern California, while activity along the Intermountain seismic belt during the interval was relatively low (compare with Fig. 6-6 in Smith, this volume). Some of the more general, long-term characteristics of the seismicity pattern relevant to a discussion of structure of the lithosphere and tectonic relations in the Pacific Coast region are:

1. The deformation between plate boundaries is not confined to a single through-going fault; rather, it is distributed over a zone some 50 to 100 km wide on either side of the major branches of the San Andreas system. This is particularly apparent in southern California between the Transverse Ranges and the Salton Trough, where persistent seismic activity covers a 200- to 300-km-wide zone from the Mojave Desert to the Channel Islands (see Hileman and others, 1973). Nevertheless, during major earthquakes a major portion of the displacement between plates occurs on single fault surfaces. This is pointed up by displacements of as much as 7 m produced across the San Andreas fault by the magnitude 8 Fort Tejon (1857) and San Francisco (1906) earthquakes.

2. Earthquakes in California south of Cape Mendocino are confined almost entirely to the upper 15 km of the crust. Notable exceptions include the magnitude 6 Point Mugu earthquake (1973) located at a depth of 17 km (Ellsworth and others, 1973) and two small earthquakes (magnitude 1.5) located at depths of 40 km beneath Oroville in the western foothills of the Sierra Nevada (Marks and Lindh, 1978). In the vicinity of Cape Mendocino and the Puget Sound, however, earthquakes at depths of 30 to as much as 60 km are common (Seeber and others, 1970; Crosson, 1972; Smith, 1976).

3. The level of seismic activity between Cape Mendocino and the Puget Sound region to the north is low. In particular, the Cascade Range, in contrast to other chains of andesitic volcanoes around the Pacific basin, lacks a band of earthquake hypocenters dipping beneath the continent along an active Benioff zone. This suggests that there is at this time little if any relative motion between the Juan de Fuca and North American plates, even though the magnetic anomalies indicate that a displacement rate of 2.5 to 3 cm/yr must have continued to a time between 4 to 7 m.y. ago (Atwater, 1970).

4. Focal mechanism for earthquakes throughout the Pacific Coast States are generally consistent with relative right-lateral motion of the North American plate with respect to the Pacific plate in a direction parallel with the northwest trend of the San Andreas fault system. Notable exceptions to this pattern include the 1 August 1975 Oroville earthquake ($M = 5.7$) and the 60-km-deep, 1965 Puget Sound earthquake ($M = 6.5$), both of which had normal fault-plane solutions, suggesting that local deformation is dominantly an east-west extension.

5. The size distribution of earthquakes in both time and space provide a measure of the displacement history (seismic slip rate) along a fault zone in terms of the cumulative seismic moments (Hanks and others, 1975; Thatcher and others, 1975). (Seismic moment is defined as $M_0 = \mu dA$, where μ is the elastic shear modulus, and d is the average displacement on a fault surface of area A.) Applications of this technique to the San Jacinto fault, historically the most active branch of the San Andreas fault system at the magnitude 6 level, indicate an average seismic slip of 1.5 cm/yr since 1912 (Brune, 1968). The average slip rate on the San Jacinto fault over the past few thousand to a million years, based on geologic evidence, is 0.3 cm/yr (Sharp, 1967; Clark and others, 1972), and the displacement rate between the North American and Pacific plates, based on the average opening rate of the mouth of the Gulf of California over the past 4 m.y. is 6 cm/yr (Atwater, 1970). The differences in slip rates presumably reflect, in part at least, episodic movement on different branches of the San Andreas fault system, with perhaps some aseismic slip or creep that would not be detected by seismic methods.

CRUSTAL STRUCTURE

The primary evidence for the structure of the Earth's crust and upper mantle in the Pacific Coast States comes from long-range seismic refraction profiles using either explosions or regional earthquakes as sources. Additional evidence is provided by relative traveltime differences of P-waves from distant earthquakes or from surface-wave dispersion. Accordingly, structural models of the crust are presented in terms of the distribution of seismic compressional (P) wave velocities with depth in the Earth. The following two points deserve emphasis in this regard:

1. Most models of the crust are presented in terms of several homogeneous "layers" with discrete P-wave velocities separated by sharp boundaries. The simplicity of these models reflects the generally coarse spacing and limited resolution of the available data. Where detailed, high-resolution data exist, as in the Soviet Union (see, for example, Sollogub, 1972) or in Hardin County, Texas, where a deep crustal reflection survey was recently completed (Oliver and others, 1976), it is clear that the fine-scale structure of the crust and upper mantle is just as complicated as geologic structures observed in surface exposures.

2. Apart from the necessity of having reliable velocity models of the crust for the accurate location of earthquake hypocenters, the P- or S-wave velocity structure is in itself of interest only in that it provides critical information on the physical properties and composition of rocks in the crust and upper mantle. The crucial link is provided by laboratory measurements of elastic velocities in rocks thought to be samples of the deep crust and upper mantle brought to the surface as inclusions in volcanic eruptions or by major tectonic displacement. Velocity measurements on a variety of crustal and mantle rock types at confining pressures as high as 10 kb are available in the literature (for example, Press, 1966, or Christensen and Fountain, 1975). Significant new measurements at both pressures and temperatures appropriate for conditions in the crust and upper mantle are in progress in several laboratories (Peselnick and Stewart, 1975). These measurements make clear that the composition of the crust is considerably more varied (and complicated) than was suggested by earlier concepts in which the compositions of the 6.0 km/s "upper layer" and the 6.7 km/s "intermediate layer" were regarded as granitic and basaltic, respectively.

Locations of long-range crustal refraction profiles in the Pacific Coast States are given in Figure 7-2. The published interpretations of these profiles provide the basis for much of the material in this review. Also plotted in Figure 7-2 are the locations of several recently established profiles. Although the interpretations of these profiles are not complete, some of the preliminary indications of the new data will be mentioned where appropriate.

We will consider the crustal structure together with significant aspects of local seismicity, beginning with the Salton Trough in southern California, moving progressively northward, and concluding with the Columbia Plateau and western Snake River Plain in Washington and Idaho.

Salton Trough, Mojave Desert, and Transverse Ranges

The oblique opening of the Gulf of California resulting from crustal spreading of the East Pacific Rise is generally supposed to extend into the Salton Trough at least as far as the Salton Sea in a series of right-stepping transform-fault segments separated by small spreading centers (Lomnitz and others, 1970; Elders and others, 1972). This supposition is supported by detailed seismicity showing right-stepping offsets in the epicenter patterns from the Cerro Prieto to the Imperial faults in the vicinity of the Cerro Prieto geothermal field in Mexico

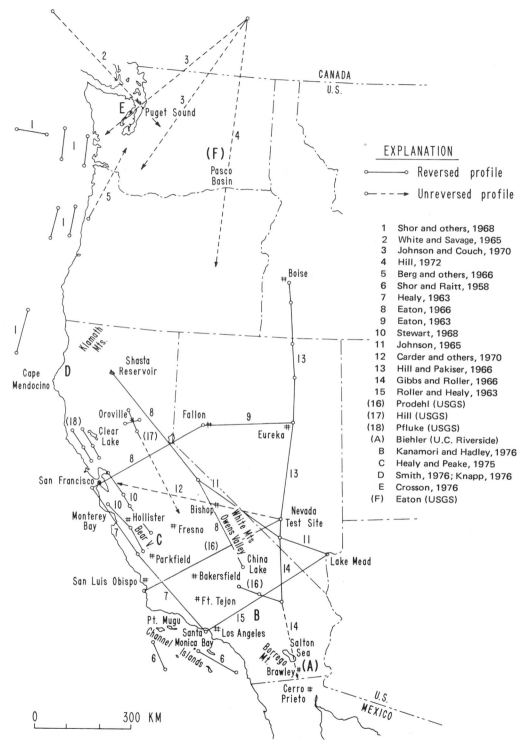

Figure 7-2. Index map showing location of seismic refraction profiles and regional seismic network studies based on local earthquake and quarry blast data. Unpublished results indicated by parentheses.

(Albores and others, 1977) and from the Imperial to the Brawley faults in the vicinity of the Brawley geothermal area in the Imperial Valley (Hill and others, 1975; Johnson and Hadley, 1976). A suggestion of this seismicity pattern can be seen in Figure 7-1.

The deep structural depression of the basement floor that forms the section of the Salton Trough between the Salton Sea and the International Boundary (the Imperial Valley) is filled by 6 km of late Tertiary and Quarternary alluvium and lake deposits. Results of seismic-refraction surveys in the Imperial Valley described by Biehler and others (1964) show that the P-wave velocities increase with depth in the sediments from about 2.0 to 4.5 km/s and that the P-wave velocity of the basement varies from 5.5 to 6.4 km/s. Analysis of gravity data over the Imperial Valley, together with unpublished results from additional seismic-refraction measurements, indicates that the crust is only about 20 km thick beneath the Imperial Valley (S. R. Biehler, 1976, oral commun.; Gibbs and Roller, 1966). On the basis of this same evidence, it appears that the crust probably thickens rather abruptly to approximately 30 km beneath the Peninsular Ranges and to between 26 and 28 km beneath the igneous and metamorphic terrane of the Mojave Desert. The deep structural depression and thin crust of the Imperial Valley are generally consistent with current concepts that the continental crust is being obliquely rifted apart by an active spreading center (Elders and others, 1972).

Hamilton's analysis (1970) of seismic refraction in the Borrego Mountain region in the Peninsular Ranges just west of the Imperial Valley indicates an upper crust with a 6.0-km/s upper layer underlain by a relatively shallow 7.1-km/s intermediate layer at a depth of 14 km. Early first arrivals on unreversed profile in the Peninsular Ranges of southern California by Hadley and Kanamori (1977) also suggest a shallow depth (about 15 km) to intermediate velocities (6.7 km/s), and their data suggest a total crustal thickness of about 32 km.

The Mojave Desert is generally regarded as a relatively stable crustal block between the Garlock fault on the north and the main San Andreas fault system on the southwest. Widespread seismic activity in the south-central section of the Mojave Desert, however, indicates that significant deformation is occurring in this section of the block. Focal mechanisms of these earthquakes show strike-slip patterns with right-lateral slip on faults striking N15°W to N45°W, subparallel to the San Andreas fault, or left-lateral striking N70°E to N90°E, subparallel to either the Garlock fault on the north or the Transverse Range faults on the south (G. Fuis, 1976, written commun).

The crust of the Mojave Desert is between 26 and 32 km thick with an upper mantle P-wave velocity of 7.8 km/s (Roller and Healy, 1963; Kanamori and Hadley, 1975). The data obtained by Kanamori and Hadley (1975) from recordings of quarry blasts in southern California indicate a relatively uniform structure within the crust that can be characterized by a 5.5-km/s layer 4 km thick over a 6.3-km/s layer about 23 km thick, with the possibility of a 5-km thick, 6.8-km/s layer at the base of the crust. This compares with results obtained by Roller and Healy (1963) from reversed seismic-refraction profiles between Santa Monica Bay and Lake Mead that show a velocity of 6.1 km/s from near the surface to about 20 km, and a velocity of 6.8 to 7.0 km/s at depths between 20 and 26 km. In both cases, evidence for the 6.8- to 7.0-km/s layer at the base of the crust is based on secondary arrivals, and there remains some uncertainty regarding both the proper identification of these later wave patterns and the structure of the lower crust. The crustal structure of the Mojave Desert is generally similar to that of the southern Basin and Range province (Johnson, 1965; Langston and Helmberger, 1974), although P-wave velocities of 6.1 to 6.4 km/s in the upper crust of the Mojave Desert are somewhat higher than the 6.0-km/s value reported in the Basin and Range province.

The east-trending Transverse Ranges, which cut across the dominant northwest structural grain in California, are a result of north-south crustal compression and underthrusting associated with the westward bend in the San Andreas fault between San Bernardino and the Garlock fault (Anderson, 1971). Focal mechanisms of the magnitude 6.6 San Fernando earthquake (1971) and magnitude 6.0 Point Mugu earthquake (1973) show predominantly thrust movement on north-dipping planes reflecting this mode of deformation. Yet, maximum focal depths determined for earthquakes in this region of thrusting and crustal shortening are only marginally deeper (17 to 18 km) than elsewhere on the San Andreas fault (10 to 15 km), where the two sides of the crust slip past each other in pure strike-slip motion (Friedman and others, 1976; Fuis and others, 1977).

Roller and Healy (1963) suggested that, on the basis of reflected arrivals, the crust beneath the Transverse Ranges may be as much as 40 km thick. Kanamori and Hadley (1975), however, see no clear evidence for an increase in crustal thickness under either the Transverse Ranges or the San Bernardino Mountains in their analysis of regional traveltime data from quarry blasts and earthquakes. Rather, they suggest that the crustal structure in the Transverse Ranges is transitional between the Mojave Desert and Peninsular Ranges structure and that an 8.3-km/s body underlies the Transverse Ranges within the mantle at a depth of 40 km (Hadley and Kanamori, 1977).

Sierra Nevada and Great Valley

The east flank of the Sierra Nevada Range provides impressive evidence for displacement on steeply dipping normal faults. Structural relations with dated lava flows and glacial deposits indicate that this faulting began approximately 2 m.y. ago and has continued through recent time (Christensen, 1966). Contemporary seismic activity along the eastern flank of the Sierra Nevada (see Fig. 7-1) appears to be dominated by a stress field similar to that of the San Andreas system. Surface displacements of the magnitude 8 Owens Valley earthquake (1872) were predominantly strike slip, with a small component of normal faulting, and although there remains some uncertainty as to the sense of displacement on the fault, most of the evidence points to right-lateral motion (Bateman, 1961). Fault-plane solutions obtained by Pitt and Steeples (1975) for earthquakes near the northern end of Owens Valley (the northern extent of the continuous band of seismicity along the eastern flank of the Sierras in Fig. 7-1) also show predominately strike-slip solutions with right-lateral nodal planes subparallel with the Sierra Nevada frontal faults. One earthquake in their study, however, had a clear dip-slip mechanism with relative northeast extension. All earthquakes in their study were located with focal depths less than 18 km.

Seismic activity in the Great Valley is characterized by localized groups of earthquakes near the margins (Fig. 7-1). The magnitude 5.9 Oroville earthquake of 1 August 1975 (Bufe and others, 1976; Lahr and others, 1976) is a well-studied example of the seismic activity in the Sierra foothills at the eastern margin of the Great Valley. This earthquake and its aftershocks define a normal fault dipping steeply to the west (valley-side down) extending from the surface to a depth of about 10 km. Two small earthquakes ($M = 1.5$) late in the aftershock sequence, however, are reliably located at a depth of 40 ± 3 km (Marks and Lindh, 1978). These are the deepest earthquakes located in California south of the Mendocino region.

The P-wave velocity structure of the Great Valley, Sierra Nevada, and western Basin and Range is illustrated by transverse and longitudinal cross sections in Figures 7-3 and 7-4, respectively. These crustal models are based on Eaton's (1966) analysis of seismic-refraction profiles between shot points near San Francisco, San Luis Obispo, Shasta Reservoir, China Lake, and Fallon, Nevada, (Fig. 7-2; Eaton, 1963, 1966; Bateman and Eaton, 1967). The results show that the crust thickens from some 30 km beneath the western Basin and Range province to more than 50 km beneath the high Sierra Nevada, Owens Valley, and White Mountain region (also see Johnson, 1965) and gradually thins to 20 km or less beneath the western margin of the Great Valley. Much of the crustal root beneath the Sierra Nevada is accounted for by an increase in thickness of the 6.7- to 6.9-km/s intermediate layer. Some evidence exists for an increase in P-wave velocity from 6.0 to 6.4 km/s at approximately 10 km beneath the central Sierra Nevada. Carder and others (1970) interpreted an unreversed profile from the Nevada Test Site to Monterey Bay as indicating no substantial crustal thickening beneath the Sierra Nevada. Pakiser (1974), however, has demonstrated that minimum P-wave traveltime paths for this profile are, in fact, consistent with a structure similar to that in Figure 7-3.

Bateman and Eaton (1967) argued that the extensive Sierran root, which coincides more closely with the batholith than with the topography of the range, was formed contemporaneously with the batholith in a large Mesozoic synclinorium rather than during late Cenozoic uplift of the range as suggested by Christensen (1966) or Hamilton (1969). Resolution of this question and the nature of the process leading to the late Cenozoic uplift of the Sierra Nevada depends on a more complete definition of the structure and composition of the underlying deep crust and upper mantle. Some constraint on the composition of the lower crust is provided by gravity data, which require densities of at least $3.0 \text{ g}/\text{cm}^3$ at depths below 25 to 30 km (Thompson and Talwani, 1964; Bateman and Eaton, 1967; Oliver, 1977). This high density suggests that the intermediate velocities in the lower crust represent mafic rocks such as amphibolites or metagabbros (Bateman and Eaton, 1967) rather than less dense and more acidic granulites, which are likely candidates for the intermediate layer in many other regions (Christensen and Fountain, 1975).

The crustal structure between the southern end of the Cascade Range and the Sierra Nevada batholith (the first 200 km south of Shasta Reservoir in Fig. 7-4) is distinctly different from that under the high Sierra Nevada. Here the depth to the top of the intermediate layer, which has a P-wave velocity of 6.8 km/s, is less than 10 km beneath a 5.9-km/s upper crust. The depth to the base of the crust is not well resolved but appears to be 35 to 40 km, which is significantly shallower than the Sierran root.

Preliminary analysis of an unreversed refraction profile (obtained in 1976 by the U.S. Geological Survey) that extends some 200 km south of Oroville, California (see Fig. 7-2), together with

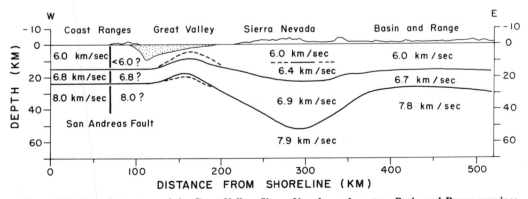

Figure 7-3. Crustal structure of the Great Valley, Sierra Nevada, and western Basin and Range province along a section passing through Fresno, California, and approximately 20 km north of Bishop, California, based on Eaton's interpretation of seismic refraction measurements (Eaton, 1963, 1966; Bateman and Eaton, 1967). Solid and dashed lines show alternate interpretations of eastern Great Valley structure.

the band of high magnetic and gravity anomalies along the eastern margin of the Great Valley (Cady, 1975), indicates that the greenstone belt of the Sierra foothills is also underlain at shallow depth by intermediate *P*-wave velocities (Fig. 7-3). Accordingly, the northern Sierra Nevada structure in Figure 7-4 may be part of a continuous stretch of Mesozoic metamorphic crust from the foothills greeenstone belt through the northern Sierras to the Klamath Mountains, which locally are covered by a veneer of Cenozoic lava flows. Hamilton (1969) has suggested that the northern Sierra Nevada structure represents a much younger Cenozoic volcanic crust common to the Cascades and high volcanic plateaus formed by massive crustal rifting. The *P*-wave velocity of 5.9 km/s in the upper crust is significantly higher than velocities of 5.2 km/s associated with 5- to 10-km-thick accumulations of Cenozoic lava flows as in the Columbia Plateau or the Snake River Plain.

Unfortunately, the available data provide poor resolution of the deep crustal structure beneath the Great Valley and its junction with the Coast Ranges to the west.

Central California Coast Ranges and the San Andreas Fault Zone

The San Andreas fault zone cuts through the central California Coast Ranges at low angle, juxtaposing a basement of Mesozoic eugeosynclinal rocks of the Franciscan sequence on the east against Mesozoic crystalline basement on the west. Seismicity in this region is concentrated along the San Andreas fault and its major branches along the east side of San Francisco Bay, the Hayward and Calaveras faults (Wesson and others, 1973). A diffuse pattern of earthquakes is evident on either side of these major faults in the central Coast Ranges (Fig. 7-1). Earthquakes in the Coast Ranges almost invariably have strike-slip focal mechanisms consistent with right-lateral slip on planes subparallel to the San Andreas fault and focal depths less than 15 km. The most intense seismic activity in this region over the past 10 yr has been concentrated along the segment of the San Andreas fault between Parkfield and Hollister (Eaton and others, 1970; Ellsworth, 1975; Fig. 7-1).

Early work by Byerly (1939) established that the basement *P*-wave velocity (P_g) east of the San Andreas fault zone in Franciscan rocks is about 5.7 km/s. This is significantly lower than the values of 6.0 km/s commonly found for P_g velocities in most other continental areas. More recent work (Eaton, 1963, 1966; Stewart, 1968) has substantiated this low *P*-wave velocity for Franciscan basement rocks.

Some general aspects of the crustal velocity structure of the central California Coast Ranges

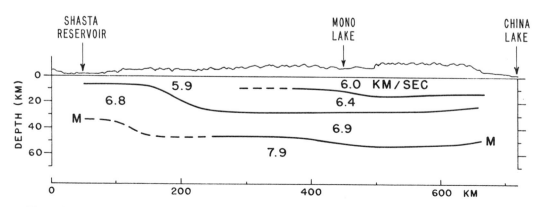

Figure 7-4. Crustal structure of the Sierra Nevada along longitudinal section from Shasta Reservoir, California, to China Lake, California (Eaton, 1966).

are schematically illustrated in Figure 7-5. The upper 15 km of the structure is adapted from a detailed study by Healy and Peake (1975) of the Bear Valley region immediately south of the point where the San Andreas fault zone splays into its several branches extending through the San Francisco Bay area (Figs. 7-1, 7-2). The lower crustal structure is adapted from the analysis of regional profiles by Healy (1963), Eaton (1966), and Stewart (1968). Work in the same general region by Stewart and O'Neill (1972), Wesson and others (1973), Mayer-Rosa (1973), Boore and Hill (1973), Aki and Lee (1976), and McEvilly and Clymer (1975) emphasizes the complicated lateral variations in the structure of the upper crust along the San Andreas fault zone in central California.

Several aspects of the structure summarized in Figure 7-5 are noteworthy in terms of current patterns of seismic activity and tectonic processes along the San Andreas fault zone in the central California Coast Ranges:

1. Rocks in the rift zones between major branches of the San Andreas fault have exceptionally low seismic velocities extending to depths of 8 km or more (Mayer-Rosa, 1973; Healy and Peake, 1975). These low velocities are thought to be largely due to sedimentary fill and the highly fractured state of rocks in the rift zones. The possibility remains, however, that factors such as high fluid pressures or dilatancy induced by high tectonic stress differences may contribute to these low velocities, particularly at depths greater than 6 km where overburden pressures exceed 2 kb.

2. P-wave velocities in the upper 10 km of the crust are systematically lower in the Franciscan terrane east of the fault than in the crystalline terrane of the Salinian block west of the fault. The P_g (basement) velocities of 6.0 to 6.4 km/s west of the fault are typical of granitic

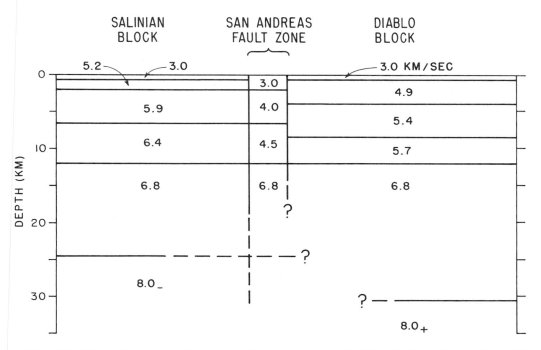

Figure 7-5. Schematic representation of crustal structure in central California Coast Ranges. Upper 15 km is based on an interpretation by Healy and Peake (1975) of data in the vicinity of the San Andreas fault zone in the Bear Valley region about 40 km south of Hollister. Deeper structure is based on analysis of refraction profiles by Healy (1963), Eaton (1966), and Stewart (1968).

rocks, and the 5.5 to 5.7 km/s P_g-velocities east of the fault are consistent with recent laboratory measurements of ultrasonic velocities in samples of Franciscan graywackes (Stewart and Peselnick, 1977). The velocity structure of the top 3 to 4 km is variable and more complicated than suggested by the schematic representation in Figure 7-5. Details of the shallow velocity structure are governed by such factors as the local thickness of the Mesozoic Great Valley thrust sheet overlying the Franciscan basement west of the fault and local thickness of Tertiary sedimentary and volcanic rocks on both sides of the fault (Irwin and Barnes, 1975).

3. The strong variations in P-wave velocities across the fault zone make accurate location of earthquake hypocenters difficult. Routine hypocenter-location methods, which rely on plane-layered velocity models, systematically place the hypocenters on vertical planes parallel with, but offset on either side of, the fault zone by as much as 3 to 5 km (see Mayer-Rosa, 1973; Ellsworth, 1975; Fig. 7-1). Efforts to account for the lateral variations in velocity across the fault zone in locating earthquakes in the Bear Valley region indicate that the hypocenters may actually lie within the fault zone (Boore and Hill, 1973). Techniques for handling lateral variations in velocity structure are not easily incorporated into routine hypocenter-location programs, although some progress is being made in this regard (Aki and Lee, 1976). Meanwhile, the precise locations of earthquake hypocenters relative to the fault zone remain uncertain by several kilometres.

4. Available evidence suggests that the depth to the 6.8-km/s layer may be roughly the same (12 to 15 km) on both sides of the San Andreas fault. The maximum depth for earthquakes in the Coast Ranges corresponds with the top of this layer. Although the change in composition at this depth may account for the absence of deeper earthquakes, laboratory studies show that factors such as high fluid pressures or temperatures are likely to be more important than composition in controlling the transition from failure as a series of stick-slip events (earthquakes) to stable sliding (Brace and Byerlee, 1970; Sommers and Byerlee, 1977). Moreover, the composition of the intermediate layer may be different on either side of the fault. Bailey and others (1964) and Ernst (1965) suggest that the intermediate layer in the Franciscan terrane consists of oceanic crust thickened by imbricate thrusting and folding during subduction. The intermediate layer beneath the Salinian block could consist of relatively acidic granulites (Christensen and Fountain, 1975) or perhaps more mafic amphibolites and metagabbros thought to form the lower crust beneath the Sierra Nevada.

5. On the basis of data from the relative delays of P waves from distant earthquakes recorded on seismograph networks in central California, much of the increase in crustal thickness from 23 to 25 km beneath the Salinian block to 30 km beneath the Diablo block occurs in the general vicinity of the San Andreas fault zone (McEvilly, 1966; Kind, 1972; Husebye and others, 1976). Preliminary analysis of data from the 1974 deep-reflection survey on either side of the San Andreas fault in Bear Valley, however, indicates a local crustal thickness of 24 km, with no evidence for significant change in thickness across the fault zone itself (McEvilly and Clymer, 1975). The deep-reflection data also indicate that the Mohorovičić discontinuity is a 2- to 4-km-thick transition zone beneath the granitic crust and a somewhat sharper boundary beneath the Franciscan crust.

6. The average P-wave velocity in the uppermost mantle (the P_n-velocity) beneath the central Coast Ranges of 8.0 km/s (Kind, 1972) is slightly higher than the 7.8- to 7.9-km/s values found under most areas in the Western United States. P_n-velocities of 8.0 to 8.4 km/s are typical of tectonically stable regions with low to normal heat flow of 1.5 HFU or less (Roy and others, 1972). Relatively high heat-flow values (2.0 HFU) are found in the central Coast Ranges (Lachenbruch and Sass, 1973), however, and in this respect P_n-velocities of 8.0 km/s appear anomalous. Lachenbruch and Sass (1973) have suggested that the high heat flow can

be accounted for by a "normal" mantle heat flow (0.8 HFU) entering the lower crust, the excess heat being generated by deformation in the seismogenic upper 15 km of the crust spread over the 100-km-wide band of seismicity that includes the San Andreas fault zone (Fig. 7-1).

7. Analysis of both explosion and earthquake data from regional sources suggests that P_n-velocities east of the San Andreas fault are somewhat lower than to the west (Kind, 1972). The work by Husebye and others (1976) on the systematic inversion of P-wave teleseismic sources provide preliminary indications that a velocity difference of 2% to 4% may persist to depths of 50 to 80 km into the mantle beneath the San Andreas fault zone.

Northern California Coast, Cape Mendocino

North of San Francisco Bay, the San Andreas fault parallels the coast, running just onshore as far as Point Arena and just offshore beyond to Cape Mendocino. Here the Coast Ranges lie almost entirely east of the fault in Franciscan terrane. As is evident in Figure 7-1, the main branch of the San Andreas fault in this region, which broke in 1906, is now seismically quiet. Most of the seismic activity in the Coast Ranges north of the San Francisco Bay area is associated with the Healdsberg fault and with a zone of diffuse epicenters extending to the east and north. Data from a recently installed seismic network in the Geysers–Clear Lake area 100 km north of San Francisco shows persistent microearthquake activity in the Geysers geothermal area (Bufe and Lester, 1975). Little is known about the crustal structure of the Coast Ranges north of the bay area; a moderate explosion seismology program carried out by the U.S. Geological Survey in September 1976 in the Geysers–Clear Lake area should begin to resolve the structure of at least the upper crust here (see entry 18 in Fig. 7-2).

The level of seismic activity is relatively high in the Cape Mendocino region where the San Andreas fault apparently makes a sharp bend seaward into the Mendocino Fracture Zone. Studies of the seismic activity in this region by Seeber and others (1970) and Smith (1976) using temporary arrays of seismic stations show a bimodal depth distribution of earthquakes with a shallow concentration of activity from roughly 0 to 10 km (similar to normal depth range for earthquakes on the San Andreas fault to the south) and a deeper zone from 15 to 35 km. Focal mechanisms indicate predominately right-lateral, strike-slip displacement on northwest-trending fault planes that radiate from Cape Mendocino northwest into the Gorda Basin; earthquakes located on the Gorda escarpment (western end of the Mendocino Fracture Zone) show both right-lateral, strike-slip, and thrust solutions on planes striking nearly east-west (Simila and others, 1975; Seeber and others, 1970; Bolt and others, 1968; McEvilly, 1968). Smith (1976) found evidence in focal mechanisms and aftershock distributions for earthquakes north of Cape Mendocino that indicates left-lateral, strike-slip displacement on fault planes that cut across the regional structural trend in a northeast direction. These focal mechanisms are generally consistent with a predominant north-south compression in this tectonically complex region where the Gorda Basin is evidently undergoing internal deformation while being obliquely thrust beneath the North American and Pacific plates (McEvilly, 1968; Seeber and others, 1970; Silver, this volume).

An initial effort to infer the crustal structure in the Cape Mendocino area by Knapp (1976), using data from local earthquakes recorded on a 16-station array, suggests a general similarity to the crustal structure of the Franciscan terrane east of the San Andreas fault in central California. In particular, he found evidence for a P_g-velocity of 5.8 km/s, an intermediate layer with a P-wave velocity of 7.0 km/s at a depth of 17 km, a total crustal thickness of about 25 km, and a P_n-velocity of 7.8 km/s

Coast Ranges, Cascade Range, and the Volcanic Plateaus of Oregon and Southern Washington

The chain of andesitic volcanoes forming the Cascade Range separates Tertiary eugeosynclinal rocks of the Coast Ranges on the west from the extensive high-alumina basalt flows of the volcanic plateaus of the east. Considering the evidence for intense Cenozoic volcanic activity in the Cascades (White and McBirney, this volume) the level of seismic activity in Oregon and southern Washington is surprisingly low (see Fig. 7-1). This low seismic activity, together with the waning volcanic activity, however, provides strong evidence that the Juan de Fuca plate is no longer being actively subducted beneath the North American plate along the Oregon-Washington coast.

Data on the crustal structure in Oregon and southern Washington are sparse. Dehlinger and others (1965) used local earthquakes recorded on as many as 15 seismograph stations located throughout the Pacific Northwest to establish regional traveltime curves. On the basis of these traveltime curves, they concluded that the crust west of the Cascades is 25 to 30 km thick, with an upper mantle P_n velocity of about 7.8 km/s; east of the Cascades the crust is 35 to 40 km thick with a P_n-velocity of about 8.0 km/s. The velocity structure within the crust is not resolved from these traveltime curves, although there is evidence for P-wave velocities of about 6.6 km/s in the lower crust on both sides of the Cascade Range. Sparse data along an unreversed profile described by Berg and others (1966) suggest a depth of 10 km or less to intermediate velocities (6.6 to 7.4 km/s) beneath the Coast Ranges in Oregon and southern Washington. McCollum and Crosson (1975), using data recorded on a more extensive set of regional stations, also found evidence for a lower P_n-velocity west of the Cascades (7.7 km/s) than to the east (8.1 km/s). Results from a series of marine refraction profiles off the Washington and Oregon coasts (Shor and others, 1968) indicate that low P_n-velocities (7.7 to 7.9 km/s) also underlie the thin oceanic crust (8 to 9 km thick) of the Gorda plate. The crust of the Pacific plate south of the Mendocino Fracture Zone and west of the Juan de Fuca rise appears more typically oceanic with a 10- to 12-km thickness and P_n-velocities of 8.1 to 8.3 km/s (Shor and others, 1968; Dehlinger and others, 1965).

Limited seismic-refraction data from two unreversed profiles across the northern Cascades (Johnson and Couch, 1970), together with the regional P_n data (McCollum and Crosson, 1975), indicate that the Cascade Mountains lack a significant crustal root. This is consistent with analysis of the gravity data over individual Cascade volcanoes (see Pakiser, 1964; LaFehr, 1965), which suggests that the topographic load of the volcanoes is to a large extent locally compensated by mass deficiencies in the upper 10 km of the crust.

Puget Sound

The Puget Sound in western Washington lies at the northern end of the Puget-Williamette structural depression between the Tertiary eugeosynclinal rocks forming the Coast Ranges on the west and the andesitic volcanoes forming the Cascade Range on the east. Compared with the rest of the Pacific Northwest, seismic activity in this region is relatively high (see Fig. 7-1). The destructive earthquakes of magnitude 7.1 in 1949 and 6.5 in 1965 at depths between 60 and 70 km beneath the Puget Sound region served to emphasize this activity and to stimulate interest in the tectonic processes operating in this region.

The resolution afforded by the seismograph network installed in the Puget Sound region by the University of Washington in 1970 clearly shows that earthquakes here occur at depths as great as 80 km (Crosson, 1972 1976). This is considerably deeper than maximum focal depths of earthquakes elsewhere in the conterminous United States. Figure 7-6 summarizes

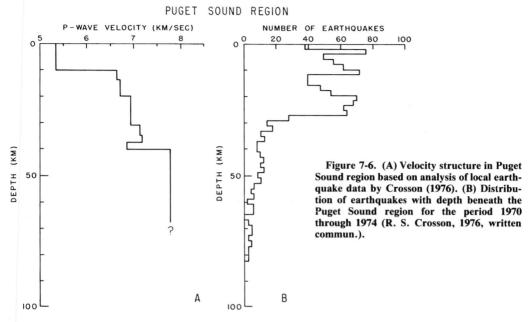

Figure 7-6. (A) Velocity structure in Puget Sound region based on analysis of local earthquake data by Crosson (1976). (B) Distribution of earthquakes with depth beneath the Puget Sound region for the period 1970 through 1974 (R. S. Crosson, 1976, written commun.).

the distribution of earthquakes with depth beneath the Puget Sound region for the interval 1970 to 1974 (Crosson, 1972, and 1976, written commun.). Although most earthquakes occurred at depths less than 40 km during this period, the magnitude of these shallower events are small (magnitude 4 or less), and a major proportion of the elastic energy released during this century occurred at depths of 60 to 70 km, with the magnitude 6.5 and 7.1 earthquakes in 1965 and 1949 (Crosson, 1972). The hypocenters of earthquakes located within the Puget Sound network are scattered throughout the area and show no clear lineations with recognized structures (Crosson, 1972).

Fault-plane solutions determined for earthquakes in the upper 20 km of the crust show both thrust and strike-slip mechanisms consistent with north-south compression (Crosson, 1972; and 1976, written commun.). Composite solutions for earthquakes in the lower crust and upper mantle, as well as solutions for the 60-km-deep 1965 earthquake, show normal mechanisms on north-striking planes consistent with relative east-west extension (R. S. Crosson, 1976, written commun.). The apparent change in orientation of the contemporary greatest principal stress from north-south to vertical at a depth of about 20 km is particularly intriguing. One possible explanation is that the deeper stress field may be related to the gravitational sinking of a residual section of the Juan de Fuca plate beneath the Puget Sound region (Isacks and Molnar, 1971). Further resolution of the distribution of earthquake hypocenters and focal mechanisms will be required to distinguish between this and other possibilities.

Crosson (1976) has used data recorded on the Puget Sound seismograph network from local earthquakes with a span of focal depths between 5 to 50 km to infer the velocity model illustrated by the solid line in Figure 7-6A. This plane-layer model has a 40-km-thick crust with a shallow intermediate layer beginning with 6.6 km/s at 10 km and increasing to 7.1 km/s at about 38 km. Crosson found evidence for a slight low-velocity zone at the base of the crust and a 7.8 km/s P-wave velocity in the uppermost mantle. Langston and Blum (1977) have proposed a different model for the crust, based on their analysis of wave forms from the 1965 Puget Sound earthquake recorded at teleseismic distances. Their model has an exceptionally thin crust (only 13 km thick) and a pronounced low-velocity zone in the mantle between depths of 40 to 55 km.

Crosson's model is generally consistent with regional traveltime curves across the Puget Sound region (White and Savage, 1965; Berg and others, 1966), which show apparent velocities of about 6.7 km/s to distances of 300 km, implying a thick intermediate layer. These regional traveltime curves are unreversed, however, and a more complete resolution of the structure in this region, where gravity data suggest strong lateral structural variations (Danes and others, 1965), will require a series of well-placed, reversed refraction profiles.

Columbia Plateau and Western Snake River Plain

East of the Cascade Range in Washington and northern Oregon lie the extensive flood basalts of the Columbia Plateau. These massive tholeiitic basalt flows of Miocene to early Pliocene age lap against the Paleozoic granitic and metamorphic terrane to the north, the Idaho batholith to the east, and the metamorphic core of the Blue Mountain uplift to the south (Waters, 1962; Swanson, 1967). The Pliocene-Pleistocene flood basalts of the Snake River Plain to the southeast are commonly included with the Columbia Plateau as part of a major flood-basalt geologic province.

Data on seismic activity in the Columbia Plateau has been provided by a network of seismic stations installed by the U.S. Geological Survey in the central section of the plateau (the Pasco Basin area) in 1969. The number of stations in this network was gradually increased from 6 in 1969 to 24 in 1974. This network reveals the rather persistent occurrence of small earthquakes (maximum magnitude 4) at depths of 5 km or less throughout the Pasco Basin (see Malone and others, 1975). Focal mechanisms of some of the larger events (magnitudes 3 to 4) have thrust fault-plane solutions consistent with north-south compression. The network also reveals a diffuse north trend of small earthquakes with focal depths of 18 to 25 km beneath the western margin of the basin. These earthquakes also have thrust focal mechanisms consistent with north-south compression (M. Pitt, 1976, oral commun.).

J. P. Eaton (1976, oral commun.), in analyzing data from regional quarry blasts recorded on the Pasco Basin seismograph network to obtain a crustal model for the central part of the Columbia Plateau, found that the crustal structure can best be approximated by a 5.2-km/s layer over a 6.1-km/s layer, with the thickness of the 5.2 km/s material increasing from 5 km beneath the margins of the Pasco Basin to about 10 km beneath the center of the basin. These data also indicate a high P_n-velocity (between 8.0 to 8.2 km/s) and a minimum crustal thickness of about 25 km beneath the center of the Pasco Basin. Data from a 600-km-long unreversed refraction profile extending due south from the Canadian border (the shot point was 200 km north of the border) across the Columbia Plateau in eastern Oregon (Hill, 1972), together with the constraints provided by Eaton's analysis of the quarry blast data, indicate that the crust thins from 35 km beneath the granitic terrane in northeastern Washington to about 25 km beneath the Pasco Basin, then thickens again farther south beneath the Blue Mountain uplift (see Fig. 7-7).

The bimodal basalt-rhyolite volcanism of the Snake River Plain is younger (Pliocene and Pleistocene), less voluminous, and compositionally distinct from the Columbia River Basalt Group (see Armstrong, this volume; Christiansen and McKee, this volume). Figure 7-8 illustrates the crustal structure of the western Snake River Plain in terms of two independent interpretations of a series of reversed seismic-refraction profiles between Eureka, Nevada, and Boise, Idaho (Hill and Pakiser, 1966; Prodehl, 1970). These two interpretations serve to emphasize the differences in detailed velocity structure that can result from different assumptions and analysis

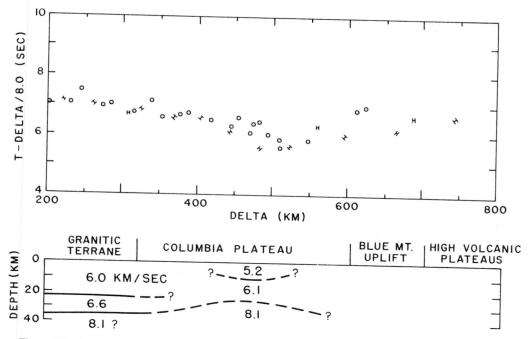

Figure 7-7. Crustal structual of the Columbia Plateau along north-south profile from the Canadian border into east Oregon (see Fig. 7-2). Structure within the Columbia Plateau is constrained by Eaton's analysis of quarry blasts recorded on the Pasco Basin seismograph network (J. P. Eaton, 1976, oral commun.). Upper plot is reduced traveltime curve of first arrivals (P_n) on which variations in structure between granitic terrane of northern Washington and Columbia Plateau is based (Hill, 1972). Symbols O and H indicate the two types of instrumentation used to record the data (see Hill, 1972).

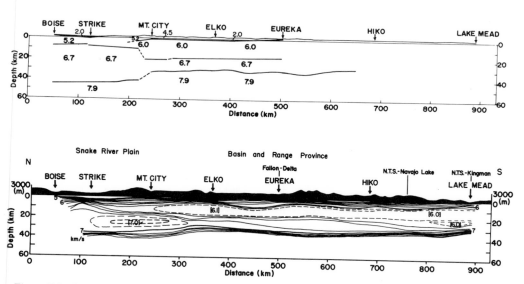

Figure 7-8. Crustal structure of the western Snake River Plain and northern Basin and Range province comparing two independent interpretations of the same data set. Upper section shows boundaries between "layers" of assumed uniform P-wave velocity (Hill and Pakiser, 1966). Lower section shows contours of equal P-wave velocity at 0.2 km/s intervals assuming a continuous variation in velocity with depth (Prodehl, 1970).

techniques applied to the same data set. Both interpretations, however, show generally close agreement for the gross structure.

In particular, the crust under the western Snake River Plain is 40 to 45 km thick, with a thick 6.6- to 7.0-km/s intermediate layer and a thin, low-velocity (5.2 km/s) upper crust. The upper mantle P-wave velocity is 7.9 km/s. This velocity structure stands in sharp contrast to the thin, low-velocity crust of the Columbia Plateau (see Fig. 7-8). Both the Columbia Plateau and western Snake River Plain are commonly regarded as resulting from massive crustal rifting (Hamilton and Myers, 1966; Hill and Pakiser, 1966); yet their structural differences indicate that they either developed through different tectonic processes or represent two stages in the development of similar tectonic processes.

UPPER–MANTLE STRUCTURE

The variations in P_n-velocities described above indicate that some of the complexity in crustal structure of the Pacific Coast States extends into the upper mantle. Generally sparse data, however, provide only a few tantalizing hints on possible structural variations in the upper mantle and their relation to tectonic processes.

A reasonable picture of the average velocity structure of the upper mantle beneath the Western United States has been developed by Helmberger (1973) and Wiggins and Helmberger (1973). They have relied primarily on long-period body-wave data recorded on LRSM (Long Range Seismic Measurement) and WWSN (World-Wide Seismic Station Network) stations from regional earthquakes and Nevada Test Site explosions. Because most of these stations are located east of long 118°W, the data provide only a weak sampling of the upper mantle beneath the Pacific Coast States.

Their results, summarized in Figure 7-9, show a pronounced low-velocity zone (LVZ) at depths between roughly 80 and 150 km, followed by a rather smooth increase in velocity with depth to at least 200 km. Attenuation of seismic waves due to the partial loss of elastic energy as heat (anelasticity) tends to be more pronounced in the LVZ than other sections of the upper mantle. Helmberger (1973) found evidence that the LVZ beneath the Western United States strongly attenuates seismic body waves ($Q \sim 50$) relative to the surrounding mantle (Q between 500 and 2,000). (Q is proportional to the maximum elastic energy in a seismic wave divided by the fraction of elastic energy loss to heat during one cycle of oscillation.) A low Q (high attenuation) in the low-velocity zone is consistent with the generally accepted assumption that the upper mantle LVZ is a result of a small percentage of partial melting (Anderson and Sammis, 1970), and that it is coincident with the weak layer (asthenosphere) over which the lithospheric plates move. The configuration of the LVZ is thus undoubtedly related to regional tectonic processes, and knowledge of lateral variations in the LVZ from the average structure shown in Figure 7-9 becomes important.

In this regard, York and Helmberger (1973) have used the traveltime difference between the P wave reflected from the bottom of the LVZ and the PL wave (a guided wave propagating in the crust with a relatively stable velocity between 4.8 and 5.2 km/s) to provide some indication of variations in the thickness of the LVZ beneath the Western United States. The most striking result from their analysis is evidence for a relatively thick LVZ extending from the Gulf of California and southern California into the eastern Basin and Range province. Their analysis provides rather weak evidence for an "average" Western United States upper mantle (model A in Fig. 7-9) beneath the Pacific Coast States, with perhaps some thickening of the LVZ beneath the Cascade Range.

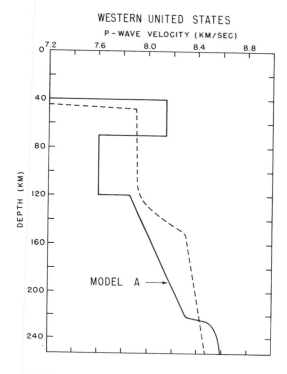

WESTERN UNITED STATES

Figure 7-9. Average *P*-wave velocity structure of the upper mantle beneath the Western United States after Helmberger (1973). The low-velocity zone coincides with a low *Q* (high attenuation) zone. Dashed model suggests that the high-velocity "lid" above the low-velocity zone may be absent in southern California and the Basin and Range province (Archambeau and others, 1969).

Thickness of the LVZ in the context of the time difference between *PL* and *P* waves refers to some combination of actual thickness and the extent of the velocity decrease within the LVZ. The presence or absence of a high-velocity layer in the upper mantle above the LVZ is one of the critical factors contributing to apparent thickness. Low P_n-velocities (7.7 to 7.9 km/s) in the Basin and Range province and southern California, for example, suggest the absence of a significant high-velocity layer over the LVZ in the region where York and Helmberger (1973) find evidence that the LVZ is "thicker" than in model A in Figure 7-9.

Additional evidence for variations in detailed structure of the upper mantle are emerging from the study of *P* waves from distant earthquakes (teleseismic events) recorded on local seismic networks. Raikes (1976), for example, described evidence in *P*-wave traveltime differences (residuals) for an elongate, high-velocity body in the upper mantle in southern California beneath the Transverse Ranges. Hadley and Kanamori (1976) saw evidence in reversed refraction data obtained from regional quarry blasts that this high-velocity body has a *P*-wave velocity of 8.3 km/s with a convex upper surface at a depth of 40 km beneath the Transverse Ranges. They inferred that this high-velocity body may be as much as 50 km thick on the basis of early *PKP* arrivals (*P* waves transmitted through the core). The pattern of anomalous traveltimes associated with this upper mantel high-velocity body does not appear to be offset by the San Andreas fault.

The evidence from teleseismic *P*-wave residuals for a small *P*-wave velocity difference across a downward extension of the San Andreas fault zone to depths of 50 km or more in the upper mantle beneath the central California Coast Ranges was mentioned earlier (Husebye and others, 1976). Solomon and Butler (1974) have interpreted teleseismic *P*-wave residuals recorded on stations in northern California and Oregon as providing evidence for a "dead slab" (that is, a remnant of a once-active subduction zone) beneath this region dipping to the east. McKenzie and Julian (1971) found evidence for a similar structure in the mantle beneath western Washington in the distribution of P-wave residuals from the 1965 Puget Sound earthquake.

Such structures carry important implications regarding the tectonic development of the Pacific Coast States, but because of ambiguities inherent with the interpretation of P-delay data, the existence and configuration of these structures need to be further tested using independent seismic techniques.

DISCUSSION

Our understanding of the gross crustal structure for the major tectonic units of the Pacific Coast States is schematically summarized by Figure 7-10. The high proportion of dashed lines and question marks in the figure emphasizes the limited resolution afforded by available data. The most obvious pattern that emerges from Figure 7-10 is a general correlation of P-wave velocities in the upper 5 to 10 km of the crust with dominant rock types in the upper crust. P-wave velocities of 5.9 to 6.3 km/s, for example, are associated with the Mesozoic granitic or crystalline blocks such as the Sierra Nevada, the Peninsular Ranges, or the Salinian block of the Coast Ranges. Somewhat lower velocities ranging from 5.5 to 5.8 km/s are associated with the Mesozoic and early Tertiary eugeosynclinal rocks of the Franciscan terrane in the Coast Ranges. Still lower velocities (about 5.2 km/s) are associated with the massive basalt flows of the Columbia Plateau and western Snake River Plain; the lowest velocities (2.0 to 5.0 km/s) are confined to deep sedimentary basins such as the Great Valley or the Imperial Valley and sections of the San Andreas fault zone.

Beyond this, few clear patterns emerge in the gross structural relations between tectonic units. The intermediate layer shows a wide variety of depths and thickness, and total crustal thickness is poorly correlated with either topographic elevation, depth to the intermediate layer, or variations of the P-wave velocities (P_n) in the uppermost mantle. Although diversity may diminish as new data improve and expand the resolution of crustal and upper mantle structure, much of it is probably real, reflecting the variety of tectonic processes operating in the vicinity of the active plate boundaries of the Pacific Coast region.

This diversity extends to exceptions in common relations between other geophysical parameters, such as the inverse relation between P_n-velocities and heat flow that characterizes broad sections of the conterminous United States (Blackwell, 1971; Sass and others, 1971). In particular, although most of southern California shares the low P_n-velocities (<8.0 km/s) and high heat flow (>1.5 HFU) characteristic of the Basin and Range province, both the central California Coast Ranges and the Columbia Plateau have high P_n-velocities and high heat flow, whereas the Sierra Nevada, the Peninsular Ranges, the northern Coast Ranges, and the Puget Sound have both low P_n-velocities and low heat flow. Lachenbruch and Sass (this volume) review dynamic processes such as mass transport or frictional heating that may explain some of these anomalous heat flow–P_n-velocity patterns.

The diversity in structure illustrated in Figure 7-10 raises a question regarding the extent to which the various models are in isostatic balance. Average gravity anomalies over broad sections of the Pacific Coast region indicate that an approximate state of isostatic balance is realized in the major tectonic units (Eaton and others, this volume). Thus a rough check on the models is provided by the isostatic condition that the total mass per unit area between the surface and the depth of compensation (a depth below which lateral density variations are insignificant) be nearly the same for all major tectonic units.

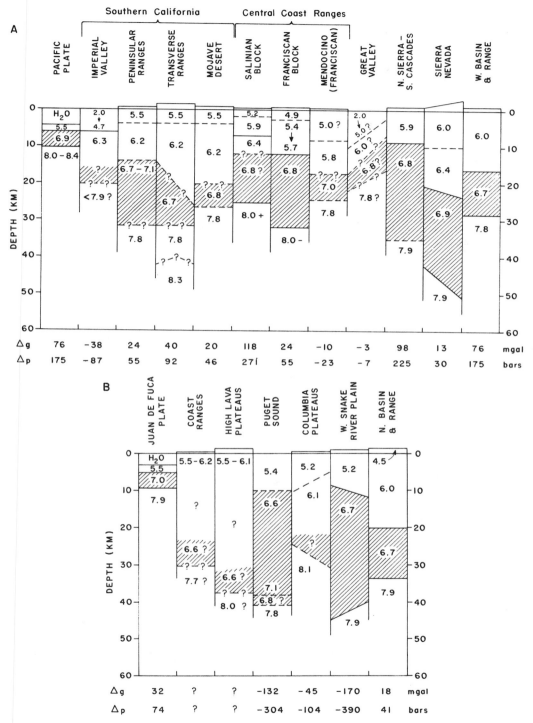

Figure 7-10. Summary of gross crustal structure for major tectonic units in the Pacific Coast States. Numbers indicate *P*-wave velocity in kilometres per second for major layers. Intermediate layer is indicated by shading, *M*-discontinuity by a heavy line. Structures based on reversed refraction profiles have solid lines, those based on unreversed profiles and regional traveltime data have dashed lines. Queries indicate poor resolution. An isostatic gravity anomaly, Δ*g*, and pressure difference, Δ*p*, computed for each structure assuming a 60-km depth of compensation given below each model (see text). (A) California. (B) Oregon, Washington, and Idaho.

The extent to which this condition is met by the layered models in Figure 7-10 can be estimated by calculating the mass per unit area, M, for each crustal model as $M = \Sigma \rho_i\, h_i$. Here ρ_i and h_i are the density and thickness of the ith layer, and the sum includes all layers from the regional surface elevation to an assumed depth of compensation of 60 km. This depth of compensation is chosen to be just deeper than the base of the thickest crust. (Although regional density variations probably extend below 60 km in the upper mantle beneath the Pacific Coast States, available data are inadequate to warrant a more elaborate assumption regarding the true depth of compensation.)

The densities ρ_i used in those calculations are related to P-wave velocities in the crustal models through the velocity-density curve plotted in Figure 7-11. This somewhat sinuous curve represents an attempt to account for variations in both rock type and lithostatic pressure in the crust and uppermost mantle. At P-wave velocities greater than 6.0 km/s this curve is drawn to approximate Eaton's curve for crystalline rocks under confining pressures from 2 to 10 kb (Bateman and Eaton, 1967); at P-wave velocities less than 6.0 km/s it is drawn to approximate the Nafe-Drake curve for saturated sediments and sedimentary rocks at room pressure (Nafe and Drake, 1963). Bear in mind, however, that variations in factors such as fluid pore pressure, temperature, and composition can result in density differences of ± 0.2 g/cm^3 or so from values predicted for a given P-wave velocity by a single curve such as that in Figure 7-11.

Relative differences in the mass per unit area calculated from the crustal models can be expressed in more familiar terms as gravity and lithostatic pressure differences, or

$$\Delta g\,(\text{gal}) = 2\pi y \left(\sum_i \rho_i\, h_i - \bar{M} \right)$$

$$\Delta p\,(\text{bars}) = g \left(\sum_i \rho_i\, h_i - \bar{M} \right).$$

Here γ is the gravitational constant (6.67×10^{-8} cgs), g is the gravitational acceleration of the Earth's surface (980 gal), and $\bar{M}$ is the average mass per unit area for all models in Figure 7-10. Note that if all the crustal models were in perfect isostatic balance at the assumed depth of compensation, then

$$\sum_i \rho_i\, h_i = \bar{M}$$

for each model and both Δg and Δp would be zero for all models. In this sense, Δg represents an isostatic gravity anomaly for the respective crustal models, and Δp represents a deviation in lithostatic pressure at the depth of compensation.

The calculated values of Δg and Δp, tabulated beneath the respective crustal models in Figure 7-10, scatter about zero with standard deviations of ± 88 mgal and ± 203 bars, respectively. Much of this scatter reflects the approximate nature of such calculations. If, for example, the actual density of rocks over a 10-km-depth interval in a specific model differs by 0.1 g/cm^3 from the value predicted by the curve in Figure 7-11, the calculated value of Δg and Δp will be off by 42 mgal and 98 bars, respectively. This, together with the approximate nature of the concept of isotasy in a laterally heterogeneous Earth with finite strength (Artyushkov, 1974), and the likelihood that some density difference persists below the assumed 60-km depth of compensation suggest that computed values of Δg and Δp within roughly

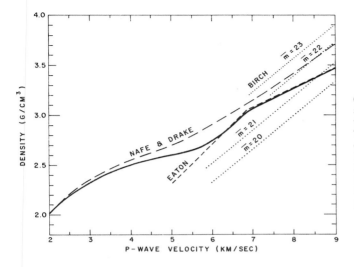

Figure 7-11. Velocity-density curve (solid line) used for computing gravity anomalies and pressure differences for the gross structural models in Figure 7-10. Broken lines indicate velocity density curves proposed by Nafe and Drake (1963) and Eaton (Bateman and Eaton, 1967). Birch's (1960) curves for rocks with mean atomic weight, $\overline{M}$, 20 through 23 at 10-kb confining pressure are also plotted.

±100 mgal and ±200 bars of zero should be regarded as satisfying an approximate condition of isostatic balance.

Most of the gross crustal structures summarized in Figure 7-10 satisfy this necessarily rather loose condition. The exceptions, when compared with regional gravity anomalies, may indicate possible complications in velocity and density structure not accounted for by the existing interpretations of available seismic data or a single velocity-density curve such as that in Figure 7-11.

Two exceptions are the models for the Puget Sound and the western Snake River Plain with computed gravity anomalies of −130 and −170 mgal, respectively. The computed value for the Puget Sound is generally consistent with gravity data in western Washington, which suggests an isostatic anomaly of −50 to −100 mgal for the Puget Sound region (Daneš, 1969). The computed value for the western Snake River Plain, however, is significantly less than the near zero to slightly positive isostatic anomaly indicated by regional gravity data over the western Snake River Plain (Hill, 1963; Mabey, 1976). This discrepancy may be explained by rocks somewhere in the crust or upper mantle beneath the western Snake River Plain with densities higher than those predicted by the curve in Figure 7-11 or perhaps by the existence of relatively high P-wave velocities (probably located in the lower part of the crust) not resolved in interpretations of the existing seismic data.

Another exception is the model for the Salinian block in the central California Coast Ranges with a computed gravity anomaly of 118 mgal, nearly 100 mgal greater than that computed for the immediately adjacent Franciscan block. Gravity measurements over the central Coast Ranges indicate that both blocks are approximately compensated (Byerly, 1966). The model for the Salinian block in Figure 7-10 thus overestimates the mass somewhere in the crust or upper mantle. This discrepancy could be mitigated if the intermediate layer were significantly thinner than inferred by Healy and Peake (1975) or perhaps by the presence of low-velocity zones within any of the major layers of the Salinian crust. Neither possibility is excluded by existing data, and, in fact, the anomalously rapid decrease in amplitude of waves refracted by the 6.4-km/s layer (Stewart, 1968) suggests that the P-wave velocity may decrease markedly within this layer (Hill, 1971).

Considerable latitude exists for rationalizing the apparent discrepancies in the mass balance between models for the various tectonic units. Unfortunately, existing data provide few constraints on the various possibilities. These discrepancies emphasize the need for more detailed data to resolve the fine structure of the crust and upper mantle.

Although we are beginning to make progress in resolving the fine structure of the crust in the Pacific Coast States (Hill, 1971; McEvilly and Clymer, 1975; Prodehl, 1970), much remains to be done in defining even the gross structure of the lithosphere. Most of the velocity models of the lithosphere summarized here can be comfortably worked into the current concepts of the development of the Pacific Coast region in terms of plate tectonics. This is true, however, largely because the velocity models provide little constraint rather than strong support for the various hypotheses. Nevertheless, existing knowledge of the gross velocity structure of the lithosphere and the current patterns of seismic activity do suggest a number of critical questions regarding tectonic processes in the Pacific Coast region that can, in principle, be answered by further definition of both the gross and fine structure of the lithosphere. A few of the more critical questions include:

1. What controls the maximum depth of earthquake occurrence along the San Andreas fault system in California? This is closely related to the question of the thickness of the lithosphere in California. In particular, does the earthquake activity stop at a depth of about 15 km because the lithosphere is only 15 km thick and rides over a weak decoupling layer (a low-velocity zone?) within the crust as suggested by Anderson (1971) and Tocher (1975)? Or does stick-slip failure give way to stable sliding on the vertical slip surface between thick lithospheric plates? If additional data on the velocity structure of the deep crust and upper mantle show that differences in velocity persist across the San Andreas fault zone to depths of 50 km or more (as is weakly suggested by the results of Husebye and others, 1976), the concept of a shallow decoupling layer loses considerable credibility.

2. Are the upper-mantle earthquakes beneath the Cape Mendocino and Puget Sound regions related to stresses in a remnant of a partially subducted Juan de Fuca plate? An affirmative answer seems likely, but details regarding the configuration of such a plate relative to accurate hypocenter locations and reliable focal mechanisms for the earthquakes will be required for an understanding of current tectonic processes.

3. What is the velocity structure beneath the Cascades and volcanic plateaus of eastern Oregon? Hamilton and Myers (1966) have suggested an evolution of the Pacific Northwest region that involves massive crustal rifting in which a once continuous belt of Cretaceous granitic rocks including the Sierra Nevada and Idaho batholith was disrupted and moved to its present configuration. If this is correct, we may not find a typical continental crustal structure beneath the intervening volcanic plateaus but perhaps a structure similar to that found just south of the Snake River Plain with a thick intermediate layer.

4. What is the composition of the intermediate layer beneath the various tectonic units? Current ideas suggest that the intermediate layer beneath eugeosynclinal terranes such as the Franciscan block consists of oceanic crust thickened by imbricate thrusting during subduction (Ernst, 1965), that beneath Cretaceous granitic terranes it consists of amphibolites and metagabbros (Bateman and Eaton, 1967) or perhaps more acidic granulites (Christensen and Fountain, 1975), and that beneath the Snake River Plain it consists of massive intrusions of basalt (Hill and Pakiser, 1966). Resolution of this question will require an integration of high-resolution seismic velocity data (both P- and S-wave velocities) with laboratory ultrasonic velocity measurements, the inversion of heat-flow data for geothermal gradients, and gravity data.

5. Are extensive low-velocity zones present within either the upper crust (the 5.8- to 6.4-km/s "layer") or the intermediate layer in any of the major tectonic units? Such low-velocity zones can, in principle, be detected by refraction profiles with dense station spacing together with

new interpretation methods based on complete wave theory. Where low-velocity zones can be shown to exist, they carry important tectonic implications in terms of high geothermal gradients, zones of high pore pressure, or perhaps variation in composition within major crustal layers.

6. What is the configuration and composition of the high-velocity bodies in the upper mantle that recent data suggest may be present beneath the northern Coast Ranges? A remnant of a partially subducted oceanic lithosphere should by all accounts be present beneath the northern Coast Ranges and Cascades, and it is important to resolve its depth, dip, and thickness as reliably as possible. The nature of the 8.3-km/s body beneath the Transverse Ranges suggested by Hadley and Kanamori (1976) remains a mystery.

7. What is the configuration of the upper-mantle, low-velocity zone beneath the Pacific Coast region?

8. And, finally, does significant velocity anisotropy exist in any of the major crustal layers or upper mantle in the Pacific Coast region? Anisotropy is difficult to detect in regions with strong structural variations, and existing data are certainly inadequate to the task. If a significant and systematic velocity anisotropy is present over a sufficiently broad region, however, it could require major revisions in interpretations of seismic velocity structures based on the common assumption of isotropic elasticity.

REFERENCES CITED

Aki, K. and Lee, W.H.K., 1976, Determination of three-dimensional velocity anomalies under a seismic array using first *P* arrival times from local earthquakes; 1, A homogeneous initial model: Jour. Geophys. Research, v. 81, p. 4381–4399.

Albores, A., Reyes, A., Brune, J., Day, S., Gonzales, J., Nava, A., Garoilazo, L., and Vaillard, I., 1977, Microearthquake study at the Cerro Prieto, Mexico, geothermal field [abs]: EOS (Am. Geophys. Union Trans.), v. 58, p. 1187.

Anderson, D. L. 1971, The San Andreas fault: Sci. American, v. 225, p. 52–67.

Anderson, D. L., and Sammis, C., 1970, Partial melting in the upper mantle: Physics Earth and Planetary Interiors, v. 3, p. 41–50.

Archambeau, C. B., Flinn, E. A., and Lambert, D. G., 1969, Fine structure of the upper mantle: Jour. Geophys. Research, v. 74, p. 5825–5865.

Armstrong, R. L., 1978, Cenozoic igneous history of the U.S. Cordillera from lat 42° to 49°N, *in* Smith, R. B., and Eaton, G. P., eds., Cenozoic tectonics and regional geophysics of the western Cordillera: Geol. Soc. America Mem 152 (this volume).

Artyushkov, E. V., 1974, Can the Earth's crust be in a state of isostasy?: Jour. Geophys. Research, v. 79, p. 741–752.

Atwater, T., 1970, Implications of plate tectonics for the Cenozoic tectonic evolution of western North

America: Geol. Soc. America Bull., v. 81, p. 3515–3536.

Bailey, E. H., Irwin, W. P., and Jones, D. L., 1964, Franciscan and related rocks, and their significance in the geology of western California: California Div. Mines and Geology Bull. 183, 177 p.

Bateman, P. C., 1961, Willard D. Johnson and the strike-slip component of fault movement in the Owens Valley earthquake of 1872: Seismol. Soc. America Bull., v. 51, p. 483–493.

Bateman, P. C., and Eaton, J. P., 1967, Sierra Nevada batholith: Science, v. 158, p. 1407–1417.

Berg, J. W., Jr., Trembly, L., Emilia, D. A., Hutt, J. R., King, J. M., Long, L. T., McKnight, W. R., Sarmah, S. K., Souders, R., Thiruvathukal, J. V., and Vessler, D. A., 1966, Crustal refraction profile, Oregon coast range: Seismol. Soc. America Bull., v. 56, p. 1357–1362.

Biehler, S. R., Kovach, R. L., and Allen, C. R., 1964, Geophysical framework of the northern end of Gulf of California structural province, *in* Van Andel, T. H., and Shor, G. G., eds, Marine geology of the Gulf of California: Am Assoc. Petroleum Geologists Mem. 3, p. 126–156.

Birch, F., 1960, The velocity of compressional waves in rocks to 10 kilobars, 1: Jour. Geophys. Research, v. 56, pp. 1083–1102.

Blackwell, D. D., 1971, The thermal structure of

the continental crust, *in* Heacock, J. G., eds., The structure and physical properties of the Earth's crust: Am. Geophys. Union Geophys. Mon. 14, p. 169–184.

Bolt, B. A., Lomnitz, C., and McEvilly, T. V., 1968, Seismological evidence of the tectonics of central and northern California and the Mendocino escarpment: Seismol. Soc. America Bull., v. 58, 1725–1767.

Boore, D. M., and Hill, D. P., 1973, Wave propagation characteristics in the vincinity of the San Andreas fault, *in* Kovach, R. L., and Nur, A., eds., Conf. on tectonic problems San Andreas fault system proc.: Stanford Univ. Pubs. Geol. Sci., v. 13, p. 215–224.

Brace, W. F., and Byerlee, J., 1970, California earthquakes: Why so shallow focus?: Science, v. 168, p. 1573–1575.

Brune, J. N., 1968, Seismic moment, seismicity and rate of slip along major fault zones: Jour. Geophys. Research, v. 73, p. 777–784.

Bufe, C. G., and Lester, F. W., 1975, Seismicity of the Geysers–Clear Lake region, California [abs]: EOS, v. 56, p. 1020.

Bufe, C. G., Lester, F. W., Lahr, K. M., Lahr, J. C., Seekins, L. C., and Hanks, T. C., 1976, Oroville earthquakes: Normal faulting in the Sierra Nevada foothills: Science, v. 192, p. 72–74.

Byerly, P., 1939, Near earthquakes in central California: Seismol. Soc. America Bull., v. 29, p. 427–462.

Byerly, P. E., 1966, Interpretations of gravity data from the central Coast Ranges and San Joaquin Valley, California: Geol. Soc. America Bull., v. 77, p. 83–94.

Cady, J., 1975, Magnetic and gravity anomalies in the Great Valley and western Sierra Nevada metamorphic belt, California: Geol. Soc. America Spec. Paper 168, 56 p.

Carder, D. S., Qamar, A., and McEvilly, T. V., 1970, Trans-California seismic profile—Pahute Mesa to San Francisco Bay: Seismol. Soc. America Bull., v. 60, p. 1829–1846.

Christensen, M. N., 1966, Late Cenozoic crustal movements in the Sierra Nevada of California: Geol. Soc. America Bull., v. 77, p. 168–182.

Christensen, N. I., and Fountain, D. M., 1975, Constitution of the lower continental crust based on experimental studies of seismic velocities in granulite: Geol. Soc. America Bull., v. 86, p. 227–236.

Christiansen, R. L., and McKee, E. H., 1978, Late Cenozoic volcanic and tectonic evolution of the Great Basin and Columbia intermontane region, *in* Smith, R. B., and Eaton, G. P., eds., Cenozoic tectonics and regional geophysics of the western Cordillera: Geol. Soc. America Mem. 152 (this volume).

Clark, M. M., Grantz, A., and Rubin, M., 1972, Holocene activity of the Coyote Creek fault as recorded in sediments of Lake Cohuilla: U.S. Geol. Survey Prof. Paper 787, p. 5–86.

Coney, P. J., 1978, Mesozoic-Cenozoic Cordilleran plate tectonics, *in* Smith, R. B., and Eaton, G. P., eds., Cenozoic tectonics and regional geophysics of the western Cordillera: Geol. Soc. America Mem. 152 (this volume).

Crosson, R. S., 1972, Small earthquakes, structure, and tectonics of the Puget Sound region: Seismol. Soc. America Bull., p. 1133–1172.

——1976, Crustal structure modeling of earthquake data, 2, Velocity structure of the Puget Sound region, Washington: Jour. Geophys. Research, v. 81, p. 3047–3054.

Daneš, Z. F., 1969, Gravity results in western Washington: EOS, v. 50, p. 548–550.

Daneš, Z. F., Bonno, M., Brau, E., Gilham, W. D., Hoffman. R. F., Johansen, D., Jones, M. H., Malfait, B., Masten, J., and Teague, G. O., 1965, Geophysical investigation of the southern Puget Sound area, Washington: Jour. Geophys. Research, v. 70, p. 5573–5580.

Dehlinger, P., Chiburis, E. F., and Collver, M. M., 1965, Local travel-time curves and their geologic implications for the Pacific Northwest States: Seismol. Soc. America Bull., v. 55, p. 587–608.

Eaton, G. P., Wahl, R. R., Prostka, H. J., Mabey, D. R., and Kleinkopf, M. D., 1978, Regional gravity and tectonic patterns: Their relation to late Cenozoic epeirogeny and lateral spreading of the western Cordillera, *in* Smith, R. B., and Eaton, G. P., eds., Cenozoic tectonics and regional geophysics of the western Cordillera: Geol. Soc. America Mem. 152 (this volume).

Eaton, J. P., 1963, Crustal structure from San Francisco, California, to Eureka, Nevada, from seismic-refraction measurements: Jour. Geophys. research, v. 69, p. 5789–5806.

——1966, Crustal structure in northern and central California from seismic evidence, *in* Geology of northern California: Calif. Div. Mines Geol. Bull. 190, p. 419–426.

Eaton, J. P., O'Neill, M. E., and Murdock, J. N., 1970, Aftershocks of the 1966 Parkfield-Cholame, California, earthquake: A detailed study: Seismol. Soc. America Bull., v. 60, p. 1151–1197.

Elders, W. A., Rex, R. W., Meidav, T., Robinson, P. T., and Biehler, S., 1972, Crustal spreading in southern California: Science, v. 178, p. 15–24.

Ellsworth, W. L., 1975, Bear Valley, California, earthquake sequence of February-March 1972: Seismol. Soc. America Bull., v. 65, p. 483–506.

Ellsworth, W. L., Campbell, R. H., Hill, D. P., Page, R. A., Alewine, R. W., Hanks, T. C., Heaton, T. H., Hileman, J. A., Kanamori, H., Minster, B., and Whitcomb, J. H., 1973, Point Mugu, California earthquake of 21 February 1973 and its aftershocks: Science, v. 182, p. 1127–1129.

Ernst, W. G., 1965, Mineral paragenesis in Franciscan metamorphic rocks, Panoche Pass, California: Geol. Soc. America Bull., v. 76, p. 879–914.

Friedman, M. E., Whitcomb, J. H., Allen, C. R., and Hileman, J. A., 1976, Seismicity of the southern California region, Jan. 1972–31 Dec. 1974: Pasadena, California Inst. Technology Seismol. Lab., 934 p.

Fuis, G. S., Friedman, M. E., and Hileman, J. A., 1977, Preliminary catalog of earthquakes in southern California, July 1974-Sept. 1976: U.S. Geol. Survey Open-File Rept., 90 p.

Gibbs, J. F., and Roller, J. C., 1966, Crustal structure determined by seismic-refraction measurements between the Nevada Test Site and Ludlow, California, in Geological Survey Research 1966: U.S. Geol. Survey Prof. Paper 550-D, p. 0125–0131.

Hadley, D., and Kanamori, H., 1977, Seismic structure of the Transverse Ranges, California: Geol. Soc. America Bull., v. 88, p. 1469–1478.

Hamilton, R. M., 1970, Time-term analysis of explosion data from the vicinity of the Borrego Mountain, California, earthquake of 9 April 1968: Seismol. Soc. America Bull., v. 60, p. 367–381.

Hamilton, W., 1969, Mesozoic California and the underflow of Pacific mantle: Geol. Soc. America Bull., v. 80, p. 2409–2430.

Hamilton, W., and Myers, W. B., 1966, Cenozoic tectonics of the Western United States: Rev. Geophysics, v. 4, p. 509–549.

Hanks, T. C., Hileman, J. A., and Thatcher, W., 1975, Seismic moments of the larger earthquakes of the southern California region: Geol. Soc. America Bull., v. 86, p. 1131–1139.

Healy, J. H., 1963, Crustal structure along the coast of California from seismic refraction measurements: Jour. Geophys. Research, v. 68, p. 5777–5787.

Healy, J. H., and Peake, L. G., 1975, Seismic velocity structure along a section of the San Andreas fault near Bear Valley, California: Seismol. Soc. America Bull., v. 65, p. 1177–1198.

Helmberger, D. V., 1973, On the structure of the low-velocity zone: Royal Astron. Soc. Geophys. Jour., v. 29, p. 367–372.

Hileman, J. A., Allen, C. R., and Nordquist, J. M., 1973, Seismicity of the southern California region 1 January 1932 to 31 December 1972: California Inst. Technology Earthquake Research Assoc. Contr. 2385, 83 p.

Hill, D. P., 1963, Gravity and crustal structure in the western Snake River Plain, Idaho: Jour. Geophys. Research, v. 68, p. 5807–5819.

———1971, Velocity gradients and anelasticity from crustal body-wave amplitudes: Jour. Geophys. Research, v. 76, p. 3309–3325.

———1972, Crustal and upper mantle structure of the Columbia Plateau from long range seismic-refraction measurements: Geol. Soc. America Bull., v. 83, p. 1639–1648.

Hill, D. P., and Pakiser, L. C., 1966, Crustal structure between the Nevada Test Site and Boise, Idaho, from seismic-refraction measurements, in Steinhart, J. S., and Smith, T. J., eds., The earth beneath the continents (Merle A. Tuve volume): Am. Geophys. Union Geophys. Mon. 10, p. 391–419.

Hill, D. P., Mowinckel, P., and Peake, L. G., 1975, Earthquakes, active faults, and geothermal areas in the Imperial Valley, California: Science, v. 188, p. 1306–1308.

Husebye, E., Christoffersson, A., Aki, K., and Powell, C., 1976, Preliminary results on the two-dimensional seismic structure of the lithosphere under the U.S.G.S. central California seismic array: Royal Astron. Soc. Geophys. Jour., v. 46, p. 319–340.

Irwin, W. P., and Barnes, I., 1975, Effects of geologic structure and metamorphic fluids on seismic behavior of the San Andreas fault system in central and northern California: Geology, v. 3, p. 713–716.

Isacks, B., and Molnar, P., 1971, Distribution of stresses in the descending lithosphere from a global survey of mantle earthquakes: Rev. Geophysics and Space Physics, v. 9, p. 103–174.

Johnson, C. E., and Hadley, D. M., 1976, Tectonic implications of the Brawley swarm, Imperial Valley, California, January-February, 1975: Seismol. Soc. America Bull., v. 66, p. 1155–1158.

Johnson, L. R., 1965, Crustal structure between Lake

Mead, Nevada, and Mono Lake, California: Jour. Geophys. Research, v. 70, p. 2863–2872.

Johnson, S. H., and Couch, R. W., 1970, Crustal structure in the north Cascade Mountains of Washington and Bristish Columbia from seismic-refraction measurements: Seismol. Soc. America Bull., v. 60, p. 1259–1270.

Kanamori, H., and Hadley, D., 1975, Crustal structure and temporal velocity change in southern California: Pure and Appl. Geophysics, v. 113, p. 257–280.

Kind, R., 1972, Residuals and velocities of P_n waves recorded by the San Andreas seismograph network: Seismol. Soc. America Bull., v. 62, p. 85–100.

Knapp, J. S., Jr., 1976, Velocity changes associated with the Ferndale earthquake [M.Sc. thesis]: Seattle, Univ. Washington, 93 p.

Lachenbruch, A. H., and Sass, J. H., 1973, Thermo mechanical aspects of the San Andreas fault system, in Kovach, R. L., and Nur, A., eds., Conf. on tectonic problems of the San Andreas fault system Proc.: Stanford Univ. Pubs. Geol. Sci., v. 13, p. 192–205.

——1978, Models of an extending lithosphere and heat flow in the Basin and Range province, in Smith, R. B., and Eaton, G. P., eds., Cenozoic tectonics and regional geophysics of the western Cordillera: Geol. Soc. American Mem. 152 (this volume).

LaFehr, T. R., 1965, Gravity isostasy and crustal structure in the southern Cascade Range: Jour. Geophys. Research, v. 70, p. 5581–5598.

Lahr, K. M., Lahr, J. C., Lindh, A. G., Bufe, C. G., and Lester, F. W., 1976, The August Oroville earthquakes: Seismol. Soc. America Bull., v. 66, p. 1085–1099.

Langston, C. A., and Blum, D. E., 1977, The April 29, 1965, Puget Sound earthquake and the crustal and upper mantle structure of western Washington: Seismol. Soc. America Bull., v. 67, p. 693–711.

Langston, C. A., and Helmberger, D. V., 1974, Interpretation of body and Rayleigh waves from NTS to Tucson: Seismol. Soc. America Bull., v. 64, p. 1919–1930.

Lomnitz, C., Mooser, F., Allen C., Brune, J., and Thatcher, W., 1970, Seismicity and tectonics of the northern Gulf of California region, Mexico—Preliminary results: Geofísica Internac., v. 10, p. 37–48.

Mabey, D. R., 1976, Interpretation of a gravity profile across the western Snake River Plain, Idaho: Geology, v. 4, p. 53–55.

Malone, S. D., Rothe, G. H., and Smith, S. W., 1975, Details of microearthquake swarms in the Columbia Basin, Washington: Seismol. Soc. America Bull., v. 65, p. 855–864.

Marks, S. M., and Lindh, A. G., 1978, Regional seismicity of the Sierran foothills in the vicinity of Oroville, California: Seismol. Soc. America Bull. (in press).

Mayer-Rosa, D., 1973, Traveltime anomalies and distribution of earthquakes along the Calaveras fault zone, California: Seismol. Soc. America Bull., v. 63, p. 713–729.

McCollom, R. L., and Crosson, R. S., 1975, An array study of upper mantle velocity in Washington State: Seismol. Soc. America Bull., v. 65, p. 467–482.

McEvilly, T. V., 1966, Crustal structure estimation with a large scale array: Royal Astron. Soc. Geophys. Jour., v. 11, p. 13–17.

——Seafloor mechanics north of Cape Mendocino, California: Nature, v. 224, p. 125–133.

McEvilly, T. V., and Clymer, R. W., 1975, A deep crustal reflection survey on the San Andreas fault: EOS, v. 56, p. 1021.

McKenzie, D., and Julian, B., 1971, Puget Sound, Washington, earthquake and the mantle structure beneath the northwestern United States: Geol. Soc. America Bull., v. 82, p. 3519–3524.

Nafe, J. W., and Drake, C. L., 1963, Physical properties of marine sediments, in Hill, M. N., ed., The sea, Vol. 3: New York, Interscience, p. 794–815.

Oliver, H. W., 1977, Gravity and magnetic investigations of the Sierra Nevada batholith, California: Geol. Soc. America Bull., v. 88, p. 445–461.

Oliver, J., Dobrin, M., Kaufman, S., Meyer, R., and Phinney, R., 1976, Continuous seismic reflection profiling of deep basement; Hardeman County, Texas: Geol. Soc. America Bull, v. 87, p. 1537–1546.

Pakiser, L. C., 1964, Gravity, volcanism, and crustal structure in the southern Cascade Range, California: Geol. Soc. America Bull., v. 75, p. 611–620.

——1974, root of the Sierra Nevada: Geol. Soc. America Abs. with Programs, v. 6, p. 234.

Pakiser, L. C., and Steinhart, J. S., 1964, Explosion seismology in the Western Hemisphere, in Odishaw, H., ed., Research in geophysics, Vol. 2, Solid earth and interface phenomena: Cambridge, Massachusetts Inst. Technology Press, p. 123–147.

Peselnick, L., and Stewart, R. M., 1975, a sample assembly for velocity measurements in rocks at elevated temperatures and pressures: Jour. Geophys. Research, v. 73, p. 3765.

Pitt, A. M., and Steeples, D. W., 1975, Microearthquakes in the Mono Lake–northern Owens Valley region from September 28 to October 18, 1970: Seismol. Soc. America Bull., v. 65, p. 835–844.

Press, F., 1966, Seismic velocities, in Clark, S. P., ed., Handbook of physical constants (rev. ed.): Geol. Soc. America Mem. 97, p. 195–218.

Prodehl, D., 1970, Seismic-refraction study of crustal structure in the Western United States: Geol. Soc. America Bull., v. 81, p. 2629–2646.

Raikes, S. A., 1976, The azimuthal variation of teleseismic P-wave residuals for stations in southern California: Earth and Planetary Sci. Letters, v. 29, p. 367–372.

Roller, J. C., and Healy, J. H., 1963, Seismic-refraction measurements of crustal structure between Santa Monica Bay and Lake Mead: Jour. Geophys. Research, v. 68, p. 5837–5849.

Roy, R. F., Blackwell, D. D., and Decker, E. R., 1972, Continental heat flow, in Robertson, E. C., ed., Nature of the solid earth: New York, McGraw-Hill Book Co., p. 506–543.

Sass, J. H., Lachenbruch, A. H., Munroe, R. J., Green, G. W., and Moses, T. H., 1971, Heat flow in the Western United States: Jour. Geophys. Research, v. 76, p. 6376–6413.

Seeber, L., Barazongi, M., and Nowroozi, A., 1970, Microseismicity and tectonics of coastal northern California: Seismol. Soc. America Bull., v. 60, p. 1669–1700.

Sharp, R. V., 1967, San Jacinto fault zone in the Peninsular Ranges of southern California: Geol. Soc. America Bull., v. 78, p. 705–730.

Shor, G. G., Jr., and Raitt, R. W., 1958, Seismic studies in the southern California continental borderland, in Geofísica aplicada: Internat. Geol. Cong., 20th, México, D. F. 1956, Sec. 9, tome 2, p. 243–259.

Shor, G. G., Jr., Dehlinger, P., Kirk, H. K., and French, W. S., 1968, Seismic refraction studies off Oregon and northern California: Jour. Geophys. Research, v. 73, p. 2175–2194.

Silver, E. A., 1978, Geophysical studies and tectonic development of the continental margin off the Western United States, lat 34° to 48°N, in Smith, R. B., and Eaton, G. P., eds., Cenozoic tectonics and regional geophysics of the western Cordillera: Geol. Soc. America Mem 152 (this volume).

Simila, G. W., Peppin, W. H., and McEvilly, T. V., 1975, Seismotectonics of the Cape Mendocino,

California, area: Geol. Soc. America Bull., v. 86, p. 1399–1406.

Smith, R. B., 1978, Seismicity, crustal structure, and intraplate tectonics of the interior of the western Cordillera, in Smith, R. B., and Eaton, G. P., eds., Cenozoic tectonics and regional geophysics of the western Cordillera: Geol. Soc. America Mem. 152 (this volume).

Smith, S. W., 1976, Seismicity north of Cape Mendocino: Earthquake Notes, v. 47, p. 20.

Sollogub, V. B., and Chekunow, A. V., 1972, The results of DSS measurements in the cooperating countries; The Soviet Union; The Soviet, in Gyorgy, S., ed., The crustal structure of central and southeastern Europe based on the results of explosion seismology: Budapest, Geofiz. Koezlem, Spec. Ed., p. 43–68.

Solomon, S. C., and Bulter, R. G., 1974, Prospecting for dead slabs: Earth and Planetary Sci. Letters, v. 21, p. 421–430.

Sommers, R., and Byerlee, J., 1977, A note on the effect of fault gouge composition on the stability of frictional sliding: Internat. Jour. Rock Mechanics, v. 14, p. 155–160.

Stewart, R., and Peselnick, L., 1977, Velocity of compressional waves in dry Franciscan rocks to 8 kilobars and 300°C: Jour. Geophys. Research, v. 82, p. 2027–2039.

Stewart, S. W., 1968, Preliminary comparison of seismic traveltime and inferred crustal structure adjacent to the San Andreas fault in the Diablo and Gabilan Ranges of central California, in Dickinson, W. R., and Grantz, A., eds, Geologic Problems of San Andreas fault system Conf. Proc.: Stanford Univ., Publ. Geol. Sci., v. 11, p. 218–230.

Stewart, S. W., and O'Neill, M. E., 1972, Seismic traveltimes and near-surface crustal velocity structure bounding the San Andreas fault zone near Parkfield California: U.S. Geol. Survey Prof. Paper 800-C, p. C117–C125.

Swanson, D. A., 1967, Yakima Basalt of the Tieton River area, south-central Washinton: Geol. Soc. America Bull., v. 78, p. 1077–1110.

Thatcher, W., Hileman, J. A., and Hanks, T. C., 1975, Seismic slip distribution along the San Jacinto fault zone, southern California, and its implications: Geol. Soc. America Bull., v. 86, p. 1140–1146.

Thompson, G. A., and Talwani, M., 1964, Crustal structure from Pacific basin to central Nevada: Jour. Geophys. Research, v. 69, p. 4813–4838.

Tocher, D., 1975, On crustal plates: Seismol. Soc. America Bull., v. 65, p. 1495–1500.

Warren, D. H., and Healy, J. H., 1973, Structure of the crust in the conterminous United States: Tectonophysics, v. 20, p. 203–213.

Waters, A. C., 1962, Basalt magma types and their tectonic associations—Pacific Northwest of the United States *in* The crust of the Pacific basin: Am. Geophys. Union Geophys. Mon. 6, p. 158–170.

Wesson, R. L., Roller, J. C., and Lee, W.H.K., 1973, Time-term analysis and geological interpretation of seismic traveltime data from the Coast Ranges of central California: Seismol. Soc. America Bull., v. 63, p. 1447–1472.

White, C. M., and McBirney, A. R., 1978, Some quantitative aspects of orogenic volcanism in the Oregon Cascades, *in* Smith, R. B., and Eaton, G. P., eds., Cenozoic tectonics and regional geophysics of the western Cordillera: Geol. Soc. America Mem. 152 (this volume).

White, W. R., and Savage, J. C., 1965, A seismic refraction and gravity study of the Earth's crust in British Columbia: Seismol. Soc. America Bull., v. 55, p. 463–486.

Wiggins, R., and Helmberger, D. V., 1973, Upper mantle structure of Western United States: Jour. Geophys. Research, v. 78, p. 1870–1880.

York, J. E., and Helmberger, D. V., 1973, Low-velocity zone variations in the southwestern United States: Jour. Geophys. Research, v. 78, p. 1882–1886.

MANUSCRIPT RECEIVED BY THE SOCIETY AUGUST 15, 1977

MANUSCRIPT ACCEPTED SEPTEMBER 2, 1977

Printed in U.S.A.

Geological Society of America
Memoir 152

8

Heat flow and energy loss in the Western United States

DAVID D. BLACKWELL
Department of Geological Sciences
Southern Methodist University
Dallas, Texas 75275

ABSTRACT

Recent heat-flow studies in the Western United States, especially the Cordillera, are discussed and summarized and a new heat-flow map is presented. The major features of the map have already been described: high heat flow in the Northern Rocky Mountains, Columbia Plateau, High Cascades, and Basin and Range provinces (the Cordilleran thermal anomaly zone); high heat flow along the San Andreas–Gulf of California transform system; high heat flow in the Southern Rocky Mountains; moderate heat flow in part of the Colorado Plateau; and low heat flow along the Sierra Nevada and the coastal provinces of Oregon and Washington. In addition, much detail is apparent in the Cordilleran thermal anomaly zone. Very high heat flow (greater than 2.5 HFU) is found in the Cascades, and part of the Brothers fault zone in Oregon, part of the Snake River Plain in Idaho, Yellowstone in Wyoming, the Battle Mountain "high" in Nevada, the Geysers area and the Imperial Valley in California, and the Rio Grande rift in New Mexico. Areas of low heat flow are associated with part of the Columbia Basin in Washington, the eastern part of the Snake River Plain in Idaho, and the Eureka "low" in Nevada. The heat-flow map is very complicated because it includes the effects of crust and mantle radioactivity and magmatic heat sources, regional hydrology, and thermal refraction due to structurally related thermal conductivity contrasts.

In the active tectonic areas there are energy losses associated with volcanism, intrusion, and hydrothermal convection. Such losses may not be measured in a typical heat-flow survey. These losses are evaluated and shown to be a significant part of the total heat flow in many areas. Heat flow is compared to the geographical distribution of volcanism, plutonism, hydrothermal activity, and average topography. The regions of high heat flow correlate well with the areas of Cenozoic volcanism, plutonism, and thermal spring activity. The lack of a one-to-one correlation of areas of active plutonism to regional concentrations of thermal springs is shown. The relationship of topography to heat flow is complicated, and it appears that in general the average composition of the crust changes during a major continental thermal event so that the relationships between topography and heat flow may change during the evolution of the thermal event.

Detailed heat-flow interpretation relies on the relationship between heat flow (Q) and radioactive heat generation (A). The Basin and Range plot of Q versus A applies only to areas where the most recent volcanic event is older than 17 m.y. In areas of younger thermal events, the reduced heat flow (that is, the heat flow measured at the surface less heat production from crustal radioactive sources) is generally higher than 1.4 HFU, and hydrothermal convection and volcanism are major mechanisms involved in the total energy transfer. Furthermore, transitions between thermal provinces are narrow (generally less than 20 km); therefore, the sources directly responsible for the surface-measured variations in heat flow must be in the crust. Thermal boundaries also usually appear in areas of contemporary seismicity.

As a synthesis of the discussion, a map of energy release (as opposed to heat flow measured at the surface) and a simple model of a Cordilleran thermal event are presented. The generalized map of energy release shows total thermal energy transfer from the mantle, including nonconductive energy losses. This map shows highest heat flow along the eastern and western borders of the Cordilleran thermal anomaly zone and a smoother variation of heat flow within the zone than does the heat-flow map. During a continental thermal event typical of the Cordilleran thermal anomaly zone, which may be infinitely more varied in intensity and duration than an oceanic spreading event, the conductive heat flow will be only a part of the total energy loss, and the dominant heat-transfer mechanisms change with time over the life of the event. For the areas where a thermal event is young, volcanism and plutonism may be the prime energy-loss mechanisms, but in spite of the high overall energy loss, large areas of young volcanism may have very low conductive heat flow because of the dominant effect of hydrothermal convection as a mechanism for plutonic energy loss. As the thermal event decays, regional hydrothermal convection and heat conduction become the dominant heat-transfer mechanisms. For areas where thermal events are older than 17 m.y., heat conduction is the dominant heat-transfer mechanism although hydrothermal convection may be locally significant.

INTRODUCTION

Research into the thermal characteristics of the Western United States has never been as active as at the present time. Of particular interest are studies of such subjects as the volcanic history, nature of hydrothermal circulation in the crust in geothermal areas, and regional geophysical investigations. These studies have been given added impetus and urgency by the need to explore for and evaluate the impact of geothermal resources on the energy needs of the United States. Most of these studies have been begun in the past 5 years, and their full results and impact have yet to reach the scientific literature. For example, there are probably twice as many heat-flow measurements in preparation for publication for the Western United States as have been published up to 1976. The number of active investigators has increased also, and at the present time there are probably 100 to 200 new heat-flow measurements made in the Western United States every year. Private exploration groups are drilling holes in geothermal areas at the rate of 1,000/yr. It is not uncommon at the present time to have 20 to 50 drill holes in a single geothermal prospect.

In a recent tabulation of data in the United States, Sass and others (1976) included slightly over 400 values for the Western United States. An additional 100 or so heat-flow values are in the final stages of preparation or in press for each western state. The results of these studies have revealed a complexity and detail of heat-flow variation previously unknown. In the geologically complicated Western United States, the spatial frequency of variation of the heat flow at the surface is on the order of 1 to 20 km; therefore, pre-existing data, often at a spacing of 50 to 100 km and in biased locations, are not adequate for understanding the variations now seen so prominently.

The predominant thermal feature of the Cordillera is the zone of high heat flow 500 to 1,000 km wide that includes the Northern Rocky Mountains, Cascade Range, Columbia Plateau, and Basin and Range provinces in the United States. I have noted (Blackwell, 1969) that the thermally anomalous region extends north to the Canadian border and is not confined to the Basin and Range province, previously the focus of most attention (Menard, 1960; Pakiser and Zietz, 1965; Cook, 1966). This thermal feature was called the Cordilleran thermal anomaly zone (Blackwell, 1969) to emphasize the fact that the high heat flow is widespread and not confined to a single physiographic or tectonic province. Subsequent studies have extended the zone along the Cordillera south to the Mexican volcanic belt (Mooser, 1972; Smith, 1974; Blackwell, unpub. data) and north into central British Columbia (Judge, 1975; Hyndman, 1976).

The Cordilleran thermal anomaly zone is bounded discontinuously along the west by a narrower zone of anomalously low heat flow and on the east by the Great Plains and the Colorado Plateau, which have somewhat lower heat flow. In New Mexico the Cordilleran thermal anomaly zone intersects the Rio Grande rift and the Southern Rocky Mountains. The nature of the relationship of these two features to the zone is not well understood.

The pattern of an area of anomalously low heat flow oceanward of a broad zone of high heat flow and the association of that pattern with andesitic volcanism can uniquely be related to the presence in the recent past of a subduction zone (Blackwell, 1971, 1974; Roy and others, 1972; Hyndman, 1976). The heat-flow studies thus furnish evidence of a dominance of subduction-zone tectonics on the Cenozoic history of the Cordillera; the results of the studies imply a history and location of the Cenozoic subduction-zone activity uniquely compatible with the tectonic model deduced by Atwater (1970) from the distribution of ocean-floor magnetic anomalies.

Parts of the paper deal with the whole Western United States. The emphasis, however, will be on the Cordillera, and little will be said about the Middle and Southern Rocky Mountains or about the Rio Grande rift. Furthermore, the object of this paper is not the discussion of the relationship of heat flow to plate tectonics in the Cordillera, but a description and evaluation of the distribution of energy loss and the thermal evolution of the crust and upper mantle of the Cordillera. The discussion of this paper does not supersede that of Roy and others (1972), and reference to that paper should be made for general descriptions of the thermal character of all of the Western United States.

As a result of the increasingly detailed knowledge of the heat-flow pattern, whole new chapters on the relation of heat flow to the tectonic features and the geologic history of the Cordillera are being written. Because so many of these data and results are in a state of preparation and interpretation at the present time, this paper will summarize only in a generally qualitative way some of these recent findings, describe the characteristics of the heat-flow data, and discuss some of the correlations of the heat-flow data with other investigations.

In the first section of the paper a heat-flow map of the Western United States will be presented and described. Then the various mechanisms contributing to energy loss (volcanism, and so forth), in addition to the heat flow measured during regional surveys, will be discussed and evaluated in a preliminary way to see if they contribute significantly to the energy loss. Next the heat-flow pattern will be compared to the distribution of these other types of energy-release phenomena. In the following section the relationship between heat flow measured at the surface, crustal heat production, and structural-volcanic provinces will be described.

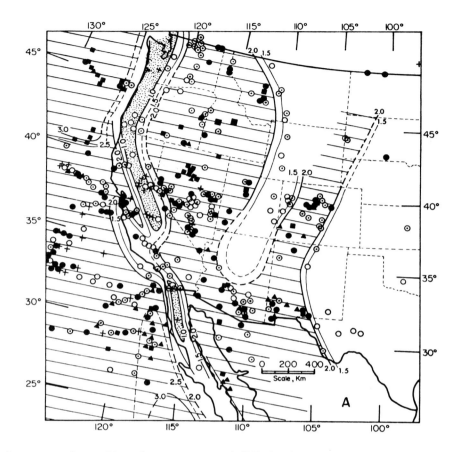

Then the nature of transitions between areas of differing heat flow will be briefly discussed in the light of recent studies. In the final section, a quantitative model for the thermal evolution of the Western United States will be suggested and a synthetic map discussed which describes in a more complete way than previous heat-flow maps the distribution and magnitude of energy release.

For continuity of the present discussion with the large body of data in the literature, cgs units will be used throughout. Conversion factors of the most commonly used units to SI units are given below for reference.[1]

HEAT-FLOW DISTRIBUTION

Two heat-flow maps (Roy and others, 1972; Sass and others, 1971) have been available for tectonic and geologic interpretation. These maps attempt to illustrate different qualities and thus warrant some discussion. Roy and others (1972) and Blackwell (1971) presented variations of the same heat-flow map (Fig. 8-1A). The map was contoured, and areas of high and low heat flow were shaded. As defined, the shaded areas indicate regions of high

[1] 1 heat-flow unit (HFU) = 1×10^{-6} cal/(cm^2·s) = 41.8 mW/m^2; 1 heat-generation unit (HGU) = 1×10^{-13} cal/(cm^3·s) = 0.418 μW/m^3.

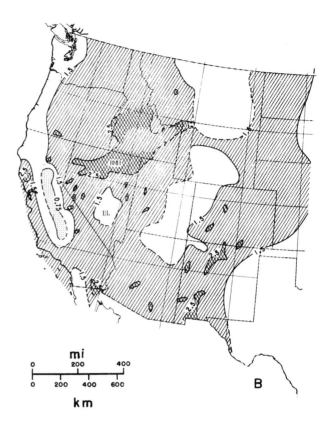

Figure 8-1. (A, facing page) Heat-flow map of the Western United States (Roy and others, 1972) with data through 1974 (Brott and others, 1976). Plus signs represent heat-flow values in the range of 0 to 0.99 HFU; open circles, 1.0 to 1.49 HFU; dotted circles, 1.5 to 1.99 HFU; solid circles, 2.0 to 2.49 HFU; solid triangles, 2.5 to 2.99 HFU; solid rectangles, >3.0 HFU. (B) Heat-flow map of the Western United States (Sass and others, 1976). BMH is the Battle Mountain high, and EL is the Eureka low. Contours are in heat-flow units.

(greater than 2 HFU) and low (less than 1 HFU) heat flow *"with average values of flux that would be measured in rocks with surface radioactivity of within the range of granodiorite"* (Roy and others, 1972, Fig. 13). This map was not an attempt to show heat flow at the surface (that is, measurable in 100- to 300-m-deep drill holes); instead the map was an attempt to generalize on the conditions in the upper mantle as reflected above an average continental crust. There was no attempt to subdivide the regions of high heat flow in a detailed way, and contouring of individual anomalies was not attempted, even though it was certainly recognized that many different geologic provinces were included in the regions of high heat flow and that regional variations in heat flow would occur within these differing geologic provinces (Roy and others, 1968).

A different approach to the representation of the heat-flow data (see Fig. 8-1B) was presented and discussed by Sass and others (1976, Fig. 8-3B; an updated version of a map published by Sass and others, 1971). This map is an attempt to present a contour map of heat flow as measured at the surface, where "surface" is roughly defined as the outermost 100 to 300 m of the Earth.

The main problem with a heat-flow map such as the one by Sass and others (1976) is that the effects of variations in crustal heat production, large-scale hydrologic effects, structural complications, local heat sources, and mantle effects are all included without differentiation. Furthermore, any new heat-flow value may reflect a previously unknown anomaly and require a change in the map. In view of the lateral scale of heat-flow variations, such a map cannot be complete until heat-flow data for all the Western United States have been gathered at a spacing of 1 to 20 km. Thus, in using such a map, it should be kept clearly in mind that contours in data gaps are subject to change as more data become available.

A problem with both maps is that it is difficult to separate one kind of anomaly (mantle or deep crustal) from the other (near surface or upper crustal) without closely spaced data and without a good understanding of the geologic setting, history, and regional hydrology of the determination site. Furthermore, the user of a heat-flow map may be interested in local anomalies (geothermal areas, for example) that are not shown on a regional heat-flow map. So the first kind of map is simpler to interpret but omits much information of importance, whereas the second is complicated and requires more data to contour properly.

Figure 8-2 shows a contour map of the heat flow in the Western United States. This map is similar to the map of Sass and others (1976) in California, Nevada, and Arizona. The remainder of the map has been contoured to include much new data in the Pacific Northwest. In the Pacific Northwest the contouring is based on the data of Blackwell and Robertson (1973), Blackwell (1974), Bowen and others (1977), Brott and others (1976, 1978), Hull and others (1977), and Morgan and others (1977). Local anomalies probably related to hydrothermal systems have been omitted because the consistency with which they can be shown given the present density of heat-flow data is uneven. The 1.0- and 2.0-HFU contours have been added in some areas.

The contouring of a heat-flow map is entirely subjective because only in a few locations are the data sufficiently dense for detailed analysis. Thus, given the gaps in the data, individual points that are anomalous relative to their immediate surroundings can either be ignored or be connected together to make the map much more complicated. For example, Reiter and others (1975) connected a series of high heat-flow values to form a long narrow ribbon along the Rio Grande rift. Alternatively, such values can be considered local anomalies and ignored in the contouring. The approach in drawing Figure 8-2 has been in general to consider each point significant and to connect several such anomalous points into discrete anomalies. This decision should be kept in mind by anyone using Figure 8-2.

The origin of most of the major features of the map (Fig. 8-2) is well known and is discussed in detail by Roy and others (1972) and Sass and others (1971). For example, the low heat flow along the Pacific Coast in Oregon and Washington extending into the Sierra Nevada of California is due to present and recent (past 10 m.y. or so) subduction of the Juan De Fuca plate under western North America (Roy and others, 1972; Blackwell, 1971, 1974; Hyndman, 1976). The high heat flow of the Cordilleran thermal anomaly zone (Blackwell, 1969), which consists of the Northern Rocky Mountains, Columbia Plateau, and Basin and Range province, is still a major feature of the map, but there is considerable character to the heat-flow map in these provinces (see below). The high heat flow along the San Andreas (Henyey and Wasserberg, 1971; Lachenbruch and Sass, 1973) and in the vicinity of the spreading center in the Gulf of California (Elders and others, 1972; Henyey and Bischoff, 1973; Smith, 1974; Lawver and others, 1973) shows clearly. The high heat flow in the Southern Rocky Mountains was pointed out by Decker (1969). The work of Reiter and others (1975) and Decker and Smithson (1975) in New Mexico has resulted in the identification of a major anomaly with very high heat flow associated with the Rio Grande rift. Work is in progress to solve the long-standing problem of the nature of the Basin and Range–Colorado Plateau transition (David Chapman, 1977, personal commun.). Additional data from the northern Colorado–Wyoming area (E. R. Decker, 1976, personal commun.) are in the process of analysis.

The newest data on the map are those in the Pacific Northwest, and a brief description of the anomalies will be given. East of the area of low heat flow along the coast, the heat flow increases beneath the Cascade Range, a currently active continental andesite arc. Farther to the east in Washington, northern Idaho, and western Montana, the heat flow is generally high, and the variations are primarily due to variations in crustal radioactive heat generation.

Recent data indicate that the heat-flow pattern in Oregon and southern Idaho is considerably more complicated. The heat-flow values at the western margin of the Oregon High Cascades average about 2.5 HFU, and an extremely complicated pattern of generally high heat flow is found associated with the Snake River Plain in Idaho (Brott and others, 1976, 1978). These variations in heat flow are too large to be due to variations in crustal radioactivity alone, and a combination of crustal thermal sources and large-scale thermal refraction due to crustal

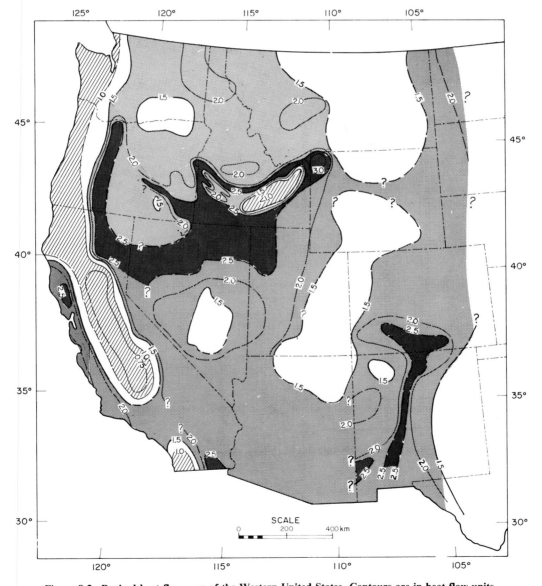

Figure 8-2. Revised heat-flow map of the Western United States. Contours are in heat-flow units.

inhomogeneities is necessary to explain the data. The low heat flow in eastern Idaho is caused by the Snake Plain aquifer. Water recharges the aquifer at the east and travels very rapidly to the west to discharge at Thousand Springs in south-central Idaho. The flow is so large that temperature gradients are near zero in the aquifer, which is over 300 m thick. No heat-flow values are available for wells that penetrate the aquifer, but Brott and others (1978) have proposed that the heat flow increases to the east to average values of 3.0 to 5.0 HFU near Island Park. The main evidence for this increase is a systematic increase in elevation eastward (see Fig. 8-5). Furthermore, average heat-flow values for a large area along the southeast margin of the Snake River Plain average 5.0 HFU (17 values).

Thus a significant thermal boundary can now be recognized between the Columbia Plateau–Northern Rocky Mountains and the Brothers fault zone–Snake River Plain. The boundary is not well located, but runs somewhere through the Blue Mountains and the central part of the Idaho batholith. In the Cascade Range the heat flow decreases northward by about 1 HFU at the Columbia River, where the young volcanism decreases in intensity.

To the south the high heat flow associated with the Snake River Plain and the Brothers fault zone (Walker, 1969) appears to merge with the Battle Mountain high described by Sass and others (1971, 1976). The origin of the high heat flow is attributed by Sass and others (1971, p. 6409) to "the transient effects of a fairly recent crustal intrusion." A second anomaly occurs farther south in Nevada, the Eureka low (Sass and others, 1971, 1976). The low heat flow is observed over an area of several thousand square kilometres and is attributed to either large-scale hydrologic phenomena or to anomalously low heat flow from the mantle.

There is high heat flow observed well to the east of the Great Plains–Northern Rocky Mountains boundary in eastern Montana and western North Dakota (Blackwell, 1969; Combs and Simmons, 1973). This region of high heat flow is somewhat enigmatic, and the origin of the excess heat flow is not known. The high heat flow may be a mantle effect, or it may be due to hydrologic effects in the Great Plains aquifers (see Adolphson and LeRoux, 1968).

NONCONDUCTIVE ENERGY RELEASE

Conventional heat-flow measurements do not measure the total energy loss from the interior of the Earth in an area of active tectonism (Shimazu, 1964; Blackwell, 1967). Significant sources such as volcanism, intrusive activity, hydrothermal convective heat transfer, and so forth, may or may not be represented in the measurements. So a map of heat flow such as Figure 8-2 does not necessarily represent the energy transfer from the upper mantle. The object of this section is to consider such energy losses for the Cordillera to see if, compared to the conductive heat-flow values, they are significant.

Stored Thermal Energy

One phenomenon that absorbs energy from the mantle without being readily apparent at the surface is heating of the lithosphere. If a conservative average warming of 200°C from the surface to 200 km is assumed for the thermally anomalous areas of the Cordillera and the heat capacity is assumed to be 0.25 cal/(g·°C), the total energy stored for a density of 3.3 g/cm^3 is 3.3×10^9 cal/cm^2. If this energy were to be supplied at a uniform rate for 80 m.y., a flux of 1.3 HFU would be required. This figure is nearly equal to the total heat flow from the mantle proposed for the Cordilleran thermal anomaly zone. Perhaps the gap in volcanic activity in the Great Basin region between the end of the Sierra Nevada event about 70 m.y. ago (Kistler and others, 1971) and the beginning of Cenozoic volcanism about 40 m.y. ago (Armstrong and others, 1969; Stewart and others, 1977) is due to the period required for heating and melting of the upper mantle and crust after shallowing of the subducting block at the end of the Mesozoic. Of course, eventually the section will cool off and the heat will be released, but in the context of the time scale of Cordilleran tectonism, the stored heat represents a significant component of the energy budget.

Thermal metamorphism of the crust could also absorb a significant amount of energy. Since the initial metamorphic grade of the crust is not known, it is difficult to estimate how much energy might be absorbed by such a mechanism. Hence no estimate of this effect is attempted.

Energy Flow from Volcanism

A second possible source of energy loss from the interior of the Earth that is not measured in heat-flow surveys is the heat lost at the surface during the eruption and cooling of volcanic rocks. The parameters needed are the volume, the heat loss per unit volume, and the time span of volcanism. For the purposes of this discussion the average heat lost during cooling of an average volcanic or intrusive rock is assumed to be about 845 cal/cm^3 (325 cal/g times 2.6 g/cm^3). The volume of volcanic rocks is the most elusive parameter because the products are quickly removed by erosion. In order to compare the energy flux to the heat flow, knowledge of the area of the source is needed, but is unknown. It is certainly larger than the intrusive center (see Marsh and Carmichael, 1974). For purposes of the discussion I will assume that the area covered by the volcanic rocks is equivalent to the source area. The results are summarized in Table 8-1.

White and McBirney (this volume) have estimated a volume flux of volcanic rocks in the Oregon Cascades during the past 20 m.y. of 8.8×10^{-6} km^3/yr per 1 km of length (about

TABLE 8-1. MAGNITUDES OF ENERGY TRANSFER FOR VARIOUS MECHANISMS

Locality	Area (km^2)	Time span (m.y.)	Volume flux (km^3/yr)	Total energy flux (cal/s)	Total heat flow (HFU)
Volcanism					
Oregon Cascades	17,500	20	3.1×10^{-3}	4.1×10^9	0.47
Andes	. .	10	3.0–4.2×10^{-6}*	. .	0.16–0.23
Columbia Plateau	250,000	3	1.2×10^{-1}	3.1×10^9	1.3
Great Basin ignimbrites	180,000	20	1.0×10^{-2}	2.6×10^8	0.15
San Juan volcanics	25,000	3	1.3×10^{-2}	2.7×10^8	1.1
Plutonism					
Yellowstone–Snake River Plain	2,500	15	5.2×10^{-2}	1.2×10^9	48
Andes	. .	. .	2.9–9.9×10^{-6}*	. .	0.16–0.54
Hydrothermal convection					
Western United States	2,300,000	. .	. .	$>2.4 \times 10^8$†	>0.01§
Southern Idaho batholith	32,500	. .	. .	6.5×10^7	0.2
Stored thermal energy					
Cordilleran thermal anomaly zone	2,000,000	40–80	. .	2.4–4.8×10^{10}	1.2–2.4

*Per 1 km of length.
†Excluding Yellowstone. Actual value is probably 2 to 10 times this amount.
§Average over areas of high heat flow.

350 km). The width of the volcanic zone is about 50 km, so the average heat-flow rate would be 0.5 HFU over the area of late Cenozoic volcanism (see Table 8-1). Francis and Rundle (1976) estimated the volume of volcanic rocks for the continental arc volcanism of the central Andes. Their estimates give a heat flow about one-half that estimated for the Oregon Cascades.

In a completely different setting, Baksi and Watkins (1973) estimated a volcanic production rate for the Miocene basalts of the Columbia Plateau, including southeastern Oregon, of 1.2×10^{-1} km^3/yr for about 3 m.y. This value corresponds to an energy flux of 3.1×10^9 cal/s and an equivalent heat-flow value for the approximately 250,000 km^2 underlain by the basalt of 1.3 HFU.

Lipman and others (1970) estimated the eruption rate for the San Juan volcanic rocks of Colorado at about 1.0×10^{-2} km^3/yr for the rocks of intermediate composition and for the ash-flow tuffs. This eruption rate corresponds to an energy flux of 1.1 HFU for the duration of the volcanism. The activity occurred between about 32 to 34 m.y. ago for the intermediate rocks and 28 m.y. ago for the ash-flow tuffs.

As a final example, the average heat flow during the voluminous ignimbrite eruptions of mid-Cenozoic time (approximately 20 to 39 m.y. ago, Stewart and others, 1977) in the Great Basin is considered. Mackin (1960) estimated the volume of the ignimbrites at 205,000 km^3. The area covered is about 180,000 km^2, so the average heat flow for the period was 0.16 HFU. This value is probably a minimum, but it is comparable to the other figures.

Energy from Intrusive Activity and Hydrothermal Convection

If plutons cool by heat conduction alone and if the heat-flow measurements are appropriately spaced, then the energy loss would be adequately represented in the heat-flow data. However, it appears that such conductive cooling seldom if ever occurs in plutons emplaced in the middle to upper crust, because these bodies cool primarily by hydrothermal convection. Thus the energy will be lost in a possibly complex way that may be missed by conventional heat-flow studies. For the purposes of this discussion, the effects of intrusive heat loss and hydrothermal convection will be considered together. As more attention has been focused on geothermal energy, some thermal studies of intrusive-related geothermal systems have been made (Lachenbruch and others, 1976a, 1976b; Morgan and others, 1977), but much remains to be done.

Francis and Rundle (1976) estimated the Mesozoic intrusive activity in the central Andes based on the area of granitic rocks of the Coastal batholith of Peru (see Table 8-1). The inferred energy flux is 0.2 to 0.5 HFU per 1 km of length if the zone is assumed to be 50 km wide. They noted that similar estimates apply to the Sierra Nevada and other regions of the circum-Pacific batholith chains; thus this heat-loss value is apparently typical of andesitic arc plutonism.

In order to obtain a different perspective, the currently most active volcanic zone in the Cordillera, the Snake River Plain–Yellowstone region is considered. Fournier and others (1976) have measured the current rate of heat loss for the hydrothermal convection system in the Yellowstone caldera. They estimated an energy-loss rate of 1.2×10^9 cal/s (Table 8-1) and a heat-flow value of 48 HFU. The current rate of energy flux in Yellowstone is probably typical of the flux rate of the hot spot for the past 15 m.y. or so. As another perspective on the energy transfer, the total conducted heat for the Cordilleran thermal anomaly zone between the Canadian border and the southern limits of the Great Basin (about 1.2×10^6 km^2) is about 2.6×10^{10} cal/s, so that the convective heat transfer in Yellowstone is equal to 5% of the total conductive heat loss in the Cordillera!

Yellowstone is the giant among the active convective systems in the Western United States.

The energy-flow rate calculated for all geothermal areas in the Western United States from expected subsurface temperatures and flow rates tabulated by Renner and others (1975) is 2.4×10^8 cal/s. This value is probably a considerable underestimate, however.

Some, perhaps most, of this energy flow represents cooling shallow magma chambers, but some merely represents redistribution of the regional conductive heat flow. For example, in the southern part of the Idaho batholith where the last major thermal event was in Eocene time (Armstrong, 1974) and where there is no evidence for contemporary volcanic or intrusive activity, fully 10% of the average regional heat flow, 0.2 HFU, is transported to the surface by hot-spring activity. The country rock is granite, which is usually considered a relatively impermeable rock type.

There is no doubt that hydrothermal convection plays an important role in energy transfer in the Cordillera. In local areas the percentage of the energy transported in this manner may be larger than the total heat flow outside such areas. Comparison of the energy flux for the convective system to that from volcanism and conductive heat flow suggests that over 90% of the total heat during the life of the system is transferred to the surface by the convective system. As an overall average it would appear that 10% or more of the average conductive heat flow in the Cordilleran thermal anomaly zone is carried to the surface by this mechanism, and energy-flux averages should reflect this additional energy loss.

Miscellaneous Energy Sources

Other phenomena that may play a part in the energy budget include the energy lost in seismic wave radiation during an earthquake and the energy required to make the fracture. Both of these effects can be shown to be less than 0.02 HFU (Blackwell, 1967). The radiated energy can be calculated by using the recurrence relationship for earthquakes in the various areas of the Cordillera (Ryall and others, 1966) and the energy versus magnitude relationship (Richter, 1958). The increase in potential energy required during uplift is also small.

In areas near active faults there may be significant contributions to the conductive heat flow from friction-generated heat along the fault zone. Such a mechanism may explain the high heat flow along the San Andreas in California (Brune and others, 1969; Henyey and Wasserburg, 1971; Lachenbruch and Sass, 1973). Such heat will in general be measured by regional heat-flow studies and does not need to be considered here.

Summary

The discussion of this section has shown that numerous actual or potential nonconductive energy-loss mechanisms are operative in regions of active tectonism and that these mechanisms must be included in any attempt to estimate total heat flow or energy budget for a given area. Estimated rates of energy loss by volcanism during active periods may range from 0.1 to 1.5 HFU. Estimated rates of energy loss by intrusive activity are as much as one order of magnitude more. The smaller values relate to activity averaged over larger areas and times, whereas the larger values relate to areas of active intrusion and volcanism. The average energy loss from the upper mantle in the past 40 m.y. or so for the Cordilleran thermal anomaly zone has been from 0.5 to 1.5 HFU greater than the Q_0 difference of 0.6 HFU between the heat flow in the Basin and Range province and the Eastern United States (Roy and others, 1968). This extra energy loss must be accounted for by any tectonic model for the development of the Cordillera. The object of the next section is to investigate the geographic distribution of these various mechanisms of energy loss in the Cordillera.

CORRELATION OF HEAT FLOW AND OTHER ENERGY-LOSS MECHANISMS

Distribution of Cenozoic Volcanism

Figure 8-3 shows a generalized map of the distribution of Cenozoic volcanic rocks in the Western United States. This map was compiled from the "Geologic Map of North America" (1965), the various state geologic maps, and recent summaries of the age of Cenozoic volcanism published by Stewart and others (1977) and Stewart and Carlson (this volume). The map (Fig. 8-3) represents an attempt to show the time of initiation of volcanism, its duration, and the time of the last major volcanism in broad areas. Such a map does not convey an idea of the volume of volcanism represented in the various areas or the intensity of the thermal event associated with the volcanic episodes. This problem will be discussed in the final section. The 1.5-HFU contour from Figure 8-2 is shown on this and several subsequent maps for reference, as it is the contour that most clearly outlines the regions of high and normal or low heat flow.

Not surprisingly there is a close correlation between the areas of Cenozoic volcanism and the areas of high heat flow. Also the areas of highest heat flow correlate reasonably well with areas in which volcanism has occurred within the past 17 m.y. The notable exception to this generalization is the area of the Columbia Plateau basalts (dated at 15 to 17 m.y. B.P.) in Washington (Baksi and Watkins, 1973) where the surface heat flow is between 1.2 and 1.5 HFU. The source of the basalts is near the common boundary of the Washington, Idaho, and Oregon where heat-flow data are sparse.

Lipman and others (1972) and Christiansen and Lipman (1972) have discussed the overall compositional variations with time and space of the Cenozoic volcanism. They discussed the general change with time of the volcanism from andesite to the basalt-rhyolite association at different locations in the Western United States. In the preparation of Figure 8-3, some areas of minor young basaltic volcanism have been included with older age groups. It is known that the upper mantle is partially molten under the whole Cordilleran thermal anomaly zone, and it seems more likely that the heat flow will relate to province-wide silicic or combined silicic-basaltic volcanic activity rather then to relatively isolated basalt occurrences.

Distribution of Late Cenozoic Intrusive Activity and Hydrothermal Systems

The geographic distribution of late Cenozoic intrusive activity and current major hydrothermal activity is shown in Figure 8-4. Included are the locations of "identified volcanic systems" in the Western United States where an "identified volcanic system" is one "showing evidence from the presence of silicic eruptives that a high-level magma chamber formed in the recent past or is forming at the present time" (Smith and Shaw, 1975, p. 73). Most of the centers shown in Figure 8-4 are the centers of extensive Holocene or Pleistocene volcanism associated with, in many cases, extensive hydrothermal systems. Some examples are the Cascade stratocone volcanoes, Yellowstone, the Jemez Mountains, and Long Valley. The locations of these centers, of course, agree well with the areas of youngest volcanism shown in Figure 8-3. Nonetheless, it is interesting to see where contemporary magma chambers may be. The postulated magma chambers are concentrated in narrow bands along the Cascade Range, the Sierra Nevada–Great Basin transition (the western boundary of the Cordilleran thermal anomaly zone), the Intermountain seismic belt (Smith and Sbar, 1974; the eastern boundary of the Cordilleran thermal anomaly zone), the Rio Grande rift zone, the San Andreas–Gulf of California system, Basin and Range–Colorado Plateau boundary in Arizona, and the Snake River Plain–Brothers fault zone feature in Oregon and Idaho.

The locations of hydrothermal convection systems with estimated aquifer temperatures (based on geochemical thermometers) of 90 to 150°C and in excess of 150°C (Renner and others, 1975) are also shown in Figure 8-4. In general these systems are the largest ones. A peculiar feature of this map is that, with a few notable exceptions, the largest and hottest hydrothermal systems are not associated with the young magma chambers. A map of all the hot springs in the United States (Waring, 1965) is similar, but has many warm springs outside the 1.5-HFU contour.

There are numerous correlations of the heat-flow data (Fig. 8-2) with these data. There

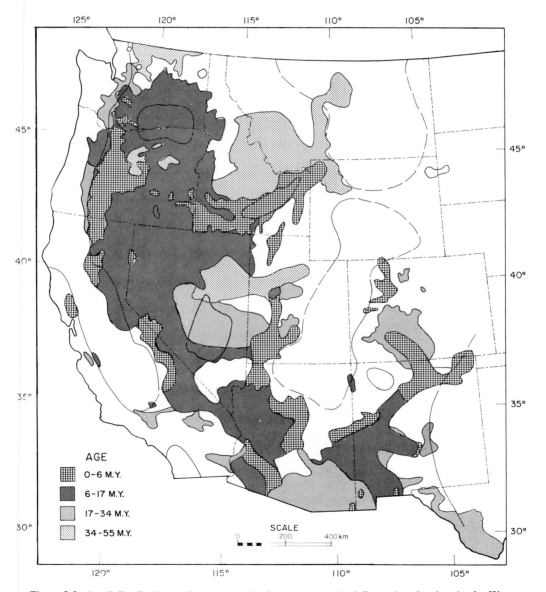

Figure 8-3. Areal distribution and age range (see key to patterns) of Cenozoic volcanism in the Western United States. Areas of no large-scale Cenozoic volcanism are blank. The 1.5-HFU contour from Figure 2 is shown for reference.

are areas of unusually high heat flow associated with the Oregon Cascade intrusive zone, the Rio Grande rift, the Snake River Plain, and the San Andreas–Gulf of California region. In contrast, on the basis of present data, the heat flow does not seem to be unusually high along the Sierra Nevada–Great Basin transition. There are not yet enough data to evaluate the heat flow along the Basin and Range–Colorado Plateau transition where there are also a series of young intrusions.

Probably one of the most striking correlations is between the location of the major hydrothermal systems and the area of very high heat flow included in the Battle Mountain–Snake River

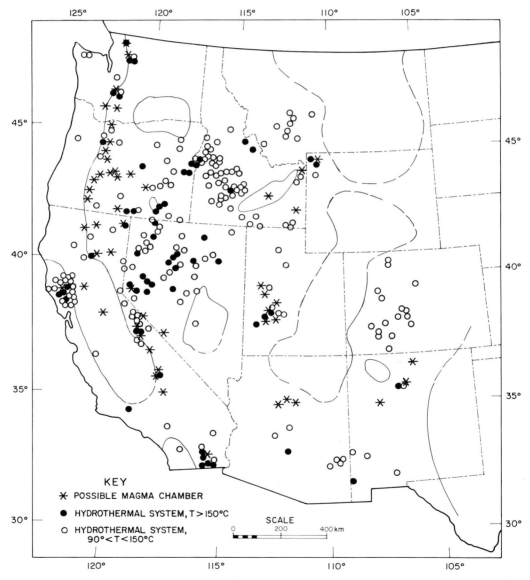

Figure 8-4. Location of possible contemporary magma chambers and major hydrothermal systems. The 1.5-HFU contour from Figure 8-2 is shown for reference.

Plain–Brothers fault zone region. This area has also been characterized by volcanism (and presumably intrusive activity) between 17 and 6 m.y. ago (see Fig. 8-3). The volcanism younger than 6 m.y. is basaltic and relatively small in volume, and the main zones of silicic volcanism have shifted to the east (Armstrong and others, 1975) and west (MacLeod and others, 1976). Thus many of the most extensive hydrothermal systems are restricted to areas of older centers of volcanism. Most of the remainder of the area of high heat flow (Cordilleran thermal anomaly zone and Southern Rocky Mountains) is associated with volcanism, but only in the middle to early Cenozoic (more than 17 m.y. ago). Evidence of young magma chambers and extensive hydrothermal systems is lacking in these areas.

Although detailed heat-loss budgets for the possible magma chambers are few, on the basis of the inferences in the previous section, much more heat is probably being lost from these systems than is indicated from the amount of volcanism alone. Data on the heat loss from these centers are necessary before the current rates of energy loss can be accurately determined.

Regional Topography

Topography is another characteristic feature that is related to the various mechanisms of energy release. Suppe and others (1975) have discussed the relation of Quaternary volcanism with topography. Strange and Woollard (1964) published a map of averaged topography for the Western United States that is useful for this type of comparison, although the size of the averaging area is probably too large (1° of latitude × 1° of longitude). This map is shown in Figure 8-5 and has been generalized in that the original contour interval of 500 ft (152 m) has been replaced by contours at 500-m intervals. The map of regional topography is virtually identical to the regionally averaged Bouguer gravity map (Mabey, 1966; Eaton and others, this volume), and both reflect the same crust and mantle effects.

Comparison of Figures 8-5 and 8-2 quickly illustrates that there are only gross similarities between the two sets of data. The major influence on the topography is the average density of the crust-mantle section down to some depth. The average density is of course more affected than the temperature by variations in crustal thickness, for example (see Crouch and Thompson, 1976), so that the map of Figure 8-5 includes the effects of compositional density variations as well as density variations due to temperature (phase changes or thermal expansion). The geographic variations of crustal thickness and density have been discussed elsewhere (Eaton and others, this volume; Hill, this volume), but consideration of these other effects should be kept in mind during this discussion.

In general, the Western United States is characterized by high heat flow and high topography, but there are many exceptions to this generalization and there is no simple one-to-one correlation. For example, some areas of high elevation have high heat flow, such as the Southern Rocky Mountains and the Yellowstone region. On the other hand the Gulf of California and the Rio Grande rift have high heat flow and are low in elevation, whereas the Sierra Nevada has very low heat flow and high average elevation. Suppe and others (1975) argued that the region of highest elevation has the highest crust and mantle temperatures (essentially outlined by the 2.0-km elevation contour in the central Western United States). This region cuts across most of the geologic and geophysical trends of the Cordillera, and such a simple correlation seems to make little tectonic sense.

Two areas that seem to show a relatively straightforward relationship between heat flow and topography are the Oregon Cascades and Coast Ranges, and the Snake River Plain–Yellowstone region. In western Oregon the heat flow is low between the coast and the edge of the High Cascade province (the locus of Pliocene to Holocene volcanism). At that point the heat flow increases dramatically from about 1 HFU to about 2.5 HFU over a distance of 20 km or less (Hull and others, 1977). The topography also rises from about 100 to 300 m to over 1.0 km, but rather more gradually. The gradual change is exaggerated even more because of the averaging grid size on Figure 8-5. There appears to be no significant change in crustal structure across the heat-flow transition (Johnson and Couch, 1070); therefore, it is possible that the elevation difference is related fairly directly to temperature effects.

In the Snake River Plain, Armstrong and others (1975) documented a progressive eastward-moving episode of silicic volcanism culminating in the contemporary activity at Yellowstone. The volcanism began about 17 m.y. ago near the Idaho-Oregon border and progressed eastward at a velocity of about 3.5 cm/yr. Brott and others (1978) showed that the heat-flow data in the western Snake River Plain can be explained by the volcanic-age model of Armstrong and others (1975) and presented a thermal model for the Snake River Plain based on the volcanic-age model. They predicted heat-flow values that increase systematically eastward toward Yellowstone. These heat-flow values reach 3 HFU or more along the axis of the Snake River Plain (shown in Fig. 8-5 for reference) in the east. Along the axis of the Snake River Plain, elevation increases smoothly from about 700 m in the west to about 2,500 m on the Yellowstone Plateau. This systematic change in elevation is attributed by Brott and others (1978) to the progressive decrease in heat flow (and lithospheric temperatures) to the west from Yellowstone (the present position of the hot spot) to the Oregon-Idaho border (the location of the hot spot 17 m.y. ago).

This example also points out one of the enigmatic features of the correlations of heat flow with topography, that is, although the active center of volcanism stands high, slightly older parts of the Snake River Plain, although still much hotter than the surroundings, are lower in elevation. This apparently anomalous behavior is common—for example, the Great Basin is lower than but hotter than the Colorado Plateau and Sierra Nevada; the Rio Grande rift has the same relationship to its surroundings, and so forth. Suppe and others (1975) pointed out this problem and suggested that in the case of the Snake River Plain, the change in elevation before and after the volcanic episode was due either to loss of crustal material by explosive volcanism that removed the material from the area and/or subcrustal erosion. The net result of the process is a crust-mantle section that is more mafic in composition, and hence its higher average density more than compensates for the increased temperature.

There are several other processes that can lead to the same result. While the volcanic center remains high, material may be removed from the center by normal erosion. This results in a subsequent increase in density of the overall crustal and upper-mantle section after cooling. Secondly, there may be spreading or rifting associated with the volcanic event. The rifting process thins the granitic part of the crust with essentially no chance for its replacement. This results in a much more mafic and thus denser crust and upper-mantle section. The action of this process is most clearly shown in North America in the case of the Gulf of California (Elders and others, 1972). Each stage of the progressive disruption of the continental crust is shown between the Mojave desert and the Gulf of California itself. In this process little silicic volcanism occurs even though the continental crust is completely split, in contrast to the extensive silicic volcanism of the Snake River Plain.

If any or all of these processes occur and the topography becomes inverted upon cooling, then the deposition of sediments back into the resulting basin could lead to subsidence from loading. For example, the western Snake River Plain has about 2 km of sediments on top of most of the volcanic rocks (Newton and Corcoran, 1963). Whatever the mechanism within the Snake River Plain itself, the process may have operated relatively uniformly since the various parts show a systematic correlation between heat flow and elevation.

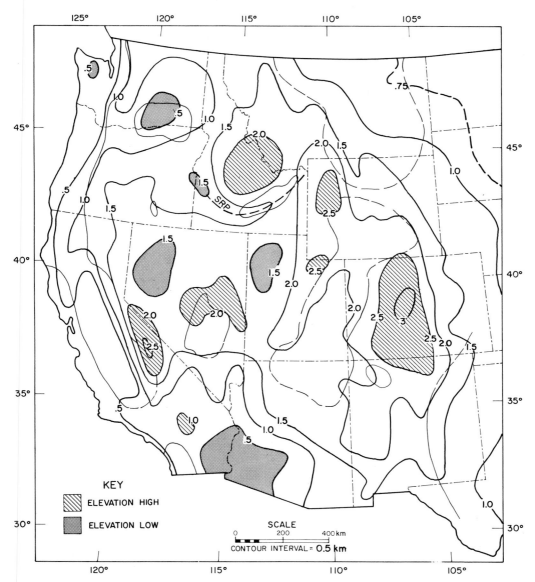

Figure 8-5. Average elevation (1° × 1° squares; Strange and Woollard, 1964) in the Western United States. Closed high and low areas are shaded, and the axis of the Snake River Plain (SRP) is indicated. The 1.5-HFU contour from Figure 8-2 is shown for reference.

The collapse of topography at some time in a volcanic and/or rifting cycle so that eventually higher heat flow may coincide with lower topography can explain many of the apparently anomalous relationships between heat flow and topography in the Western United States. For example, the Basin and Range province sits lower than the bounding Colorado Plateau or Sierra Nevada in spite of its higher heat flow. On the other hand, the Southern Rocky Mountains with high heat flow and some of the highest crustal temperatures in the United States, but no extensive young volcanism and rifting, stand well above all the surrounding provinces.

RELATIONSHIPS BETWEEN HEAT FLOW AND HEAT PRODUCTION

Interpretation of conventional heat-flow data such as shown in Figure 8-2 relies heavily on the relationship of the heat flow to the radioactive heat generation of the crust. So in order to carry the discussion of the heat-flow data forward, this section will discuss the relationship of heat flow to radioactive heat generation in the Cordilleran thermal anomaly zone.

Birch and others (1968) found that the heat flow (Q) and the radioactive heat production (A) in plutonic rocks in New England were linearly related by an equation of the form

$$Q = Q_0 + bA \tag{1}$$

where Q_0 is a constant that can be interpreted as the heat flow from below a radioactive layer of thickness b, the slope of the line relating Q and A. Similar results were later reported for parts of the Western United States (Roy and others, 1968, 1972; Lachenbruch, 1968, 1970). Roy and others (1968) identified two thermal provinces in the Western United States on the basis of their Q-A relationships: the Sierra Nevada and the Basin and Range province. The Sierra Nevada curve has a slope of 10.1 ± 0.1 km and an intercept of 0.40 ± 0.03 HFU; the Basin and Range curve has a slope of 9.4 ± 1.3 km and an intercept of 1.40 ± 0.09 HFU (standard errors are given for all the parameters). Lachenbruch (1968, 1970) discussed the Q-A relationship for the Sierra Nevada and the meaning and interpretation of the linear relationship.

Subsequently, it was found that the Basin and Range Q-A relationship applied fairly well to the Northern Rocky Mountains (Blackwell and Robertson, 1973; Blackwell, 1974) and to the Southern Rocky Mountains (E. R. Decker in Roy and others, 1972). Sass and others (1976) noted the difficulty in applying the same line to the heat flow observed in Nevada.

The correlation of heat flow and heat production in granite has led to great advances in heat-flow interpretation. However, in the Western United States the conventional Q-A relationships cannot be obtained and used in interpretation because in most of the area no granite is exposed and, in numerous instances, even the nature of the basement is unknown. In Nevada, for example, only 2.8% of the exposed rocks are granite (Archbold, 1972), and many of the exposures are small and thus are subject to uncertainties of interpretation relating to the thermal conductivity and heat generation of radioactive elements of the surrounding country rock. In Oregon the percentage of granite outcrop is even less than in Nevada, and in fact the entire southeastern quarter of the state has no granite exposures suitable for heat-flow measurements. Thus, heat-flow measurements must be made in areas that are not ideal for interpretation and at sites that may be severely biased relative to the regional heat flow.

The significance of the Basin and Range Q-A relationship is the subject of some discussion and will be briefly reviewed. Roy and others (1968) noted that heat-flow values too high to be explained by the Basin and Range Q-A relationship and by reasonable variations in crustal radioactivity were commonly observed in the thermally anomalous regions of the Cordillera. These heat-flow values were presumed to be related to crustal intrusive events or nearby hydrothermal anomalies. Furthermore, Roy and others (1968) pointed out that melting would occur at shallow depths in the mantle given the temperature-depth curve implied by the Basin and Range Q-A relationship and that only a small rise in temperature and heat flow could cause crustal melting. Thus they concluded that regional heat-flow values much in excess of an average of 2.5 HFU were not likely to occur except in regions of active crustal disruption and/or silicic volcanism.

If the Q-A data in a given area fit a straight line, then the intercept value may be interpreted as the heat flow from below the radioactive layer. In the Western United States, the slopes of the two Q-A relationships are about 10 km, and Roy and others (1972) defined "reduced heat flow" as the observed heat flow minus the heat generation in a 10-km-thick layer (or b, if known) of the observed A value. To a first approximation, then, Q_0 may be calculated for isolated Q-A pairs. This heat-flow value is called a reduced heat flow to emphasize the assumptions needed for its calculation. In an area of active tectonism and volcanism, the scatter in heat flow is far too extreme to be explained by variations in radioactive heat production and too rapid to be caused by mantle sources. In such an area, calculation of reduced heat flow will remove the first-order crustal radioactivity effects so that these other types of effects can be more readily examined.

Figure 8-6 shows the locations of measurements converted to reduced heat-flow values for the Western United States and the generalized areas of Cenozoic volcanism younger than 17 m.y. and of Cenozoic volcanism older than 17 m.y. Heat flow and heat production in these two areas are shown in Figure 8-7. The values from California have been omitted because of the complexity of thermal and tectonic forces at work there.

There is a significant correlation between the volcanic age data and the reduced heat flow. In general, values of reduced heat flow in the range 1.0 to 1.5 HFU are found in areas where the age of the last major volcanism is older than about 17 m.y. These data approximately fit the Basin and Range Q-A line of Roy and others (1968). The only exception to the systematic correlation of the Basin and Range Q-A relationship with areas characterized by older volcanism is the Columbia Plateau. Part of the Columbia Basin in Washington covered by the voluminous Miocene basalts has a consistent reduced heat flow of 1.2 to 1.4 HFU (Blackwell, 1974; Hull and others, 1977).

In areas where volcanism is younger than 17 m.y. the Q-A relationship is much more complex, and indeed it is questionable if one exists in the conventional sense. Major subprovinces in the Great Basin (the Battle Mountain high and the Eureka low) have been identified by Sass and others (1971). These provinces more or less correspond to the areas in Nevada where volcanism occurred from 17 to 6 m.y. ago. In east-central Nevada there is an area of older volcanism, and most of the points from Nevada that fall near the Basin and Range curve are found there (see Fig. 8-6).

In the areas of most voluminous young volcanism—the High Cascades of Oregon, the Snake River Plain, and the southeastern fourth of Oregon—reduced heat-flow values cannot be readily determined because no granite is exposed. Only along the western margins of the Snake River Plain are there extensive exposures of granite suitable for calculation of conventional reduced heat flow. Five data points from the margins of the western Snake River Plain are shown in Figure 8-6. Three of the Q-A points are discussed by Brott and others (1976), and Urban and Diment (1975) have published heat-flow values for the two others. Radioactivity values for the last two heat-flow values have been obtained from regional radioactivity measurements (Swanberg and Blackwell, 1973; Blackwell, unpub. data). The average reduced value for these points, all within 10 to 20 km of the margins of the plain, is 2.8 HFU, yet the age of most of the voluminous silicic volcanic activity is 10 to 15 m.y. (Armstrong and others, 1975). Brott and others (1978) have shown that this high value of heat flow along the margins of the Snake River Plain can be explained by the combination of a small residual thermal anomaly resulting from the volcanic event and large-scale thermal refraction due to the thermal conductivity contrast between the sedimentary and volcanic rocks beneath the Snake River Plain and the granites of the Idaho batholith along the margins.

Of course, in contemporarily active Yellowstone Park, the average convective heat flow

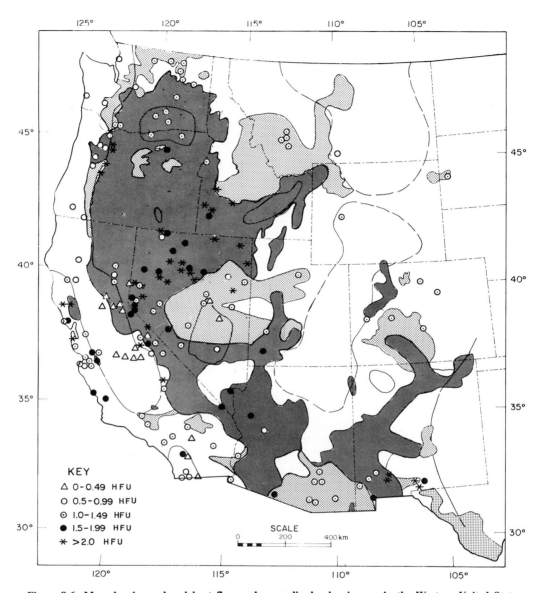

Figure 8-6. Map showing reduced heat flow and generalized volcanic ages in the Western United States. Areas of volcanism less than 17 m.y. and between 17 and 55 m.y. old are shown in dark and light shading, respectively. The 1.5-HFU contour from Figure 8-2 is shown for reference.

over the 2,500 km^2 of the caldera is 48 HFU (Fournier and others, 1976), and over the portion of the park covered by Yellowstone Lake, the conductive heat flow from the underlying convective system is about 20 HFU (Morgan and others, 1977). The "reduced heat flow" for this 50- to 75-km-wide section of the crust is only 0.3 to 0.5 HFU less than the observed heat flow. Certainly this heat flow must represent an upper limit for heat transfer per unit area for a continental area.

In southeastern Oregon, the heat-flow values are highly variable; they average 1.4 to 3.0

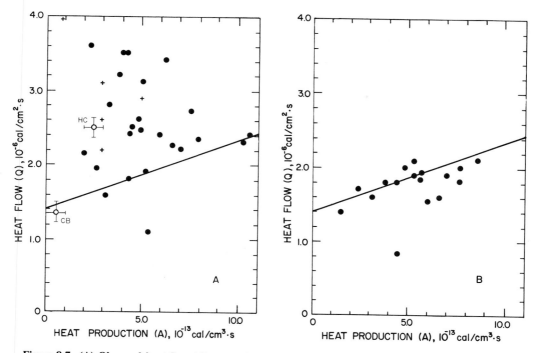

Figure 8-7. (A) Observed heat flow (Q) versus heat generation (A) in the upper crust for the Cordilleran thermal anomaly zone (excluding California) for areas where provincial volcanism is less than 17 m.y. old. HC is for the western border of the High Cascades, and CB is for the Columbia Basin portion of the Columbia Plateau. The plus signs represent data along the margin of the western Snake River Plain. The Basin and Range Q-A line is shown for reference. (B) Observed heat flow versus heat generation in granitic rocks for areas where provincial volcanism is greater than 17 m.y. old. The Basin and Range Q-A line is shown for reference.

HFU over areas of several tens of kilometres square (Hull and others, 1977). The nature of the crustal radioactivity is not known because no basement rocks are exposed. However, on the basis of measurements on the granitic rocks of the Blue Mountains (Swanberg and Blackwell, 1973) and the probable nature of the crust, a value of 3 HGU is probably an upper limit to the heat production, and lateral variations are unlikely to cause heat-flow variations of more than 0.2 ± 0.1 HFU. Thus, most of the variations must be due to structure, shallow heat sources, and/or large-scale hydrologic effects.

An east-to-west thermal event in southeastern Oregon, mirroring in many ways the Snake River Plain trend, has been described by MacLeod and others (1976). No elevation variation coincides with the Oregon volcanic trend, however, and the energy loss associated with the event is probably 1 or 2 orders of magnitude less than that of the Yellowstone event. Still the average regional heat flow is much above the value of 1.5 to 1.7 HFU predicted by the Basin and Range Q-A relationship. A simple unweighted average of heat-flow measurements for southeastern Oregon, excluding the western Snake River area, is 2.5 ± 0.4 HFU (standard error shown).

In a recent study of heat flow in the Oregon Cascades, Hull and others (1977) measured heat-flow values of 2.5 HFU in the volcanic rocks at the west margin of the High Cascades. The heat production of the few small stocks exposed in the western Cascades ranges from 1 to 3 HGU (Gosnold, 1977). The volcanic rocks are probably underlain by rocks like these stocks, and if their values are typical of the crust, then the reduced heat-flow values are greater than 2 HFU. The energy loss by volcanism in the Oregon Cascades is given in Table 8-1.

The present data make it clear that there are large regions in the Western United States that have reduced heat-flow values much above the intercept value of the Basin and Range Q-A relationship. The main characteristic of these regions is the presence of extensive volcanism within the past 17 m.y. Thus the scatter of points on Figure 8-7A is caused by the complexity of the heat-source distribution in areas of young thermal events, the complexity of heat-transfer mechanisms, and large-scale variations in the thermal conductivity of the crust. Given these factors, a simple linear Q-A relationship cannot be expected to hold for large areas in the provinces with young tectonism and volcanism.

CHARACTERISTICS OF HEAT–FLOW TRANSITION ZONES

One of the early objectives of detailed heat-flow studies was the investigation of the nature (magnitude, width, and so forth) of the transition zones between the various thermal provinces. Roy and others (1972) found that the Sierra Nevada–Great Basin heat-flow transition at lat 39°N occurred over a distance of less than about 50 km, and they measured only one point in the transition zone. These data are shown in Figure 8-8A. The Great Plains–Basin and Range transition was found to occur over a distance of less than 50 km with no points in the transition zone (see also Warren and others, 1969). Subsequent work has been done in both transition zones [see Decker and Smithson (1975) and Reiter and others (1975) for data on the Basin and Range–Great Plains transition]. A cross section is shown in Figure 8-8A for the Sierra Nevada–Basin and Range transition at lat 39°N including data from Sass and others (1971) and Henyey and Lee (1976). In addition, Lachenbruch and others (1976a; also Lachenbruch, 1968) have published data on the Sierra Nevada–Great Basin transition between lat 37°N and 38°N, south of the area studied by Roy and others (1972) (see Fig. 8-8B).

The main conclusion that can be drawn from the results shown in Figure 8-8 is that the transition zones are much more complicated than suggested by earlier data. In general, variations of heat flow of 50% to 100% occur across distances of 10 to 20 km, and a single zone where the province-wide average heat-flow values change in some smooth way does not occur. The rapid spatial variations observed must be caused by crustal, not mantle, effects. The two main crustal effects that might be involved are large-scale thermal refraction and magmatic and/or hydrothermal heat sources. Lateral variations of radioactive heat sources can be ruled out except as local complications (within the Sierra Nevada, for example) because the observed effects are much too large.

From geophysical studies it is known that the physical property differences among heat-flow provinces (seismic velocity, for example) extend to depths of 150 to 200 km or more (Archambeau and others, 1969; Pakiser and Zietz, 1965). It can also be argued from regionalized topography (see Fig. 8-5, the systematic variation in elevation in the Great Plains, for example) and gravity (Thompson and Talwani, 1964) that the upper-mantle temperature differences persist deep into the mantle. No boundary extending to these depths can cause the sharp surface boundary. In the Sierra Nevada, the reduced heat flow is constant at 0.4 HFU from the western foothills to the crest. High heat-flow values occur within 20 km of this low reduced heat flow at both lat 39°N and lat 37° to 38°N. The major crustal boundary where the crust thins from about 45 km in the Sierra Nevada to about 25 km in the Great Basin (Eaton,

1966; Johnson, 1965) occurs well to the east of the first high heat-flow values (see Roy and others, 1972). There is a systematic change in elevation from west to east across the Sierra Nevada with the east side having an elevation more than 2 km above the west even though the heat flow is equally low on both sides. The elevation changes must be due to recent (past 5 m.y.) lateral or bottom heating of the cold block, although the heat has not yet been conducted to the surface. However, magmas are rising to the surface in advance of the thermal conduction, and thus a very sharp apparent heat-flow transition or several apparent transitions are generated by the crustal magmatism even though the mantle transition may be more gradual and the actual source of the high heat flow may be in the upper mantle.

Thus the Sierra Nevada boundary is a very complex thermal transition, and the heat flow (Fig. 8-8) is responding in a complex way to the mechanisms by which the energy is transferred and to the tectonic and structural influences on the heat flow. Furthermore, the measured heat flow may be only part of the energy transfer involved. Most of the heat-flow transition zones are similarly complex. Thus the nature of the heat-flow pattern in such a transition zone may relate more to details of the shallow crustal structure and the mechanisms of shallow crustal heat transfer than to the presumed mantle sources of energy loss.

DISCUSSION

As a result of the additional data available at the present time, it is apparent that the heat-flow pattern in the Western United States, and particularly the Cordillera, is very complicated. The discussion in this section represents an attempt to generalize the results presented and to discuss a simple regional model for the high heat flow in the Cordilleran thermal anomaly zone. The heat flow in the coastal regions has been discussed previously (Blackwell, 1971; Roy and others, 1972) for the areas of low heat flow and for the San Andreas system (Henyey and Wasserburg, 1971; Lachenbruch and Sass, 1973).

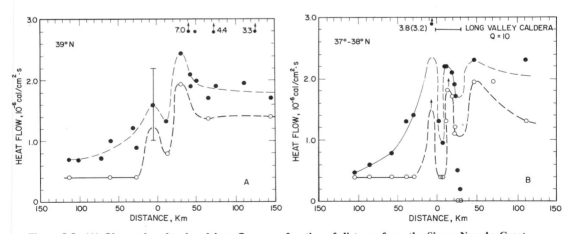

Figure 8-8. (A) Observed and reduced heat flow as a function of distance from the Sierra Nevada–Great Basin transition at lat 39°N. The observed heat-flow values are shown as dots, and the reduced values are shown as circles. (B) Observed and reduced heat flow as a function of distance from the Sierra Nevada–Great Basin transition at lat 37° to 38°N. The observed heat-flow values are shown as dots, and the reduced values are shown as circles.

Energy Flux in the Western United States

Building on the results of the previous discussions, an energy-loss (heat-flow) map is shown in Figure 8-9 that is different in concept from previous maps of heat flow (see Figs. 8-1 and 8-2) in the Western United States. This map represents a semiquantitative attempt to evaluate the actual amount of energy reaching the surface after removal of the effects of heat-flow perturbations and nonconductive energy losses related to structure, volcanism, regional aquifer systems, and hydrothermal convection. The map is based on the distribution of energy loss at crustal levels (although the source of the energy is in the mantle), and so no transition zones are shown (on this scale) between most of the areas of contrasting heat flow. Outside the Cordilleran thermal anomaly zone, the energy-loss map is virtually identical to a heat-flow map, although some features are different. For example, the high heat flow in the northern Great Plains (Fig. 8-2) has been left off because it may be a hydrologic feature. In the thermally anomalous regions, more character is shown than in the heat-flow map of Roy and others (1972), but the distribution of energy loss is considerably different than shown by a conventional heat-flow map (Figs. 8-1B and 8-2). In Figure 8-9 the high energy-loss areas are divided into three categories shown by different map patterns.

The areas that fit the Basin and Range Q-A relationship (and where no heat-flow data exist, those areas with volcanism no more recent than 17 m.y. ago) fall in the energy-loss range of 1.5 to 1.99 HFU. Here the reduced heat flow is about 1.4 HFU; the surface heat flow ranges from 1.4 to 2.4 HFU and averages 1.8 ± 0.1 HFU, and nonconductive energy transfer is minor. These areas include the Northern Rocky Mountains, the Columbia Basin, the central part of the Great Basin, and part of the Basin and Range province in Arizona. A small area of this characteristic heat flow is postulated in extreme southern Nevada, based on the absence of regional volcanism there.

The second category within the areas of high heat flow represents areas characterized by energy loss ranging from 2.0 to 3.0 HFU, major volcanic episodes greater than several million years old, and extensive hydrothermal systems. The major regions in this category are the northern Great Basin, the western Snake River Plain, the High Lava Plains in Oregon, part of the Basin and Range province in Arizona, and the Rio Grande rift zone. In this interpretation, the Battle Mountain high of Sass and others (1971, 1976) is included as part of this intermediate category of the Cordilleran thermal anomaly zone and not as a distinct province. The volcanism in that part of Nevada ended approximately 10 m.y. ago, and there are extensive hydrothermal systems active at the present time. There are no obvious tectonic or volcanologic reasons to separate this area from the surrounding areas.

The third category of high heat flow includes areas with total energy-loss values greater than 3 HFU (much greater in some areas). This category includes the areas of currently active volcanism and tectonism such as Yellowstone and the eastern Snake River Plain, the Oregon Cascade Range, the eastern and western boundaries of the Great Basin, the southern boundary of the Colorado Plateau, parts of the San Andreas system (the Geysers region), the Rio Grande rift (the Valles caldera and possibly other areas), and the Imperial Valley. In most of these areas the conductive heat flow is a fraction of the total heat loss and the main energy-loss mechanisms are volcanism and hydrothermal convection. The energy loss may range from values of 3 to 5 HFU across tens of thousands of square kilometres to 40 to 50 HFU across thousands of square kilometres in the cases of most extreme crustal disruption.

There are a number of correlations of the energy-loss map in Figure 8-9 with certain geologic and geophysical features that are not so clearly shown in conventional maps of heat flow.

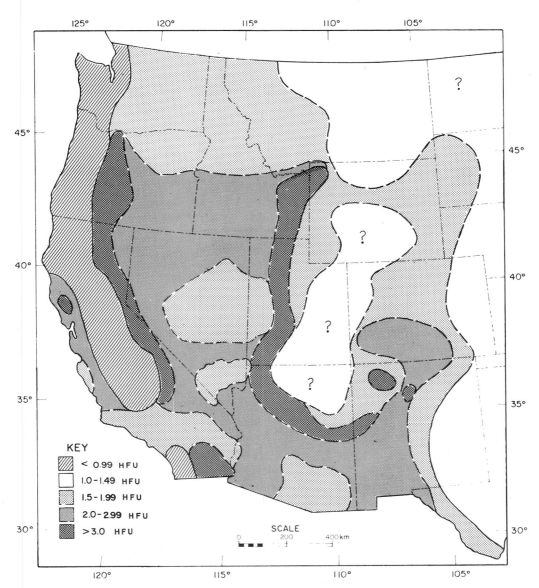

Figure 8-9. Energy-flux map of the Western United States. Contours are in heat-flow units.

For example, the map of energy loss indicates a number of thermal boundaries inside the Cordilleran thermal anomaly zone that are related to zones of seismicity, crustal structure changes, and electrical resistivity variations. In particular the transition from the areas of high heat flow of southern Oregon and Idaho to the areas of lower heat flow in northern Oregon, Washington, northern Idaho, and Montana is an upper-mantle electrical resistivity boundary (Gough, 1974), a crustal structure boundary (Hill, this volume), and a seismic belt (Smith, 1977, and this volume). Similarly, the areas of possible transitions in heat flow in southern Nevada correlate with major changes in crustal and upper-mantle characteristics (Eaton and others, this volume).

Thus, in the Western United States, most of the tectonically and volcanically active belts are thermal boundaries and/or areas of unusually high energy loss. Recognition of this important property of these belts will help in understanding the nature, origin, and mechanism of formation of these important features. The reasons for this concentration of energy loss are not known. Some of the possibilities are that a melting event is moving outward from the center of the Basin and Range province (Armstrong and others, 1969), the transition zones are the locations of mantle heating above that in the rest of the thermally anomalous regions, or the transitions act as stress concentrators because of the change in mechanical properties across these zones and thus extension concentrates at these locations. The correlation of earthquake zones, thin crust, and electrical resistivity variations with the eastern and western boundaries of the Cordilleran thermal anomaly zone has been pointed out (Blackwell, 1969; Roy and others, 1972; Smith and Sbar, 1974; Smith and others, 1975; Smith, 1977, and this volume).

Origin of the Thermal Pattern in the Cordilleran Thermal Anomaly Zone

The accumulated data discussed in this paper reinforce the conclusion that the thermal features of the Cordillera are dominantly shaped by the effects of the Mesozoic and Cenozoic subduction (Blackwell, 1971; Roy and others, 1972). The average energy anomaly above the normal continental heat flow of the Cordilleran thermal anomaly zone throughout the past 40 m.y. at least has been 1.5 to 3.0 HFU over an area averaging 800 km in width. This amount of energy for this period of time can only come from active plate interactions, and the thermal patterns, including the areas of low heat flow, can best be explained by subduction-zone interactions. Furthermore, energy of this order of magnitude for this length of time can most likely come from frictional heat generated along the subduction zone.

Although this overall mechanism explains most of the various features a general way, many important details of the tectonic and volcanic development must be explained by details of the process and its possible interaction with hot spots, subducted ridge locations, and so forth (Atwater, 1970; Lipman and others, 1972; Christiansen and Lipman, 1972; Smith, 1977).

The most satisfactory thermal models of subduction zones have been developed by Hasebe and others (1970) in order to explain the heat flow over the Japanese arcs, the area most similar to the Western United States in size. Over the part of the subduction zone with high heat flow, energy is continuously pumped into the upper mantle from friction along the top of the downgoing slab. This energy may go into heating the upper mantle, crust, and so forth. Eventually if the subduction zone does not change its position, the mantle wedge begins to convect, either as a solid or as a liquid, and the heat is transferred rapidly upward. At any given time, large portions of the mantle (100 to 300 km thick, solid to partially liquid) may be metastable. If an instability develops in the upper mantle (caused by a change in stress state, subduction characteristics, high concentrations of magma, or any one of many other possible mechanisms), then heat and perhaps magma will be concentrated at some point and will result in a thermal event in the crust (see Marsh and Carmichael, 1974). The instability may propagate in an irregular fashion through the hot, partially melted mantle and cause systematic spatial variations in the crustal thermal event. This mechanism is envisioned for the Western United States as an explanation for the volcanic progression patterns.

Whether the concept of a mantle diapir (Karig, 1971; Scholtz and others, 1971) has any validity is problematical. It seems more likely that the upper mantle above the subduction zone is in a continuous state of convection. Volcanic events have occurred at times and places of sufficient heat input to cause significant crustal melting and the possibility of some (probably small) amount of silicic material input from the mantle. These events have probably been strongly influenced by the residual effects of much older (Mesozoic and Paleozoic) thermal events that left a thinner and weaker lithosphere in places (such as the Southern Rocky Mountains, for example) in the same way that current spreading is focused in the area of highest energy loss from the mantle (hottest upper mantle?) The relationship of the volcanic events to the nature of the Pacific–North American plate interaction remains to be completely understood. The crustal spreading in the Great Basin might be related to such events, but similar crustal structure is present in the Pacific Northwest far north of the postulated interactions responsible for the Great Basin rifting (Blackwell, 1969).

The heat-flow patterns observed in the Cordilleran thermal anomaly zone cannot be uniquely related to the mantle effects that give rise to the surface pattern, and many different models are compatible with the heat-flow data. The heat-flow data set energy limits, but the volcanic and tectonic history may be more informative in determining the mechanisms.

Thermal Model for the Cordilleran Thermal Anomaly Zone

The results of the previous sections demonstrate that the heat flow along the Cordillera is high and variable. There appear to be many complicated correlations of the heat flow with other geologic and geophysical parameters, but the clearest correlation is with the Cenozoic volcanic history. The heat flow may also correlate with past spreading or normal faulting, as it seems to correlate with the contemporary seismic zones. The history of normal faulting is much less well known than the age of the volcanism, however (Stewart, this volume).

The correlation between the Basin and Range Q-A relationship and the areas with volcanic activity older than 17 m.y. has been shown. In areas of younger volcanism the heat transfer is very complicated, and numerous nonconductive mechanisms contribute to the total energy transfer. These types of energy loss must be evaluated to obtain the true heat flow in a given area. The composition and volume of the volcanism and plutonism also relate to the energy budget and to the nature of the thermal event. In turn the various parameters are related to the driving mechanism(s) of the tectonism and volcanism.

In an ocean setting, a relatively simple model gives a good representation of the thermal event associated with sea-floor spreading (McKenzie, 1967). On the continents, thermal events are much more complicated, however. In the oceans, topography appears to be closely related to heat flow and helps in determining the thermal models where the heat-flow pattern is disturbed by hydrothermal convection near the ridge crests [see Parsons and Sclater (1977) for discussion of these points]. Such a simple relationship between heat flow and elevation does not work in general for the continents because of the common changes in crustal density associated with thermal events. Thus much detailed information on crustal structure is necessary before correlations between heat flow and elevation can be made for continents. The only places in the United States where elevation correlates with heat flow in a relatively straightforward way are along the Snake River Plain, western Oregon, and the Great Plains–Southern Rocky Mountains.

In the Western United States, thermal events might range from a primarily heating event where a section of the lithosphere is heated to its melting point over some depth range to a rifting event where the continental crust is completely attenuated without significant volcanism, to a thermal event associated with plate-plate interaction along a transform fault. An example of the extreme type of melting thermal event is of course Yellowstone; of the rifting event, the Imperial Valley–Gulf of California; and of the transform fault, the San Andreas. The rifting event could resemble the volcanic event if the spreading occurred primarily by emplacement of new crustal material. In the Imperial Valley it appears that the rifting has occurred primarily by attenuation of the continental crust. In the Basin and Range province, spreading has been accompanied by silicic volcanism as well as crustal thinning; thus, presumably there have been much higher crustal temperatures. In both cases the mantle material must flow into the extending area in the mantle to accommodate the increase in surface area.

Undoubtedly once an intense thermal event has affected an area, the nature of the future thermal events, particularly the volcanic events, is restricted. Therefore, subsequent volcanic events differ in character even if the source and energy loss remain the same. The transition from the andesite to the basalt-rhyolite volcanic association (Lipman and others, 1972; Christiansen and Lipman, 1972) may reflect such an irreversible process. The progression of volcanism outward from the Idaho-Oregon border in the past 10 to 15 m.y. may be a typical volcanic event in which a thermal instability developed and progressed into other metastable regions in a manner similar to the migration of salt domes or salt walls. The Yellowstone system may represent the superposition of an additional heat source such as a plume (Morgan, 1972). The intensity of energy transfer at Yellowstone is certainly not out of the realm of subduction zones, however, as virtually identical size and energy loss are associated with the Taupo graben (Elder, 1965) of New Zealand.

During an active spreading or melting event (or any combination of the two), temperatures would rise with time (continuous thermal model). Eventually the temperature of the lithosphere would approach a new steady-state value if the activity persisted for long enough. During the period of activity the effect of lithospheric heating, volcanism, and hydrothermal convection heat losses would have to be added to the measured heat-flow to obtain the true mantle energy flux. Once activity ceases and cooling starts, then the subsequent thermal behavior of the crust and upper mantle will be essentially independent of the mechanism of origin of the high temperatures. It appears that in most of the Cordilleran thermal anomaly zone there have been both volcanism and extension. Unfortunately, it is very difficult to date extensional events except as they are associated with volcanism. With more data in the future it may be possible to subdivide the zone more completely on the basis of the date of the last major thermal event.

The cooling history of a widespread thermal event may be modeled as a one-dimensional instantaneous heat-conduction problem where the temperature at each depth at the end of the thermal event (beginning of the cooling) minus the background temperature (approximately linear) is the initial temperature anomaly and simple subsequent cooling is assumed. This type of model approximates the ocean-ridge cooling problem as well (Parsons and Sclater, 1977). Unfortunately, there is no way to know in detail the temperature in the crust and upper mantle after any given thermal event. Furthermore, in an active volcanic province there may be a random distribution of magma-emplacement depths and times, and the time history of the temperature at some depth will be a complex superposition of vertical and lateral conduction effects. Such a complexity does not cloud the ocean-ridge cooling model.

I consider as a very simple model for decay of a thermal event the instantaneous cooling of a semi-infinite half space at a constant initial temperature (Carslaw and Jaeger, 1959, p. 62) of 1,140°C. This model represents the maximum possible variation in heat flow with time because no background temperature or heat flow is assumed except that a heat flow of 0.35 HFU is added to account for the average radioactivity of the surface layer. The heat flow is produced in a 10-km-thick layer with a heat generation of 3.5 HGU. The temperature increment due to the radioactive layer is 60°C so that the initial temperature in the calculation is 1,200°C overall. This model corresponds to a thermal event that results in the heating of the crust to the melting point of basalt throughout. The cooling history of this model (V1) is shown in Figure 8-10 for a thermal conductivity of 6×10^{-3} cal/(cm·s·°C) and a

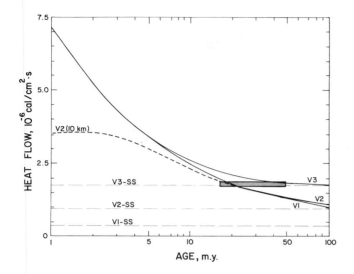

Figure 8-10. Thermal models for regional volcanic events in the Cordilleran thermal anomaly zone. The models are explained in the text. SS indicates the steady-state asymptotic heat-flow values.

thermal diffusivity of 0.01 cm^2/s. In this model, most of the change in heat flow occurs during the first 20 m.y., and the heat flow decreases with time from a value of 7.2 HFU after 1 m.y. to about 1.8 HFU after 15 m.y. and 1.3 HFU after 50 m.y.

In the V2 model shown in Figure 8-10, the initial temperature is assumed to be at 1,200°C throughout at zero time, and to cool to the steady-state temperature appropriate for a heat flow of 0.95 HFU at the surface, a mantle heat flow of 0.6 HFU, and a radioactive layer with a heat generation of 3.5 HGU, at infinite time. The anomaly was calculated using equation 29 of Parsons and Sclater (1977). The value of 0.6 HFU was chosen as a reasonable mantle heat flow for a cratonic or shield area. The main result is that V2 cannot be distinguished from the previously discussed model (V1) for times less than 100 m.y. at least.

As a variant of the V2 model, the anomaly was assumed to extend up to a depth of 10 km. The effect of this modification is also shown in Figure 8-10. The heat flow after 1 m.y. is much lower, of course, and the two models become indistinguishable after about 10 to 15 m.y. In the case of a continental volcanic event, the intensity can vary tremendously, so that the theoretical regional heat flow at some subsequent time will represent a superposition of many spatially and temporally varied relatively local events. After a few million years, the anomalies will become indistinguishable as their effects merge. The time required is related to the spatial relations of the various components that make up the thermal event. Thus, the two cases for model V2 that are shown are only two possibilities of an infinite number of one-, two-, and three-dimensional anomalies that could be components of a major volcanic event.

A third model is shown in Figure 8-10 (V3). This model is identical to V2 except that the mantle heat flow is constrained to be 1.4 HFU. This model is only distinguishable from the previous two at times greater than 10 m.y. The range of age and heat flow for the Basin and Range Q-A curve, given the error of the intercept and assuming an average A of 3.5 HGU (so that this model is comparable to V1 and V2), is shown by the box in Figure 8-10. The heat-flow results do not exclusively favor either model. Either a thermal history with a mantle heat flow of 0.6 HFU (or less) or with a mantle heat flow of about 1.3 HFU is consistent with the data. A distinction is important, however, because the implications of the models are very different.

If the parts of the Cordilleran thermal anomaly zone where the thermal events are old are cooling off to a shield type of mantle heat flow, then there has been no continued input of heat into the mantle after the end of the thermal event. If the thermal decay is toward a mantle heat flow of 1.3 HFU or so, then the mantle heat flow has remained anomalous even after the end of the thermal event most obvious at the surface. This latter explanation is favored. With the heat-flow and heat-generation data now in the process of being analyzed and continuing detailed geochronologic studies of volcanic and intrusive rocks, it should be possible soon to analyze the geographic variations of the mantle heat flow in the Cordilleran thermal anomaly zone in detail and investigate in detail the correlation of the mantle heat flow with age of the thermal event. Such studies will result in a significant improvement in the understanding of the thermal history of the Cordilleran thermal anomaly zone in particular and the continental lithosphere in general.

ACKNOWLEDGMENTS

Many people have been involved in the studies described here. Particularly, I acknowledge my colleagues R. G. Bowen, C. A. Brott, R. A. Chadwick, W. D. Gosnold, J. C. Mitchell, P. Morgan, E. C. Robertson, J. E. Schuster, and J. L. Steele. D. S. Chapman and W. P. Nash read the manuscript and made numerous helpful comments. The paper would not have been completed without the encouragement, assistance, and insistence of R. B. Smith. The studies described have been supported in large part by the National Science Foundation through grants GA-11351 and AER 76-00108. The paper was written during a sabbatical leave from Southern Methodist University spent at the University of Utah. Both institutions are thanked for their support.

REFERENCES CITED

Adolphson, D. G., and LeRoux, E. F., 1968, Temperature variations of deep flowing wells in South Dakota: U.S. Geol. Survey Prof. Paper 600-D, p. 60–62.

Archambeau, C. B., Flinn, E. A., and Lambert, P. G., 1969, Fine structure of the upper mantle: Jour. Geophys. Research, v. 74, p. 5835–5866.

Archbold, N. L., 1972, Modified geologic map of Nevada: Nevada Bur. Mines and Geology, map 44.

Armstrong, R. L., 1974, Geochronometry of the Eocene volcanic-plutonic episode in Idaho: Northwest Geology, v. 3, p. 1–14.

Armstrong, R. L., Ekren, E. B., McKee, E. H., and Noble, D. C., 1969, Space-time relations of Cenozoic silicic volcanism in the Great Basin of the western United States: Am. Jour. Sci., v. 267, p. 478–490.

Armstrong, R. L., Leeman, W. P., and Malde, H. E., 1975, Quaternary and Neogene volcanic rocks of the Snake River Plain, Idaho: Am. Jour. Sci., v. 275, p. 225–251.

Atwater, T., 1970, Implications of plate tectonics for the Cenozoic tectonic evolution of western North America: Geol. Soc. America Bull., v. 81, p. 3513–3536.

Baksi, A. K., and Watkins, N. D., 1973, Volcanic production rates: Comparison of oceanic ridges, islands, and the Columbia Plateau basalts: Science, v. 180, p. 493–496.

Birch, F., Roy, R. F., and Decker, E. R., 1968, Heat flow and thermal history in New England and New York, in Zen, E., White, W. S., Hadley, J. B., and Thompson, J. B., Jr., eds., Studies of Appalachian geology—Northern and maritime: New York, Interscience, p. 437–451.

Blackwell, D. D., 1967, Terrestrial heat-flow determinations in the northwestern United States [Ph.D. thesis]: Cambridge, Mass., Harvard Univ.

——1969, Heat-flow determinations in the northwestern United States: Jour. Geophys. Research, v. 74, p. 992–1007.

——1971, The thermal structure of the continental crust, *in* Heacock, J. G., ed., The structure and physical properties of the earth's crust: Am. Geophys. Union Geophys. Mon. 14, p. 169–184.

——1974, Terrestrial heat flow and its implications on the location of geothermal reservoirs in Washington: Washington Div. Mines and Geology Inf. Circ., no. 50, p. 21–33.

Blackwell, D. D., and Robertson, E. C., 1973, Thermal studies of the Boulder batholith and vicinity, Montana: Soc. Econ. Geologists Guidebook, Butte Field Mtg., August 18–21, p. D-1–D-8.

Bowen, R. G., Blackwell, D. D., and Hull, D. A., 1977, Geothermal exploration studies in Oregon: Oregon Dept. Geology and Mineral Industries Misc. Paper 19, 50 p.

Brott, C. A., Blackwell, D. D., and Mitchell, J. C., 1976, Heat flow study of the Snake River Plain, Idaho: Idaho Dept. Water Resources Water Inf. Bull. 30, Pt. 8, 195 p.

Brott, C. A., Blackwell, D. D., and Mitchell, J. C., 1978, Tectonic implications of the heat flow of the western Snake River Plain, Idaho: Geol. Soc. America Bull., v. 89 (in press).

Brune, J. N., Henyey, T. L., and Roy, R. F., 1969, Heat flow, stress, and rate of slip along the San Andreas fault, California: Jour. Geophys. Research, v. 74, p. 3821–3827.

Carslaw, H. S., and Jaeger, J. C., 1959, Conduction of heat in solids (2nd ed.): Oxford, Clarendon Press, 510 p.

Christiansen, R. L., and Lipman, P. W., 1972, Cenozoic volcanism and plate-tectonic evolution of the western United States. II. Late Cenozoic: Royal Soc. London Philos. Trans., v. 271, p. 249–284.

Combs, J., and Simmons, G., 1973, Terrestrial heat flow determinations in the north-central United States: Jour. Geophys. Research, v. 78, p. 441–461.

Cook, K. L., 1966, Rift system in the Basin and Range province, *in* Irving, T. N., ed., The world rift system: Canada Geol. Survey Paper 66–14, p. 246–279.

Crouch, S. T., and Thompson, G. A., 1976, Thermal model of continental lithosphere: Jour. Geophys. Research, v. 81, p. 4857–5862.

Decker, E. R., 1969, Heat flow in Colorado and New Mexico: Jour. Geophys. Research, v. 75, p. 550–559.

Decker, E. R., and Smithson, S. B., 1975, Heat flow and gravity interpretation across the Rio Grande rift in southern New mexico and West Texas: Jour. Geophys. Research, v. 80, p. 2542–2552.

Eaton, G. P., Wahl, R. R., Prostka, H. J., Mabey, D. R., and Kleinkopf, M. D., 1978, Regional gravity and tectonic patterns: Their relation to late Cenozoic epeirogeny and lateral spreading of the western Cordillera, *in* Smith, R. B., and Eaton, G. P., eds., Cenozoic tectonics and regional geophysics of the western Cordillera: Geol. Soc. America Mem. 152 (this volume).

Eaton, J. P., 1966, Crustal structure in northern and central California from seismic evidence, *in* Geology of northern California: California Div. Mines and Geology Bull. 190, p. 419–426.

Elder, J. W., 1965, Physical processes in geothermal areas, *in* Lee, W.H.K., ed., Terrestrial heat flow: Am. Geophys. Union Geophys. Mon. 8, p. 211–240.

Elders, W. A., Rex, R. W., Meidev, T., Robinson, P. T., and Biehler, S., 1972, Crustal spreading in southern California: Science, v. 178, p. 15–24.

Fournier, R. O., White, D. E., and Truesdell, A. H., 1976, Convective heat flow in Yellowstone National Park, *in* Proc. 2nd United Nations Symposium on the development and use of geothermal resources, Vol. 1: Washington, D.C., U.S. Govt. Printing Office, p. 731–740.

Francis, P. W., and Rundle, C. C., 1976, Rates of production of the main magma types in the central Andes: Geol. Soc. America Bull., v. 87, p. 474–480.

Geologic map of North America, 1965: U.S. Geol. Survey, scale 1:5,000,000.

Gosnold, W. D., 1977, A model for uranium and thorium assimilation by intrusive magmas and crystallizing plutons through interaction with crustal fluids [Ph.D. thesis]: Dallas, Texas, Southern Methodist Univ.

Gough, D. I., 1974, Electrical conductivity under western North America in relation to heat flow, seismology, and structure: Jour. Geomagnetism and Geoelectricity, v. 26, p. 105–123.

Hasebe, K., Fujii, N., and Uyeda, S., 1970, Thermal processes under island arcs: Tectonophysics, v. 10, p. 335–355.

Henyey, T. L., and Bischoff, J. L., 1973, Tectonic elements of the northern part of the Gulf of California: Geol. Soc. America Bull., v. 84, p. 315–330.

Henyey, T. L., and Lee, T. C., 1976, Heat flow in Lake Tahoe, California–Nevada, and the Sierra Nevada–Basin and Range transition: Geol. Soc. America Bull., v. 87, p. 1179–1187.

Henyey, T. L., and Wasserburg, G. J., 1971, Heat flow near major strike-slip faults in California:

Jour. Geophys. Research, v. 76, p. 7924–7946.

Hill, D. P., 1978, Seismic evidence for the structure and Cenozoic tectonics of the Pacific Coast States, *in* Smith, R. B., and Eaton, G. P., eds., Cenozoic tectonics and regional geophysics of the western Cordillera: Geol. Soc. America Mem. 152 (this volume).

Hull, D. A., Bowen, R. G., Blackwell, D. D., and Peterson, N. V., 1977, Preliminary heat flow map and evaluation of Oregon's geothermal energy potential: Ore Bin, v. 39, p. 109–123.

Hyndman, R. D., 1976, Heat flow measurements in the inlets of south-western British Columbia: Jour. Geophys. Research, v. 81, p. 337–349.

Johnson, L. R., 1965, Crustal structure between Lake Mead, Nevada, and Mono Lake, California: Jour. Geophys. Research, v. 70, p. 2863–2872.

Johnson, S. H., and Couch, R. W., 1970, Crustal structure in the Northern Cascade Mountains of Washington and British Columbia, from seismic refraction measurements: Seismol. Soc. America Bull., v. 60, p. 1259–1270.

Judge, A., 1975, Terrestrial heat flow determinations in the southern Canadian Cordillera [abs.]: EOS (Am. Geophys. Union Trans.), v. 56, p. 245.

Karig, D. E., 1971, Origin and development of marginal basins in the western Pacific: Jour. Geophys. Research, v. 76, p. 2542–2561.

Kistler, R. W., Evernden, J. F., and Shaw, H. R., 1971, Sierra Nevada plutonic cycle: Pt. 1, Origin of composite granitic batholiths: Geol. Soc. America Bull., v. 82, p. 853–868.

Lachenbruch, A. H., 1968, Preliminary geothermal model of the Sierra Nevada: Jour. Geophys. Research, v. 73, p. 6977–6989.

——1970, Crustal temperature and heat production: Implications of the linear heat-flow relation: Jour. Geophys. Research, v. 75, p. 3291–3300.

Lachenbruch, A. H., and Sass, J. H., 1973, Thermomechanical aspects of the San Andreas fault system: Stanford Univ. Pub., Geol. Sci., v. 13, p. 192–205.

Lachenbruch, A. H., Sass, J. H., Munroe, R. J., and Moses, T. H., Jr., 1976a, Geothermal setting and simple heat conduction models for the Long Valley caldera: Jour. Geophys. Research, v. 81, p. 769–784.

Lachenbruch, A. H., Sorey, M. L., Lewis, R. E., and Sass, J. H., 1976b, The near-surface hydrothermal regime of Long Valley caldera: Jour. Geophys. Research, v. 81, p. 763–768.

Lawver, L. A., Sclater, J. G., Henyey, T. L., and Rogers, J., 1973, Heat flow measurements in the southern portion of the Gulf of California:

Earth and Planetary Sci. Letters, v. 12, p. 198–208.

Lipman, P. W., Steve, T. A., and Mehnert, H. H., 1970, Volcanic history of the San Juan Mountains, Colorado, as indicated by potassium-argon dating: Geol. Soc. America Bull., v. 81, p. 2329–2352.

Lipman, P. W., Prostka, H. J., and Christiansen, R. L., 1972, Cenozoic volcanism and plate-tectonic evolution of the western United States. I, Early and middle Cenozoic: Royal Soc. London Philos. Trans., ser. A, v. 271, p. 217–248.

Mabey, D. R., 1966, Relation between Bouguer gravity anomalies and regional topography in Nevada and the eastern Snake River Plain, Idaho: U.S. Geol. Survey Prof. Paper 550-B, p. 108–110.

Mackin, J. H., 1960, Structural significance of Tertiary volcanic rocks in southwestern Utah: Am. Jour. Sci., v. 258, p. 81–131.

MacLeod, N. S., Walker, G. W., and McKee, E. H., 1976, Geothermal significance of eastward increase in age of late Cenozoic rhyolite domes in southeastern Oregon, *in* Proc. 2nd United Nations symposium on the development and use of geothermal potential Vol. 1: Washington, D.C., U.S. Govt. Printing Office, p. 465–474.

Marsh, B. D., and Carmichael, I.S.E., 1974, Benioff zone magmatism: Jour. Geophys. Research, v. 49, p. 1196–1206.

McKenzie, D. P., 1967, Some remarks on heat flow and gravity anomalies: Jour. Geophys. Research, v. 72, p. 6261–6273.

Menard, H. W., 1960, The East Pacific Rise: Science, v. 132, p. 1737–1746.

Mooser, F., 1972, The Mexican volcanic belt: Structure and tectonics: Geofísica Internac., v. 12, p. 55–70.

Morgan, P., Blackwell, D. D., Spafford, R. E., and Smith, R. E., 1977, Heat flow measurements in Yellowstone Lake and the thermal structure of the Yellowstone caldera: Jour. Geophys. Research, v. 82, p. 3719–3732.

Morgan, W. J., 1972, Plate motions and deep mantle convection: Geol. Soc. America Mem. 132, p. 7–22.

Newton, V. C., and Corcoran, R. E., 1963, Petroleum geology of the western Snake River Basin: Oil and gas investigation no. 1: Oregon Dept. Geology and Mineral Industries, 67 p.

Pakiser, L. C., and Zietz, I., 1965, Transcontinental crustal and upper mantle structure: Rev. Geophysics, v. 3, p. 505–520.

Parsons, B., and Sclater, J. G., 1977, An analysis of the variation of ocean floor bathymetry and heat flow with age: Jour. Geophys. Research, v. 82, p. 803–827.

Reiter, M., Edwards, C. L., Hartman, H., and Weidman, D., 1975, Terrestrial heat flow along the Rio Grande rift, New Mexico and southern Colorado: Geol. Soc. America Bull., v. 86, p. 811–818.

Renner, J. L., White, D. E., and Williams, D. L., 1975, Hydrothermal convection systems: U.S. Geol. Survey Circ. 726, p. 5–57.

Richter, C. F., 1958, Elementary seismology: San Francisco, W. H. Freeman & Co. 768 p.

Roy, R. F., Blackwell, D. D., and Birch, F., 1968, Heat generation of plutonic rocks and continental heat-flow provinces: Earth and Planetary Sci. Letters, v. 5, p. 1–12.

Roy, R. F., Blackwell, D. D., and Decker, E. R., 1972, Continental heat flow, in Robertson, E. C., ed., The nature of the solid Earth: New York, McGraw-Hill Book Co., p. 506–543.

Ryall, A., Slemmons, D. B., and Gedney, L. D., 1966, Seismicity, tectonism, and surface faulting in the western United States during historic time: Seismol. Soc. America Bull., v. 56, p. 1105–1135.

Sass, J. H., Lachenbruch, A. H., Munroe, R. J., Greene, G. W., and Moses, T. H., Jr., 1971, Heat flow in the western United States: Jour. Geophys. Research, v. 76, p. 6379–6413.

Sass, J. H., Diment, W. H., Lachenbruch, A. H., Marshall, B. V., Monroe, R. J., Moses, T. H., Jr., and Urban, T. C., 1976, A new heat-flow contour map of the conterminous United States: U.S. Geol. Survey Open-File Rept. 76-756, 24 p.

Scholz, C. H., Barazangi, M., and Sbar, M. L., 1971, late Cenozoic evolution of the Great Basin, western United States, as an ensialic interarc basin: Geol. Soc. America Bull., v. 82, p. 2979–2990.

Shimazu, Y., 1964, Energy analysis of geological phenomena, an aid to obtain a quantitative information of crustal evolution: Nagoya Univ. Jour. Earth Sci., v. 12, p. 85–101.

Smith, D. L., 1974, Heat flow, radioactive heat generation, and theoretical tectonics for northwestern Mexico: Earth and Planetary Sci. Letters, v. 23, p. 43–52.

Smith, R. B., 1977, Intraplate tectonics of the western North American plate: Tectonophysics, v. 37, p. 323–336.

——1978, Seismicity, crustal structure, and intraplate tectonics of the interior of the western Cordillera, in Smith R. B., and Eaton, G. P., eds., Cenozoic tectonics and regional geophysics of the western Cordillera: Geol. Soc. America

Mem. 152 (this volume).

Smith, R. B., and Sbar, M. L., 1974, Contemporary tectonics and seismicity of the western United States with emphasis on the Intermountain seismic belt: Geol. Soc. America Bull., v. 85, p. 1205–1218.

Smith, R. B., Braile, L. W., and Keller, G. R., 1975, Upper crustal low-velocity layers: Possible effect of high temperatures over a mantle upwarp at the Basin Range–Colorado Plateau transition: Earth and Planetary Sci. Letters, v. 28, p. 197–204.

Smith, R. L., and Shaw, H. R., 1975, Igneous-related geothermal systems: U.S. Geol. Survey Circ. 726, p. 58–83.

Stewart, J. H., 1978, Basin-range structure in western North America: A review, in Smith, R. B. and Eaton, G. P., eds., Cenozoic tectonics and regional geophysics of the western Cordillera: Geol. Soc. America Mem. 152 (this volume).

Stewart, J. H., and Carlson, J. E., 1978, Generalized maps showing distribution, lithology, and age of Cenozoic igneous rocks in the western United States, in Smith, R. B., and Eaton, G. P., eds., Cenozoic tectonics and regional geophysics of the western Cordillera: Geol. Soc. America Mem. 152 (this volume).

Stewart, J. H., Moore, W. J., and Zietz, I., 1977, East-west patterns of Cenozoic igneous rocks, aeromagnetic anomalies, and mineral deposits, Nevada and Utah: Geol. Soc. America Bull., v. 88, p. 67–77.

Strange, W. E., and Woollard, G. P., 1964, The use of geologic and geophysical parameters in the evaluation, interpretation, and production of gravity data: Aeronautical Chart and Inf. Center, U.S. Air Force, St. Louis, Missouri, Contract AF 23 (601)-3879 (Phase 1).

Suppe, J., Powell, C., and Berry, R., 1975, Regional topography, seismicity, Quaternary volcanism, and the present-day tectonics of the western United States: Am. Jour. Sci., v. 275–A, p. 397–436.

Swanberg, C. A., and Blackwell, D. D., 1973, Areal distribution and geophysical significance of heat generation in the Idaho batholith and adjacent intrusions in eastern Oregon and western Montana: Geol. Soc. America Bull., v. 84, p. 1261–1282.

Thompson, G. A., and Talwani, M., 1964, Crustal structure from Pacific basin to central Nevada: Jour. Geophys. Research, v. 69, p. 4813–4837.

Urban, R. C., and Diment, W. H., 1975, Heat flow on the south flank of the Snake River rift: Geol. Soc. America Abs. with Programs, v. 7, p. 648.

Walker, G. W., 1969, Geology of the High Lava Plains province, *in* Mineral and water resources of Oregon: Oregon Dept. Geology and Mineral Industries Bull. 64, p. 109–119.

Waring, G. A., 1965, Thermal springs of the United States and other countries of the world—A summary: U.S. Geol. Survey Prof. Paper 492, 383 p.

Warren, R. E., Sclater, J. G., Vacquier, V., and Roy, R. F., 1969, A comparison of terrestrial heat flow and transient geomagnetic fluctuations in the southwestern United States: Geophysics, v. 34, p. 463–478.

White, C. M., and McBirney, A. R., 1978, Some quantitative aspects of orogenic volcanism in the Oregon Cascades, *in* Smith, R. B., and Eaton, G. P., eds., Cenozoic tectonics and regional geophysics of the western Cordillera: Geol. Soc. America Mem. 152 (this volume).

MANUSCRIPT RECEIVED BY THE SOCIETY AUGUST 15, 1977

MANUSCRIPT ACCEPTED SEPTEMBER 2, 1977

Printed in U.S.A.

Geological Society of America
Memoir 152

9

Models of an extending lithosphere and heat flow in the Basin and Range province

Arthur H. Lachenbruch
J. H. Sass
U.S. Geological Survey
345 Middlefield Road
Menlo Park, California 94025

ABSTRACT

Reduced heat flow in the Basin and Range province is characteristically greater by 50% to 100% than that in stable regions; in the hotter subprovinces like the Battle Mountain High, it is greater by 300%. Evidence for distributed tectonic extension and magmatism throughout the province suggests that much of the anomalous heat is transferred from the asthenosphere by convection in the lithosphere, in the solid state by stretching, and in the magmatic state by intrusion. Simple steady-state thermomechanical models of these processes yield relations among reduced heat flow, asthenosphere flux, lithosphere thickness, extension rate, and basalt production by the asthenosphere. Thermal effects in an extending lithosphere lead to decreased estimates of temperature and increased estimates of lithosphere thickness in the Basin and Range province. Moderate extension rates can account for high heat flow in the province without calling on anomalous conductive flux from the asthenosphere. The heat and mass budgets of bimodal volcanic centers suggest that they occur at points where the lithosphere is pulling apart rapidly, drawing up basalt to fill the void. Intrusion can probably facilitate lithosphere extension at low stress levels either by brittle "hydrofracturing" by basaltic dikes or by warming and thinning caused by basaltic underplating. Whether lithosphere extension occurs in the distributed mode or in the plate-tectonic mode might depend largely upon whether the lateral divergence of mass can be supplied by asthenosphere basalt, or whether it must be supplied by the ascent of very viscous ultramafic material which requires wide conduits separated by large distances. For a range of plausible models of distributed extension, the anomalous heat flow increases roughly 1 HFU (10^{-6} cal·cm^{-2}·s^{-1}) for every 1% to 2%/m.y. increase in extension rate; the relation suggests extension rates in the Great Basin consistent with estimates from structural evidence. It also suggests much more rapid local extension in the hotter subprovinces, an inference supported by limited evidence from other sources.

INTRODUCTION

Heat flow from the Earth's surface plays a central role in discussions of the tectonics of active regions. Although it is the principal source of information regarding temperature in the lithosphere and thermal manifestations of flow in the asthenosphere, the implications of heat flow are more ambiguous than is sometimes realized. The ambiguities are introduced at two distinct interpretive steps: (1) abstracting a generalized or characteristic heat flow from the observational data for a region and (2) selecting a thermal model for downward continuation of this generalized surface heat flow. In a previous paper (Lachenbruch and Sass, 1977), we discussed the first step for the Basin and Range province in some detail. In this paper (which is a sequel to that one), we present an updated heat-flow map and review the basis for selecting heat flows to characterize different regions in the province and the basis for the simple conduction model used to construct crustal geotherms. We then consider simple thermal models that are thought to be more appropriate to an extending lithosphere.

Relations between reduced heat flow and thickness of the continental lithosphere were discussed in terms of heat-conduction models by Crough and Thompson (1976) and Pollack and Chapman (1977). In their models the anomalous heat flow at the surface (corrected for radioactivity) is the same as anomalous flux conducted into the base of the lithosphere; no provision is made for convective transport in the lithosphere. Although each of these models is reasonable for the purpose for which it was used, the models cannot be applied to a lithosphere undergoing distributed extension; within such a lithosphere, material must be moving vertically, and vertical convection of heat must be taking place. In this paper we examine relations among surface heat flow, extension rate, lithosphere thickness, and asthenosphere flux for the simplest thermomechanical models of extension. Because of the complexity of the problem, we consider only steady-state models. The validity of the steady-state assumption must be judged a posteriori, according to the resulting lithosphere thickness and the supposed duration of the geologic processes responsible for the thermal condition. Although the assumption is a serious limitation, it yields a useful limiting case that provides insight from simple analytical results. The models are discussed in terms of heat-flow regimes believed to be characteristic of the Basin and Range province; they indicate some constraints and options for the interpretation of heat flow in regions of extensional tectonics.

A list of symbols is given in Appendix 1.

REVIEW OF THE STATUS OF HEAT–FLOW STUDIES IN THE
BASIN AND RANGE PROVINCE

Regional Heat Flow

The number of heat-flow determinations suitable for a regional analysis in the Western United States has now increased to more than 500 (from 2 in the 1950s), but still only gross regional generalizations are possible because of the irregular coverage and local variability. Figure 9-1 is an updated version of the heat-flow map presented recently by Lachenbruch and Sass (1977), to which paper the reader is referred for a background discussion. With notable exceptions, the generalized heat flow throughout most of the Western United States, and throughout the Basin and Range province in particular, is greater than 1.5 HFU, the approximate average for the continents. A histogram for the Basin and Range province (Fig.

9-2) shows that in addition to having a large mean, the heat-flow values in the province are widely dispersed. We have suggested elsewhere (Lachenbruch and Sass, 1977) that the large dispersion is caused primarily by convective transfer associated with ground-water circulation and that the large mean is probably caused primarily by convective transfer by ascending magma. In the earlier paper (Lachenbruch and Sass, 1977), we considered the effects on regional heat flow of circulatory hydrothermal convection; in this paper we shall consider some effects of magmatic movements and solid-state convection in the lithosphere.

The maps (Fig. 9-1) show subregions of extremely high heat flow (>2.5 HFU). Many, like the Long Valley volcanic center in eastern California (LV, Fig. 9-1a), are only a few tens of kilometres across. However, two large subprovinces, the Battle Mountain High (BMH, Fig. 9-1a) in northern Nevada (Sass and others, 1971) and the Rio Grande Rift (RGR, Fig. 9-1a) in central New Mexico and Colorado (Reiter and others, 1975), are taking shape as regional features, although their boundaries are not yet well defined. We have drawn the map in such a way as to suggest a possible connection between the Battle Mountain High and the Yellowstone volcanic region in Wyoming (Y, Fig. 9-1a); the two are separated by the Snake River Plain (SRP, Fig. 9-1a), a volcanic region in which the measurement of heat flow is complicated by hydrologic circulation (Brott and others, 1976). The Snake River Plain also separates the Battle Mountain High from a region of very high conductive heat flow and hydrothermal activity in the Idaho batholith (IB, Fig. 9-1a).

Heat Production

Several of the heat-flow measurements in the Basin and Range province were made in granitic rock where determinations were also made of the radioactive heat production (A_o) of core and outcrop samples. The heat-flow–heat-production (q–A_o) data for the Basin and Range province are displayed with similar data from the Sierra Nevada province and the Eastern United States (that is, east of the Great Plains; see Fig. 9-1b) in Figure 9-3. The data for the Sierra Nevada province and the Eastern United States follow the well-known linear relation first discovered in radioactive rocks of New England by Birch and others (1968).

$$q = q_r + DA_o. \tag{1}$$

In this relation the intercept q_r is the "reduced heat flow" or heat flow for zero surface radioactivity. The slope D is a "characteristic depth" for the vertical distribution of heat production (10 km for the Sierra Nevada; 7½ km for the Eastern United States). Although this relation has been widely discussed (for example, Birch and others, 1968; Roy and others, 1968, 1972; Lachenbruch, 1968, 1970; Blackwell, 1971; Lachenbruch and Sass, 1977), we consider it briefly here, as it forms the basis for discussion to follow.

The simplest interpretation, consistent with the requirement that equation 1 survive differential erosion, is that heat production $A(z)$ be distributed vertically (in the upper crust at least) approximately as

$$A(z) = A_o e^{-z/D}. \tag{2}$$

Where equation 1 applies throughout large provinces, the reduced heat flow q_r is uniform, and it is reasonable to assume that the more local effects expected from hydrologic and magmatic convection are unimportant and that q_r represents uniform conductive flux from

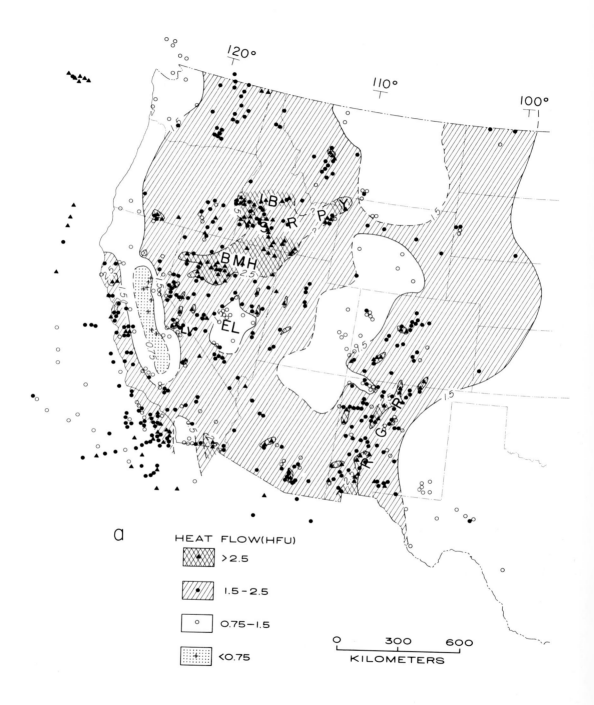

a

HEAT FLOW(HFU)

>2.5

1.5–2.5

0.75–1.5

<0.75

0   300   600
KILOMETERS

Figure 9-1. (a) A contour map of surface heat flow in the Western United States adapted from Figures 1 and 2 of Lachenbruch and Sass (1977) with additional U.S. Geological Survey unpublished data and new data in Idaho from Brott and others (1976). Abbreviations are BMH for Battle Mountain High, EL

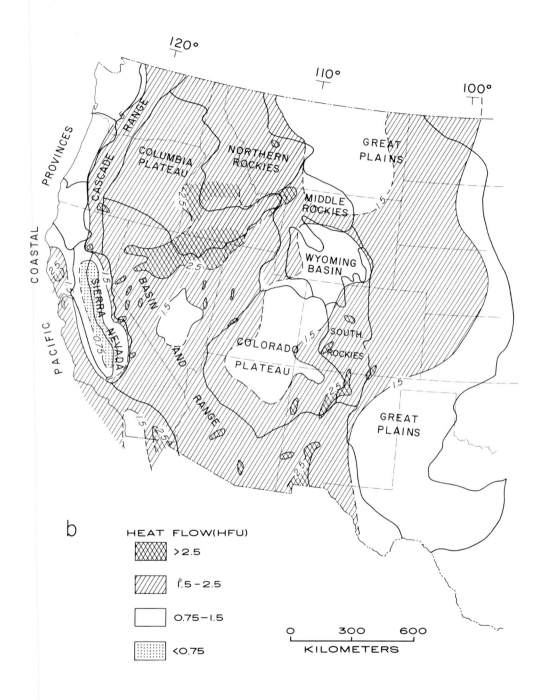

for Eureka Low, LV for Long Valley volcanic center, IB for Idaho Batholith, SRP for eastern and Central Snake River Plain, Y for Yellowstone thermal area, and RGR for Rio Grande Rift. (b) Heat-flow contours superimposed on major physiographic units of the Western United States.

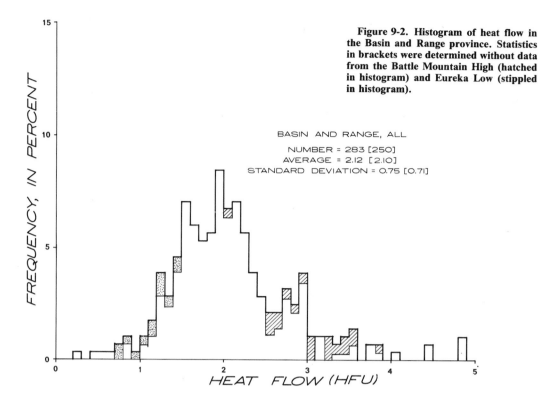

Figure 9-2. Histogram of heat flow in the Basin and Range province. Statistics in brackets were determined without data from the Battle Mountain High (hatched in histogram) and Eureka Low (stippled in histogram).

BASIN AND RANGE, ALL

NUMBER = 283 [250]
AVERAGE = 2.12 [2.10]
STANDARD DEVIATION = 0.75 [0.71]

the deep crust or mantle. If we assume uniform thermal conductivity K, the steady temperature θ under these circumstances is given by

$$\theta(z) = \frac{1}{K} [q_r z + D^2 A_o (1 - e^{-z/D})].$$

(3)

The depth to which equation 3 or similar relations based on equation 1 are valid depends upon (1) the depth to which heat transfer is predominantly by steady conduction and (2) the depth to which equation 2 or similar relations between $A(z)$ and A_o are valid.

When only a few heat-flow–heat-production pairs were available from the Basin and Range province, they were described reasonably well by a regression line (dashed, Fig. 9-3) proposed by Roy and others (1968). However, from the accumulated data shown in Figure 9-3, it now appears that the linear relation does not apply throughout the province as a whole; q_r is not uniform, and D is not defined. However, it is still possible to define the "non-radiogenic" contribution q_r for each site (Roy and others, 1972) in a manner consistent with equation 1:

$$q_r \equiv q - DA_o.$$

(4)

The distribution of reduced heat flow q_r for the Basin and Range province (Fig. 9-4b) has been computed from equation 4 by assuming $D = 10$ km, the value established in related rocks of the adjacent Sierra Nevada province (Fig. 9-3). From the data in Figure 9-4, we

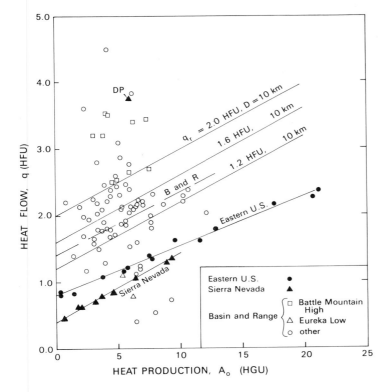

Figure 9-3. Heat flow and radioactive heat production in crystalline rock from three provinces. Original least-squares lines, labeled Sierra Nevada, Eastern U.S., and B and R, are from Roy and others (1968); they predate most of the Basin and Range observations. DP is adjacent to the Long Valley volcanic center.

concluded that variations in surface heat flow caused by convective processes and their related transients are greater by a factor of perhaps 3 or 4 than those caused by lateral variations in crustal radioactivity (Lachenbruch and Sass, 1977).

Crustal Temperatures: A Static Model

Almost half the values of q_r in Figure 9-4b lie in the modal range 1.2 to 1.6 HFU. We take this to be the characteristic range of q_r for the province, based on the belief that most of the departures, both positive and negative, result from circulatory hydrothermal convection. However, in hotter subregions like the Battle Mountain High, q_r is consistently greater than 2.0, and it averages about 2.5 HFU (the actual heat flows, including those for which radioactivity was not available, average about 3 in the Battle Mountain High, Fig. 9-2). The generalized crustal temperature profiles for the province (curves C, D, E, and F, Fig. 9-5) were obtained from equation 3 by introducing the four values of q_r (1.2, 1.6, 2.0, and 2.5 HFU) and the near-average value of 5 HGU for A_o (Fig. 9-4c). From the intercept value of the line for the Eastern United States (Fig. 9-3), we may infer that 0.8 HFU is a characteristic value for q_r in stable regions (Roy and others, 1968). Curve B, Figure 9-5, shows the temperatures one should expect in the Basin and Range province if it were underlain by such a stable lithosphere. Shown also in Figure 9-5 (curve A) is the generalized curve for the Sierra Nevada province, where q_r is only 0.4 HFU (see Fig. 9-3). As this is probably a transient condition, curve A probably underestimates temperatures at depth (see, for example, Blackwell, 1971).

Figure 9-5 is about as far as we can go at present in constructing geotherms with a steady conduction model and observations of heat flow and radioactivity at the surface. In the more

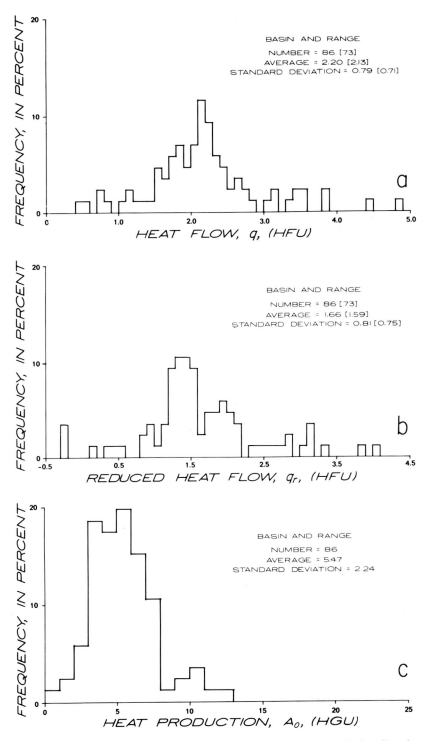

Figure 9-4. Histograms of (a) heat flow, (b) reduced heat flow, and (c) radioactive heat production for sites at which both q and A_o were measured in granitic rocks of the Basin and Range province. Statistics in brackets in (a) and (b) were determined without data from Battle Mountain High and Eureka Low.

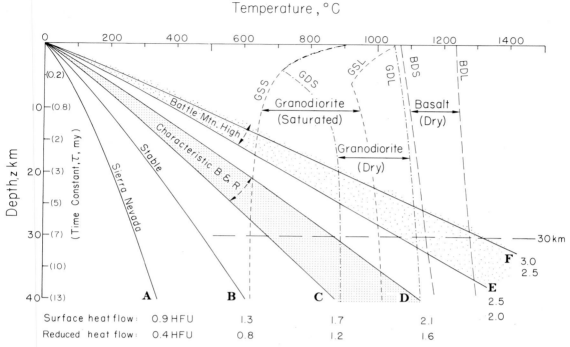

Figure 9-5. Generalized conductive temperature profiles for the Sierra Nevada crust (A), a stable reference crust (B), the characteristic Basin and Range crust (C to D), and lower limiting (E) and average (F) conditions in the crust of the Battle Mountain High. All curves are drawn for a surface heat production (A_o) of 5 HGU, characteristic depth (D) of 10 km, and thermal conductivity (K) of 6 mcal/(cm·s·°C) (eq. 3). Surface heat flow and reduced heat flow are shown at the bottom of each curve. Melting relations (after Wyllie, 1971) are shown for intermediate crustal rock by the curves GSS (granodiorite saturated solidus), GSL (granodiorite saturated liquidus), GDS (granodiorite "dry" solidus, Wyllie, type II), and GDL (granodiorite dry liquidus) and for basalt by BDS (basalt dry solidus), and BDL (basalt dry liquidus). In parentheses on the depth axis is shown the time constant for an overburden of thickness z (see text).

stable regions, curves constructed on this basis are probably reasonable throughout the crust; in active areas they may be useful down to those depths where convective transport or transients become important. Although the applicability of Figure 9-5 to near-surface conductive regimes is based upon several uncertain assumptions, it represents a virtual certainty in comparison to the conjectures required for discussion of thermal processes deeper in the lithosphere in tectonically active regions like the Basin and Range province.

DISCUSSION OF FIGURE 9-5

Major Assumptions

The scheme applied for downward extrapolation of temperatures from surface data in Figure 9-5 contains several assumptions that should be considered before we attempt to generalize it to accommodate convection. The more important ones are discussed below:

1. Systematic convective loss at the surface is neglected. We have assumed that after correction

for the contribution of crustal radioactivity, the average heat flux through the Earth's surface is q_r, determined from measurements of conductive flux. We have therefore neglected heat delivered to the surface convectively by volcanoes and hot and warm springs. The thickness of post-Oligocene volcanic rocks in the Great Basin probably averages on the order of 100 m (E. H. McKee, 1977, oral commun.). Averaged over time, the contribution of their heat to the combined surface flux is negligible ($\sim 10^{-2}$ HFU; see Fig. 9-15). With the exception of the Yellowstone system, the hotter known hydrothermal systems presently are discharging heat at a rate that would not significantly affect the regional crustal heat budget (Lachenbruch and Sass, 1977). However, details of the total hydrologic heat loss to the surface drainage from less conspicuous systems is unknown; this loss is neglected although it could be significant.

2. The materials are homogeneous, the geometry is quasi–one dimensional, and the heat production is given by equation 2. The first two are obvious simplifications that usually permit bracketing models without intuitive difficulty by selecting extreme values of the parameters. The assumption for radioactivity may be reasonable to crustal depths, although values for the lower crust are uncertain. If extended deeper, equation 2 underestimates heat production in the upper mantle and results in an underestimation of lithosphere thickness and overestimation of temperature. The effect is easily bracketed by reducing the assumed flux from the asthenosphere by the estimated heat production of the upper mantle and the possible underestimate by equation 2 for the lower crust; plausible adjustments are not likely to exceed about 0.3 HFU.

3. The materials are in a quasi-stationary thermal state. By this we mean that at any depth the temperature (and flow for convective models to be discussed) has only short-term variations about a steady mean value. To introduce the time context, we have shown on the ordinate axis (Fig. 9-5) the conductive time constant ($\tau = z^2/4\alpha$, α = thermal diffusivity) for the layer of thickness z. For conduction models the quasi-stationary condition is approached in the layer one or three time constants after a rapid change in temperature or heat flux (respectively) at depth z (Lachenbruch and Sass, 1977). (For example, the temperature is nearly steady throughout a 30-km crust 7 m.y. after a rapid temperature change at its base [Fig. 9-5].) If conditions on z change slowly over time periods large relative to the time constant, the stationary model can be a good approximation even though neither temperature nor flux is constant on z.

4. Heat transfer to the depth of extrapolation is predominantly by thermal conduction. This may be the most significant limitation of Figure 9-5 in tectonically active provinces like the Basin and Range. There we have abundant evidence of heat transfer by mass movement in the forms of hydrothermal activity, magmatism, and crustal extension. Under such conditions convection should not be neglected; it probably contains the explanation for the high heat flow, so laboriously documented.

Over periods $\gtrsim 10^6$ years the integrated effect on regional heat flow of circulatory hydrothermal convection in the upper few kilometres of the crust is probably unimportant (Lachenbruch and Sass, 1977), and we presume that this effect has been adequately filtered out by selecting modal values of q_r in constructing Figure 9-5. Hence, we focus on how magmatism and tectonic extension might affect the high heat flow observed.

Simple Magmatic Model for Elevated Surface Heat Flow

Figure 9-5 can be applied to a simple semiquantitative model for high heat flow. Suppose basalt rises through the lithosphere to some level H in the crust where it is impeded, perhaps by a viscous siliceous melt generated as the basalt exchanges its heat with indigenous rocks

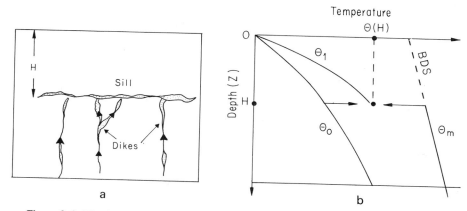

Figure 9-6. The simple intrusive model. A sill at depth H is maintained at average temperature $\theta(H)$ by dikes at temperature θ_m in a medium initially at temperature θ_o. The steady temperature in the conductive overburden is θ_1.

(Fig. 9-6a). Let the initial temperature of the lithosphere be θ_o, and the temperature of the rising basalt be θ_m (Fig. 9-6b). Suppose θ_o is curve C and θ_m is BDS, Figure 9-5. If the basalt supply were adequate to maintain the sill-like melt at a time-averaged temperature $\theta(H)$, for a time $\tau(H)$, a stationary conductive condition θ_1 would be established in the overburden (Fig. 9-6b). For example, if $H = 20$ km, $\theta(H) = 900°C$, then after 3 m.y. (τ (20 km), Fig. 9-5) θ_1 would be the curve F, Figure 9-5, typical of the Battle Mountain High. Downward extrapolation from surface observations would correctly identify the magmatic temperature at depth H, but extrapolation to greater depth (for example, for curve F) would not be valid because there the temperature could range, according to position, between θ_m and θ_o. To maintain the silicic melt at 900°C, the rising basalt would, on the average, need to convect heat to the 20-km depth at the rate of 1.3 HFU (the difference between q_r for curves C and F) to supply the upward heat loss, and in addition such heat as might be required to approach a quasi-stationary state beneath H. The quasi-stationary state beneath H would develop if the intrusive mass flux, though intermittent, were uniform over times on the scale of the time constant τ at each depth. Under these conditions the average temperature at each depth would assure that the combined conductive and convective fluxes at that depth were just sufficient to supply the 2.5 HFU required to sustain curve F (plus the requirements for lateral convection if there were extension). We shall consider an analytical model of this quasi-stationary intrusive condition later. For the present, it is sufficient to note that in the stationary convecting zone the average gradient and temperatures will be lower than indicated by the curves of Figure 9-5. This is because the vertical gradient is proportional to the conductive component which is only a fraction of the combined flux.

Steady-State Conducting Lithosphere as a Limiting Case

There is one more point to be made from Figure 9-5 before proceeding with tectonic models of high heat flow. For such models we shall define the lithosphere in a thermal sense so that the temperature at its base is sufficiently high to generate substantial amounts of melt if the composition permits. Hence, we take the base of the lithosphere R to be at the dry solidus temperature. Thus,

$$\theta(R) = \theta_m(R) \tag{5a}$$

where

$$\theta_m(z) \equiv \theta_o + mz, \tag{5b}$$

$$\theta_o = 1050°C, \tag{5c}$$

and

$$m = 3°C/km. \tag{5d}$$

The parameters are chosen to coincide approximately with BDS, Figure 9-5, and we assume that this adequately represents the peridotite solidus in the upper mantle as well (see Fig. 8-4, Yoder, 1976). It is emphasized that we are defining the lithosphere R as the region with temperatures below the dry solidus because the large amounts of basalt required by the magmatic models (to be discussed) probably could not originate above the base of such a region (Wyllie, 1971). The lithosphere defined in terms of seismic behavior may be substantially thinner than R because of possible effects in the upper mantle of melting below the dry solidus, high temperature, or contamination by basaltic intrusion.

A formal extrapolation of the curves of Figure 9-5 to the solidus, θ_m, is shown in Figure 9-7. These curves underestimate the thickness R of the lithosphere in a stationary thermal state for three reasons: (1) They neglect convective transport, which as just discussed, would generally reduce gradients (for a given surface heat flow) and increase lithosphere thickness; (2) they are based upon an assumed uniform conductivity of 6 mcal/(cm·s·°C); the actual conductivity of the mantle is probably larger and the gradients would be correspondingly smaller; and (3) they are based upon the distribution of heat production given by equation 2, the effect of which has been discussed. That 9-7 gives an underestimate is supported by the value for the "stable" curve which is $R = 94$ km, substantially less than the values usually estimated for such regions. (Increasing K by one conductivity unit or decreasing q_r by 0.1 HFU to accommodate upper-mantle conditions would each increase the estimate of the stable R by about 20 km; these effects are less important for the thinner extending lithospheres that we shall be considering.)

SIMPLE MODELS OF A STEADILY EXTENDING LITHOSPHERE

Conditions in the Lithosphere

It is well known from the pattern of normal faulting observed at the surface that the Basin and Range province is extending.

If we view the extension as stretching in the homogeneous plastic sense (like pulling taffy), we expect the lithosphere to become thinner in the process. For a lithosphere whose base is defined by a temperature condition (eq. 5), thinning will increase the gradient and hence the heat loss to the surface; this loss, in turn, will be limited by the rate at which processes in the asthenosphere can supply heat to the lithosphere. If the extension rate of the lithosphere and the processes in the upper asthenosphere remain relatively uniform, the lithosphere will eventually reach a stationary thickness; the rate of accretion (that is, of descent of the basalt-melting isotherm) at its base will just compensate for the thinning. Two limiting cases of this model are illustrated in Figure 9-8a and 8b. In one (STR, Fig. 9-8a) the accreting material is all crystalline, and in the other (UPL, Fig. 8b) it is all liquid basalt which gives up its latent heat as it crystallizes at the base of the lithosphere. In a third model (INT, Fig. 9-8c) the extension is accommodated by intermittent intrusion of dikes that over long periods of time are homogeneously distributed throughout the lithosphere.

In the simple steady conduction model leading to Figure 9-7, the heat flow at the surface q is simply the sum of the heat conducted into the base of the lithosphere and the radioactive heat generation within the lithosphere. The latter, obtained from equation 2, is

$$\int_{0}^{R} A_o e^{-z/D} \, dz = A_o D - A_o D \, e^{-R/D}. \tag{6}$$

The second term on the right is the heat generated by the distribution (eq. 2) at depths greater than R. Hence, for the model of Figure 9-7, heat conducted into the base of the lithosphere is

$$q_r + A_o D \, e^{-R/D} \simeq q_r \quad , \quad R >> D = 10 \text{ km.} \tag{7}$$

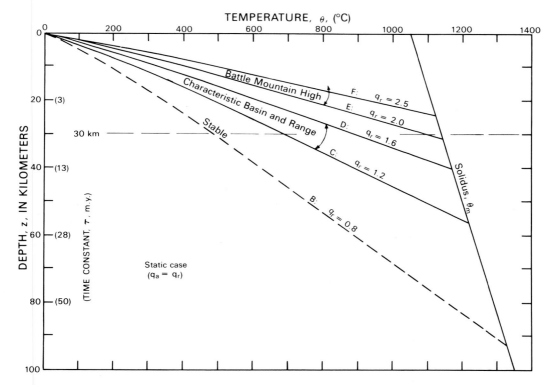

Figure 9-7. Model of the static conducting lithosphere obtained by extending the curves of Figure 5 (eq. 3) to the solidus (θ_m, eq. 5).

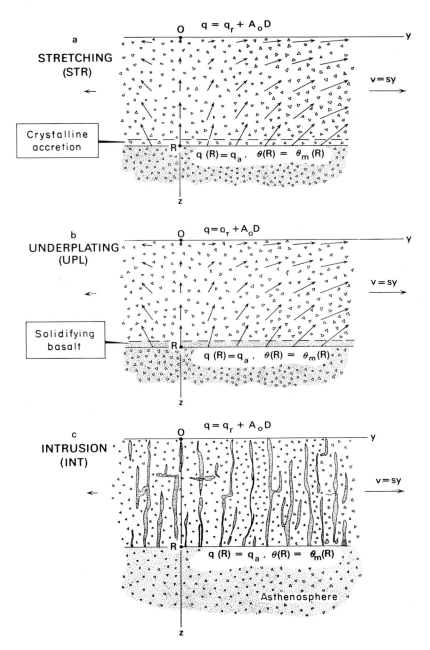

Figure 9-8. Simple thermomechanical models of a steadily extending lithosphere of thickness R with base at solidus temperature, $\theta_m(R)$. (a) Homogeneous stretching with accretion of crystalline material at the base (STR). (b) Homogeneous stretching with accretion of solidifying basalt at the base (UPL). (c) Extension by distributed dike intrusion (INT). Arrows in (a) and (b) are velocity vectors (v, w), and s is horizontal strain rate.

Thus, for the simple static model the heat conducted into the base of the lithosphere is, for practical purposes, the value of q_r observed at the surface. In tectonically active areas, however, q_r will generally contain contributions from heat both conducted and convected into the base of the lithosphere, and neither will be identifiable from thermal observations at the surface. (Their sum will not equal q_r because some heat is convected laterally in an extending lithosphere.) For completely specified mechanical models like those illustrated in Figure 9-8, the extension rate at the surface will determine the vertical velocity field in the lithosphere, and with equation 5 the convective input to a lithosphere of any thickness will be known. However, the conductive flux into the base of the lithosphere is still unspecified; we shall denote it by

$$q_a + A_o D e^{-R/D} \simeq q_a \quad , \quad R >> D = 10 \text{ km.} \tag{8}$$

(The radioactive distribution [eq. 2] is being retained to permit a simple comparison of results with the static model.) For convenience we shall refer to the conductive flux into the base of the lithosphere as q_a, although in the analysis the more complete expression on the left side of equation 8 will be used. In general, q_a will be less than q_r; the two quantities will approach one another as the extension rate goes to zero.

To determine the temperature distribution through an extending lithosphere and the relations among spreading rate, surface heat flow, and lithosphere thickness, we must know q_a. It is the only term in the thermal budget not determined by processes in the lithosphere. Hence, the surface heat flow contains information about asthenosphere processes only to the extent that it is sensitive to q_a and to the mechanical coupling at $z = R$ associated with extension.

Processes in the Asthenosphere and Limits to q_a

Although the lithosphere of the Great Basin is extending, it is in isostatic balance, and hence the mass per unit area is not decreasing (Thompson and Burke, 1974). It follows that mass must be moving upward in the asthenosphere above some depth at which it is replenished by lateral inflow. The flow pattern, which is fundamental to several of the tectonic models discussed in this volume, could take many forms. Consider three extreme cases: (1) buoyant rise of a massive diapir or "plume" producing an active asthenosphere in which the horizontal strain rate diminishes upward near the base of the lithosphere providing the viscous stress to pull the lithosphere apart; (2) a passive asthenosphere, dragged apart and stretched as the lithosphere falls apart, perhaps from regional gravitational stress, or (Scholz and others, 1971; Christiansen and McKee, this volume) from changing boundary stress at the plate margins; and (3) penetrative convection of the asthenosphere by the melted fraction through a relatively static crystalline framework; one source of melt might be the remnant of subducted material at depth. It is likely that (3) would operate simultaneously with (1) or (2), and the modes could change as the system evolves. For example, (1) and (3) could raise the temperature, and the topographic changes caused by thermal expansion or thinning of the lithosphere might generate gravitational stress operating in mode (2). The time for approach to a stationary thermal state could be on the order of 10^8 years. However, substantial heat transfer could occur in transient states, and the processes could have been in progress long before the initiation of basin-range faulting. Each of these modes has rather different implications for heat, mass, and momentum transfer in the asthenosphere, but quantitative modeling is difficult as it is sensitive to uncertain assumptions regarding chemical composition, mechanical properties, thermal conductivity, heat production, and choice of boundary conditions (for example,

Froideveaux and Schubert, 1975); we shall not pursue it. Nevertheless, it is clear that the ultimate source of anomalous heat flow in the Basin and Range province must be the heat carried by mass moving upward in the asthenosphere, augmented perhaps by viscous dissipation in the flow.

Because of convection (Tozer, 1967), the coexistence of liquid and crystalline phases (Kay and others, 1970), and the role of radioactivity (Clark and Ringwood, 1964), the thermal gradient in the asthenosphere must generally be small relative to that in the lithosphere. Thus, a substantial proportion of the heat convected upward through the asthenosphere could be converted to internal energy by melting (and to a lesser extent the work of volume expansion). To take an extreme example, the excess heat convected across a layer with 100°C temperature drop could all be consumed by melting a 30% fraction (if such a large meltable fraction were present); the heat delivered to the base of the lithosphere might be only that conducted along a melting point gradient. For the numerical values we have assumed, that condition leads to $q_a \approx 0.18$ HFU, which is a lower limiting value for the models of Figure 9-8a and 8c; if q_a were less than 0.18, the geotherm would not intersect the melting curve as required by equation 5a. In any case, the upward movement of mass through the asthenosphere is likely to result in an increasing proportion of melt near its top (Green, 1973). The heat stored in this melt cannot be delivered to the surface unless the melt is allowed to cool to subsolidus temperatures, a process permitted by the "magmatic" models, Figure 9-8b and 8c. The rate of basalt production required by these models may be one of the more important constraints on asthenosphere flow imposed by heat-flow observations.

Additional limits to the value of q_a are imposed by the requirement that it reduce to q_r (or slightly less if we allow for the uncertain radioactivity of the lower crust and upper mantle) as the extension rate goes to zero, in which case the static model must obtain. If we believe that the high heat flow and lithosphere extension are causally related, then the value of q_a should approach the value characteristic of stable regions; $q_r \approx 0.8$ (or perhaps as little as 0.5, when we allow for neglected radioactivity). If we believe that high heat flow is not causally related to extension, q_a cannot exceed q_r in any case. Beyond that, it is difficult to generalize, and it is not obvious whether q_a would increase or decrease with increasing extension rate. With a thinner lithosphere characteristic of rapid extension, flowing material occurs higher in the upper mantle, probably reducing the gradient and conductive flux as discussed above. However, the estimate of q_a is complicated by the possibility of larger conductive flux in a boundary layer at $z = R$; this is likely to be controlled by the latent heat release, an extreme case of which is accommodated by the model of Figure 9-8b.

The Steady-State Assumption

For the idealized models of Figure 9-8, we assume that the extending layer is in a thermal and mechanical steady state. In terms of geological applications, this assumption is a mathematical fiction justified by the insight it yields through simple analytical results, and by the fact that alternative time-dependent models permit an enormous range of possibilities poorly constrained by our present knowledge. Surface manifestations such as faulting and extrusion tell us that movements of mass in the crust are likely to be episodic and nonuniformly distributed; each change in the velocity field will result in a time-dependent or "transient" response in the temperature field. We have assumed that short term variations in the velocity components average out to uniform values over periods of millions of years, although even average values will, in general, vary on all time scales. The time necessary for approach to a steady state in a layer extending at an average uniform rate depends upon the conductive time constant for the layer (τ, see Figs. 9–5, 9–7), the mechanical mode and rate of extension, and upon the history of thermal conditions at the lower boundary. This history depends upon little-known processes of heat and mass transfer at great depth, and they would have to be specified before we could describe the approach to equilibrium in the extending layer.

Nevertheless, it is possible to make rough estimates of geologic conditions under which steady-state results might reasonably be applied. For example, in the intrusive mode (Fig. 9-8c) it can be shown quite generally (Lachenbruch, unpub.) that if the basal temperature in a layer of thickness z' is held constant after the time ($t = 0$) that extension is initiated at a uniform strain rate s, then the transient temperature disturbance in the layer will decay as $\exp\{-[s\tau(z') + (\pi/2)^2]\, t/\tau(z')\}$. Similarly, if conductive flux at the base is held constant, the leading term in the Fourier expansion for the transient decays as $\exp\{-[s\tau(z') + (\pi/4)^2]\cdot t/\tau(z')\}$. In each case the boundary value may undergo a large initial change when extension commences. According to these relations the temperature change will generally exceed about 85% of its equilibrium value when the expressions following the minus signs exceed about 1.9. In the static case ($s = 0$), this occurs when $t \gtrsim 3/4\,\tau(z')$ for constant boundary temperature and when $t \gtrsim 3\,\tau(z')$ for constant boundary flux, confirming the rule-of-thumb given in an earlier section. The first term in brackets, which represents the effects of convective transport, hastens the approach to equilibrium for both boundary conditions; the effect is important only for larger extension rates. For extension in the other modes the situation is more complicated; convection hastens the approach to a steady state for extension at constant basal temperature and retards it for extension at constant basal flux. However, for the range of conditions expected to apply regionally over periods of millions of years, the static rule-of-thumb applies reasonably well to all modes.

The adjustment of lithosphere thickness to its stationary value might significantly retard equilibrium, depending on the mechanism and initial condition assumed, and plausible boundary conditions that would cause the thermal steady state to be approached more slowly than in the constant flux case, or more promptly than in the constant temperature case, can be imagined. In general, however, if the time constant of the extending layer exceeds the period over which extension can reasonably be considered uniform, the applicability of steady-state results should be considered with some caution. The present episode of extension in the Great Basin has probably been in progress for about 17 m.y.; in other parts of the Basin and Range province, it probably started as much as 10 m.y. earlier (Christiansen and Lipman, 1972; Stewart, 1971). The layer thicknesses associated with time constants of such durations are 45 km and 60 km, respectively. As these are of the same general magnitude as the expected lithosphere thickness in the Basin and Range province, it is not unreasonable to expect steady-state results to give useful insight into average conditions there. Numerical details of the results, particularly for the thicker lithospheres, should, of course, not be taken too literally. [In another paper (Lachenbruch, in press), the theory is applied to only the crustal portion of an extending lithosphere, thereby avoiding the uncertainty associated with longer time constants.].

ANALYTICAL RESULTS

The model of Figure 9-8a (abbreviated STR) represents homogeneous stretching of a lithosphere whose thickness is maintained constant by accretion of crystalline material on its base. The model of Figure 9-8b is the same except that the accreting material is all liquid basalt which gives up its latent heat, a process we shall call "underplating" (abbreviated UPL). In these models convective transport in the lithosphere is in the solid state. In the model of Figure

9-8c (abbreviated INT), vertical convective transport is by fluid basalt in dikelike intrusions which, over long time periods, generate an average stationary thermal state. The intrusion accommodates lithosphere extension without vertical solid-state convection. Models similar to this one have been used by Bodvarsson (1954) and Palmason (1973) to describe conditions in Iceland.

We let w be the average upward velocity (in the negative z direction) and let v be the horizontal velocity (in the y direction) at any level in the lithosphere. The horizontal strain rate, denoted by s, is assumed to be uniform in the lithosphere. Thus for an incompressible lithosphere, the two-dimensional continuity condition is

$$\frac{\partial w}{\partial z} = \frac{\partial v}{\partial y} = s. \tag{9}$$

Taking the origin at an arbitrary point ($y = 0$) on the surface $z = 0$, we obtain for the velocity components

$$w = sz \tag{10a}$$

$$v = sy. \tag{10b}$$

For the intrusive model, w is average upward velocity of basalt at any depth in the lithosphere, that is, the vertical volume flux of basalt. Although we have given equations 9 and 10 in two-dimensional form, the thermal models are quasi–one dimensional, and s can be viewed more generally as horizontal areal strain rate.

For simplicity in these models, we assume the thermal conductivity [$K = 6 \text{ mcal}/(\text{cm} \cdot \text{s} \cdot {}^\circ\text{C})$] and volumetric specific heat [$\rho c = 0.6 \text{ cal}/(\text{cm}^3 \cdot {}^\circ\text{C})$] of all materials, liquid and solid, are the same. Effects of these assumptions are unimportant for the present purpose; they can easily be evaluated. The latent heat parameter L/c is assumed to be 350°C.

As in the simple conductive model, we assume the distribution of heat production is given by equation 2 with $A_o = 5$ HGU, $D = 10$ km. A minor modification of the following results would allow for the finite heat production of the lower lithosphere, and this modification, and an adjustment of mantle conductivity should be made if interest centers on lithosphere thickness for small extension rates. An additional modification of the following equations can account for mechanically generated heat. However, for the strain rates ($\lesssim 10^{-15} \text{ s}^{-1} \simeq$ 3%/m.y.) and deviatoric stresses ($\lesssim 1$ kb) of interest, the effect is unimportant.

Stretching with Crystalline Accretion (STR) or Underplating (UPL)

The differential equation of heat transfer for both cases is

$$K\frac{d^2\theta}{dz^2} + w\rho c\frac{d\theta}{dz} = -A_o e^{-z/D}. \tag{11}$$

The terms from left to right represent, respectively, conduction, convection, and radioactive heat production. We introduce the notation

$$\beta^2 = \frac{K}{s\rho c} \tag{12a}$$

$$\varepsilon(z) = \frac{z^2}{2\beta^2}, \tag{12b}$$

where β is a characteristic length. In general, conductive or convective transfer will predominate according to whether β is respectively large or small relative to R. (In our applications the two quantities are generally of the same order of magnitude, and both forms of transport are important.) The quantity ε is introduced into the exponent in equation 11 (with no important physical consequence) to obtain a simple analytical solution. Introducing equations 10a, 12a, and 12b into equation 11, we obtain

$$\frac{d^2\theta}{dz^2} + \frac{z}{\beta^2}\frac{d\theta}{dz} = -\frac{A_o}{K}e^{-z/D-\varepsilon(z)}. \tag{13}$$

Boundary conditions are

$$\theta = 0 \quad , \quad z = 0 \tag{14a}$$

$$K\frac{d\theta}{dz} = q_a + A_o D e^{-R/D-\varepsilon(R)} + \frac{LR}{c\beta^2}K \quad , \quad z = R. \tag{14b}$$

The temperature as a function of depth z for lithosphere thickness R, spreading rate s (related to β through eq. 12a), and basal conductive flux q_a is

$$\theta(z;R,\beta,q_a) = e^{R^2/2\beta^2}\left[q_a + \frac{LR}{c\beta^2}K\right]\frac{\beta}{K}\left(\frac{\pi}{2}\right)^{1/2} erf\left(\frac{z^2}{2\beta^2}\right)^{1/2}$$

$$+ \frac{A_o D^2}{K}[1 - e^{-z/D} + 0(\varepsilon(D))]. \tag{15}$$

The heat flow is

$$q(z;R,\beta,q_a) = \left[q_a + \frac{L}{c}\frac{R}{\beta^2}K\right]e^{(R^2-z^2)/2\beta^2} + A_o D e^{-z/D-\varepsilon(z)}, \tag{16a}$$

and at the surface $z = 0$

$$q(0;R,\beta,q_a) = \left[q_a + \frac{LR}{c\beta^2}K\right]e^{R^2/2\beta^2} + A_o D. \tag{16b}$$

Comparing equations 16b and 4, we see that the first term (including the brackets) in 16b represents reduced heat flow q_r. Equation 15 is a generalization of the static conduction model

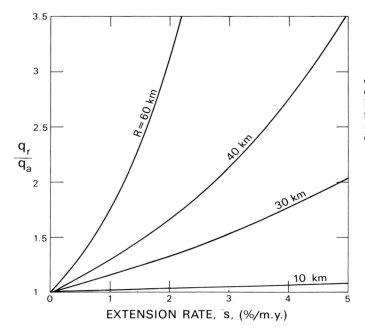

Figure 9-9. Effect on heat flow of homogeneous stretching, STR, of a lithosphere of thickness R. Reduced heat flow is q_r, conductive flux from the asthenosphere is q_a, and extension strain rate is s (see eq. 16b).

(eq. 3) to which it reduces when $s \to 0$ (that is, when $\beta \to \infty$). Equations 15 and 16 represent the case STR when $L = 0$ and the case UPL otherwise. The enhancement of q_r relative to q_a caused by stretching of the lithosphere is shown graphically in Figure 9-9 for the case $L = 0$.

Stationary Intrusive Model (INT)

In this case basalt rises through dikelike fractures at the temperature θ_m (eq. 5). The differential equation for the average temperature θ of the lithosphere at depth z is

$$K \frac{d^2 \theta}{dz^2} = -A_o e^{-z/D} + s\rho c \left[\theta - \left(\theta_m + \frac{L}{c} \right) \right] - wp c \frac{d\theta_m}{dz}. \tag{17}$$

The second term on the right (including the brackets) represents heat given up by solidification and cooling of the magma in place. The last term represents heat given up by the magma during its rise; the term can be modified to accommodate extrusion, but that has been judged to be unimportant to the heat budget in the Great Basin. Introducing equations 5b, 10a, and 12a into equation 17, we obtain

$$\frac{d^2 \theta}{dz^2} = -\frac{A_o}{K} e^{-z/D} + \frac{1}{\beta^2} \left[\theta - \left(\theta_o + \frac{L}{c} \right) \right] - \frac{2m}{\beta^2} z. \tag{18}$$

Boundary conditions are

$$\theta = 0 \quad , \quad z = 0 \quad , \tag{19a}$$

$$K\frac{d\theta}{dz} = q_a + A_o D e^{-R/D} \quad , \quad z = R. \tag{19b}$$

The average temperature at any depth z for given R, s (obtained from β), and q_a is then

$$\theta(z;R,\beta,q_a) = A_o \frac{D^2}{K}\left(1 - \frac{D^2}{\beta^2}\right)^{-1}\left(\frac{\cosh\dfrac{R-z}{\beta}}{\cosh\dfrac{R}{\beta}} - e^{-z/D}\right)$$

$$+ \left(\theta_o + \frac{L}{c}\right)\left(1 - \frac{\cosh\dfrac{R-z}{\beta}}{\cosh\dfrac{R}{\beta}}\right)$$

$$+ \frac{\beta}{K}\left[q_a - A_o\frac{D^3}{\beta^2}\left(1 - \frac{D^2}{\beta^2}\right)^{-1}e^{-R/D} - 2mK\right]\frac{\sinh\dfrac{z}{\beta}}{\cosh\dfrac{R}{\beta}}$$

$$+ 2mz. \tag{20}$$

The heat flow at the surface $z = 0$ is given by

$$q(0;R,\beta,q_a) = A_o D\left(1 - \frac{D^2}{\beta^2}\right)^{-1}\left(1 - \frac{D}{\beta}\tanh\frac{R}{\beta}\right)$$

$$+ \frac{K}{\beta}\left(\theta_o + \frac{L}{c}\right)\tanh\frac{R}{\beta}$$

$$+ \left[q_a - A_o\frac{D^3}{\beta^2}\left(1 - \frac{D^2}{\beta^2}\right)^{-1}e^{-R/D} - 2mK\right]\frac{1}{\cosh\dfrac{R}{\beta}}$$

$$+ 2mK. \tag{21}$$

Reduced heat flow q_r is obtained from equation 21 by subtracting $A_o D$ (see eq. 4). Like equation 15, equation 20 is a generalization of the static conduction model (eq. 3), to which it reduces as $s \to 0$.

Procedure for Applying the Analysis

The foregoing results satisfy the conditions for heat and mass flux at the base of an extending lithosphere whose thickness R is arbitrary. We can therefore adjust R to satisfy the temperature condition (eq. 5a). We are then left with the four parameters—surface heat flow, asthenosphere flux, extension rate, and lithosphere thickness—any two of which must be known to specify the lithosphere regime completely for each spreading mode. Of these, the surface heat flow is the most observable, and the radioactive contribution to it ($A_o D$) is relatively unimportant. We shall therefore investigate the lithosphere conditions associated with each of the standard heat-flow regimes, C, D, E, and F (Fig. 9-7), representing reduced heat flows of 1.2, 1.6, 2.0, and 2.5, respectively. The actual surface heat flows for which the calculations are made are each greater than q_r by the radioactive contribution, 0.5 HFU; most of the results do not change in important ways for different radioactive contributions. With q_r specified, assigning any one of the parameters s, q_a, or R will determine the other parameters as well as the thermal regime $\theta(z)$ for each extension mode.

DISCUSSION OF THE ANALYTICAL RESULTS

Figure 9-10 shows the model-dependence of attempts to deduce lithosphere conditions from a knowledge of reduced heat flow. For example, the three curves that converge near the letter F, Figure 9-10a, show the combinations of s and q_a, for each extension mode, that are consistent with the reduced heat flow of 2.5 HFU, typical of the Battle Mountain High. As the extension rate s approaches zero, the asthenosphere flux q_a must approach the reduced heat flow q_r, and the curves converge to $q_a = q_r = 2.5$ HFU as required by the static model (eq. 3, Fig. 9-7). As spreading rate s increases, the required reduced heat flow can be supplied with a progressively smaller asthenosphere contribution q_a; the difference is made up by heat transport associated with vertical mass movement within the extending lithosphere. Lithosphere convection in the INT mode is the most effective in the sense that it can produce a given surface heat flow with the least extension rate; STR requires the greatest. The depth to the source of basaltic melt R is therefore greatest for the INT mode for a given spreading rate (Fig. 9-10b); the efficient convective transport greatly reduces thermal gradients in the lower lithosphere. For a given extension rate, STR and UPL yield the same R (Fig. 9-10b), but less asthenosphere flux is required by UPL (Fig. 9-10a or 10c) because of the contribution of latent heat at its base. The value of this contribution is the difference between the ordinates for STR and UPL in Figure 9-10a and 10c.

We shall refer to INT and UPL as the "magmatic modes." For both modes the volume increase of the extending lithosphere is accommodated by a magmatic increment W that must be supplied to the base of the lithosphere at the rate sR (eq. 10a). Hence,

$$W = sR \tag{22}$$

is the rate of basalt production (per horizontal unit area) required of the asthenosphere to sustain these modes. It is seen from Figure 9-10d that for a given asthenosphere flux the basalt production required by any surface heat flow is approximately the same for each magmatic mode; R is greater for INT, but s is greater for UPL, and the product sR (eq. 22) is approximately the same for each. (The dotted curves in Figure 9-10d represent a combination model to be discussed.)

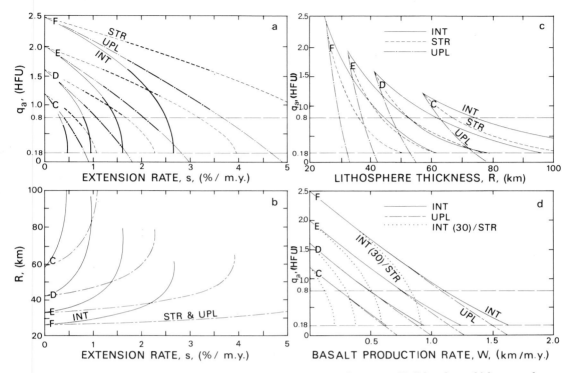

Figure 9-10. Relations among (a) asthenosphere flux and extension rate, (b) lithosphere thickness and extension rate, (c) asthenosphere flux and lithosphere thickness, and (d) asthenosphere flux and basalt production for STR (dash), UPL (dash-dot), and INT (solid). Each set of three curves labeled F represents a reduced heat flow of $q_r = 2.5$ HFU; those labeled E are for $q_r = 2.0$; D are for $q_r = 1.6$; and C are for $q_r = 1.2$. Heavy portions of curves in part (a) represent conditions satisfying relation 23.

Depth to the Basaltic Source or "Lithosphere Thickness" R

Figure 9-11 illustrates the relation between reduced heat flow and R for each of the three modes for an assumed asthenosphere flux q_a of 0.8 HFU. (Values of R for other values of q_a can be read from Fig. 9-10c.) As explained previously, 0.8 HFU is close to the value of q_a expected in stable regions, but we do not know a priori whether q_a might increase or decrease with extension rate, or even if the two quantities should be functionally related. The static case $q_a = q_r$ (dotted Fig. 9-11) represents an upper limit to q_a for all three modes; it provides a lower limit to R. Thus, steady extension of the lithosphere can result in a substantial increase in the estimate of thickness; the amount of increase is sensitive to the mode of extension and to the value of asthenosphere flux. However, even in the static case for characteristic Basin and Range province conditions (C to D, Fig. 9-11), R is at least 40 to 60 km—considerably greater than the 30 km of crust with which the lithosphere of the Basin and Range is sometimes identified. For reasons presented earlier, however, the depth to dry basalt melting might be substantially greater than the thickness of a seismically defined lithosphere, and the contradiction, if it exists, might be one of definition.

It is seen from Figures 9-10 and 9-11 that increasing reduced heat flow does not necessarily imply decreasing lithosphere thickness or increasing asthenosphere flux q_a. If the extension

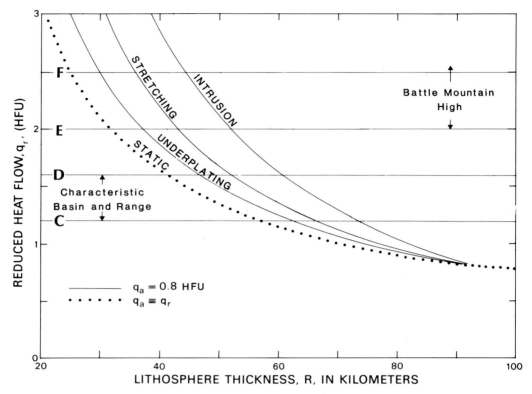

Figure 9-11. Relation between reduced heat flow and lithosphere thickness at constant asthenosphere flux (q_a = 0.8 HFU) for the three simple extension modes (solid curves). Dotted curve is limiting static case (q_a = q_r).

mode changes (for example, from UPL to INT), it is quite possible for q_r to increase while lithosphere thickness increases and/or asthenosphere flux q_a decreases.

Steady Temperatures in an Extending Lithosphere

The four solid curves in each part of Figure 9-12 show temperature profiles for each of the surface conditions C, D, E, and F in an extending lithosphere. Solid curves in Figure 9-12a, 12b, and 12c show each mode for the intermediate value q_a = 0.8 HFU. Solid curves in Figure 9-12d, 12e, and 12f show each mode for the lower limiting value q_a = 0.18, HFU, and the dashed curves on each figure represent the upper limiting case q_a = q_r, reproduced from Figure 9-7. The extension rate (s) compatible with each case is shown beside the solid curves; the value of s for each dashed curve is, of course, zero. Comparison of the dashed curves with the corresponding solid ones illustrates how steady convective transport within the lithosphere reduces estimates of average temperature (for a given q_r) at each depth and increases the depth R to the basaltic source.

The static model and the six interpretations of the extending lithosphere are compared for average conditions in the Battle Mountain High (regime F) in Figure 9-13a and for a characteristic Basin and Range condition (regime D) in Figure 9-13b. Upper crustal temperatures

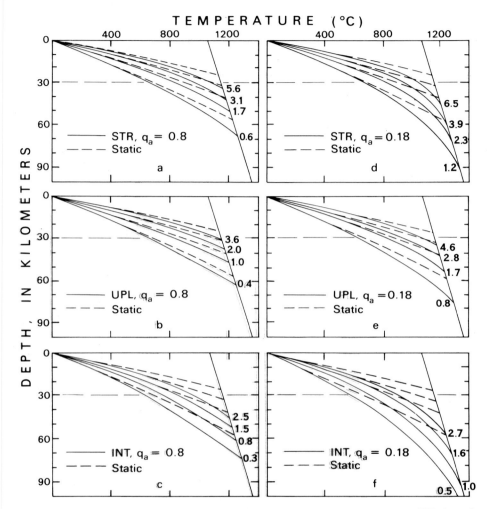

Figure 9-12. Geotherms for lithospheres extending in the three simple modes, STR (a and d), UPL (b and e), and INT (c and f). Dashed curves are for the static case ($q_a = q_r$) reproduced from Figure 9-7. Solid curves in (d), (e), and (f) are for the lower limiting value $q_a = 0.18$ HFU; in (a), (b), and (c) they are for $q_a = 0.8$ HFU. Each set of four curves represents (from top to bottom) reduced heat flows of 2.5, 2.0, 1.6, and 1.2 HFU. Numbers after each solid curve denote extension rates in percent per million years.

are reasonably similar among each set of models; variations are up to perhaps 100°C at mid-crustal levels ($z \sim 15$ km). Temperatures near the base of the crust ($z \sim 30$ km) vary almost 300°C among models for the Battle Mountain High (Fig. 9-13a) and almost 200°C among those for the characteristic Basin and Range condition (Fig. 9-13b). UPL gives higher temperatures and a thinner lithosphere than INT; for UPL the magmatic source occurs at the base; for INT it is distributed throughout the lithosphere. STR requires very large extension rates because it has no magmatic source. It is seen that for both regimes (F and D) the depth to basalt melting R varies by a factor of 2, even when the static model is excluded. Comparison of Figure 9-13 with the melting relations in Figure 9-5 shows that partial melting of crustal rocks

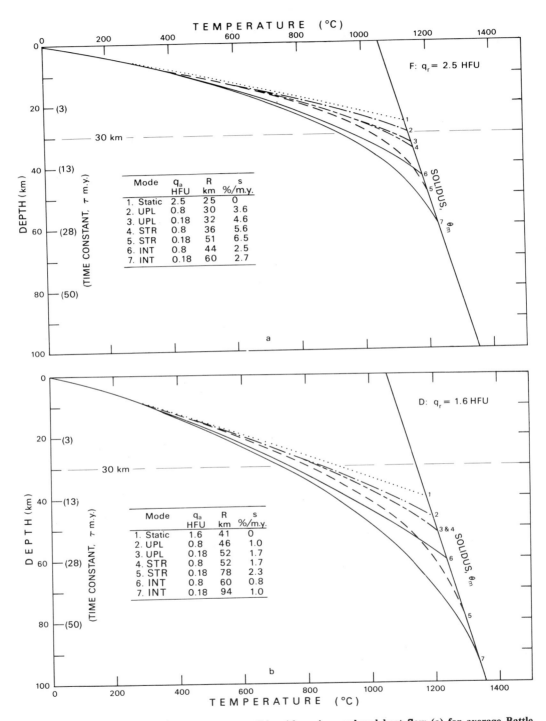

Figure 9-13. Comparison of geotherms compatible with a given reduced heat flow (a) for average Battle Mountain High conditions (F) and (b) for a characteristic Basin and Range condition (D). Number after each curve refers to location in the table (inset) where curve is described.

is possible in the Basin and Range province even under the reduced temperatures given by steady lithosphere extension.

Relation Between Reduced Heat Flow and Extension Rate

The solid curves of Figure 9-14 illustrate how extension affects reduced heat flow in a lithosphere extending under constant asthenosphere flux q_a assumed to be 0.8 HFU. The dashed curve is for the intrusive mode with q_a at its lower limiting value, 0.18 HFU. The proximity of the two curves for INT shows the insensitivity to q_a of the relation between q_r and s for that mode. This insensitivity is illustrated more generally in Figure 9-10a by the near-vertical lower portions of the INT curves. If we knew extension was in the INT mode and that q_a did not increase rapidly with s, then heat flow q_r would be a good measure of spreading rate s.

In considerations of lithosphere temperature (for example, Fig. 9-12) or thickness (Fig. 9-11) the static case gave a useful limit. This is not so in considerations of spreading rate which, of course, goes to zero as $q_a \rightarrow q_r$. In an attempt to abstract some rule-of-thumb from the range of possibilities in Figure 9-10, we have shaded the region in Figure 9-14 in which the reduced heat flow increases from a characteristic stable value of 0.8 HFU at the rate of 1 HFU for every 1% to 2%/m.y. increase in spreading rate. The relation is

$$q_r(\text{HFU}) \sim 0.8 + 1/2 \text{ to } 1 \times s(\%/\text{m.y.}). \tag{23}$$

This region encloses a substantial range of plausible models, particularly if a portion of the extension is accommodated by dike intrusion. The range of conditions falling within the region described by the relation 23 is shown by the darkened portions of curves in Figure 9-10a.

Basalt Production

The relations among reduced heat flow q_r, asthenosphere flux q_a, and basalt production W shown in Figure 9-10d for both UPL and INT can be represented quite well by the dimensional relation

$$(q_r - q_a) [\text{HFU}] \simeq 1.6 \ W \ [\text{km}/\text{m.y.}]. \tag{24}$$

This relation can be viewed simply in terms of Figure 9-15 which shows the steady heat flux Δq from the surface of the lithosphere that would be produced by a volume flux of basalt W into its base. We assume that basalt enters at $\theta_m(R)$ and loses its heat at the average ambient temperature θ'. Thus,

$$\Delta q = \rho c \left[\theta_m(R) - \theta' + \frac{L}{c} \right] W. \tag{25}$$

We assume $\theta_m(R) = 1200°C$, $\rho c = 0.6 \ \text{cal} \cdot \text{cm}^{-3} \cdot °\text{C}^{-1}$, and $L/c = 350°C$. The slope for the curve $\theta' = 700°C$ in Figure 9-15 is about 1.6 HFU per km/m.y. This is the same as the slope in equation 24, generalized for INT and UPL from Figure 9-10d. Thus, in the INT mode, for which all of the convected heat is carried by basalt, the basalt must lose its heat to the lithosphere at an average temperature of about 700°C. For UPL the basaltic heat is

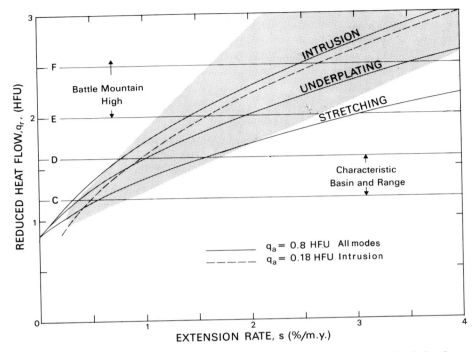

Figure 9-14. Variation of reduced heat flow q_r with extension rate s in each of the three simple modes for an assumed asthenosphere flux q_a = 0.8 HFU (solid curves), and for the INT mode for q_a = 0.18 (dashed curve). The limiting static case coincides with the ordinate axis. Shaded region corresponds to relation 23 in text.

given up at $\theta_m(R)$, a condition represented approximately by the curve $\theta' = 1200°C$ (Fig. 9-15). As the curves for UPL and INT coincide in Figure 9-10d, the difference between ordinates of the curves for 1200°C and 700°C, Figure 9-15, must represent the contribution of solid-state convection (that is, vertical motion due to stretching) in the UPL mode.

Applying equation 22, we see that equation 24 yields an increase in heat flow with spreading rate at the rate of 1%/m.y. per HFU for $R = 62.5$ km and 2%/m.y. per HFU for $R \simeq 31$ km. These are reasonable values consistent with relation 23 and with a simple version of the INT model presented elsewhere (Lachenbruch and others, 1976) and with a special case considered by Oxburgh and Turcotte (1971).

The rates of basalt production required by extension in the magmatic modes are very large. Even if we assume the asthenosphere flux is 1.2 HFU (a static condition for regime C), the average production required in the characteristic Basin and Range regime (C to D) is ~0.1 km/m.y.; for $q_a = 0.8$ it is greater by a factor of 3 (Fig. 9-10d). For the average Battle Mountain High regime (F, Fig. 9-10d), basalt production at the rate of about 1 km/m.y. is required for $q_a = 0.8$, and 0.6 km/m.y. even if q_a is large as 1.5 HFU. For comparison, the Columbia basalt flows, with an average thickness of 1 to 2 km emplaced in 2 to 3 m.y., represent a rate of supply (per unit area) of 0.3 to 1 km/m.y. (Baksi and Watkins, 1973; Swanson and others, 1975). Thus, if the high heat flow in the Basin and Range province is due primarily to basaltic additions to a steadily extending lithosphere, basalt must be supplied in the large hotter subregions at a rate comparable to that of the Columbia Plateau extrusion for many millions of years, and throughout the province as a whole at a rate less only by a factor of perhaps 3 to 5. The basalt requirement can, of course, be reduced (or eliminated) if it is assumed that part (or all) of the lithosphere is extending in the non-magmatic mode (STR).

(It is interesting to note from Figure 9-15 that magmatic heat delivered by extrusion during the Columbia Plateau events represents 1 to 3 HFU; for intrusion in the INT and UPL modes at comparable rates, the magmatic contribution to heat flow would be less by factors of about 2 and 4, respectively.)

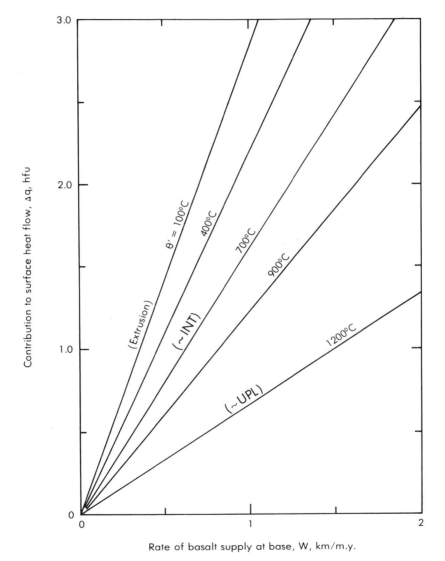

Figure 9-15. Steady surface heat loss Δq **from a layer receiving basalt through its base at the rate** W **at 1200°C; the basalt's heat is lost to the layer at the average ambient temperature** θ'**.**

Combined Models

Although the simple extension modes of Figure 9-8 yield useful limiting cases, it is of interest to combine them in various ways to describe more meaningful geologic models. There are two simple combination schemes: (1) to allow more than one mode to operate at each depth, or (2) to allow one mode to operate in a layer extending from the surface to depth H and a second to operate in the region between H and R. The first scheme was implicit in the foregoing discussion of limiting special cases. In this section we consider a few examples of the second scheme. Results are obtained by modifying the equations to match the temperature and conductive flux at the interface $z = H$. As the velocity fields are the same for all modes (eq. 10), the continuity of mass flux and convective flux at the boundary is assured. In the layered models, we can generally accommodate any estimated departure from steady conditions in the lower layer by adjusting our estimate of q_a, which is unknown in any case.

Dike Intrusion Overlain by Underplating

This model denoted by UPL(H)/INT, and illustrated in Figure 9-16d, provides one quantitative version of the "simple magmatic model" illustrated in Figure 9-6. Extension of the lower lithosphere (below H) is accommodated by dike intrusion, and above H by solid-state stretching;

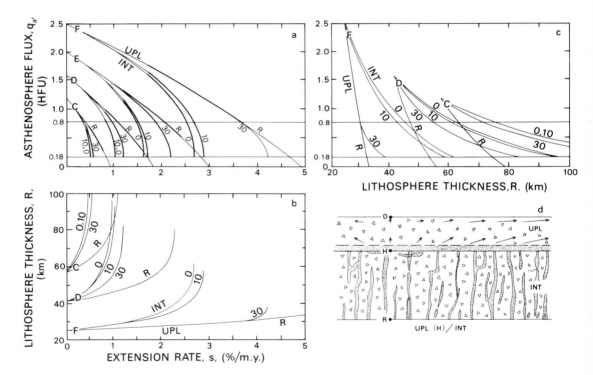

Figure 9-16. Combinations of asthenosphere flux q_a, extension rate s, and lithosphere thickness R, compatible with the characteristic reduced heat flows (regimes C, D, E, and F) for the mode UPL(H)/INT, shown schematically in part (d). Values on curves represent H in kilometres; the curves $H = 0$ and $H = R$ represent, respectively, INT and UPL reproduced from Figure 9-10. Heavy portions of curves in part (a) represent conditions satisfying relation 23.

for example, normal faulting. The thinning of the upper layer is compensated by slow sill formation at depth H. (For this model, it is necessary to replace L/c in equations 14 to 16 by $[L/c + \theta_m(H) - \theta(H)]$ to account for sensible heat released by the basaltic contribution to the sill.)

Figure 9-16a, 16b, and 16c illustrates two cases, $H = 10$ km and $H = 30$ km, both of which are bracketed in all three representations by the simple modes INT and UPL, to which the combined mode reduces when $H = 0$ and $H = R$, respectively. As both modes are magmatic, basalt production (Fig. 9-10d) is not sensitive to the value of H.

The case $H = 10$ approximates the model of the Basin and Range province by Thompson and Burke (1974) wherein extension in the upper 10 km is accommodated by normal faulting and at greater depths by dike intrusion. It is seen from the curves $H = 10$, Figure 16a, 16b, and 16c, that departures from INT caused by the layer of normal faulting are generally unimportant.

The case $H = 30$ km represents a crust extending by underplating above an upper mantle extending by dike intrusion. For the hottest regimes, it is approximated well by UPL, and for the coolest it is closer to INT. Temperatures can be estimated from Figures 9-12 and 9-13.

Dike Intrusion Underlain by Stretching

The large basalt production required of UPL and INT stems from the requirement that the volume change on extension be made up by magmatic contributions throughout the entire lithosphere of depth R. In this model (denoted by INT(H)/STR and illustrated in Fig. 9-17d), basalt, originating at R, rises through the lower region (below H), but it does not reside there. For example, the basalt might rise as lenticular dikelike bodies that pinch off on their lower ends as first suggested by Weertman (1971; see also Secor and Pollard, 1975). The basalt forms dikes in the layer above H which we shall take to represent the crust by setting $H = 30$ km. Below H the upper mantle (down to the depth R of basalt melting) extends by homogeneous stretching; this region can be viewed as part of either the lithosphere or the asthenosphere.

In this case we must distinguish between the volume flux of the rising basalt w_i and the vertical velocity associated with stretching w_s. Hence,

$$w = sz \tag{26a}$$

$$= w_i + w_s \tag{26b}$$

where

$$w_i = sz, \quad 0 < z < H \tag{27a}$$

$$= sH, \quad H < z < R \tag{27b}$$

and

$$w_s = 0, \qquad 0 < z < H \tag{27c}$$

$$= s(z - H), \quad H < z < R. \tag{27d}$$

As basalt masses rise from R to H, maintaining their melting temperature θ_m, they will cool and release heat at the rate

$$\rho c w_i \frac{d\theta_m}{dz} = \frac{K}{\beta^2} Hm, \quad H < z < R. \tag{28}$$

The influence of this constant source has been added to equations 15 and 16 to accommodate this small effect in the calculations.

For conditions illustrated in Figure 9-17a and 17b, the combined model is bracketed by STR and INT, to which the model reduces when $H \to 0$ and R, respectively. On physical grounds the curve $H = 30$ must pass to the curve INT as a progressively increasing proportion of the basalt freezes in transit through the upper mantle. Hence, the region bounded by these two curves ($H = 30$ and INT, Fig. 9-17a, 17b, and 17c) may represent a range of plausible conditions. The region satisfies the relation 23 over a broad range of parameters (darkened portions of curves, Fig. 9-17a). It is interesting that this combined model yields an equilibrium depth to basalt melting (Fig. 9-17c) even greater than INT; for $q_a = 0.8$ HFU, R varies from 48 km in regime F to 75 km in regime C. As only the crust (of thickness $H = 30$) is extending

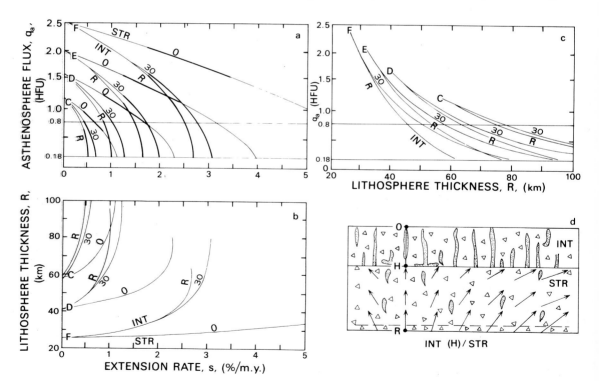

Figure 9-17. Combinations of asthenosphere flux q_a, extension rate s, and lithosphere thickness R compatible with the characteristic reduced heat flows (regimes C, D, E, and F) for the mode INT(H)/STR shown schematically in part (d). Values on curves represent H in kilometres; curves $H = 0$ and $H = R$ represent, respectively, STR and INT reproduced from Figure 9-10. Heavy portions of curves in part (a) represent conditions satisfying relation 23.

in a magmatic mode, basalt production W is given by sH, shown by the dotted curves in Figure 9-10d. Although it is still substantial in this combined model, the basalt production is considerably less than that required by the pure magmatic modes.

Temperatures for this combined model with $q_a = 0.8$ HFU are shown in Figure 9-18; they could have been anticipated fairly well from the information in Figure 9-17 and the curves of Figure 9-12. Extension rates required by the model are shown by numbers on the curves of Figure 9-18; they satisfy relation 23.

Other Modes

It has been shown that the three simple modes of Figure 9-8, individually and in combination, are sufficient to indicate the thermal effects of uniform extension of a lithosphere of constant thickness under a variety of conditions of sill and dike intrusion and stretching. However, these modes do not complete the possibilities. For example, we could consider a fourth mode, "overplating," wherein sill injection at the top of a layer is compensated by stretching of the layer, and solid-state convection is downward. This case can be treated by a simple modification of the foregoing results. Unlike stretching and dike intrusion, sill intrusion does not require extension. Thus, plausible magmatic models of elevated heat flow in a non-extending lithosphere might involve the steady supply to a sill at the rate W(km/m.y.) at some depth

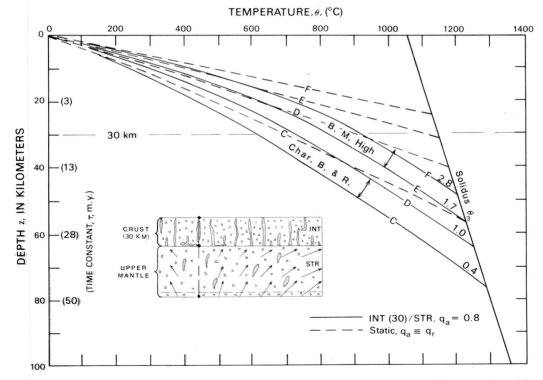

Figure 9-18. Comparison of geotherms for the static (dashed) and INT(30)/STR (solid curves) models for regimes C, D, E, and F. Numbers on solid curves are extension rate in percent per million years. Physical model is shown schematically by inset.

where the rising magma spreads laterally (perhaps because of loss of buoyancy or formation of viscous melts in the light crustal rocks). For such cases, the contribution of magmatic heat to surface heat flow can be estimated roughly from Figure 9-15, and the attenuation of asthenosphere flux from below (by downward convection in the vertically displaced lower lithosphere) can be estimated from results like equation 10 of Lachenbruch and Sass (1977). Although such stationary models, and their transient generalizations, are probably appropriate for describing some heat-flow anomalies, we have not considered them here, as we are investigating the thermal implications of tectonic extension.

DISCUSSION AND APPLICATIONS

In the stable Eastern United States, the reduced heat flow is about 0.8 HFU; in the Basin and Range province, reduced heat flow is characteristically greater by 50% to 100%, and in the Battle Mountain High subprovince by 300%. We surmise that these high heat flows are caused in some way by lithosphere extension and magmatism, for which there is abundant evidence in the province. What do the high heat flows tell us about these processes and about the temperature and thickness of the underlying lithosphere? The answer depends upon the mechanical mode of extension, the conductive flux from the asthenosphere q_a and the thermal history, all of which are unknown. We have investigated the question with simple steady-state thermomechanical models of extension and arbitrary asthenosphere flux. The lithosphere would be hottest and thinnest if the reduced heat flow q_r (with a small correction for neglected radioactivity) were equal to the asthenosphere flux q_a. Strictly speaking, however, in that case (Fig. 9-7) there could be no extension or magmatism, and it is precluded by the geologic observations. Heat convection across a steadily extending lithosphere requires that more heat be conducted out of the top (q_r) than into the bottom (q_a), unless there is significant undetected convective discharge at the Earth's surface. However, if extension were very slow and in a thermally inefficient mode (for example, solid-state stretching, STR) the difference between q_r and q_a would be small, and the large values of q_r could simply be mimicking large values of q_a. In that case heat would have to be conducted across a boundary layer at the base of the lithosphere at rates approaching q_r. Although we have not investigated heat transfer in the asthenosphere, such high values of q_a seem unlikely; for the oceanic asthenosphere they are precluded by heat-flow evidence. Furthermore, the abundant evidence for basaltic activity in the Basin and Range province suggests that extension there is occurring (in part at least) in the magmatic modes, that is, those for which the volume increase of the extending lithosphere is made up largely by magmatic additions as dikes (INT and its combinations) or sills (UPL and its combinations). They are more efficient in the sense that they produce a larger difference between q_r and q_a for a given extension rate. Hence, we consider it probable that much, if not most, of the observed regional variation in q_r results from variation in convective transport within the extending lithosphere.

The range of possibilities for lithosphere temperature and thickness, extension rate, basalt production, and asthenosphere flux permitted by the simple steady-state models underscores the ambiguity of interpretations of reduced heat flow in regions of extensional tectonics. We have attempted to present results in a form sufficiently general that they may be used to test the implications of specific models suggested by evidence from other sources. As we have made no systematic attempt to examine such evidence, it is premature to suggest a preferred model. Nevertheless, the analysis suggests some generalizations:

1. The lithosphere defined as the region overlying the basalt (or peridotite) dry solidus

in the Basin and Range province generally includes a great deal more than the crust. Any discrepancy with the lithosphere defined from seismic data may be semantic, relating to the effects on seismic propagation of melting below the dry solidus, high temperature, or contamination by basaltic intrusion in the upper mantle.

2. Temperature estimates based on heat-flow observations are reduced by the effects of lithosphere extension but are still consistent with substantial partial melting of crustal rocks in the warmer parts of the Basin and Range province.

3. If the steady volume increase of at least the crustal part of the extending lithosphere is supplied by distributed dike intrusion (INT), the surface heat flow is insensitive to asthenosphere flux q_a, so long as the latter does not exceed about 1 HFU.

4. For the foregoing case and a number of other plausible steady-state models, the anomalous heat flow increases about 1 HFU for each 1% to 2%/m.y. increase in extension rate. The high regional heat flow and its variations throughout the Basin and Range province can be accounted for by modest rates of lithosphere extension; there is no need to invoke high or variable conductive flux from the asthenosphere.

5. The ultimate source of high heat flow in the Basin and Range province must be upward convective transport in the asthenosphere.

Ambiguities regarding q_a and extension mode may not be resolved until we understand the flow in the asthenosphere. However, surface observations of relations among heat flow, volcanic activity, and extension rate may give some clues. We shall conclude with a few remarks about them.

Silicic Volcanic Centers

Since the start of Basin and Range tectonism in early Miocene time, silicic volcanic activity in the province has generally been associated with basaltic activity (Christiansen and Lipman, 1972). During periods when silicic magma chambers existed in the upper crust, basaltic extrusion often appeared around their edges (Eaton and others, 1975; Bailey and others, 1976). Such observations have led to the belief that basaltic intrusion of the lithosphere might be a primary source of heat for this "bimodal" crustal magmatic activity and that the throughflow of basalt to the surface might commonly be blocked by the viscous silicic melts that the basalt's heat creates (Smith and Shaw, 1976; Bailey and others, 1976; Christiansen and Lipman, 1972). In this connection it is enlightening to consider the heat and mass budget of these volcanic centers.

For the volcanic center at Long Valley, California (LV, Fig. 9-1a), it has been estimated (Sorey and Lewis, 1976; Lachenbruch and others, 1976) that an upper crustal magma chamber ~20 km in diameter resulted in a combined (hydrothermal and conductive) anomalous flux of at least 10 HFU in excess of the ambient regime for at least 2 m.y. The supply of this steady loss for 2 m.y. would require a heat input equivalent to that of a 14-km layer of basalt of comparable area, quenched to the magma temperature (assumed to be 800°C, see Fig. 9-15). To warm the underlying rock to magmatic temperatures would take perhaps 5 or 6 km more, and an additional 1 or 2 km would be required to supply the heat lost from extruded rock. It is difficult to imagine a mechanism for this mass influx and cooling that does not incorporate rapid local crustal spreading. Furthermore, it is difficult to imagine thermomechanical processes that could sustain such intense heat discharge other than those associated with an influx of basalt. Because of its low viscosity (Shaw, 1965; Shaw and others, 1968), high melting temperature, and large latent heat, basalt is the most efficient heat transfer fluid available. It could rise freely through narrow conduits in the lower lithosphere, and at crustal levels, give up its heat of crystallization and much sensible heat at temperatures high enough to mobilize or melt in place the indigenous silicic material (see Fig. 9-5). (The rapid heat loss from the magma chamber to the surface must, in turn, depend on vigorous hydrothermal transport.)

The most reasonable conclusion seems to be that these bimodel silicic volcanic centers exist because they are at places where the lithosphere is pulling apart rapidly, drawing up basalt from below to fill the void (Lachenbruch and others, 1976). Such a process has been suggested on the basis of structural arguments by Wright and Troxel (1968) and Carr (1974). If these processes took place directly beneath the Long Valley magma chamber, an increase in area of 35% or more would be required for a lithosphere 50 km thick. If the strain were accommodated by east-west extension, it would amount to 7 or 8 km of displacement, which could account for the observed ellipticity of the caldera structure; the displacement could, of course, be distributed over a larger area and longer time. Direct evidence for east-west extension of this sort has recently been obtained from a study of seismicity at the Coso Volcanic Center, another Quaternary silicic volcanic center about 200 km south of Long Valley (Weaver and Hill, 1978).

Lithosphere Extension and Passive Intrusion

The foregoing considerations suggest processes related to extension in the magmatic modes, and they invite speculation that the primary source of heat for high regional heat flow in the Basin and Range province, like that for its volcanic centers, might be basalt passively intruding the extending lithosphere. It has been pointed out that extension of the lithosphere and isostatic balance imply upflow of mass in the asthenosphere, and that this will generally result in an upward increase in melted fraction toward the base of the lithosphere. The pressure on the asthenosphere melt will be the overburden pressure, but at least one horizontal stress in the rigid portion of the extending lithosphere will be consistently less. This would be an unstable condition if the rigid layer behaved brittlely; the melt could passively invade the layer, in effect by "hydrofracturing" it. The through-flow of basalts to the surface might generally be impeded by solidification in transit, by the formation of viscous silicic melts as heat is exchanged with the crustal rocks, and by loss of buoyancy at crustal levels. Above some depth of intense intrusion (for example, H in mode UPL(H)/INT), extension might be mainly by normal faulting (a form of "stretching") and the heat transfer mainly by conduction, modified by localized and transient silicic intrusion and hydrothermal effects. This view is generally consistent with many structural accounts of the Basin and Range province, although they vary in other respects (see, for example, Hamilton and Myers, 1966; Pakiser and Zietz, 1965; Thompson and Burke, 1974; Scholz and others, 1971).

"Hydrofracturing" by basaltic magma differs in two important ways from the more familiar hydrofracturing by water; basalt has a viscosity ($\sim 10^3$ poise) greater by a factor of $\sim 10^5$ than water, and for basalt the ambient temperature is below the freezing temperature. The fracture width is determined primarily by the elasticity of the rigid lithosphere, the extensional tectonic stress, and the buoyancy of the basalt; widths could be only on the order of metres for cracks tens of kilometres in length. If the crack travels too fast, viscous forces will reduce the pressure in the magma and arrest propagation (see, for example, Newman and Smyrl, 1974; Lachenbruch, 1973, eq. A-14); if it travels too slowly the basalt will freeze. Estimation of these limiting propagation velocities is a very complex problem (for a recent review see Anderson and Grew, 1977), but order-of-magnitude calculations (Fedotov, 1976;

Maaloe, 1973; Lachenbruch, 1977, unpub.) suggest that for modest extensional stresses there may be a substantial "velocity window" through which basaltic dikes could traverse a brittle lithosphere without freezing. If this is true and if the yield stress of the lithosphere exceeds the stress needed for "hydrofracturing," distributed spreading of the lithosphere might be largely in the INT mode, rate-limited by the availability of basalt from the asthenosphere.

The other magmatic mode (UPL) might also facilitate distributed extension. If the basalt supply were depleted and the extension rate remained the same, the UPL mode would pass to the STR mode, and solid-state flow would remain unchanged. However, the basaltic heat contribution would vanish, and the lithosphere would thicken and cool toward the new and stronger steady-state configuration (STR) (for example, if $q_a = 0.8$ HFU and $s = 2\%/$m.y., then q_r would decrease by 0.3 HFU, Fig. 9-14, and R would increase by 30%, Fig. 9-11; equilibrium temperatures at 30 km would drop from 1000°C, Fig. 9-12b, to ~875°C). Unless the tectonic stress or q_a increased, spreading rate would decrease. This would further cool and strengthen the lithosphere and eventually arrest the extension process. Hence, either magmatic mode (INT or UPL) could facilitate distributed extension of the lithosphere; basalt production by the asthenosphere might even be a necessary condition for distributed extension.

It has been shown (Fig. 9-10d) that for a moderate asthenosphere flux the rate of basalt production by the asthenosphere required to sustain the magmatic modes can be quite substantial. For a given spreading rate, the basalt flux is proportional to the thickness of that portion of lithosphere extending in a magmatic mode (for example, R for INT or UPL, H for INT(H)/STR). If, for example, that portion should be 30 km thick, then the basalt required to produce a unit area of new surface would be greater by six times than the corresponding quantity for an ocean-spreading center where the basaltic crust is about 5 km thick. If the entire Great Basin were extending at the rate of about 1 cm/yr, the asthenosphere beneath it would have to supply basalt at the same rate as that beneath an oceanic spreading center with a spreading half rate of 3 cm/yr.

At ocean-spreading centers, lithosphere more than 60 km thick must be created, and less than 10% of it, the crustal part, is supplied by basalt. Consequently, very viscous (~10^{20} poise) ultramafic material must rise. Because of the large viscosity and small buoyancy of this material, the conduits through which it rises must be very wide (Lachenbruch, 1973, 1976) to accommodate appreciable extension rates. Furthermore, the walls of these conduits must separate rapidly to maintain the temperature of the passively rising ultramafic material appreciably above ambient. Such conduits must therefore accommodate extension over large areas and consequently there should be few of them. Hence, lithosphere production that requires ultramafic intrusion favors localized or plate-tectonic extension as distinguished from distributed tectonic extension, which we suggest might be sustained by an abundant supply of the less viscous basalt. Perhaps the localized thermal anomaly at Yellowstone Park, Wyoming (Fig. 9-1b), results from the fact that the lithosphere there is very thick (Iyer, 1975) and its spreading requires the ascent of very viscous asthenosphere material in a single wide conduit region, the opening of which must accommodate strain accumulation over a large area.

Heat Flow and Distributed Spreading in Back-Arc Basins

The tectonic setting of the Basin and Range province is often compared to that of marginal oceanic basins that occur behind island arcs and associated subduction zones (for references, see Thompson and Burke, 1974, p. 234). In each setting, lithosphere extension seems to be distributed through the province rather than being concentrated along an axis as at an oceanic spreading center. Watanabe and others (1977) have noted that newly formed back-arc basins cool toward a steady heat flow of about 2.2 HFU which they seem to maintain with a thin lithosphere (40 to 50 km) for tens of millions of years. Thereafter, a heat source is apparently removed and a second stage of cooling results in an approach to the flux typical of the old Pacific plate. Perhaps the first steady-state results from a phase of distributed extension, and the additional heat source represents convective transport through the lithosphere. For example, if q_a were constant at 0.8 HFU, distributed extension at the rate of 1.5%/m.y. in the INT mode would yield a reduced heat flow of 2.0 HFU (Fig. 9-10a, regime E); for the UPL mode, 2%/m.y. would be required. This would result in total spreading across a 500-km basin at rates of 7.5 and 10 mm/yr and lithosphere thicknesses (Fig. 9-10c) of 52 and 38 km for INT and UPL, respectively. When extension stopped, possibly because of depletion of the asthenosphere basalt required by the magmatic modes, the lithosphere would cool conductively toward the static condition associated with $q_a = q_r$, ~0.8 HFU as observed (see, for example, curve B, Fig. 9-7).

Regional Heat Flow and Rate of Extension in the Basin and Range Province

The present width of the Basin and Range province across central Nevada-Utah is about 800 km. If the crustal strain is accommodated by east-west extension and the mean reduced heat flow is 1.4 HFU (the average of regimes C and D), then relation 23 yields an extension rate of 0.6% to 1.2%/m.y. and a spreading rate for the province of 5 to 10 mm/yr. On the assumption that the present is typical of the past 17 m.y., we multiply to obtain a total extension of 80 to 160 km. Refinements such as allowing for growth by encroachment on the Sierra Nevada and Colorado Plateau provinces or for local effects at volcanic centers hardly seem warranted in view of the tenuous basis for relation 23. It is encouraging, however, that this result is generally consistent with independent estimates made from geometric considerations of fault displacements (for a discussion, see Thompson and Burke, 1974).

According to relation 23, the rate of crustal extension in high heat-flow subprovinces like the Rio Grande rift and the Battle Mountain High might be much greater than that for the Basin and Range province as a whole. This is consistent with frequently stated views of spreading in the Rio Grande rift (for example, Hamilton and Myers, 1966; Reiter and others, 1975), but the tectonic implications of the Battle Mountain High deserve comment. For the reduced heat flow of 2.5 HFU (regime F), relation 23 suggests that the area of the Battle Mountain High may be increasing ~1.7% to 3.4%/m.y. If the strain were relieved by displacement (across its 150± km width) normal to the northeasterly axis, we would obtain a total local spreading rate of 2½ to 5 mm/yr. For other directions of uniaxial strain, the total rate would be greater. Although historic seismicity in the area is low, the thermal result is supported by Slemmons' (1967) study of Pliocene and Quaternary crustal movements, which shows an intense concentration of faulting in the Battle Mountain High region throughout the recent past. The nearly north-south trend of normal faults, predominant in much of the Great Basin, seems to swing to a more northeasterly trend in the Battle Mountain High (Stewart, 1971), more or less parallel to the long axis of the Battle Mountain High and its extension through the eastern Snake River Plain to Yellowstone Park, Wyoming (Fig. 9-1b). Perhaps the distributed extension inferred from high heat flow in the Battle Mountain High and the more localized rifting in the Snake River Plain (Hill and Pakiser, 1966; Hamilton and Myers, 1966) represent contrasting responses of two different crustal provinces to the same regional tectonic system of stress.

ACKNOWLEDGMENTS

We are grateful to our colleague, John Ziagos, for his innovative efforts in programming and executing the numerical computations, and to Charles Bacon, David Chapman, Warren Hamilton, David Hill, Roy Hyndman, George Thompson, and Mary Lou Zoback, for their thoughtful comments on our manuscript.

APPENDIX 1. LIST OF SYMBOLS

θ = temperature

θ' = average ambient temperature at which basalt loses its heat

z = depth

y = horizontal distance

v = horizontal velocity

w = average vertical velocity at any depth (vertical volume flux)

W = volume flux of basalt at base of lithosphere

s = $\dfrac{\partial v}{\partial y}$, horizontal strain rate

K = thermal conductivity

ρ = density

c = specific heat

L = latent heat of melting for basalt (per unit mass)

β = $(K/s\rho c)^{1/2}$, characteristic length for lithosphere convection

$A(z)$ = radioactive heat production at depth z

A_o = radioactive heat production at surface ($z = 0$)

D = characteristic depth for distribution of radioactive heat production

$q(z)$ = conductive heat flow at depth z

q = conductive heat flow at surface

q_r = $q - A_o D$, reduced heat flow

$\theta_m(z)$ = $\theta_o + mz$, temperature of basalt dry solidus at depth z

R = "lithosphere thickness" (depth to θ_m).

q_a = conductive flux from asthenosphere (see eq. 8)

INT   = mode of lithosphere extension by distributed dike intrusion (Fig. 9-8c)

UPL   = mode of lithosphere extension by stretching and underplating (Fig. 9-8b)

STR   = mode of lithosphere extension by stretching (Fig. 9-8a)

C   = lithosphere regime for q_r = 1.2 HFU

D   = lithosphere regime for q_r = 1.6 HFU

E   = lithosphere regime for q_r = 2.0 HFU

F   = lithosphere regime for q_r = 2.5 HFU

HFU   = heat-flow unit, 10^{-6} cal$\cdot$cm$^{-2}\cdot$s^{-1} = 41.8 mW$\cdot$m^{-2}

HGU   = heat generation unit, 10^{-13} cal$\cdot$cm$^{-3}\cdot$s^{-1} = 0.418 μW$\cdot$m^{-3}

REFERENCES CITED

Anderson, O. L., and Grew, P. C., 1977, Stress corrosion theory of crack propagation with applications to geophysics: Rev. Geophysics and Space Physics, v. 15, p. 77–104.

Bailey, R. A., Dalrymple, G. B., and Lanphere, M. A., 1976, Volcanism, structure, and geochronology of Long Valley Caldera, Mono County, California: Jour. Geophys. Research, v. 81, p. 725–744.

Baksi, A. K., and Watkins, N. D., 1973, Volcanic production rates: Comparison of oceanic ridges, islands, and the Columbia Plateau basalts: Science, v. 180, p. 493–496.

Birch, Francis, Roy, R. F., and Decker, E. R., 1968, Heat flow and thermal history in New England and New York, in Zen, E-an, White, W. S., Hadley, J. B., and Thompson, J. B., Jr., eds., Studies of Appalachian geology: New York, Northern and Maritime: Interscience, p. 437–451.

Blackwell, D. D., 1971, The thermal structure of the continental crust, in Heacock, J. G., ed., The structure and physical properties of the earth's crust: Am. Geophys. Union Geophys. Mon. 14, p. 169–184.

Bodvarsson, Gunnar, 1954, Terrestrial heat balance in Iceland: Timarit Verkfraedingafelags Islands, p. 69–76.

Brott, C. A., Blackwell, D. D., and Mitchell, J. C., 1976, Heat flow study of the Snake River Plain region, Idaho: Boise, Idaho Dept. Water Resources, Geothermal Inv. in Idaho, Water Inf. Bull., v. 30, pt. 8, 195 p.

Carr, W. J., 1974, Summary of tectonic and structural evidence for stress orientation at the Nevada Test Site: U.S. Geol. Survey Open-File Rept. 74–176, 53 p.

Chapman, D. S., and Pollack, H. N., 1977, Regional geotherms and lithosphere thickness: Geology, v. 5, p. 265–268.

Christiansen, R. L., and Lipman, P. W., 1972, Cenozoic volcanism and plate-tectonic evolution of the Western United States—Pt. II, late Cenozoic: Royal Soc. London Philos. Trans., Ser. A, v. 271, p. 249–284.

Christiansen, R. L., and McKee, E. H., 1978, Late Cenozoic volcanic and tectonic evolution of the Great Basin and Columbia intermontane region, in Smith, R. B., and Eaton, G. P., eds., Cenozoic tectonics and regional geophysics of the western cordillera: Geol. Soc. America Mem. 152 (this volume).

Clark, S. P., Jr., and Ringwood, A. E., 1964, Density distribution and constitution of the mantle. Rev. Geophysics, v. 2, p. 35–88.

Crough, S. T., and Thompson, G. A., 1976, Thermal model of continental lithosphere: Jour. Geophys. Research, v. 81, p. 4857–4862.

Eaton, G. P., Christiansen, R. L., Pitt, A. M., Mabey, D. R., Blank, H. R., Zietz, Isidore, and Gettings, M. E., 1975, Magma beneath Yellowstone National Park: Science, v. 188, p. 787–796.

Fedotov, S. A., 1976, On the ascent of basic magmas in the earth's crust and the mechanism of basaltic fissure eruptions: Akad. Nauk SSSR Izv. Ser. Geol., no. 10, p. 5–23.

Froideveaux, C., and Schubert, G., 1975, Plate motion and structure of the continental asthenosphere: A realistic model of the upper mantle:

Jour. Geophys. Research, v. 80, p. 2553–2564.

Green, D. H., 1973, Contrasted melting relations in a pyrocite upper mantle under mid-ocean ridge, stable crust, and island arc environments: Tectonophysics, v. 17, p. 285–297.

Hamilton, Warren, and Myers, W. B., 1966, Cenozoic tectonics of the Western United States: Rev. Geophysics, v. 4, p. 509–549.

Hill, D. P., and Pakiser, L. C., 1966, Crustal structure between the Nevada Test Site and Boise, Idaho, from seismic refraction measurements, in Steinhart, J. S., and Smith, T. J., eds., The earth beneath the continents: Am. Geophys. Union Mon. 10, p. 391–419.

Iyer, H. M., 1975, Anomalous delays of teleseismic P-waves in Yellowstone National Park: Nature, v. 253, p. 425–427.

Kay, R., Hubbard, N. J., and Gast, P. W., 1970, Chemical characteristics and origin of oceanic ridge volcanic rocks: Jour. Geophys. Research, v. 75, p. 1585–1613.

Lachenbruch, A. H., 1968, Preliminary geothermal model of the Sierra Nevada: Jour. Geophys. Research, v. 73, p. 6977–6989.

——1970, Crustal temperature and heat production: Implications of the linear heat flow relation: Jour. Geophys. Research, v. 75, p. 3291–3300.

——1973, A simple mechanical model for oceanic spreading centers: Jour. Geophys. Research, v. 78, p. 3395–3417.

——1976, Dynamics of a passive spreading center: Jour. Geophys. Research, v. 81, p. 1883–1902.

Lachenbruch, A. H., and Sass, J. H., 1977, Heat flow in the United States and the thermal regime of the crust, in Heacock, J. G., ed., The nature and physical properties of the earth's crust: Am. Geophys. Union Geophys. Mon. 20.

Lachenbruch, A. H., Sass, J. H., Munroe, R. J., and Moses, T. H., Jr., 1976, Geothermal setting and simple heat conduction models for the Long Valley Caldera: Jour. Geophys. Research, v. 81, p. 769–784.

Maaloe, Sven, 1973, Temperature and pressure relations of ascending primary magmas: Jour. Geophys. Research, v. 78, p. 6877–6886.

Newman, John, and Smyrl, W. H., 1974, Fluid flow in a propagating crack: Mettalurgical Trans., v. 5, p. 469–474.

Oxburgh, E. R., and Turcotte, D. L., 1971, Origin of paired metamorphic belts and crustal dilation in island arc regions: Jour. Geophys. Research, v. 76, p. 1315–1327.

Pakiser, L. C., and Zietz, Isidore, 1965, Transcontinental crustal and upper mantle structure: Rev. Geophysics, v. 3, p. 505–520.

Palmason, Gudmundur, 1973, Kinematics and heat flow in a volcanic rift zone, with application to Iceland: Royal Astron. Soc. Geophys. Jour., v. 33, p. 451–481.

Pollack, H. N., and Chapman, D. S., 1977, On the regional variation of heat flow, geotherms, and lithospheric thickness: Tectonophysics, v. 38, p. 279–296.

Reiter, Marshall, Edwards, C. L., Hartman, Harold, and Weidman, Charles, 1975, Terrestrial heat flow along the Rio Grande Rift: Geol. Soc. America Bull., v. 86, p. 811–818.

Roy, R. F., Blackwell, D. D., and Birch, Francis, 1968, Heat generation of plutonic rocks and continental heat flow provinces: Earth and Planetary Sci. Letters, v. 5, p. 1–12.

Roy, R. F., Blackwell, D. D., and Decker, E. R., 1972, Continental heat flow, in Robertson, E. C., ed., The nature of the solid earth: New York, McGraw-Hill, p. 506–544.

Sass, J. H., Lachenbruch, A. H., Munroe, R. J., Greene, G. W., and Moses, T. H., Jr., 1971, Heat flow in the Western United States: Jour. Geophys. Research, v. 76, p. 6376–6413.

Scholz, C. H., Barazangi, M., and Sbar, M. L., 1971, Late Cenozoic evolution of the Great Basin, Western United States, as an ensialic interarc basin: Geol. Soc. America Bull., v. 82, p. 2979–2990.

Secor, D. T., Jr., and Pollard, D. D., 1975, On the stability of open hydraulic fractures in the earth's crust: Geophys. Research Letters, v. 2, p. 510–513.

Shaw, H. R., 1965, Comments on viscosity, crystal settling, and convection in granitic magmas: Am. Jour. Sci., v. 263, p. 120–152.

Shaw, H. R., Wright, T. L., Peck, D. L., and Okamura, R., 1968, The viscosity of basaltic magma: An analysis of field measurements in Makaopuhi lava lake, Hawaii: Am. Jour. Sci., v. 266, p. 225–264.

Slemmons, D. B., 1967, Pliocene and Quaternary crustal movements of the Basin-and-Range province, USA: Osaka City Univ. Jour. Geosciences, v. 10, p. 91–103.

Smith, R. L., and Shaw, H. R., 1976, Igneous-related geothermal systems, in White, D. E., and Williams, D. L., eds., Assessment of geothermal resources of the United States—1975: U.S. Geol. Survey Circ. 726, p. 58–83.

Sorey, M. L., and Lewis, R. E., 1976, Convective heat flow from hot springs in the Long Valley Caldera, Mono County, California: Jour. Geophys. Research, v. 81, p. 785–791.

Stewart, J. H., 1971, Basin and Range structure: A system of horsts and grabens produced by deep-seated extension: Geol. Soc. America Bull., v. 82, p. 1019–1044.

Swanson, D. A., Wright, T. L., and Helz, R. T.,

1975, Linear vent systems and estimated rates of magma production and eruption for the Yakima Basalt on the Columbia Plateau: Am. Jour. Sci., v. 275, p. 877–905.

Thompson, G. A., and Burke, D. B., 1974, Regional geophysics of the Basin and Range province: Annual Review of Earth and Planetary Sciences, v. 2, p. 213–237.

Tozer, D. C., 1967, Towards a theory of thermal convection in the mantle, *in* Gaskell, T. F., ed., The earth's mantle: London, Academic Press, p. 325–353.

Watanabe, T., Langseth, M. G., and Anderson, R. N., 1977, Heat flow in back-arc basins of the western Pacific, *in* Talwani, M., and Pitman, W., eds., Evolution of island arcs, deep sea trenches and back-arc basins: Am. Geophys. Union, Maurice Ewing Ser., I, p. 137–161.

Weaver, C. S., and Hill, D. P., 1978, Earthquake swarms and local crustal spreading along major strike-slip faults in California: PAGEOPH, (in press).

Weertman, John, 1971, Theory of water-filled crevasses in glaciers applied to vertical magma chambers beneath oceanic ridges: Jour. Geophys. Research, v. 76, p. 1171–1183.

Wright, L. A., and Troxel, B. W., 1968, Evidence of northwestward crustal spreading and transform faulting in the southwestern part of the Great Basin, California and Nevada [abs.]: Geol. Soc. America Spec. Paper 121, p. 580–581.

Wyllie, P. J., 1971, Experimental limits for melting in the earth's crust and upper mantle, *in* Heacock, J. G., ed., The structure and physical properties of the earth's crust: Am. Geophys. Union Geophys. Mon. 14, p. 279–301.

Yoder, H. S., Jr., 1976, Generation of basaltic magma: Natl. Acad. Sci., Washington, D. C., 265 p.

MANUSCRIPT RECEIVED BY THE SOCIETY AUGUST 15, 1977

MANUSCRIPT ACCEPTED SEPTEMBER 2, 1977

Geological Society of America
Memoir 152

10

Geophysical studies and tectonic development of the continental margin off the Western United States, lat 34° to 48°N

Eli A. Silver
Earth Sciences Board
University of California, Santa Cruz
Santa Cruz, California 95064

ABSTRACT

The continental margin off the Western United States changes abruptly from the subduction type north of lat 40°38′N to the translational type south of that latitude. Along the Cascadia subduction zone in the northern area a change in structural geometry at about the Columbia River separates the Oregon structural type of fore-arc basin-ridge-slope accretionary prism from the Washington type of low accretionary plateau. This change in gross morphology results from the dominantly landward-dipping thrust faults off Oregon to seaward-dipping thrusts off Washington. Gravity anomalies along the Oregon slope have steep gradients and relatively high amplitudes, whereas off Washington the gradients and amplitudes are low. Magnetic anomalies from the down-going Juan de Fuca plate can be traced inland 100 km from the base of the Washington slope but only 20 km off Oregon.

The central California translational margin is cut by several long fault zones and has five large sedimentary basins and four marginal ridges. The basins all formed in late middle Miocene, probably as a result of a change in direction of Pacific-America plate motion. Basins resting on Franciscan assemblage basement rocks show a greater structural variability than those on granitic basement. Offset of granitic rocks of the Salinian block that underlie the Farallon ridge can be restored by assuming 80 km of right lateral strike-slip along the San Gregorio fault and 600 km along the San Andreas. The San Gregorio fault may join the Hosgri–San Simeon fault zone south of Point Sur, and this fault also shows evidence for 80 to 100 km of right-lateral strike-slip. The San Gregorio–Hosgri fault system may thus be one of the longest in California. The nongranitic offshore ridges are underlain by deformed sedimentary and volcanic rocks, some of which are dated as early to middle Tertiary. A sharp change in structural style between these rocks, which were probably deformed by lower to middle Tertiary subduction, and the gently folded or undisturbed overlying strata of Miocene and younger age may represent the transition from compression associated with subduction to translational shear off central California.

251

INTRODUCTION

The continental margin off the California, Oregon, and Washington coast is of the subduction type north of lat 40°30′N and the translational type to the south (Fig. 10-1). The northern area, called the Cascadia subduction zone, is the locus of subduction of the Juan de Fuca plate (Atwater, 1970; Silver, 1969,1971a, 1972). The southern area is the shear boundary between the Pacific and North American plates. The structural development of these two areas has been quite different, but significant variations exist within each region as well. The structural development of the continental margin is interpreted from geophysical data—largely seismic, gravity, and magnetic—plus bottom sampling and some drilling. This paper presents an overview of the structure of the continental margin and uses predictive aspects of plate tectonics. Original data sources can be found in Silver (1971a, 1971b, 1971c, 1972), Kulm and Fowler (1974a, 1974b), Hoskins and Griffiths (1971), and E. A. Silver, D. S. McCulloch, and J. R. Curray (in prep.).

CASCADIA SUBDUCTION ZONE

The salient aspects of the Cascadia subduction zone are a slow subduction rate, a high sedimentation rate, and the subduction of young lithosphere. These aspects combine to produce a well-defined, relatively easily studied zone of sediment offscraping, a filled trench, and low seismicity, lacking deep earthquakes. Along the entire subduction zone, magnetic anomalies

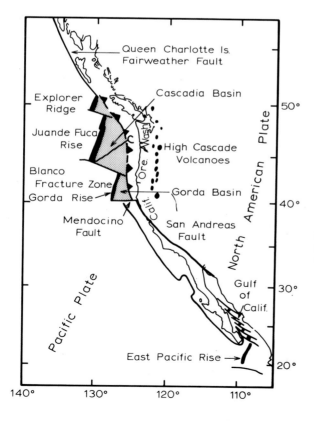

Figure 10-1. Location map of California translational margin and Cascadia subduction zone (shaded area indicates west margin of Juan de Fuca plate). Teeth on upper plate.

from source rocks on the Juan de Fuca plate can be mapped above intervening sediment that has been scraped off and accreted along the lower part of the continental slope.

The structure of the subduction zone changes markedly at the latitude of the Columbia River. To the south the slope off Oregon and northernmost California is cut by predominantly landward-side-up reverse faults that separate fold packets. The resulting structure is a steep (10° to 15°) frontal slope and a deformed outer ridge (Fig. 10-2B). Uplift of the slope and outer ridge probably proceeded by thrusting successive wedges of basin sediments below the sequence of thrust packets (Seely and others, 1974). The ridge trapped sediments in a basin on the landward side and built a marginal plateau. At the base of the continental slope a sharp fault boundary separates deformed rocks of the accreted wedge from undeformed turbidites and pelagic sediments of the Gorda and Cascadia Basins.

Magnetic anomalies formed by the spreading process on the Juan de Fuca plate can be mapped for about 20 km over the accreted wedge landward of the base of the slope, demonstrating that the ocean crust has been thrust beneath the continental margin. Magnetic modeling studies indicate that the ocean crust dips about 12° at 10 to 20 km east of the base of the slope (Silver, 1971a; Emilia and others, 1968).

Free-air gravity on the continental slope shows a sharp minimum a few kilometres east of the base of the slope and a maximum over the top of the slope (Fig. 10-3). The horizontal separation of extreme values is 20 to 30 km.

Uplift of the continental slope, outer ridge, and shelf is documented by numerous dredges, dart cores, and piston cores taken along the Oregon and northernmost California continental margin (Kulm and Fowler, 1974a; Silver, 1971a). Benthonic foraminifera of Miocene and Pliocene age are sampled from depths of as much as 900 m shallower than those in which they probably lived (Kulm and Fowler, 1974a). Uplift of strata on the slope and outer ridge is probably related to imbricate wedging associated with subduction. However, anomalous depth indicators on the continental shelf may be related, in part, to diapiric uplift of shale masses (Silver, 1972; P. Snavely, 1971, oral commun.). No samples older than Miocene have been recovered on the margin seaward of the shelf from Cape Mendocino to Vancouver Island.

In contrast to the southern part of the Cascadia subduction zone, the northern part from the Columbia River to northern Washington shows a very different style of deformation (Fig. 10-2A). Here faulting is dominantly seaward-side-up along the continental slope; these faults also separate thrust-fold packets, but the sense of offset results in a wide plateau, elevated about 1 km above the Cascadia basin, rather than a steep continental slope. Each thrust packet stacks out against the last, producing a wide, low plateau. Stacking of these thrust packets to build the plateau appears to have taken only a relatively short time. The oldest

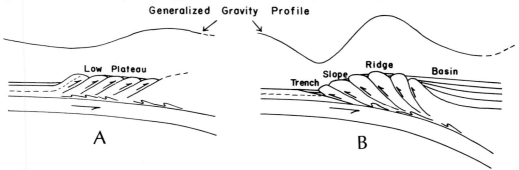

Figure 10-2. Schematic representation of the major types of accretionary prism observed off Oregon and Washington along the Cascadia subduction zone. (A) Washington-type accretionary plateau convergent continental margin. (B) Oregon-type trench-slope-ridge accretionary wedge convergent continental margin.

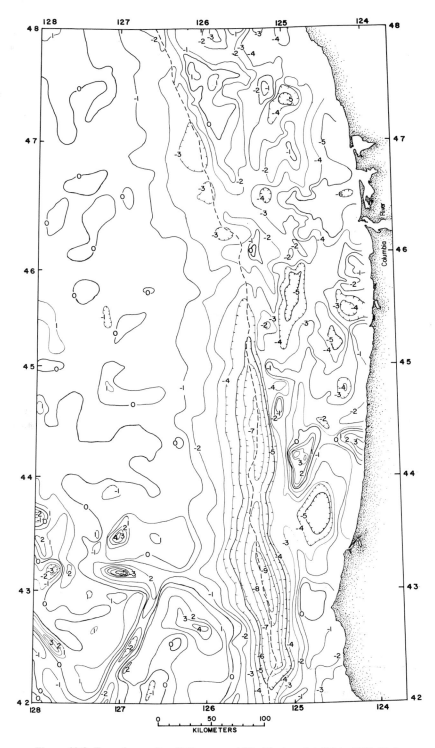

Figure 10-3. Free-air gravity off Oregon and Washington, lat 42° to 48°N. Redrawn from National Ocean Survey seamap series, North Pacific Ocean, NOS 12042-12G, 1974. Contour interval 10 mgal.

rocks sampled from this lower plateau region are mid-Pleistocene (Barnard, 1973), and the plateau is about 50 km wide. Thus the 50-km-wide plateau could have been built in 1 m.y. Assuming no volume change during deformation to give a conservative estimate, the plateau, which is 50 km wide and about 3 km thick, could be stretched out to 75 km wide and 2 km thick (the sediment thickness in the adjacent basin), thus giving a conservative horizontal shortening of 25 km in 1 m.y. or a rate of 2.5 cm/yr. This rate is close to that estimated from plate motion between Pacific–American–Juan de Fuca plates (Fig. 1; Silver, 1969; Atwater, 1970).

Barnard (1973) measured offsets along faults cutting Holocene sediment across the continental margin and concluded the Holocene rate of shortening was no more than 0.7 cm/yr. This rate does not consider Holocene thrusting in the Olympic Mountains (P. Snavely, 1975, oral commun.). It, however, does point up problems in extrapolating from long-term geologic strain rates to recent events or from short-term records to a longer time average.

The interface between deformed strata of the lower plateau and undeformed sediment of Cascadia basin is not abrupt. The seaward flank of the westernmost thrust fold is gently upturned to about 10° but truncated abruptly by a reverse fault on the landward side (Fig. 10-2A). Off southernmost Washington a series of gentle folds but no faults constitute the outer part of the lower plateau. In some areas the outermost thrust fold has been pushed up to above the general level of the plateau and a new thrust is beginning to form just to the west in Cascadia basin (Silver, 1972, Fig. 3, profile 4).

Magnetic anomalies on the Juan de Fuca plate can be traced over the deformed plateau region for up to 100 km east of the western edge of the plateau. The extent to which the anomalies can be followed eastward attests to a much shallower dip of ocean crust beneath the outer plateau off Washington (5° or less), compared to that beneath the slope to the south. Free-air gravity profiles over the Washington continental margin (Dehlinger and others, 1971) show low-gradient and low-amplitude anomalies, indicative of a broad, gentle transition between ocean and continental crust, which is also consistent with the gentle slope of the wide accretionary plateau.

Folds and faults along the leading edge of the Cascadia subduction zone invariably parallel the continental slope even though the ocean crust responsible for their formation has been subducting at oblique and temporally varying directions. This observation shows that the strike of the folds and thrust faults (but not necessarily slip vector indicators in strained rocks) is highly dependent on the initial orientation of the continental slope.

The difference between the structure of the Oregon and Washington continental margins demonstrates that vergence of thrusts and fold packets may be independent of subduction direction. Along the Cascadia subduction zone, eastward subduction is accompanied by westward vergence (eastward dips) off Oregon and eastward vergence off Washington. The Washington example—vergence opposite to the sense of overall shear in the subduction process—is rare and seems to correspond with rapid sedimentation and slow, shallow subduction. Perhaps shallow dip of the down-going slab is the key here, but cause and effect cannot easily be separated. That is, an eastward vergence of thrust packets above an eastwardly subducting slab will rapidly build out a wide, low accretionary plateau, and a shallow dip of the subduction slab will result.

Subduction probably has occurred beneath Washington, Oregon, and northern California throughout much of Cenozoic time, based on a generally accepted history of Cenozoic plate movements (Atwater, 1970; Larson and Chase, 1972). The nearly continuous Cascade volcanism throughout the Cenozoic, although not volumetrically uniform generally confirms this interpretation. Only late Cenozoic deformation, however, is observed with seismic and sampling methods on this continental margin. Much of the coastal ranges do not show deformed rocks of early to middle Tertiary subduction but do show evidence for earlier subduction. The evidence for early to middle Tertiary subduction could be buried and not yet exhumed below the shelf and marginal plateau sediments. This suggestion is supported by evidence in the Olympic Mountains where a regionally unique domal uplift exposes early to middle Tertiary subduction mélange (Tabor, 1972). The locally thick basaltic rocks of the Eocene Crescent Formation have been nicely explained as part of a seamount chain that has collided with the North American continent (Cady, 1975). The Crescent volcanic units are thick pillow basalts, rapidly varying in thickness laterally, and no ultramafic rocks or sheeted dike complexes are reported, as would be expected if this were an ophiolite complex. The Oligocene-Miocene mélange represents sediments thrust beneath the seamount chain after it collided with the continent and became detached from the down-going slab.

Subduction along the Cascadia subduction zone may have occurred as a series of short, rapid events separated by longer periods of slow convergence during the late Cenozoic. This suggestion is based on the interpretation of the magnetic anomaly pattern on the Juan de Fuca plate by Morgan (1968) and Silver (1971c). Probably as a result of stresses on the small plate from movements of the much larger Pacific and American plates, the Juan de Fuca plate is deforming internally in the small Gorda basin (Fig. 10-1; Silver, 1971b) and episodically faulting in the larger Cascadia basin (Silver, 1971c). The large northeast-trending fault in the Cascadia basin was active for less than 1 m.y. approximately 2 to 2.5 m.y. ago. It has offset all anomalies older than 2.5 m.y. by 70 km left laterally but does not offset the overlying turbidites of Pleistocene age. Plate models satisfying this information require a period of rapid subduction at that time by the triangular block between the northeast-trending fault and the Blanco Fracture Zone. Morgan (1968) noted that the projection of these faults encompasses nearly the entire Cascade Range.

Volcanism in the Cascades was volumetrically very low during the period 2 to 2.5 m.y. ago. Eruption rates, however, increased dramatically in the middle to late Pleistocene, marking the dominant building of the present High Cascade volcanoes (White and McBirney, this volume). These events may record rather precisely the lag between a subduction event and the ensuing volcanic eruption. This lag is approximately 1 m.y. Such short subduction events would be associated with high deformational strain rates in accreted sediments, probably with high seismicity (in contrast with the present time), and with voluminous volcanic eruptions lagging approximately 1 m.y. behind the subduction event.

CENTRAL CALIFORNIA TRANSLATIONAL MARGIN

Lateral shearing largely dominates the present tectonics of the central and southern California borderlands. Plate reconstructions (Atwater and Molnar, 1973) predict subduction for this region in early Tertiary time, changing to lateral shear between Pacific and American plates beginning about 29 m.y. ago at a small region off southern California. The length of shear interaction expanded with time as two triple junctions migrated north and south away from the region of initial contact to their present ephemeral locations at Cape Mendocino and the mouth of the Gulf of California (Fig. 10-1).

A broad, two-phase structural history of this continental margin can be read from seismic reflection profiles. In the first phase, during active subduction, older sedimentary rocks were highly deformed and produced a structurally irregular surface. The second phase during translational shear was a late Cenozoic episode of basin formation and marginal ridge uplift

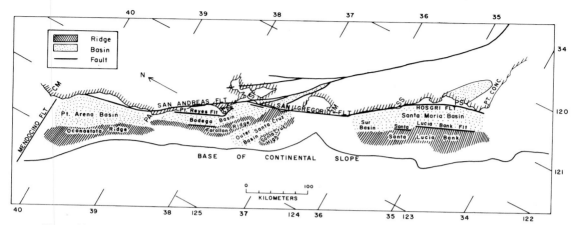

Figure 10-4. Map of basins, ridges, and faults on the central California continental margin. CM, Cape Mendocino; PA, Point Arena; PR, Point Reyes; SF, San Francisco; M, Monterey; SS, San Simeon; PS, Point Sal. The San Simeon fault is the on-shore segment of the Hosgri fault at San Simeon.

that shaped the present continental margin. During this phase younger sediments were deposited. These younger sediments are gently folded in places and locally cut by steep faults but are not highly deformed. The structure of the basins, bounded by faults with largely down-to-basin offsets, is grossly similar, despite major differences in basement rock type. Basins underlain by Franciscan rocks, however, show greater structural variability than granite-floored basins.

Santa Maria basin (Fig. 10-4) is probably underlain by Franciscan assemblage rocks. Actinolite and glaucophane schists were dredged from the east flank of Santa Lucia Bank, and immature, somewhat sheared and zeolitized graywacke was dredged from the west flank. Magnetic and gravity anomalies are mapped along the central part of the Santa Maria basin, indicating dense, magnetized rock at shallow depth below the surface. Possibly these rocks are Mesozoic ophiolites as are found onshore in this area (Hopson and others, 1973; Hall, 1975). In profile L18 (Fig. 10-5A) the basin thickens toward both bounding faults. Farther south, more complex folding is seen under the basin. To the north in Sur basin the structure simplifies (profile L6, Fig. 10-5B) to a half graben with minimum vertical relief of 3 km on the eastern boundary fault.

Shell Oil Company drilling has demonstrated that the Santa Maria basin originated in late middle Miocene time (Hoskins and Griffiths, 1971) with the uplift of Santa Lucia Bank. Significantly, they concluded that the other major basins along the central California coast—Outer Santa Cruz, Bodega, and Point Arena basins—also began to form at the same time.

Santa Lucia Bank bounds Santa Maria and Sur basins on the west and is separated from the former by the Santa Lucia Bank fault (profile L18, Fig. 10-5A). Rocks underlying the bank and continental slope are deformed. Gravity is low over the bank, which suggests no abnormally dense rocks beneath the bank, nor do magnetic anomalies appear, in contrast to the findings at shallow depth in Santa Maria basin. The older sedimentary rocks under Santa Lucia Bank may have been deformed during subduction of the Farallon plate in early to middle Tertiary time.

Rounded boulders of quartz monzonite and granodiorite were dredged from the bank, just north of the crossing of profile L18. These rocks are not significantly altered and appear petrographically similar to the granitic rocks found in the La Panza Range of the Salinian block (L. Silver, 1976, oral commun.). If the boulders were derived from the Salinian block, it must have been prior to the development of the late Tertiary basins. Alternatively, though

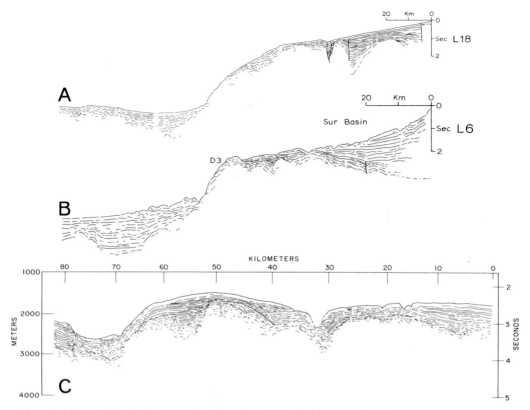

Figure 10-5. Line-drawing interpretation of reflection profiles (A) L18 (lat 35°N) across Santa Maria basin, (B) L6 (lat 36°N) across Sur basin, and (C) WX along Oconostota ridge. Vertical exaggeration approximately X10 for each section.

less likely, they may have been derived from local, and as yet undiscovered, slivers of crystalline basement rock.

Bodega and Outer Santa Cruz basins lie between the coastal fault system, the Farallon ridge, and the Santa Cruz high (Fig. 10-4). The Farallon ridge is composed of granitic rocks, and these have also been sampled by drilling beneath Bodega basin. A multichannel profile across Bodega basin (Fig. 10-6) shows 2.5 km of sediment above basement and an unconformity in the central part of the section. The comparison of the drilling results reported by Hoskins and Griffiths (1971) with this section shows the lower half of the section to be late middle Miocene and the upper half Pliocene-Pleistocene. Lower Miocene sediments thicken over the Farallon ridge, indicating a reversal of structure between lower and late middle Miocene. Bodega basin is bounded by down-to-basin, dominantly reverse faults.

Point Arena basin changes structural style northward from Point Arena. Just north of Point Arena basin, deposits and unconformities are folded. But to the northwest (profile WX, Fig. 10-5C), undisturbed sediments rest on an irregular surface of older basement rocks. This irregular older basement appears to be early to middle Tertiary Franciscan assemblage rocks, because a dredge haul recovered weakly metamorphosed and sheared graywacke containing a sparse middle Eocene to Oligocene coccolith fauna. As to the south on Santa Lucia bank, deformation of the basement rocks probably resulted from compression related to subduction. Thus, subduction appears to have been active until the mid-Tertiary.

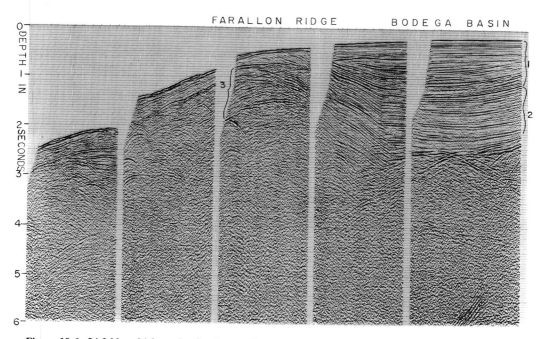

Figure 10-6. 24-fold multichannel reflection profile across part of Bodega basin and Farallon ridge (profile runs southwest from Bodega Head). 1, Holocene to upper Pliocene; 2, upper Miocene; 3, lower to middle Miocene.

The apparently synchronous development of the large shelf basins in late middle Miocene time may have an explanation in changing plate motions during Neogene time. Atwater and Molnar (1973) deduced changes in the motions of the Pacific and American plates over the past 38 m.y. by summing a global plate circuit using spreading-center data between the Pacific–Antarctic–Indian–African–North American plates for five finite time intervals. This history shows an acceleration of Pacific-America movement for the central California region over the past 20 m.y., with a major increase in slip rate about 10 m.y. ago; this change in slip rate is similar to the history of accelerating slip on the San Andreas fault (Dickinson and others, 1972; Hein, 1973).

The change in plate motion may have been the fundamental cause of the development of the late Tertiary basins on the shelf. E. A. Silver, D. S. McCulloch, and J. R. Curray (in prep.) performed a similar plate circuit for the same time intervals as Atwater and Molnar (1973). They showed changes in the direction of Pacific-America movement for the central California region of 339° for the interval 21 to 10 m.y. ago, 328° for 10 to 4.5 m.y. ago, and 321° for 4.5 to 0 m.y. ago (measured clockwise from north). The change in direction between the two plates should have produced an extensional component across the strike-slip plate boundary approximately 10 m.y. ago. Thus the extensional component of motion across the translational faults was initiated at about the time when mid-Tertiary sediments started to collect in basins bounded by high-angle faults.

In addition to basin development, very large horizontal movements have occurred along some of the major faults in central California. Right-lateral slip of 300 km in the past 22 m.y. has been documented along the San Andreas fault for the segment between the Transverse Ranges and San Francisco (Huffman, 1972; Matthews, 1976). No offset of this segment occurred during 45 to 22 m.y. ago, because middle Eocene rocks have been offset 300 km also (Nilsen and Clarke, 1975). Post-Cretaceous offset along the San Andreas north of San Francisco is 550 to 600 km, based on offset of granitic basement terranes (Silver and others, 1971) and offset sources for Cretaceous and Paleocene conglomerates at Point Arena (Wentworth, 1968; Ross, 1972).

This difference in offset history can be explained in part by movement on the fault system composed of the San Gregorio (Greene and others, 1973) and Hosgri faults (Hoskins and Griffiths, 1971); this system intersects the San Andreas at San Francisco and extends south along the coast to Point Conception. Right-lateral offset of 90 km on the San Gregorio was first suggested, based on apparent offset of granitic basement rocks across Monterey Bay (Silver, 1974). Graham and Dickinson (1978) now believe this offset is 115 km. Hall (1975) presented evidence for 80 to 100 km right slip on the Hosgri fault, on the basis of matching stratigraphy of Jurassic to Miocene rocks at Point Sal and San Simeon.

Johnson and Normark (1974) suggested that the entire 550 to 600 km of slip along the San Andreas occurred in the past 30 m.y., with 300 km offset on the San Andreas and the rest along faults west of the San Andreas but cutting granitic rocks of the Salinian block. We can now explain up to 115 km offset on the San Gregorio–Hosgri fault zone in addition to the 300 km on the San Andreas but still require at least 135 km of slip on other faults. Part of the remaining offset may have occurred in latest Cretaceous or Paleocene time (Silver and others, 1971; Nilsen and Clarke, 1975), although the need to postulate significant pre-Miocene offset seems to be declining as the other faults of the San Andreas "system" are studied in detail.

CONCLUSIONS

The continental margin off the Western United States is characterized by subduction north of lat 40°30′N and translational shear to the south. The northern region, named the Cascadia subduction zone, shows both landward- and seaward-verging folds and thrust faults in sediment scraped off the descending plate and accreted to the continental slope. The Oregon type has seaward vergence (landward dips), which results in undertucking of successive thrust sheets and uplift of a prominent outer ridge (often referred to as the "trench-slope break" for lack of a better term). Behind the ridge is a basin, usually largely full of ponded turbidites. Gravity anomalies generally show a pronounced maximum over the ridge and minimum over the lower slope or trench. In contrast, landward vergence of folds and thrusts (Washington-type) results in a wide but low plateau composed of successive thrust sheets stacked sequentially seaward, all reaching approximately the same height, roughly 1 km above the original surface of deposition. Off Washington a lower plateau 50 km wide was formed during the Pleistocene. Gravity anomalies due to the edge effect of the continental slope are significantly smaller in amplitude than for the Oregon type.

Subduction characterized the central California continental margin until the middle Tertiary (Atwater, 1970), and early Tertiary rocks are deformed. Subsequently, movement on large vertical faults produced large lateral offsets of basement terranes (on the San Andreas and San Gregorio–Hosgri fault zones), and large shelf basins developed. The basins all began collecting thick marine sections in the late middle Miocene and, therefore, probably opened as a result of a tectonic event affecting the entire continental margin. The basin-forming event coincides fairly closely with a time during which Pacific-America plate motions apparently were reorienting from a more northward to a more northwestward trend and may have developed an extensional component of slip across the more northward-trending faults.

ACKNOWLEDGMENTS

I am grateful to Dave McCulloch for many valuable suggestions. He and R. W. Couch critically reviewed the manuscript.

REFERENCES CITED

Atwater, T., 1970, Implications of plate tectonics for the Cenozoic tectonic evolution of western North America: Geol. Soc. America Bull., v. 81, p. 3513–3536.

Atwater, T., and Molnar, P., 1973, Relative motion of the Pacific and North American plates deduced from sea-floor spreading in the Atlantic, Indian, and South Pacific Oceans, in Kovach, R. L., and Nur, A., eds., Tectonic Problems of the San Andreas Fault System Conf.: Stanford Univ. Pubs. Geol. Sci., v. 13, p. 136–148.

Barnard, W. D., 1973, Late Cenozoic sedimentation on the Washington continental margin [Ph.D. thesis]: Seattle, Univ. Washington, 255 p.

Cady, W. M., 1975, Tectonic setting of the Tertiary volcanic rocks of the Olympic Peninsula, Washington: U.S. Geol. Survey Jour. Research, v. 3, p. 573–582.

Dehlinger, P., Couch, R. W., McManus, D. A., and Gemperle, M., 1971, Northeast Pacific structure, in Maxwell, A. E., ed., The sea, Vol. 4: New York, John Wiley & Sons, Inc., pt. 2, p. 133–189.

Dickinson, W. R., Cowan, D. S., and Schweickert, W. A., 1972, Test of new global tectonics: Discussion: Am. Assoc. Petroleum Geologists Bull., v. 56, p. 375–384.

Emilia, D. A., Berg, J. W., and Bales, W. E., 1968, Magnetic anomalies off the northwest coast of the United States: Geol. Soc. America Bull., v. 79, p. 1053–1062.

Graham, S., and Dickinson, W. R., 1978, Evidence for 115 kilometres of right slip on the Gregorio-Hosgri fault trend: Science, v. 199, p. 179–181.

Greene, G., Lee, W.H.K., McCulloch, D. S., and Brabb, E. E., 1973, Faults and earthquakes in the Monterey Bay region, California: U.S. Geol. Survey Misc. Field Studies Map MF-518, scale 1:200,000.

Hall, C. A., Jr., 1975, San Simeon–Hosgri fault system, coastal California: Economic and environmental implications: Science, v. 190, p. 1291–1294.

Hein, J. R., 1973, Deep-sea sediment source areas: Implications of variable rates of movement between California and the Pacific plate: Nature, v. 241, p. 40–41.

Hopson, C. A., Franco, C. J., Pessagno, G., and Mattinson, J. M., 1973, Late Jurassic ophiolite at Point Sal, Santa Barbara County, California: Geol. Soc. America Abs. with Programs, v. 5, p. 58.

Hoskins, E. G., and Griffiths, J. R., 1971, Hydrocarbon potential of northern and central California offshore: Am. Assoc. Petroleum Geologists Mem. 15, v. 1, p. 212–228.

Huffman, O. F., 1972, Lateral displacement of upper Miocene rocks and the Neogene history of offset along the San Andreas fault in central California: Geol. Soc. America Bull., v. 83, p. 2913–2946.

Johnson, J. D., and Normark, W. R., 1974, Neogene tectonic evolution of the Salinian block, west-central California: Geology, v. 2, p. 11–14.

Kulm, L. D., and Fowler, G. A., 1974a, Oregon continental margin structure and stratigraphy: A test of the imbricate thrust model, in Burk, C. A., and Drake, C. L., eds., The geology of continental margins: New York, Springer-Verlag, p. 261–283.

———1974b, Cenozoic sedimentary framework of the Gorda–Juan de Fuca plate and adjacent continental margin—A review, in Dott, R. H., Jr., and Shaver, R. H., eds., Modern and ancient geosynclinal sedimentation: Soc. Econ. Paleontologists and Mineralogists Spec. Pub. 19, p. 212–229.

Larson, R. L., and Chase, C. G., 1972, Late Mesozoic evolution of the western Pacific Ocean: Geol. Soc. America Bull., v. 83, p. 3627–3644.

Matthews, V., III, 1976, Correlation of Pinnacles and Neenach volcanic formations and their bearing on San Andreas fault problem: Am. Assoc. Petroleum Geologists Bull., v. 60, p. 2128–2141.

Morgan, W. J., 1968, Rises, trenches, great faults, and crustal blocks: Jour. Geophys. Research, v. 73, p. 1959–1982.

Nilsen, T. H., and Clarke, S. H., Jr., 1975, Sedimentation and tectonics in the early Tertiary continental borderland of central California: U.S. Geol. Survey Prof. Paper 925, 64 p.

Ross, D. C., 1972, Petrographic and chemical reconnaissance study of some granitic and gneissic rocks near the San Andreas fault from Bodega

Head to Cajon Pass, California: U.S. Geol. Survey Prof. Paper 698, 92 p.

Seely, D. R., Vail, P. R., and Walton, G. G., 1974, Trench slope model, *in* Burk, C. A., and Drake, C. L., eds., The geology of continental margins: New York, Springer-Verlag, p. 249–260.

Silver, E. A., 1969, Late Cenozoic underthrusting of the continental margin off northernmost California: Science, v. 166, p. 1265–1266.

——1971a, Transitional tectonics and late Cenozoic structure of the continental margin off northernmost California: Geol. Soc. America Bull., v. 82, p. 1–22.

——1971b, Tectonics of the Mendocino triple junction: Geol. Soc. America Bull., v. 82, p. 2965–2978.

——1971c, Small plate tectonics in the northeastern Pacific: Geol. Soc. America Bull., v. 82, p. 3491–3496.

——1972, Pleistocene tectonic accretion of the continental slope off Washington: Marine Geology, v. 13, p. 239–249.

——1974, Structural interpretation from free-air gravity on the California continental margin, 35° to 40°N: Geol. Soc. America Abs. with Programs, v. 6, p. 253.

Silver, E. A., Curray, J. R., and Cooper, A. K., 1971, Tectonic development of the continental margin off central California, *in* Lipps, J. H., and Moores, E. M., eds., Geologic guide to the northern Coast Ranges, Point Reyes region, California: Geol. Soc. Sacramento Ann. Field Trip Guidebook, p. 1–10.

Tabor, R. W., 1972, Age of Olympic metamorphism, Washington—K-Ar dating of low-grade metamorphic rocks: Geol. Soc. America Bull., v. 83, p. 1805–1816.

Wentworth, C. M., 1968, Upper Cretaceous and lower Tertiary strata near Gualala, California, and inferred large right slip on the San Andreas fault, *in* Dickinson, W. R., and Grantz, A., eds., Geologic Problems of the San Andreas Fault System Conf. Proc.: Stanford Univ. Pubs. Geol. Sci., v. 11, p. 130–143.

White, C. M., and McBirney, A. R., 1978, Some quantitative aspects of orogenic volcanism in the Oregon Cascades, *in* Smith, R. B., and Eaton, G. P., eds., Cenozoic tectonics and regional geophysics of the western Cordillera: Geol. Soc. America Mem. 152 (this volume).

MANUSCRIPT RECEIVED BY THE SOCIETY AUGUST 15, 1977

MANUSCRIPT ACCEPTED SEPTEMBER 2, 1977

Geological Society of America
Memoir 152

11

Generalized maps showing distribution, lithology, and age of Cenozoic igneous rocks in the Western United States

JOHN H. STEWART
JOHN E. CARLSON
U.S. Geological Survey
345 Middlefield Road
Menlo Park, California 94025

Two maps (Pls. 11-1 and 11-2, in pocket) have been prepared to show the general distribution, lithology, and age of Cenozoic igneous rocks in the Western United States. Plate 11-1 shows igneous rocks that range in age from 65 m.y. (start of the Cenozoic Era) to 17 m.y., and Plate 11-2 shows rocks that are 17 m.y. or younger. The 17-m.y. age is a geologically significant division point. This age is approximately the low point of the mid-Miocene lull in igneous activity that has been recognized in the Great Basin (McKee and others, 1970), in the Basin and Range province as a whole (Damon, 1971), in Colorado (Marvin and others, 1974), and in New Mexico (Chapin and Seager, 1975). In many regions, few, if any, rocks of this approximate age are known. A marked change in the character of volcanism also occurred at approximately this time. Rocks older than 17 m.y. are predominantly calc-alkalic types ranging in composition from rhyolite to andesite, whereas rocks younger than 17 m.y. are predominantly basalt, or bimodal assemblages of basalt and rhyolite (Christiansen and Lipman, 1972).

Igneous rocks shown on the maps are divided into four general categories: (1) rhyolitic, (2) andesitic, (3) basaltic, and (4) intrusive. Rhyolitic rocks consist of widespread welded tuffs as well as local flows and generally range in composition from rhyolite to quartz latite. Andesitic rocks consist mostly of flows and breccias of intermediate composition, although locally large volumes of more silicic volcanic rocks are included in this category on Plate 11-1 and locally large areas of basaltic rocks are included in it on Plate 11-2. Basaltic rocks include both alkali and tholeiitic flows. Intrusive rocks include all compositional types.

Eocene basalts in the Oregon-Washington coast ranges are not shown. These rocks consist of thick sequences of spilite and tholeiite erupted on the ocean floor in contrast to the largely calc-alkalic continental volcanic rocks shown on Plate 11-1. The basalts are omitted because they were erupted in a distinctly different geologic setting and because their inclusion on Plate 11-1 would mask the distribution patterns of the more widespread continental volcanic rocks. Also omitted are Laramide intrusive rocks in California, Nevada, Arizona, New Mexico,

Colorado, and Montana. In the Basin and Range province, these intrusive rocks range in age from about 50 to 75 m.y. and have a modal age of about 65 m.y. (Damon, 1968). The inclusion of only those Laramide plutons younger than 65 m.y. is impractical as well as misleading in that it would show the distribution of only half of the igneous rocks formed during this particular pulse of igneous activity.

Between lat 35°30' and 42°N in California, Nevada, and western Utah, the maps are based on data from Stewart and Carlson (1976) and Stewart and others (1977). Elsewhere they are based largely on the geologic map of the United States (King and Beikman, 1974). The generalized radiometric dates shown on Plate 11-1 north of lat 42°N are based on a compilation by Armstrong (this volume). In Colorado, they are from Marvin and others (1974). Elsewhere south of lat 42°N they are based on unpublished compilations prepared by W. S. Snyder, W. R. Dickinson, and M. L. Silberman (published in part in a summary form, Snyder and others, 1976). Because of the relatively short time span involved (0 to 17 m.y.), no radiometric dates are shown on Plate 11-2.

REFERENCES CITED

Armstrong, R. L., 1978, Cenozoic igneous history of the U.S. Cordillera from lat 42° to 49°N, *in* Smith, R. B. and Eaton, G. P., eds., Cenozoic tectonics and regional geophysics of the western Cordillera: Geol. Soc. America Mem. 152 (this volume).

Chapin, C. E., and Seager, W. R., 1975, Evolution of the Rio Grande rift in the Socorro and Las Cruces areas, *in* Seager, W. R., Clemons, R. E., and Callender, J. F., eds., Guidebook of the Las Cruces country: New Mexico Geol. Soc. Guidebook 26, p. 297–321.

Christiansen, R. L., and Lipman, P. W., 1972, Cenozoic volcanism and plate-tectonic evolution of the western United States. II, Late Cenozoic: Royal Soc. London Philos. Trans., v. 271, p. 249–284.

Damon, P. E., 1968, Application of the potassium-argon method to the dating of igneous and metamorphic rock within the Basin Ranges of the southwest, *in* Southern Arizona guidebook 3—Geol. Soc. America Cordilleran Sec., 64th Ann. Mtg., Tucson, 1968: Tucson, Ariz., Arizona Geol. Soc., p. 7–20.

——1971, The relationship between late Cenozoic volcanism and tectonism and orogenic-epeirogenic periodicity, *in* Turekian, K. K., ed., The late Cenozoic glacial ages: New Haven, Conn., Yale Univ. Press, p. 15–35.

King, P. B., and Beikman, H. M., 1974, Geologic map of the United States: U.S. Geol. Survey Map, scale 1:2,500,000.

Marvin, R. F., Young, E. J., Mehnert, H. H., and Naeser, C. W., 1974, Summary of radiometric age determinations on Mesozoic and Cenozoic igneous rocks and uranium and base-metal deposits in Colorado: Isochron/West, no. 11, p. 1–41.

McKee, E. H., Noble, D. C., and Silberman, M. L., 1970, Middle Miocene hiatus in volcanic activity in the Great Basin area of the Western United States: Earth and Planetary Sci. Letters, v. 8, p. 93–96.

Snyder, W. S., Dickinson, W. R., and Silberman, M. L., 1976, Tectonic implications of space-time patterns of Cenozoic magmatism in the Western United States: Earth and Planetary Sci. Letters, v. 32, p. 91–106.

Stewart, J. H., and Carlson, J. E., 1976, Cenozoic rocks of Nevada—Four maps and a brief description of distribution, lithology, age, and centers of volcanism: Nevada Bur. Mines and Geology Map 52, scale 1:1,000,000.

Stewart, J. H., Moore, W. J., and Zietz, Isidore, 1977, East-west patterns of Cenozoic igneous rocks, aeromagnetic anomalies, and mineral deposits, Nevada and Utah: Geol. Soc. America Bull., v. 88, p. 67–77.

MANUSCRIPT RECEIVED BY THE SOCIETY AUGUST 15, 1977

MANUSCRIPT ACCEPTED SEPTEMBER 2, 1977

Geological Society of America
Memoir 152

12

Cenozoic igneous history of the U.S. Cordillera from lat 42° to 49°N

RICHARD LEE ARMSTRONG
Department of Geological Sciences
University of British Columbia
Vancouver, British Columbia Canada, V6T 1W5

ABSTRACT

Except for volcanoes in the Adel Mountains and near the Black Hills, the first 10 m.y. of the Cenozoic was a time of igneous quiescence. Basaltic volcanism in the eugeosyncline west of the present-day Cascade volcanic arc began in early Eocene time, was most intense between 54 and 44 m.y. ago, and tapered off slowly, with injection of basaltic and alkali trachyte dikes continuing until the Oligocene.

In Oligocene time the eugeosynclinal rocks became welded to the continental margin, shorelines shifted westward, and volcanic activity west of the Cascade arc virtually ceased.

Igneous activity began about 55 m.y. ago over a broad region in Washington, northern Idaho, and Montana, and during the early Eocene this activity swept southward across all the Northwestern United States. Volcanism, plutonism, regional high heat flow with associated geothermal circulation systems, block faulting, and ductile deformation at depth associated with tectonic denudation as a consequence of diapiric rise of plutonic rocks occurred synchronously during an intense culmination between 50 and 43 m.y. ago (Challis episode). Volcanism and deformation then died out abruptly as the locus of igneous eruption centers shifted south into Nevada-Utah and west into the Cascade arc, where volcanic activity persisted through the rest of the Cenozoic.

Following widespread tectonic and igneous quiescence between about 38 and 18 m.y. ago, volcanic activity suddenly commenced over a large region with rapid eruption of Columbia River and related basalts. Volcanism, with bimodal chemistry, began during this episode (Columbia, 13 to 16 m.y. ago) in southwestern Idaho and over all of southeastern Oregon. Between 13 m.y. ago and today, bimodal igneous activity migrated eastward across Idaho to produce the Snake River Plain-Yellowstone volcanic field. In Oregon, at the same time, siliceous volcanic centers retreated westward. As a consequence rhyolitic volcanic centers today are most active only in Yellowstone and close to the Cascade Range. Striking synchroneity is shown by pulses of more intense igneous activity in the Cascade and Snake River Plain-Yellowstone regions.

INTRODUCTION

In the world today there is a close association of active volcanism with consuming plate boundaries (Miyashiro, 1975) and within plates at melting spots that show variable degrees of historical persistence (Wilson, 1973). One of the more important constraints on plate patterns in the past is provided by volcanic rocks—their age, distribution, volume, and chemistry being matters of significance. More than a century of geologic mapping and stratigraphic studies and one and a half decades of applied geochronometry (mostly K-Ar dating) have established a reasonably precise Cenozoic igneous history for the Western United States. In this paper I will review the story for the Northwestern United States (north of lat 42°N), an area I have been concerned with since the mid-1960s.

For purposes of discussion the volcanic history of the Northwestern United States can be subdivided according to natural breaks in the volume and extent of volcanic rocks. This *episodic* character of the volcanism has been gradually recognized and documented by many studies that will be cited later. The outstanding volcanic climaxes and pauses are the Challis episode of intense and widespread volcanism during Eocene time (Armstrong, 1974), relative quiescence between 38 and 18 m.y., the Columbia episode of middle Miocene time (Baksi and Watkins, 1973; Watkins and Baksi, 1974), a quiescent period about 12 to 13 m.y. ago, a less well defined later Miocene and early Pliocene climax or composite culmination in activity, a Pliocene hiatus, and a Pleistocene (High Cascade) culmination (McBirney and others, 1974; White and McBirney, this volume; Armstrong, 1975a). These episodes appear to be of worldwide significance, as recorded in the ash record of DSDP cores (Kennett and Thunell, 1975; Vogt, 1976). I have subdivided the volcanic history into chapters as follows: 64 to 55 m.y. ago (end of Cretaceous until almost the end of the Paleocene), 55 to 36 m.y. ago (late Paleocene to early Oligocene), 36 to 20 m.y. ago (most of the Oligocene and early Miocene), 20 to 13 m.y. ago (early to middle Miocene), and 13 m.y. ago to the present (middle Miocene and later). The boundaries do not coincide with the current arbitrary subdivisions of the geologic time scale as reviewed by Berggren (1972), who placed the European epoch boundaries at 1.8, 5.0, 22.5, 37.5, 53.5, and 65 m.y. ago.

Concurrent with the theme of widespread volcanic episodes is the concept of time-transgressive volcanic patterns. The volcanic histories of areas that are widely separated are usually not precisely correlated. Regions of volcanic activity have shifted systematically across the landscape so that their history is one of moving, ameboid patterns, gradually enlarging and contracting, starting and stopping without clear-cut terminations, and yielding a story with overlapping episodes.

In general, I cite only the most recent and comprehensive references relevant to topics under discussion. Prior literature can be sought in those sources—the ultimate supporting bibliography for this review would exceed the length of the entire paper!

IMPORTANT REVIEWS AND COMPILATIONS

Early attempts to synthesize the Cenozoic volcanic history of the northwestern Cordillera were made by Waters (1962) and Van Houten (1961), but they lacked the advantage of isotopic age determinations for age assignments and correlations between distant areas. A decade later a series of papers (Lipman and others, 1971, 1972; Christiansen and Lipman, 1972) provided a review and synthesis of volcanic age, distribution, and chemical data for the Western United States that has been the basis for all subsequent discussion and tectonic speculation. They recognized that Eocene and Oligocene volcanism (typically intermediate in silica content) was distinct in character from the Miocene and later, predominantly basaltic and bimodal (basaltic-rhyolitic), volcanism that is associated with later Cenozoic rifting and related crustal disruption. This fundamental twofold subdivision of the record has been supported by all later work. Armstrong and Higgins (1973) presented two maps that emphasize the time-transgressive nature of the two volcanic associations. Many references and useful descriptions of the different volcanic provinces of the Northwest may be found in McKee (1972).

The abundant geochronometry of Cenozoic volcanic rocks of the Northwest is currently in a well-organized state as the result of publication of regional and state reviews and compilations. Laursen and Hammond (1974) compiled published data for Washington and Oregon. Washington data were presented in greater detail and considerably augmented by Engels and others (1976) and Watkins and Baksi (1974). Additional data for Washington were found in Turner (1970), Easterbrook (1975), Armstrong, and others (1976a), Whetten (1976), and Swanson and Wright (1976). For Oregon a great deal of new data is provided by Walker and others (1974a), Watkins and Baksi (1974), Armstrong and others (1975b), and McKee and others (1976). A few more dates for Oregon were gleaned from Denison (1970), Ito and Fuller (1970), Peterson and McIntyre (1970), Snavely and others (1973), Niem and Cressy (1973), McKee and others (1975), and Armstrong and others (1976b). The geochronometry of Idaho is reviewed by Armstrong (1975b, 1976), and a few additional data are available in Leonard and Marvin (1975), Neill (1975), and R. L. Armstrong, W. M. Neill, H. Prostka, and P. Williams (in prep.). Chadwick (1972) reviewed the volcanic history of Montana. Additional data for Montana were found in McDowell (1971), Mehnert and Schmidt (1971), Elliott and others (1974), Marvin and others (1973, 1974), Zen (1975), and Williams and others (1976). The geochronometry of Wyoming must be assembled piecemeal from scattered literature—Evernden and others (1964), Blackstone (1966), Nelson and Pierce (1968), Oriel and Tracey (1970), McDowell (1971), Smedes and Prostka (1972), Christiansen and Blank (1972), Love and others (1974), Pekarek and others (1974), and Hill and others (1975).

Data from all these references were compiled on a base map before abstraction and generalization onto Figures 12-2 to 12-6 of this paper. These figures show the current distribution of dated igneous rocks and thus the control available for the story. Figure 12-1 is provided as a guide to geographic names that occur in the text but not on the paleotectonic maps.

THE BEGINNING OF THE CENOZOIC

At the beginning of the Cenozoic (Fig. 12-2) and for most of Paleocene time (65 to 53.5 m.y. ago), the Northwestern United States was largely above sea level undergoing erosion, and volcanic activity was exceedingly restricted in volume and extent. Orogeny, which had been building to a climax during the latest part of Mesozoic time, was still proceeding vigorously, especially in western Montana and Wyoming and in eastern Idaho. Folding, thrust faulting, and deep-seated ductile deformation were creating mountains until nearly the end of the Paleocene; these mountains were being eroded and the debris deposited in foredeep basins of Montana and Wyoming and on the adjacent depressed and partially submerged stable craton (Ryder and Ames, 1970; Ryder and Scholten, 1973; Armstrong and Oriel, 1965; Armstrong, 1968). The end of compressive deformation (the end of the Sevier orogeny) came during Paleocene and earliest Eocene time before the beginning of the Challis volcanic episode. The latest movements along thrust faults, amounting at most to hundreds of metres of displacement, disrupted Paleocene and very early Eocene strata and are postdated by middle(?)

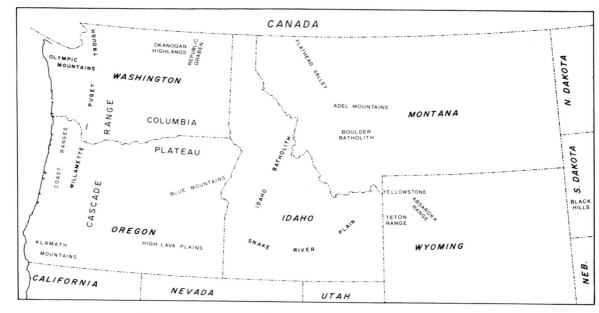

Figure 12-1. Geographic index map for the Northwestern United States.

early Eocene (~50 m.y. old) strata in southwestern Wyoming (Oriel, 1969). In the thrust belt in Montana, relatively late thrusts are postdated by a sill 58.3 m.y. old (Mehnert and Schmidt, 1971), and the period of thrusting has been bracketed between 72 and 56 m.y. ago by dating of reconstituted Cretaceous bentonites (Hoffman and others, 1976). The stratigraphic brackets for Laramide basement uplifts in Montana and Wyoming likewise indicate an end to deformation there in early Eocene time (Keefer and Love, 1963; Roberts, 1972; Love and others, 1975).

The intense Cretaceous igneous activity of the Idaho-Boulder batholith region had largely ceased by the end of the Cretaceous. The only record of volcanism at the end of the Cretaceous in the fold and thrust belt is preserved in the Adel Mountain volcanics (Chadwick, 1972) that have not been isotopically dated and in the nearby Paleocene sill already mentioned. Also active during later Paleocene and earliest Eocene time were the igneous centers of the northern Black Hills and vicinity, including Devils Tower in Wyoming (McDowell, 1971; Hill and others, 1975). Dates for these rocks indicate a time span of activity from approximately 59 to 50 m.y. ago—a unique, geographically restricted, and isolated igneous episode. Along the west coast of the Northwestern United States, the Paleocene record is poorly preserved. In California the coastal Franciscan sedimentary and tectonic environment of the later Mesozoic evidently persisted until some time in the Eocene (Blake and Jones, 1974; Evitt and Pierce, 1975), but a similar situation north of the Klamath Mountains is not known. Continental Paleocene sediments occur in and near the Cascade Mountains in northwestern Washington (Misch, 1966; Livingston, 1969). The marine geosynclinal record of the Coast Mountains of Oregon and Washington does not begin until the end of the Paleocene (Beaulieu, 1971; Cady, 1975). In general, whether or not Paleocene rocks are present is uncertain, and the oldest rocks in most local sections are thought to be lower Eocene (younger than 53.5 m.y.) and thus would not appear on Figure 12-1 at all. Where present in southwestern Oregon, and possibly in the Olympic Mountains, Paleocene strata are graywacke-turbidite marine sediments associated

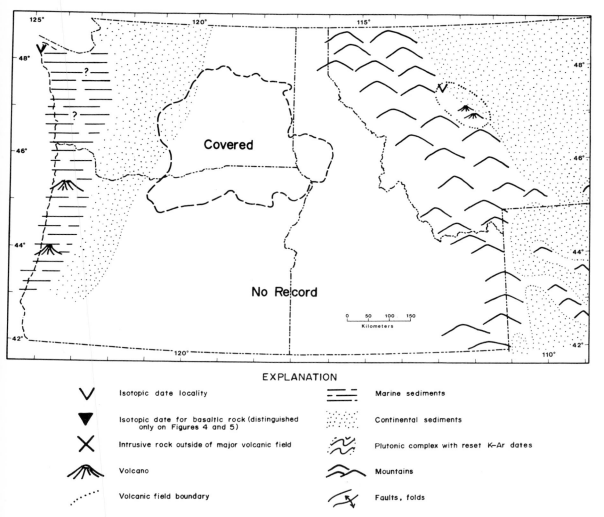

Figure 12-2. Distribution of igneous activity from 64 to 55 m.y. ago (end of Cretaceous until almost the end of the Paleocene); a relatively quiescent period.

with submarine tholeiitic basalts. There is insufficient evidence to reconstruct the paleogeography of Paleocene time along the coast in any detail.

THE CHALLIS VOLCANIC EPISODE

Beginning about 55 m.y. ago, volcanic-plutonic activity spread southward across the Canadian boundary, and by 50 m.y. ago (late Wasatchian = late early Eocene time) volcanic centers were active over the entire Northwestern United States and in the adjacent marine realm (Fig. 12-3). This activity persisted for about 5 to 10 m.y. in any given area, and then from about 43 to 36 m.y. ago it faded, nearly as swiftly as it began, by a further southward transgression leaving most of the Northwest quiescent again. In the wake of this intense igneous episode were the accumulations of volcanic rock and associated sediments, scars of tectonic disruption (fault-bounded basins, tilted blocks, and collapsed calderas), dike swarms, plutons ranging up to mesozonal batholiths in size, and large volumes of regional metamorphic and granitic rock with reset K-Ar dates, meteoric-hydrothermal modified O and H isotopic compositions, and, locally, new metamorphic fabrics.

The recognition of this episode and the correlation of phenomena attributed to it has been a recent development as a consequence of geochronometric studies. Much of the volcanic rock attributed to it was once thought to be Oligocene, and many of the intrusive bodies were undated.

In Montana-Wyoming an Eocene age (and postorogenic relation to Sevier-Laramide structures) was long established for the Absaroka volcanic rocks (Chadwick, 1972; Smedes and Prostka, 1972). As each separate volcanic field or intrusive complex in Montana and Wyoming was dated the same time span, mostly dates between 45 and 54 m.y. ago turned up with remarkable regularity (for example, Smedes and Thomas, 1965; Zen, 1975; Marvin and others, 1974; Williams and others, 1976; Marvin and others, 1973; Pekarek and others, 1974), so that we now know that almost all intrusive complexes and older Cenozoic volcanic sections can be attributed to this single igneous episode. The abundant supply of volcanic ash and erosion products is evident in the sedimentary basins of Montana and Wyoming (Roberts, 1972; Love, 1960). The middle Eocene, in particular, is richly volcanic and volumetrically large. Topography, left over from Sevier-Laramide tectonism, was gradually buried by the Eocene volcanic and sedimentary rocks so that by the end of the Eocene topography was subdued, basins shallow, and external drainage probably well developed.

The oldest deposits in the Rocky Mountain Trench–Flathead Valley of northern Montana, the Kishenehn Formation of latest Eocene–early Oligocene age (Johns, 1970), probably represent clastic filling of a rift developed during the period of Eocene volcanic activity. Many of the numerous Cenozoic sedimentary basins of western Montana, likewise, have histories beginning as fault-bounded depressions in Eocene time with subsequent episodes of fault rejuvenation, times of sediment accumulation, and intervals for which a stratigraphic record is missing (Robinson, 1960). In Idaho, as in Montana and Wyoming, all older Cenozoic volcanic sections have been found to belong to the same igneous episode as the Absaroka volcanic rocks (Axelrod, 1966; Armstrong, 1974). This volcanic assemblage, referred to as the Challis Volcanics in many areas, is more extensive, voluminous, and stratigraphically complex than the rocks of similar age in Montana and Wyoming. Contemporaneous with volcanism was emplacement of dike swarms and granitic plutons. Some of these plutons are quite large, a few were once considered part of the Mesozoic Idaho batholith. Others, of more epizonal character—especially the pink, miarolitic, alkali-rich granites and rhyolite stocks—have always been considered as Challis in age. The structural and stratigraphic relationships of Challis rocks are complex, because the prevolcanic relief was high and volcanism was accompanied by block faulting in many areas and by caldera collapse (for example, Leonard and Marvin, 1975).

The Republic graben of northern Washington is filled by Eocene volcanic and sedimentary rocks (Staatz, 1960; Muessig, 1962) that are of Absaroka-Challis age and only a fragment of a once more extensive volcanic cover that extended from the Cascades to Idaho where it probably merged with the Challis volcanic field. Dates for many volcanics in and near the graben and for contemporaneous intrusive bodies in the Cascades and Okanagan highlands are summarized by Engels and others (1976). In the central and southern Cascades, subaerial volcanic rocks and volcaniclastic sediments interfinger with deltaic and marine sediments of volcanic provenance (Foster, 1960; Wolfe and others, 1961; Snavely and Wagner, 1963; Mackin and Cary, 1965; Dott, 1966).

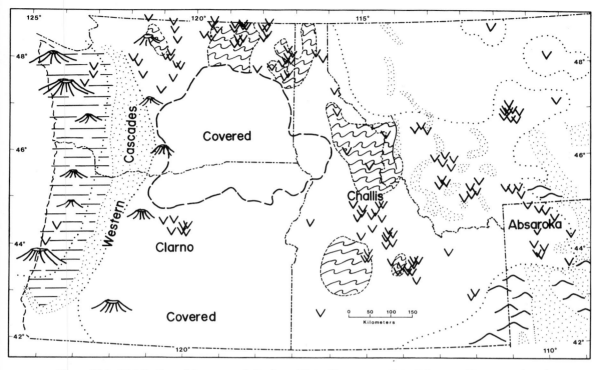

Figure 12-3. Distribution of igneous activity from 55 to 36 m.y. ago (late Paleocene, Eocene, and early Oligocene); widespread plutonic-volcanic activity of the Challis episode. See Figure 12-2 for explanation of symbols.

The oldest sequence of Cenozoic volcanic rocks in central Oregon is called the Clarno Group (Oles and Enlows, 1971), and both paleontology (middle and late Eocene) and geochronometry (41 to 46 m.y. ago) on the volcanic rocks and related andesitic plutons correlate them with the Challis-Absaroka volcanic episode. The upper part of the Clarno Group may be a bit younger than most Challis volcanic sequences but this is not yet firmly established.

At the same time as intense volcanism was occurring on land, large volumes of basalt were erupted in the submarine environment offshore in western Oregon and Washington (Beaulieu, 1971; Snavely and others, 1968; Snavely and Wagner, 1963; Cady, 1975). The greatest volume of basalt was erupted in early and middle Eocene time (~44 to 54 m.y. ago), coincident with the culmination of volcanism far inland. Large volumes of graywacke-turbidite sediment are associated with these volcanic rocks, and sediment accumulation continued after submarine volcanism had tapered off. Cessation of volcanism in the Coast Ranges was not abrupt, however; sills and dikes of gabbro, diorite, and alkali trachyte were emplaced over a considerably later time span (late Eocene through much of the Oligocene), but volumetrically these intrusive bodies are small (Snavely and Wagner, 1961; Tatsumoto and Snavely, 1969).

Chemical composition of Challis episode rocks ranges from tholeiitic basalt in the coastal geosyncline to a varied calc-alkalic assemblage of intermediate average silica content over central Oregon, northern Washington, Idaho, southwestern Montana, and northwestern Wyoming to unusual alkalic volcanic associations in many of the smaller igneous centers on the stable craton in Montana and Wyoming (Lipman and others, 1972).

In several areas (shown with a special pattern on Fig. 12-3), basement rocks, regardless of geologic age, yield K-Ar dates of 45 to 50 m.y. (Armstrong, 1974; Miller and Engels,

1975; Engels and others; 1976). Similar areas of Eocene-reset K-Ar dates occur in British Columbia (Medford, 1975; Mathews, 1976) where they are regionally associated with Challis-age igneous rocks, but neither Eocene-reset basement nor widespread Eocene volcanic rocks occur south of lat 42°N. The areas of K-Ar reset basement south of the Snake River Plain give Oligocene-Miocene K-Ar dates, and the regionally associated igneous rocks are of similar age. The regional correlations are suggestive of a genetic association. Furthermore, Taylor and Magaritz (1976) have found that the rocks in central Idaho with Eocene-reset K-Ar dates have been partially to extensively equilibrated with meteoric waters—presumably as a consequence of enormous geothermal systems operating at the time of Challis igneous activity. Heat conveyed into the crust by magmas would have propelled the deep circulation of waters and accelerated chemical reactions with host rocks. This was thus a time of concentration and formation of ore deposits.

The exposures of reset basement are, in general, composed of medium- to high-grade metamorphic rocks or meso- to catazonal igneous plutons. The association of resetting and exposures of plutonic rocks is probably not a coincidence. It is possible to speculate that these areas were sufficiently warm and buoyant to rise upward toward the surface during the Eocene and that in some places part of the metamorphic fabric of the plutonic rocks—in particular, cataclastic foliation and lineation associated with tectonic denudation during diapiric rise—is of Eocene age. This model would apply particularly to such areas as the Bitterroot mylonitic gneiss zone near Missoula, Montana (Wherenberg, 1972; Chase, 1973), and the mylonites associated with the Colville gneiss dome in Washington (Waters and Krauskopf, 1941; Fox and Rinehart, 1970). Thus, I would emphasize the association of diverse phenomena, occurring at a variety of structural levels in the crust, with the Challis episode.

OLIGOCENE–EARLY MIOCENE

As the Eocene came to an end, volcanic-plutonic activity over most of the Northwest declined drastically, and the magmatic hearth shifted southward into Nevada and Utah and contracted into the Cascade region (Fig. 12-4). In southwestern Montana and the Absaroka region, weak volcanism persisted well into Oligocene time. Ash beds are common within the numerous basins that were still slowly filling with sediment. Some of the ash may have been transported by air and water from the intensely active volcanic fields of the Great Basin to the south or from the Cascade volcanic arc to the west. By Miocene time the sedimentary basins of the Rocky Mountain region had been filled, and the area was one of gentle relief and sediment-choked drainage.

There is virtually no Oligocene record preserved in Idaho. The state must have been a region of low relief blanketed by Eocene volcanic rocks, or covered by a thin veneer of ash-rich Oligocene sediment, or a low-relief terrain of pre-Cenozoic rocks slowly undergoing erosion. The only exception was the Albion Range near the southern border where a granitic pluton crystallized about 30 m.y. ago (Armstrong, 1976), the surrounding plutonic rocks had their K-Ar dates reset to approximately 17 to 26 m.y., and lineation and a mylonitic foliation were developed in some Mesozoic and Cenozoic granitic rocks and associated high-grade metamorphosed Paleozoic sediments. This association of phenomena was contemporaneous with like processes occurring in Nevada and northwestern Utah, but it was also a repetition of the infrastructural processes associated with Challis volcanism in central Idaho, northern Idaho, and Washington.

In Oregon and Washington volcanic-intrusive activity largely contracted into the Cascades

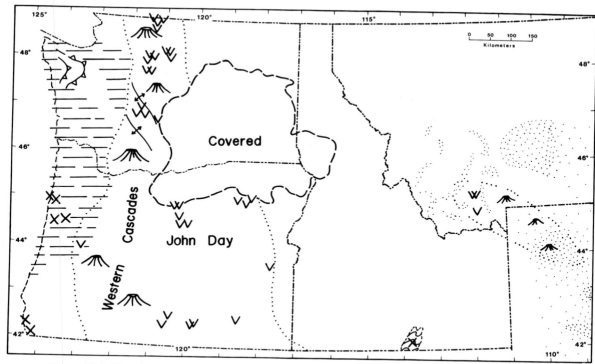

Figure 12-4. Distribution of igneous activity from 36 to 20 m.y. ago (early Oligocene to early Miocene). See Figure 12-2 for explanation of symbols.

in Oligocene time. This was the beginning of a Cascade volcanic arc of location, extent, and character comparable to the Cascades of relatively recent time. East of the Cascades in Washington the Oligocene is lacking or deeply buried beneath younger volcanic rocks. In eastern Oregon the widespread John Day Formation is composed of deposits laid down during Oligocene and early Miocene time (Hay, 1963; Fisher and Rensberger, 1972). John Day Formation is predominantly sediment, with considerable reworked volcanic ash, beds of vitric tuff, and several extensive welded ash flows from sources most typically to the west, in or near the Cascades.

The coastal geosyncline continued to fill with clastic sediment and gradually contracted in extent during Oligocene–early Miocene time. As already mentioned, a trickle of intrusive igneous activity continued from Eocene into Oligocene time, but this was no more than the dying stages of once-prolific magmatic source regions. The Olympic Mountains were undergoing tectonic imbrication (Cady, 1975) and, in deeper structural units, low-grade metamorphism was occurring during the Oligocene (about 29 m.y. ago, according to Tabor, 1972).

THE COLUMBIA VOLCANIC EPISODE

Baksi and Watkins (1973) and Watkins and Baksi (1974) have pointed out the exceptional rate of production of tholeiitic flood basalt from several centers over a short time span (mostly 13 to 16 m.y. ago) in the Columbia Plateau region (Fig. 12-5). These rocks had already been extensively studied by Waters (1961) and others (for example, Mackin, 1961; Wright and

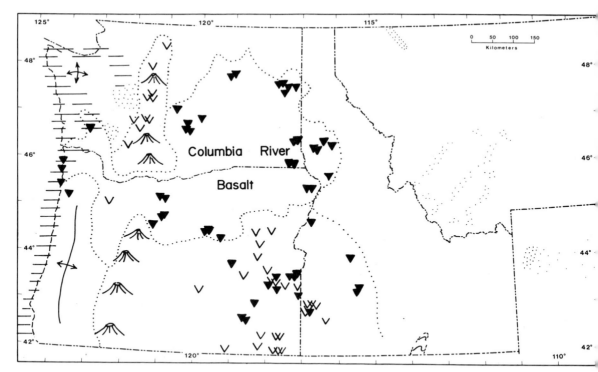

Figure 12-5 Distribution of igneous activity from 20 to 13 m.y. ago (early to middle Miocene). The extent of the Columbia episode and the first regional bimodal volcanism are shown. See Figure 12-2 for explanation of symbols.

others, 1973), but the short time span represented by the thick accumulations of basalt was not recognized until detailed geochronometric and magnetic stratigraphic work was done. The beginning of basalt eruptions, about 17 to 18 m.y. ago, coincides with major changes in regional tectonic patterns and volcanic associations that are discussed by Lipman and others (1971, 1972), Christiansen and Lipman (1972), and McKee (1971).

In Montana and Wyoming the early and middle Miocene were times of quiescence—nondeposition or, less commonly, slow filling and topping out of basins inherited from Eocene time. The region was characterized by low relief, sluggish drainage, and tuffaceous, fine-grained sediments. In western Idaho Columbia River flood basalts buried a mature topography, interfingering with sediments (Latah Formation) along the edges of the basin of accumulation. A number of dikes related to the flood basalts also occur in western Idaho and a few of these have been dated. In the Silver City area of southwestern Idaho, a bimodal volcanic suite was erupted about 15 to 16 m.y. ago and contemporaneous geothermal waters deposited rich silver ores.

South of the Columbia River Basalt field in Oregon was a large region of bimodal volcanic eruptions—both rhyolite and basalt that are dated at numerous localities. The volcanic rocks were contemporaneous with the basalt floods farther north but distinctly different in chemical character (Walker, 1970). This was the first Cenozoic bimodal volcanic association of regional extent in the Northwestern United States.

A chain of basalt-andesite-rhyolite volcanic centers continued to be active along the Cascade trend at the same time as the basalt floods and bimodal volcanism to the east (Peck and

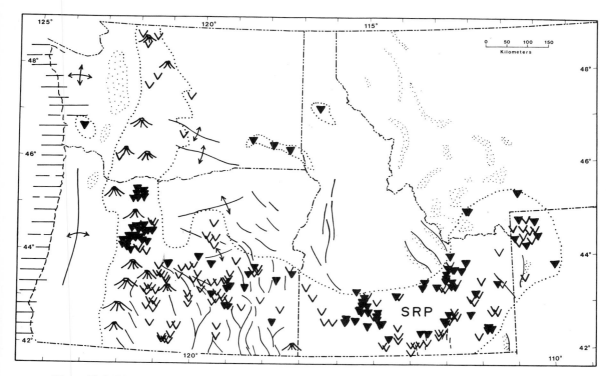

Figure 12-6. Distribution of igneous activity from 13 m.y. ago to the present (middle Miocene and later). See Figure 12-2 for explanation of symbols.

others, 1964; McBirney and others, 1974). Near the Columbia River structural low, along the Oregon-Washington boundary, the flood basalts crossed the Cascades to enter the coastal marine realm. Along the northwestern coast of Oregon are a few intrusive basalt localities associated with the Columbia episode.

During the Miocene, deformation continued in the central Olympic Mountains at the same time as deposition was occurring in peripheral areas (Tabor, 1972; Cady, 1975), including the Puget depression. Marine deposition occurred in several coastal embayments but the volume of sediment accumulated was small. Most sedimentation and much deformation had shifted to areas farther offshore (Kulm and Fowler, 1974), while the Coast Ranges of Oregon and Washington were being uplifted and deformed into their presently observed state.

LATE MIDDLE MIOCENE TO RECENT

Following the Columbia episode, the volcanic and tectonic patterns of the Northwest gradually evolved into the situation observed today (Fig. 12-6). Eruption rates varied significantly so that the record might be subdivided into as many as three separate episodes of increased activity punctuated by quieter periods (McBirney and others, 1974; White and McBirney, this volume; Armstrong, 1975a).

Volcanic activity in the Cascade arc continued, with culminations of eruptive activity about 5 to 7 and 8 to 10 m.y. ago, and with a pause about 2 to 3 m.y. ago. Then eruption of large volumes of basalt formed the High Cascade shield volcanoes of Oregon and southern

Washington. Surmounting these shields are Pleistocene, scenic but volumetrically small, andesitic to dacitic strato-volcanoes (for example, Fiske and others, 1963; Wise, 1969). At all times the Cascade region was a prolific source of volcanic-rich detritus shed into adjacent sedimentary basins and for occasional far traveled volcanic ash, ash flows, and lava tongues.

The Coast Mountains in post-Oligocene time were largely emergent, undergoing uplift and erosion. The Puget-Willamette trough remained as a partially marine sediment trap, and marine sediments were deposited in a few remaining coastal embayments. Marine deposition and deformation related to subduction continued offshore (Byrne and others, 1966; Kulm and Fowler, 1974).

The Columbia River Basalt plateau was undergoing dissection, gradual subsidence, and deformation along its western and southern edges. A small volume of late intracanyon basalt was erupted from several sources between 8.5 and 12 m.y. (Swanson and others, 1975; Swanson and Wright, 1976), and about 3 m.y. ago isolated basalt eruptions occurred in northern Idaho (Dort, 1967; Gray and Kittleman, 1967; Bishop, 1969). Much of the rock and landscape of southeastern Oregon was created during this post–early-Miocene time interval. Walker and co-workers (Walker, 1973; Walker and others, 1974a, 1974b; McKee and others, 1975, 1976) and Kittleman and others (1965) have described the bimodal volcanic assemblage that accumulated in this region since 13 m.y. ago. In all of southeastern Oregon, bimodal volcanism has persisted since early Miocene time, but in a general way the age distribution of rocks is not uniform. The ages of basaltic and especially rhyolitic eruptions are generally younger toward the Cascades with the youngest eruptions being at and near Newberry Volcano (Higgins, 1973). This northwestward-younging trend of rhyolitic eruptions was pointed out by Walker and others (1974b); the trend has been compared, and genetic similarity implied, with the time-transgressive trend of rhyolite eruptions of the Snake River Plain–Yellowstone region.

I do not think it is valid to equate these coincident time-transgressive trends. The trend of the Snake River Plain is one of inception of a brief episode of rhyolitic volcanism, whereas in southeastern Oregon the inception cannot be shown to be time-transgressive, and the history of eruption of rhyolitic material is long. What occurs there is a gradual contraction of the area of numerous rhyolitic eruptions—extending from the Cascades to western Idaho at the time of the Columbia episode, from the Cascades to the Harney basin by about 5 m.y. ago, and restricted to areas in and near the Cascades during the past few million years. This may be viewed as the exhaustion of unreplenished rhyolite magma source regions as time passed, while a source of thermal energy remained focused on the Cascades. In the volcanic fields of southeastern Oregon, the Columbia and High Cascade volcanic episodes are clearly recognized, but it is not possible to establish that culminations in activity around 6 and 10 m.y. were separated by a remission. In fact, there are numerous reliable K-Ar dates between 6 and 10 m.y. so that, taken at face value, the geochronometry now shows only a single broad culmination of activity bracketed by the pauses 2 to 3 and approximately 13 m.y. ago.

During the past 13 m.y., volcanic activity, associated with regional uplift and an advancing wave or block faulting, transgressed eastward to form the Snake River Plain–Yellowstone volcanic field (Armstrong and others, 1975a; Christiansen and Blank, 1972; Matthews and Anderson, 1973; Suppe and others, 1975). Culminations in activity about 10, 6, and 0 to 2 m.y. ago were approximately coincident with the maximum volcanic production rates of the Cascades (Armstrong, 1975a). Within any particular area the volcanism began, following a distinct gap in the stratigraphic record, with eruption of large volumes of rhyolite—ash flows, air-fall tuff, and lava flows—associated with minor amounts of basalt and the accumulation of tuffaceous sediments and clastics derived from uplifted older rocks. This was followed by eruption of flood basalts associated with a few rhyolite domes and flows. Finally, the eruptions produced a few basalt flows that are intercalated with large volumes of fine-grained tuffaceous sediment in complex ways. Because of the time-transgressive evolution of the volcanic field, these three volcanic-sedimentary facies exist as time equivalents, laterally juxtaposed, and as successors in each local succession. Yellowstone Park, on the culmination of a regional topographic dome, is currently the site of the first facies; the second facies occurs in eastern Idaho, and the third occurs in the western part of the state.

The regional uplift associated with the Yellowstone "hot spot" resulted in rejuvenation of mountainous topography in the Northern Rocky Mountains. Valley fills were eroded, redistributed, and, to a large extent, removed; this created the mountain relief now seen and produced complex drainage adjustments and changes. In new fault-bounded basins in eastern Idaho and southern Oregon, thick clastic sequences accumulated during the latest part of Cenozoic time.

The overall tectonic pattern of the past 13 m.y. has been one of expansion of the Basin and Range province along the southern margin of the region and of bimodal volcanism–block-fault topography extending from the Cascades to the Teton Range. At the same time, uplift of the Coast Ranges has continued, and buckling of the western half of the Columbia River Basalt region has produced west-northwest–to west-southwest–trending folds. This folding and the rifting along the southern side of the Columbia Plateau resulted in a clockwise rotation of this large crustal subplate. Likewise, central Idaho rotated clockwise a few degrees as rifting progressed on both sides of the Idaho batholith. The spreading of the Basin and Range province was also accommodated in southern Oregon by strike-slip displacement along west-northwest–trending faults such as the Mitchell fault (Oles and Enlows, 1971). While all this was taking place on land, continuous and continuing subduction was occurring along the continental margin (Silver, 1971, 1972; McKenzie and Julian, 1971).

SEARCH FOR AN EXPLANATION

To construct an instantaneous plate tectonic model of the Northwestern United States, we have geophysical data to augment observed processes and the constraints provided by geologic history. The regional geophysical picture was reviewed by Hamilton and Myers (1966) and Thompson and Burke (1974). Important aspects of seismicity were discussed by Sbar and others (1972) and Smith and Sbar (1974); aspects of heat flow were discussed by Roy and others (1972). Suppe and others (1975) described an instantaneous tectonic model of subduction along the continental margin and crustal disruption driven by a Yellowstone hot spot (an idea started by Matthews and Anderson, 1973) that appears a satisfactory first approximation. The geological and geophysical evidence for an exceptional focus of thermal energy in the Yellowstone region is quite impressive (Eaton and others, 1975). The uplift associated with elevated isotherms would drive the rifting of the Basin and Range province and related rotation of the central Idaho and Columbia Plateau subplates. Meanwhile, the Cascade volcanic arc would be localized and sustained by subduction of the Juan de Fuca plate offshore.

But this model can be evolved backward in time only to the Columbia episode. We then enter into speculation. Could the Columbia episode be the explosive initiation of a subcontinental hotspot? Was this an event triggered by collision of Pacific and North American plates far to the south (Atwater, 1970) or a coincidence—a fortuitous circumstance that allowed Basin and Range development to surge forward in the United States Cordillera but not elsewhere?

Is it even reasonable to link the Columbia event to the Yellowstone melting spot? The correlation of Snake River Plain and Cascade volcanic episodes remains unexplained.

Farther back in time than the Columbia event, the tectonic situation remains obscure. I do not even have a speculative plate model to offer in explanation of the earlier Cenozoic volcanic patterns and am skeptical of all models that have been offered. There is still something missing—a lack of critical data or inadequate or false conceptions prevent the reconciliation of observed history with actualistic plate models. The salient features that are most puzzling are the broad extent of volcanic fields perpendicular to the continental margin and their irregular, but well-documented, southward time transgression. The challenge of more work and thinking remains.

ACKNOWLEDGMENTS

My geochronometry projects in the Northwestern United States have been financed by U.S. National Science Foundation Grants GA 26025 and GA 694 and Canadian National Research Council Grant A-8841. I thank K. Hayne and M. L. Bevier for their assistance.

REFERENCES CITED

Armstrong, F. C., and Oriel. S. S., 1965, Tectonic development of Idaho-Wyoming thrust belt: Am. Assoc. Petroleum Geologists Bull., v. 49, p. 1847–1866.

Armstrong, R. L., 1968, The Sevier orogenic belt in Nevada and Utah: Geol. Soc. America Bull., v. 79, p. 429–458.

——1974, Geochronometry of the Eocene volcanic-plutonic episode in Idaho: Northwest Geology, v. 3, p. 1–15.

——1975a, Episodic volcanism in the central Oregon Cascade Range: Confirmation and correlation with the Snake River Plain: Geology, v. 3, p. 356.

——1975b, The geochronometry of Idaho: Isochron/West, no. 14, p. 1–50.

——1976, The geochronometry of Idaho: Isochron/West, no. 15, p. 1–33.

Armstrong, R. L., and Higgins, R. E., 1973, K-Ar dating of the beginning of Tertiary volcanism in the Mojave Desert, California: Geol. Soc. America Bull., v. 84, p. 1095–1100.

Armstrong, R. L., Leeman, W. P., and Malde, H. E., 1975a, K-Ar dating, Quaternary and Neogene volcanic rocks of the Snake River Plain, Idaho: Am. Jour. Sci., v. 275, p. 225–251.

Armstrong, R. L., Taylor, E. M., Hales, P. O., and Parker, D. J., 1975b, K-Ar dates for volcanic rocks, central Cascade Range of Oregon: Isochron/West, no. 13, p. 5–10.

Armstrong, R. L., Harakal, J. E., and Hollister, V. F., 1976a, Late Cenozoic porphyry copper deposits of the North American Cordillera: Inst. Mining and Metallurgy Trans., v. 85, p. B239–B244.

Armstrong, R. L., Taubeneck, W. H., and Hales, P. O., 1976b, Rb-Sr and K-Ar geochronometry of granitic rocks and their Sr isotopic composition—Data for Oregon and Washington: Isochron/West, no. 17, p. 27–32.

Atwater, Tanya, 1970, Implications of plate tectonics for the Cenozoic tectonic evolution of western North America: Geol. Soc. America Bull., v. 81, p. 3513–3536.

Axelrod, D. I., 1966, Potassium-argon ages of some western Tertiary floras: Am. Jour. Sci., v. 264, 497–506.

Baksi, A. K., and Watkins, N. D., 1973, Volcanic production rates: Comparison of oceanic ridges, islands, and the Columbia Plateau basalts: Science, v. 180, p. 493–496.

Beaulieu, J. D., 1971, Geologic formations of western Oregon: Oregon Dept. Geology and Mineral Industries Bull. 70, 72 p.

Berggren, W. A., 1972, A Cenozoic time-scale—Some implications for regional geology and paleobiogeography: Lethaia, v. 5, p. 195–215.

Bishop, D. T., 1969, Stratigraphy and distribution of basalt, Benewah County, Idaho: Idaho Bur. Mines and Geology Pamph. 140, 20 p.

Blackstone, D. L., Jr., 1966, Pliocene vulcanism, southern Absaroka Mountains, Wyoming: Wyoming Univ. Contr. Geology, v. 5, p. 21–30.

Blake, M. C., Jr., and Jones, D. L., 1974, Origin

of Franciscan melanges in northern California: Soc. Econ. Paleontologists and Mineralogists Spec. Pub. 19, p. 345–357.

Byrne, J. V., Fowler, C. A., and Maloney, N J., 1966, Uplift of continental margin and possible continental accretion off Oregon: Science, v. 154, p. 1654–1656.

Cady, W. M., 1975, Tectonic setting of the Tertiary volcanic rocks of the Olympic Peninsula, Washington: U.S. Geol. Survey Jour. Research, v. 3, p. 573–582.

Chadwick, R. A., 1972, Volcanism in Montana: Northwest Geology, v. 1, p. 1–20.

Chase, R. B., 1973, Petrology of the northeastern border zone of the Idaho batholith, Bitterroot Range, Montana: Montana Bur. Mines Mem. 43, 28 p.

Christiansen, R. L., and Blank, H. R., Jr., 1972, Volcanic stratigraphy of the Quaternary rhyolite plateau in Yellowstone National Park: U.S. Geol. Survey Prof. Paper 729-B, 18 p.

Christiansen, R. L. and Lipman, P. W., 1972, Cenozoic volcanism and plate tectonic evolution of the Western United States. II: Late Cenozoic: Royal Soc. London Philos. Trans., ser. A, v. 271, p. 249–284.

Denison, R. E., 1970, Oil test cores dated: Ore Bin, v. 32, p. 184.

Dort, Wakefield, Jr., 1967, Late Cenozoic volcanism, St. Joe Valley, Idaho: Northwest Sci., v. 41, p. 141–151.

Dott, R. H., Jr., 1966, Eocene deltaic sedimentation at Coos Bay, Oregon: Jour. Geologist, v. 74, p. 373–420.

Easterbrook, D. J., 1975, Mount Baker eruptions: Geology, v. 3, p. 679–682.

Eaton, G. P., Christiansen, R. L., Iyer, H. M., Pitt, A. M., Mabey, D. R., Blank, H. R., Jr., Zietz, Isidore, and Gettings, M. E., 1975, Magma beneath Yellowstone National Park: Science, v. 188, p. 787–796.

Elliott, J. E., Naeser, C. W., and Hedge, C. E., 1974, Ages of intrusives and ore deposits, Northern Rocky Mountains: U.S. Geol. Survey Prof. Paper 900, p. 32.

Engels, J. C., Tabor, R. W., Miller, F. K., and Obradovich, J. D., 1976, Summary of K-Ar, Rb-Sr, U-Pb, Pbα, and fission-track ages of rocks from Washington State prior to 1975 (exclusive of Columbia Plateau basalts): U.S. Geol. Survey Misc. Field Studies MF-710.

Evernden, J. F., Savage, D. E., Curtis, G. H., and James, G. T., 1964, Potassium-argon dates and the Cenozoic mammalian chronology of North America: Am. Jour. Sci., v. 262, p. 145–198.

Evitt, W. R., and Pierce, S. T., 1975, Early Tertiary ages from the coastal belt of the Franciscan

complex, northern California: Geology, v. 3, p. 433–436.

Fisher, R. V., and Rensberger, J. M., 1972, Physical stratigraphy of the John Day Formation, central Oregon: California Univ. Pubs. Geol. Sci., v. 101, 33 p.

Fiske, R. S., Hopson, C. A., and Waters, A. C., 1963, Geology of Mount Rainier National Park, Washington: U.S. Geol. Survey Prof. Paper 444, 93 p.

Foster, R. J., 1960, Tertiary geology of a portion of the central Cascade Mountains, Washington: Geol. Soc. America Bull., v. 71, p. 99–126.

Fox, K. F., Jr., and Rinehart, C. D., 1970, Colville batholith is a gneiss dome: U.S. Geol. Survey Prof. Paper 700A, p. 123.

Gray, Jane, and Kittleman, L. R., 1967, Geochronometry of the Columbia River Basalt and associated floras of eastern Washington and western Idaho: Am. Jour. Sci., v. 265, p. 257–291.

Hamilton, Warren, and Myers, W. B., 1966, Cenozoic tectonics of the Western United States: Rev. Geophysics, v. 4, p. 509–549.

Hay, R. L., 1963, Stratigraphy and zeolitic diagenesis of the John Day Formation of Oregon: California Univ. Pubs. Geol. Sci., v. 42, p. 119–262.

Higgins, M. W., 1973, Petrology of Newberry Volcano, central Oregon: Geol. Soc. America Bull., v. 84, p. 455–488.

Hill, D. J., Izett, G. A., and Naeser, C. W., 1975, Early Tertiary fission track ages of sphene from Devils Tower and Missouri Buttes, Black Hills, northeastern Wyoming: Geol. Soc. America Abs. with Programs, v. 7, p. 613–614.

Hoffman, Janet, Hower, John, and Aronson, J. L., 1976, Radiometric dating of time of thrusting in the disturbed belt of Montana: Geology, v. 4, p. 16–20.

Ito, H., and Fuller, M. A., 1970, A paleomagnetic study of the reversal process of the geomagnetic field, in Runcorn, S. K., ed., Paleogeophysics: New York, Academic Press, p. 133–137.

Johns, W. M., 1970, Geology and mineral deposits of Lincoln and Flathead Counties, Montana: Montana Bur. Mines and Geology Bull. 79, 182 p.

Keefer, W. R. and Love, J. D., 1963, Laramide vertical movements in central Wyoming: Wyoming Univ. Contr. Geology, v. 2, p. 47–54.

Kennett, J. P., and Thunell, R. C., 1975, Global increase in Quaternary explosive volcanism: Science, v. 187, p. 497–503.

Kittleman, L. R., Green, A. R., Hagood, A. R., Johnson, A. M., McMurray, J.M., Russell, R. G., and Weeden, D. A., 1965, Cenozoic stratigraphy of the Owyhee region, southeastern

Oregon: Oregon Univ. Mus. Nat. History Bull. 1, 45 p.

Kulm, L. D. and Fowler, G. A., 1974, Cenozoic sedimentary framework of the Gorda–Juan de Fuca plate and adjacent continental margin—A review: Soc. Econ. Paleontologists and Mineralogists Spec. Pub. 19, p. 212–229.

Laursen, J. M., and Hammond, P. E., 1974, Summary of radiometric ages of Oregon and Washington rocks, through June 1972: Isochron/West, no. 9, p. 1–32.

Leonard, B. F., and Marvin, R. F., 1975, Temporal evolution of the Thunder Mountain caldera, central Idaho: Geol. Soc. America Abs. with Programs, v. 7, p. 623–624.

Lipman, P. W., Prostka, H. J., and Christiansen, R. L., 1971, Evolving subduction zones in the Western United States, as interpreted from igneous rocks: Science, v. 174, p. 821–825.

——1972, Cenozoic volcanism and plate tectonic evolution of the Western United States. 1, Early and middle Cenozoic: Royal Soc. London, Philos. Trans. ser. A, v. 271, p. 217–248.

Livingston, V. E., Jr., 1969, Geologic history and rocks and minerals of Washington: Washington Div. Mines and Geology Inf. Circ. 45, 42 p.

Love, J. D., 1960, Cenozoic sedimentation and crustal movement in Wyoming (Bradley volume): Am. Jour. Sci., v. 258-A, p. 204–214.

Love, J. D., Tschudy, R. H., and Obradovich, J. D., 1975, Dating a Laramide orogeny, northwestern Wyoming: U.S. Geol. Survey Prof. Paper 975, p. 51.

Love, L. L., Kudo, A. M., and Love, D. W., 1974, Petrochemical and age trends in the Absaroka-Gallatin volcanic province [abs.]: EOS (Am. Geophys. Union Trans.) v. 55, p. 487.

Mackin, J. H., 1961, A stratigraphic section in the Yakima Basalt and the Ellensburg Formation in south-central Washington: Washington Div. Mines and Geology Rept. Inv. 19, 45 p.

Mackin, J. H., and Cary, A. S., 1965, Origin of Cascade landscapes: Washington Div. Mines and Geology Inf. Circ. 41, 35 p.

Marvin, R. F., Whitkind, I. J., Keefer, W. R., and Mehnert, H. H., 1973, Radiometric ages of intrusive rocks in the Little Belt Mountains, Montana: Geol. Soc. America Bull., v. 84, p. 1977–1986.

Marvin, R. F., Wier, K. L., Mehnert, H. H., and Merritt, V. M., 1974, K-Ar ages of selected Tertiary igneous rocks in southwestern Montana: Isochron/West, no. 10, p. 17–20.

Mathews, W. H., 1976, Anomalous K-Ar dates from gneisses of the Trinity Hills, British Columbia: Geol. Assoc. Canada Program with Abs., v. 1, p. 47.

Matthews, Vincent, III, and Anderson, C. E., 1973, Yellowstone convection plume and breakup of the Western United States: Nature, v. 243, p. 158–159.

McBirney, A. R., Sutter, J. F., Naslund, H. R., Sutton, K. G., and White, C. M., 1974, Episodic volcanism in the central Oregon Cascade Range: Geology, v. 2, p. 585–589.

McDowell, F. W., 1971, K-Ar ages of igneous rocks from the Western United States: Isochron/West, no. 2, p. 1–16.

McKee, Bates, 1972, Cascadia—The geological evolution of the Pacific Northwest: New York, McGraw-Hill Book Co., 394 p.

McKee, E. H., 1971, Tertiary igneous chronology of the Great Basin of western United States—Implications for tectonic models: Geol. Soc. America Bull., v. 82, p. 3497–3502.

McKee, E. H., Greene, R. C., Foord, E. E., 1975, Chronology of volcanism, tectonism, and mineralization of the McDermitt caldera, Nevada-Oregon: Geol. Soc. America Abs. with Programs, v. 7, p. 629–630.

McKee, E. H., MacLeod, N. S., and Walker, G. W., 1976, Potassium-argon ages of late Cenozoic silicic volcanic rocks, southeast Oregon: Isochron/West, no. 15, p. 37–41.

McKenzie, Dan, and Julian, Bruce, 1971, Puget Sound, Washington, earthquake and the mantle structure beneath the Northwestern United States: Geol. Soc. America Bull., v. 82, p. 3519–3524.

Medford, G. A., 1975, K-Ar and fission track geochronometry of an Eocene thermal event in the Kettle River (west half) map area, southern British Columbia: Canadian Jour. Earth Sci., v. 12, p. 836–843.

Mehnert, H. H., and Schmidt, R. G., 1971, Dating the Eldorado thrust: U.S. Geol. Survey Prof. Paper 750A, p. 37.

Miller, F. K., and Engels, J. C., 1975, Distribution and trends of discordant ages of the plutonic rocks of northeastern Washington and northern Idaho: Geol. Soc. America Bull., v. 86, p. 517–518.

Misch, Peter, 1966, Tectonic evolution of the northern Cascades of Washington State: Canadian Inst. Mining and Metallurgy Spec. Vol. 8, p. 101–148.

Miyashiro, Akiho, 1975, Volcanic rock series and tectonic setting: Earth and Planetary Sci. Ann. Rev., v. 3, p. 251–269.

Muessig, Siegfried, 1962, Tertiary volcanic rocks of the Republic area, Ferry County, Washington: U.S. Geol. Survey Prof. Paper 450D, p. 56–58.

Neill, W. M., 1975, Geology of the southeastern Owyhee Mountains and environs, Owyhee

County, Idaho: Stanford Univ. Student Research Proj. Rept., 59 p.

Nelson, W. H., and Pierce, W. G., 1968, Wapiti Formation and Trout Peak Trachyandesite, northwestern Wyoming: U.S. Geol. Survey Bull. 1254-H, 11 p.

Niem, A. R., and Cressy, F. B., Jr., 1973, K-Ar dates for sills from the Neahkahnie Mountain and Tillamook Head areas of the northwestern Oregon coast: Isochron/West, no. 7, p. 13–15.

Oles, K. F., and Enlows, H. E., 1971, Bedrock geology of the Mitchell quadrangle, Wheeler County, Oregon: Oregon Dept. Geology and Mineral Industries Bull. 72, 62 p.

Oriel, S. S., 1969, Geology of the Fort Hill quadrangle, Lincoln County, Wyoming: U.S. Geol. Survey Prof. Paper 594-M, 40 p.

Oriel, S. S., and Tracey, J. I., Jr., 1970, Uppermost Cretaceous and Tertiary stratigraphy of Fossil basin, southwestern Wyoming: U.S. Geol. Survey Prof. Paper 635, 53 p.

Peck, D. L., Griggs, A. B., Schlicker, H. G., Wells, F. G., and Dole, H. M., 1964, Geology of the central and northern parts of the western Cascade Range in Oregon: U.S. Geol. Survey Prof. Paper 449, 56 p.

Pekarek, A. H., Marvin, R. F., and Mehnert, H. H., 1974, K-Ar ages of the volcanics in the Rattlesnake Hills, central Wyoming: Geology, v. 2, p. 283–285.

Peterson, N. V., McIntyre, J. R., 1970, The reconnaissance geology and mineral resources of eastern Klamath County and western Lake County, Oregon: Oregon Dept. Geology and Mineral Industries Bull. 66, 70 p.

Roberts, A. E., 1972, Cretaceous and early Tertiary depositional and tectonic history of the Livingston area, southwestern Montana: U.S. Geol. Survey Prof. Paper 526-C, 120 p.

Robinson, G. D., 1960, Middle Tertiary unconformity in southwestern Montana: U.S. Geol. Survey Prof. Paper 400-B, p. 227–228.

Roy, R. F., Blackwell, D. D., and Decker, E. R., 1972, Continental heat flow, in Robertson, E. C., ed., The nature of the solid earth: New York, McGraw-Hill Book Co., p. 506–543.

Ryder, R. T., and Ames, H. T., 1970, Palynology and age of Beaverhead Formation and their paleotectonic implications in Lima region, Montana-Idaho: Am. Assoc. Petroleum Geologists Bull., v. 54, p. 1155–1171.

Ryder, R. T. and Scholten, R., 1973, Syntectonic conglomerates in southwestern Montana: Their nature, origin, and tectonic significance: Geol. Soc. America Bull., v. 84, p. 773–796.

Sbar, M. L., Barazangi, Muawia, Dorman, James, Scholz, C. H., and Smith, R. B., 1972, Tectonics

of the Intermountain seismic belt, Western United States: Microearthquake seismicity and composite fault plane solutions: Geol. Soc. America Bull., v. 83, p. 13–28.

Silver, E. A., 1971, Small plate tectonics in the northeastern Pacific: Geol. Soc. America Bull., v. 82, p. 3491–3496.

——1972, Pleistocene tectonic accretion of the continental slope off Washington: Marine Geology, v. 13, p. 239–249.

Smedes, H. W., and Prostka, H. J., 1972, Stratigraphic framework of the Absaroka Volcanic Supergroup in the Yellowstone National Park region: U.S. Geol. Survey Prof. Paper 729-C, 33 p.

Smedes, H. W., and Thomas, H. H., 1965, Reassignment of the Lowland Creek Volcanics to Eocene age: Jour. Geology, v. 73, p. 508–510.

Smith, R. B., and Sbar, M. L., 1974, Contemporary tectonics and seismicity of the Western United States with emphasis on the Intermountain seismic belt: Geol. Soc. America Bull., v. 85, p. 1205–1218.

Snavely, P. D., Jr., and Wagner, H. C., 1961, Differentiated gabbroic sills and associated alkalic rocks in the central part of the Oregon Coast Range, Oregon: U.S. Geol. Survey Prof. Paper 424D, p. 156–160.

——1963, Tertiary geologic history of western Oregon and Washington: Washington Div. Mines and Geology Rept. Inv. 22, 25 p.

Snavely, P. D., Jr., MacLeod, N. S., and Wagner, H. C., 1968, Tholeiitic and alkalic basalts of the Eocene Siletz River Volcanics, Oregon Coast Range: Am. Jour. Sci., v. 266, p. 454–481.

——1973, Miocene tholeiitic basalts of coastal Oregon and Washington and their relations to coeval basalts of the Columbia Plateau: Geol. Soc. America Bull., v. 84, p. 387–424.

Staatz, M. H., 1960, The Republic graben, a major structure in northeastern Washington: U.S. Geol. Survey Prof. Paper 400B, p. 304–306.

Suppe, John, Powell, Christine, and Berry, Robert, 1975, Regional topography, seismicity, Quaternary volcanism, and the present-day tectonics of the Western United States: Am. Jour. Sci., v. 275A, p. 397–436.

Swanson, D. A., and Wright, T. L., 1976, Guide to field trip between Pasco and Pullman, Washington, emphasizing stratigraphy, vent areas, and intracanyon flows of Yakima Basalt: Pullman, Washington State Univ. Dept. Geology, 33 p.

Swanson, D. A., Wright, T. L., and Helz, R. T., 1975, Linear vent systems and estimated rates of magma production and eruption for the Yakima Basalt on the Columbia Plateau: Am. Jour.

Sci., v. 275, p. 877–905.

Tabor, R. W., 1972, Age of the Olympic metamorphism, Washington: K-Ar dating of low-grade metamorphic rocks: Geol. Soc. America Bull., v. 83, p. 1805–1816.

Tatsumoto, M., and Snavely, P. D., Jr., 1969, Isotopic composition of lead in rocks of the Coast Range, Oregon and Washington: Jour. Geophys. Research, v. 74, p. 1087–1100.

Taylor, H. P., Jr., and Magaritz, M., 1976, An oxygen and hydrogen isotope study of the Idaho batholith [abs.]: EOS (Am. Geophys. Union Trans.), v. 57, p. 350.

Thompson, G. A., and Burke, D. B., 1974, Regional geophysics of the Basin and Range province: Earth and Planetary Sci. Ann. Rev., v. 2, p. 213–238.

Turner, D. L., 1970, Potassium-argon dating of Pacific Coast Miocene foraminiferal stages: Geol. Soc. America Spec. Paper 124, p. 91–129.

Van Houten, F. B., 1961, Maps of Cenozoic depositional provinces, Western United States: Am. Jour. Sci., v. 259, p. 612–621.

Vogt, P. R., 1976, Increasing evidence for global magmatic synchronism [abs.]: EOS (Am. Geophys. Union Trans.), v. 57, p. 348.

Walker, G. W., 1970, Some comparisons of basalts of southeast Oregon with those of the Columbia River Group, in Second Columbia River Basalt Symp. Proc.: Cheney, Eastern Washington State College Press, p. 223–238.

——1973, Preliminary geologic map of eastern Oregon: U.S. Geol. Survey Misc. Field Studies Map MF-495, scale 1:500,000.

Walker, G. W., Dalrymple, G. B., and Lanphere, M. A., 1974a, Index to potassium-argon ages of Cenozoic volcanic rocks of Oregon: U.S. Geol. Survey Misc. Field Studies Map MF-569, scale 1:1,000,000.

Walker, G. W., MacLeod, Norman, and McKee, E. H., 1974b, Transgressive age of late Cenozoic silicic volcanic rocks across southeastern Oregon; Implications for geothermal potential: Geol. Soc. America Abs. with Programs, v. 6, p. 272.

Waters, A. C., 1961, Stratigraphic and lithologic variations in the Columbia River Basalt: Am. Jour. Sci., v. 259, p. 583–611.

——1962, Basalt magma types and their tectonic associations: Pacific Northwest of the United States: Am. Geophys. Union Geophys. Mon. 6, p. 158–170.

Waters, A. C., and Krauskopf, K., 1941, Protoclastic border of the Colville batholith: Geol. Soc. America Bull., v. 52, p. 1355–1418.

Watkins, N. D., and Baksi, A. K., 1974, Magnetostratigraphy and oroclinal folding of the Columbia River, Steens, and Owyhee basalts in Oregon, Washington, and Idaho: Am. Jour. Sci., v. 274, p. 148–189.

Wherenberg, J. P., 1972, Geology of the Lolo Peak area, northern Bitterroot Range, Montana: Northwest Geology, v. 1, 25–32.

Whetten, J. T., 1976, Tertiary sedimentary rocks in the central part of the Chiwaukum graben, Washington: Geol. Soc. America Abs. with Programs, v. 8, p. 420.

White, C. M., and McBirney, A. R., 1978, Some quantitative aspects of orogenic volcanism in the Oregon Cascades, in Smith, R. B., and Eaton, G. P., eds., Cenozoic tectonics and regional geophysics of the western Cordillera: Geol. Soc. America Mem. 152 (this volume).

Williams, T. R., Harakal, J. E., and Armstrong, R. L., 1976, K-Ar dating of Eocene volcanic rocks near Drummond, Montana: Northwest Geology, v. 5, p. 21–24.

Wilson, J. T., 1973, Mantle plumes and plate motions: Tectonophysics, v. 19, p. 149–164.

Wise, W. S., 1969, Geology and petrology of the Mt. Hood area: A study of High Cascade volcanism: Geol. Soc. America Bull., v. 80, p. 969–1006.

Wolfe, J. A., Gower, H. D., and Vine, J. D., 1961, Age and correlation of the Puget group, King County, Washington: U.S. Geol. Survey Prof. Paper 424-C, p. C230–C232.

Wright, T. L., Grolier, M. J., and Swanson, D. A., 1973, Chemical variation related to the stratigraphy of the Columbia River Basalt: Geol. Soc. America Bull., v. 84, p. 371–386.

Zen, E-an, 1975, Eocene volcanic rocks in the Pioneer Mountains, Montana: U.S. Geol. Survey Prof. Paper 975, p. 48.

Manuscript Received by the Society August 15, 1977

Manuscript Accepted September 2, 1977

Geological Society of America
Memoir 152

13

Late Cenozoic volcanic and tectonic evolution
of the Great Basin and
Columbia Intermontane regions

Robert L. Christiansen
Edwin H. McKee
U.S. Geological Survey
345 Middlefield Road
Menlo Park, California 94025

ABSTRACT

The Great Basin is a tectonically youthful region that shares some features with the Columbia Intermontane region but is separated from the more mature southern part of the Basin and Range province by a zone of active seismicity and geophysical contrasts. Sedimentary, physiographic, and structural features show that during the past 17 m.y., extensional linear normal faulting has been active in the Great Basin region, and extension also is indicated by numerous dikes in the High Lava Plains and the Columbia Plateau. Cumulative tectonic extension in the Great Basin is more than 100 km. Since about 14 m.y. ago, tectonic activity in the Great Basin region has tended to become progressively more concentrated toward the margins, and extension has been taken up by a wide transform zone along the High Lava Plains. Within several tens of kilometres north of the High Lava Plains of Oregon and Idaho, cumulative extension is generally less than a few kilometres and has been nearly inactive since about 14 m.y. ago.

Volcanism in the past 17 m.y. has been characterized by basaltic and bimodal rhyolite-basalt suites. Between 17 and 14 m.y. ago, the predominant volcanism was basaltic, being somewhat alkalic and of relatively small volume in the central Great Basin, more voluminous and less alkalic northward into the plateaus of southern Oregon and the High Lava Plains, and extremely voluminous and tholeiitic in the Columbia Plateau. Since about 14 m.y. ago, basaltic and bimodal volcanism has occurred throughout the Great Basin region but generally has tended to erupt in successively narrower zones near its margins, probably in direct correspondence to the increasing concentration of normal faulting toward these margins. The High Lava Plains have been characterized during this same time by two linearly propagating volcanic systems, in which major cycles of rhyolitic volcanism have been initiated successively farther northwest and northeast. These two volcanic systems have propagated away from a region in the center of the High Lava Plains at about the same rate that faulting and volcanism in the Great Basin have been concentrated toward its margins.

A model that accounts for this evolution relates tectonic extension to the regional stress fields that result from the motions and changes in the interactions of the North American, Pacific, and Farallon lithospheric plates. In this model, geophysical and volcanic features of the region are interpreted to be due to a chain of heating events caused by this extension but conditioned by the stress and thermal history of the continental plate. Stress relief at the base of the lithosphere causes basaltic magma generation of varying amounts and at varying depths in the upper mantle, depending on the thickness and history of the overlying crust. The generation of basaltic magmas and their intrusion into and through the crust during continued extension have increased regional heat flow, lowered the rigidity of the lithosphere, caused crustal thinning, produced flowage and decreased seismic velocities in the upper mantle, caused regional uplift by thermal expansion, and produced rhyolitic magmas by localized partial melting of the lower crust.

According to the model, initial rifting occurred between 17 and 14 m.y. ago when northward migration of the Mendocino triple junction caused the continental-margin subduction zone to become short enough to allow partial coupling between two zones of transform displacement of the Pacific and North American plates. The increased coupling between these two zones caused extension in the North American plate perpendicular to the continental margin. Since about 14 m.y. ago, continued tectonic extension and basaltic magma generation have (1) caused a wide zone of oblique extension to become successively hotter and less rigid near the zone's central axis, (2) increasingly concentrated brittle deformation and high-level magmatism outward toward the margins of the Great Basin region, and (3) produced concentrated zones of extension and crustal melting at the intersections of the resulting marginal zones with the transitional northern transform boundary of the extending region. This accounts for the symmetrically propagating volcanic systems of the High Lava Plains. The Yellowstone melting anomaly, whose locus was controlled initially by an old structural boundary, was favorably oriented to be augmented by shear melting at the base of the lithosphere; it has become self-sustaining because of the initiation of a thermal feedback cycle and the development of a root in the mantle by inward flow around a dense, sinking, unmelted residuum.

INTRODUCTION AND REGIONAL SETTING

The Cenozoic basin-range region of the Western United States is a region of extensional block faulting that includes the Basin and Range physiographic province and some other bordering areas. A number of models have been proposed recently for the origin of the basin-range region (for example, Menard, 1964; Hamilton and Myers, 1966; Cook, 1969; Atwater, 1970; Scholtz and others, 1971; Christiansen and Lipman, 1972; Noble, 1972; Thompson and Burke, 1974; Suppe and others, 1975). Many of these models, based in part upon time-space patterns of Cenozoic volcanism and regional structures or upon global tectonic models, have focused on the Great Basin of Nevada, western Utah, and adjacent areas. To the extent that the view has widened, it commonly has been cast southward toward the rest of the Basin and Range province from southern California to New Mexico.

In this paper we attempt to build a conceptual model of late Cenozoic volcanic and tectonic evolution by focusing on relationships between the northern part of the basin-range region, here referred to as the Great Basin region, and the largely volcanic plains and plateaus that lie farther north (Fig. 13-1), sometimes referred to as the Columbia Intermontane province (for example, see Thornbury, 1965). Volcanic and tectonic features that are common to these two regions or that mark a unique transition between them may (as suggested by Hamilton and Myers, 1966) denote a genetic linkage that is as fundamental as the relationship between the Great Basin region and the southern part of the Basin and Range province.

The most important pattern common to the Great Basin and Columbia Intermontane regions is a two-stage Cenozoic volcanic and tectonic development that seems to have been nearly synchronous. From about 50 to 17 m.y. ago this entire area was a postorogenic terrane characterized in places by predominantly calc-alkalic volcanism of intermediate to silicic composition (McKee and others, 1970; Lipman and others, 1972; Stewart and Carlson, 1976 and this volume). During that time, subduction was active along the Pacific margin west of the area although after about 29 m.y. ago, subduction ceased to the south (Atwater, 1970; Atwater and Molnar, 1973). Volcanism in the area before 17 m.y. ago probably was related to that subduction (Lipman and others, 1972). A brief lull in volcanism about 17 m.y. ago (McKee and others, 1970) was followed by progressive decline of the subduction zone to the west and by regional tectonic extension and predominantly basaltic or bimodal rhyolite-basalt volcanism (Christiansen and Lipman, 1972; Stewart and Carlson, 1976 and this volume).

During the time since about 14 to 17 m.y. ago, the typical basin-range pattern of subparallel fault-bounded linear ranges separated by alluviated basins has typified most of the Great Basin region, while the Columbia Intermontane region has evolved mainly as a series of volcanic

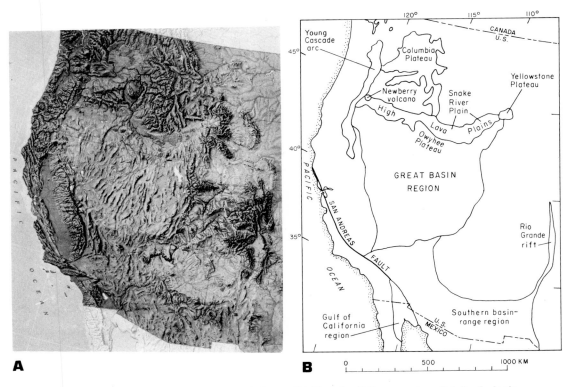

A   **B**

Figure 13-1. Physiography of the Western United States. (A) Shaded relief map; note well-defined physiographic break between the northern and southern parts of the Basin and Range province (after U.S. Geol. Survey, *National Atlas of the United States of America*, p. 56). (B) Index to some late Cenozoic tectonic and volcanic features shown on Figure 13-1A.

plains. Basin-range topography is well developed south of the High Lava Plains of central and eastern Oregon and southern Idaho (Fig. 13-1), but within several tens of kilometres north of those plains, this topographic pattern generally declines and is no longer predominant. Conversely, the volume of predominantly basaltic lavas of late Cenozoic age is greatest in the Columbia Plateau, declines southward across Oregon and Idaho into the northern Great Basin, and is relatively small in the central and southern Great Basin.

PHYSIOGRAPHY, BASIN DEPOSITS, AND STRUCTURE

Great Basin Region

The parallel linear basins and ranges of the Great Basin region are youthful fault-bounded features. The range fronts are generally steep, and fault scarps break many of the bordering alluvial fans (Slemmons, 1967). On a regional scale these characteristic features of linearity, parallelism, and conspicuous youthfulness contrast with the Basin and Range province south of a sharply defined east-trending zone through southern Nevada and California (Fig. 13-1). In the southern basin-range region the mountains commonly are lower, have less well defined fronts, and have wide surrounding complexes of pediments and alluvial fans; fresh fault scarps and active seismicity are uncommon. Eaton and others (this volume) note that this abrupt transition across southern Nevada and southeastern California is marked by a zone of seismicity and a contrast in the principal geophysical characteristics of the regions it separates. The northern basin-range region also has a higher average elevation than the southern part. Thus, it appears that the part of the basin-range region north of the southern Nevada zone is a tectonic province of active uplift and extension that is separated from a nearly stabilized region of former extension to the south.

Cenozoic sedimentary rocks in the Great Basin region most commonly occur in or on the edges of the present valleys. However, some of the older sedimentary rocks that predated or accompanied early and mid-Tertiary volcanism occur high in the block-faulted ranges (Fig. 13-2A). The prevolcanic terrane of the present Great Basin succeeded a Late Cretaceous and perhaps early Tertiary orogeny. By the time most orogenic movements had ceased, the region had considerable topographic relief, and fluvial and lacustrine sediments filled widely separated basins. The resulting sedimentary units range in age from that of the major orogenic period in Late Cretaceous time [for example, the Newark Canyon Formation of central Nevada (MacNeil, 1939; Nolan and others, 1956)], through Eocene and early Oligocene [for example, the Sheep Pass Formation of eastern Nevada (Winfrey, 1960) and the Titus Canyon Formation of the Death Valley area (Stock and Bode, 1935; Reynolds, 1969)], into the Oligocene through early Miocene age of voluminous volcanism [for example, the Horse Spring Formation of southeastern Nevada (Tschanz, 1960; Longwell and others, 1965; Armstrong, 1970; Marvin and others, 1970)]. Some local faulting continued between Cretaceous and early Miocene time, but sedimentation resulted in a progressive reduction of local relief that was not renewed by major faulting (Reynolds, 1969).

Most of the Cenozoic sedimentary rocks of the Great Basin are of late Miocene age or younger (Fig. 13-2B); this indicates that the major basins formed after mid-Miocene time. Progressive development of basin-range topography and concurrent basin filling are recorded by thick alluvial deposits (Fig. 13-2C) and numerous fault or fault-line scarps. Facies gradations are abrupt from range-front conglomerates to tuffaceous fine-grained strata; this indicates that they accumulated in basins much like those of today. The physiography and sedimentary rocks, therefore, represent a distinct geologic regime that started abruptly in mid-Miocene time and has since remained active. Where the basin-filling sedimentary rocks have been dated radiometrically or where their fossils can be calibrated to a chronology, it is clear that the large separate basins were well defined by at least about 13 m.y. ago (Robinson and others, 1968; Stewart, this volume).

It is difficult to document the age of initial basin-range faulting in detail, especially since faulting has continued virtually to the present throughout much of the region. There is, however, widespread structural evidence that basin-range topography did not exist as a regional landform prior to about 17 m.y. ago (McKee and others, 1970; McKee, 1971; Stewart, 1971; Christiansen and Lipman, 1972; Noble, 1972; McKee and Noble, 1974). Numerous sheetlike welded ash-flow tuffs of early Miocene age (25 to 20 m.y. old) in central Nevada indicate that there was little topographic relief comparable to present-day basin-range topography in that part of the province, and the relative scarcity of sedimentary rocks older than about 17 to 18 m.y. is consistent with that evidence.

Estimates of the cumulative tectonic extension across the Great Basin during the past 17 m.y. vary widely. Hamilton and Myers (1966) estimated extension of 100 to 300 km. Thompson and Burke (1974) reviewed convincing evidence for a minimum of 50 to 100 km, roughly a 10% increase in width. Stewart (this volume) reviews published evidence for cumulative extension and shows that most estimates range from less than 10% to about 35%. Although there is abundant evidence for very young basin-range faulting throughout the entire province, a major zone of seismicity occurs along the eastern (Wasatch) edge and a broader, more complex zone occurs along the western (Sierran) side of the province (Fig. 13-3). We suggest that these zones represent the loci of the most active normal faulting in the province as a whole. Stewart (this volume) and Proffett (1977) indicate a greater amount of total extension

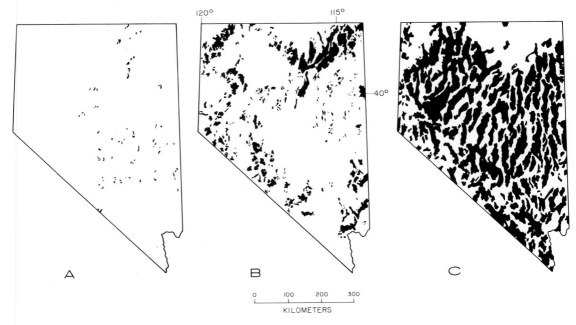

Figure 13-2. Maps showing surface distribution of Tertiary sedimentary rocks and Quaternary sediments in Nevada, taken as an example of the Great Basin region. (A) Sedimentary rocks 40 to 17 m.y. old (before basin-range faulting). (B) Tertiary sedimentary rocks less than 17 m.y. old. (C) Quaternary sediments. (In part after Stewart and Carlson, 1976.)

in the Lahontan and Bonneville basins (near the margins of the Great Basin) than in the center. Similarly, Wright (1976) demonstrated greater cumulative displacement on the western side of the Great Basin than across its central region.

Scott and others (1971) and Scholz and others (1971) suggested that basin-range faulting and extension started first near the central part of the Great Basin and spread progressively toward the present margins throughout later Cenozoic time. However, Noble (1972) gave evidence that extensional faulting began simultaneously over a wide front across the Great Basin. Noble and Slemmons (1975) noted a 12.5-m.y.-old dike emplaced into an older normal fault near the Sierran front and gave evidence that a third of the tilting of the Sierra Nevada occurred before 9.5 m.y. ago. We note further that at least some of the present basins and ranges near the edges of the province were well defined by at least 10 to 14 m.y. ago. For example, the Esmeralda Formation and related units began to accumulate before about 13 m.y. ago in basins near the western edge of the Great Basin (Robinson and others, 1968). Similarly,

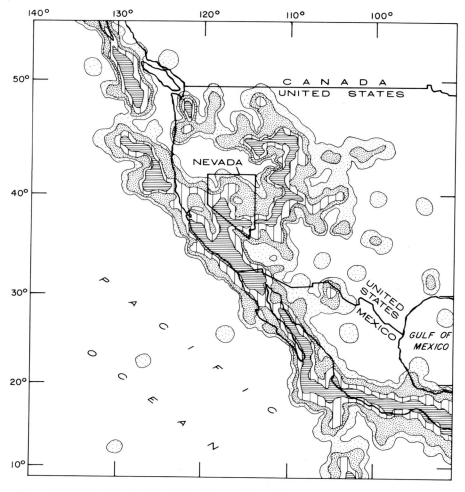

Figure 13-3. Map of seismicity of the Western United States. Based on epicenters for 1961 through 1967 (after Barazangi and Dorman, 1969); contours are 1, 2, 3, and 4 or more epicenters per circular area with radius 1° of latitude.

the Teton Range and the adjacent block-faulted basin of Jackson Hole at the northeasternmost edge of the tectonic province had formed by normal faulting oblique to Laramide structures before initial deposition of the Teewinot Formation about 10 m.y. ago (Love, 1956; Love and others, 1972).

Thus, several lines of evidence point to development of the entire province within a few million years of initial rifting. Although the present low seismicity cannot be regarded as evidence of complete tectonic inactivity in the central Great Basin (Slemmons, 1967), the present zones of greatest seismicity and cumulative extension are concentrated near the margins of the Great Basin. The pattern of tectonic extension appears to have evolved from an early stage of uniform distribution through stages of concentration into successively narrower zones toward the margins.

By analysis of first arrivals from earthquakes in the Great Basin region, Smith and Sbar (1974) and Smith (this volume) show that normal-fault mechanisms predominate and that the directions of minimum principal stress generally lie between northwest-southeast and east-west. Numerous regional geophysical studies have shown that the Basin and Range province is characterized by high heat flow (Sass and others, 1971; Lachenbruch and Sass, 1977), a thin crust and low upper-mantle seismic velocity (Hill, this volume; Smith, this volume), and a low regional gravity field (Eaton and others, this volume).

Near the symmetry axis of the Great Basin is a north-trending zone (the "quiet zone" of Stewart and others, 1977) defined by relatively low aeromagnetic intensities and a distinct lack of the high-frequency anomalies common elsewhere in the region (Fig. 13-5). This zone is especially well defined between lat 38°N and 39°N, where it appears as an aeromagnetic low between regions with a conspicuous high-frequency magnetic signature. Rocks at the surface in the low-anomaly belt are mainly middle and late Cenozoic rhyolitic ash-flow tuffs just as they are in aeromagnetic highs on either side. Thus, the low-anomaly belt probably has a deep origin, and the magnetic properties of the crust have been altered. Possibly, the Curie temperature in this belt lies at a shallower crustal level than in surrounding regions (Stewart and others, 1977; Mabey and others, this volume).

High Lava Plains and Adjacent Areas

There is a geologic transition from features typical of the Great Basin in northern Nevada and Utah to the High Lava Plains of central and eastern Oregon and southern Idaho. The pattern of youthful linear basins and ranges so conspicuous to the south continues through this transitional region, but there is a plateau-like aspect to much of the lava-covered region of central Oregon east of the Cascades.

The High Lava Plains of Oregon form a northwest-trending zone that delimits the northern edge of well-developed basin-range structure and physiography in the western part of the Basin and Range province. These plains are characterized structurally by the Brothers fault zone (Walker, 1974), a zone of northwest-trending en echelon faults of small normal to oblique displacement. No distinct parallel linear basins and ranges occur to the north (Fig. 13-1) although some normal faults occur there (Stewart, this volume, Fig. 1-1). Lawrence (1976), in an analysis of late Cenozoic faulting in Oregon, suggested that several parallel strike-slip fault zones, including the Brothers zone, bound regions of basin-range block faulting. The northern two strike-slip zones, including the Brothers, form the northern edge of basin-range block faulting.

The eastern Snake River Plain trends northeast across southwestern to east-central Idaho (Fig. 13-1). Along the southern margin of the Snake River Plain, prominent basin-range structure and topography are characteristic of the Great Basin region. The plain itself is a lava-covered region of low relief that marks a partial northern boundary to the basin-range tectonic province. One group of parallel block-faulted northwest-trending ranges (the Lemhi, Lost River, and Beaverhead Ranges) lies north of the eastern Snake River Plain, but farther northward these ranges abut the Idaho batholith terrane (Fig. 13-1), and even farther north and northeast the predominant physiographic elements are rejuvenated Mesozoic structural features of the Northern Rocky Mountains (see Pardee, 1950; Robinson, 1960, 1961). Eaton and others (1975) and Mabey and others (this volume) interpret the eastern Snake River Plain to lie along a regional structural feature of the Precambrian basement, marked by a pattern of aeromagnetic anomalies that extends beyond the plain southwestward into northern Nevada and northeastward across the Rocky Mountains and Great Plains to the Canadian border.

The western Snake River Plain is a northwest-trending basalt-filled rift that dates from mid-Miocene time (Mabey, 1976).

Seismicity along the margins of the Great Basin region continues northward nearly to the High Lava Plains; in the east, seismicity continues to the Snake River Plain (Fig. 13-3). A conspicuous east-trending belt of seismicity (the Idaho seismic belt of Smith and Sbar, 1974) marks this northward transitional boundary of the active region north of the Snake River Plain, although relatively minor seismicity continues northward to about the Canadian border. The relatively high heat flow typical of the Great Basin also characterizes the Northern Rocky Mountains (Blackwell, 1969), but the area of highest regional heat flow, like the seismically most active region, extends only a short distance north of the Snake River Plain (Lachenbruch and Sass, 1977).

Columbia Plateau

The northern part of the Columbia Intermontane region is the Columbia Plateau, a vast basaltic province that formed during middle and late Miocene time but is now quiescent. The generally flat topography of the Columbia Plateau, incised by major throughgoing drainages, contrasts markedly with the structure and topography of the Great Basin region. Large young folds characterize parts of the Columbia Plateau, but its gross structure is dominated by the large Pasco basin at its center, formed by regional subsidence during eruption of the Columbia River Group, mainly about 17 to 14 m.y. ago (Swanson and others, 1975; Watkins and Baksi, 1974). The other principal structural element of the Columbia Plateau comprises numerous more-or-less parallel north-northwest–trending basaltic dikes, the feeders for the Columbia River basaltic flows (Taubeneck, 1970; Swanson and others, 1975). These dikes represent aggregate extension of more than 1 km (Taubeneck, 1970), most of it during a short span of the mid-Miocene at about the same time as initial extension and basaltic volcanism in the Great Basin.

Thus, despite the younger folding evident in the Columbia Plateau, an initial tectonic regime of roughly east-west minimum stress was comparable to the generally east-west to northwest-southeast extension of the Great Basin region. Present seismicity of the Columbia Plateau is weak.

Regional Tectonic Pattern

In summary, the regional structure, topography, drainage, and basin deposits of the Great Basin and Columbia Intermontane regions show the following situations: (1) The late Cenozoic structures form a pattern of uplift and roughly east-west to northwest-southeast extension that is superimposed across a heterogeneous early and middle Tertiary postorogenic volcanic terrane. (2) The present pattern of structure and topography began to form about 17 m.y. ago and was essentially fully developed at least by 12 to 14 m.y. ago. (3) After an initial unified stage 17 to 14 m.y. ago, the region became divided into a zone of continuing extensional tectonism in the Great Basin region to the south and a zone of declining activity in the Columbia Plateau to the north; the High Lava Plains of Oregon and Idaho mark a transitional boundary between these two zones. (4) In contrast to the more mature-looking southern basin-range region, the entire Great Basin shows evidence of youthful tectonism. However, the most active seismicity occurs in the southern, western, and eastern parts of the Great Basin region, and the zones of greatest cumulative extension tend to be near its western and eastern margins. Thus, since 12 to 14 m.y. ago, the most active faulting appears to have been increasingly concentrated toward the margins of the Great Basin. (5) The transitional northern boundary zone along the High Lava Plains is generally aseismic (Fig. 13-3); the western branch roughly parallels the Brothers fault zone of probable strike-slip displacement, whereas the eastern branch is the subsiding eastern Snake River Plain.

VOLCANISM

Great Basin Region

Upper Cenozoic volcanic rocks, predominantly basaltic and subordinately rhyolitic, are common in the Great Basin, especially around its margins. The basalts and associated silicic rocks are most voluminous in the northern part of the province (Fig. 13-4; also see Stewart and Carlson, 1976 and this volume). Almost all the Cenozoic basalts of the Great Basin are less than about 17 m.y. old, and most are less than about 12 m.y. (McKee and others, 1970; McKee and Noble, 1974; Snyder and others, 1976).

The oldest of the upper Cenozoic basalts of Nevada are dikes and flows in north-central Nevada that have been dated by K-Ar at various places as about 17 to 14 m.y. old. These basalts form a north-northwest-trending outcrop belt that is broken by younger north-trending tectonic basins. The belt, however, is continuous beneath the basins, as shown by a prominent linear aeromagnetic high extending from central Nevada to the Idaho border (Fig. 13-5). This anomaly has been described in detail by Mabey (1966) and Robinson (1970) and is the southern part of the Oregon-Nevada lineament of Stewart and others (1975), who emphasized a direct relationship between the aeromagnetic anomaly and the mafic dikes. Although outcrops of basalt closely follow the aeromagnetic high (compare Figs. 13-4 and 13-5), analysis of the anomaly indicates that it is not caused by the surface basalts alone. The amplitude of the anomaly and its narrow linear form must also reflect a deeper zone of basaltic rocks (Mabey, 1966). This sharply defined, long (200 km), narrow (~10 km) anomaly of deep origin (10 to 15 km; Robinson, 1970) near the center of the Great Basin suggests a basaltic dike swarm close to the line of symmetry of the province although somewhat to the west. Exposed dikes in the swarm have the same trend as the belt as a whole. In the Roberts Mountains, near the south end of the belt, all the exposed basalts are dikes in a 7-km-wide swarm (Fig. 13-6). Farther northwest, most of the dikes are buried by more widespread basaltic flows, although continued presence of the dikes at depth can be inferred from the continuity of the aeromagnetic high (Fig. 13-5).

The region containing voluminous basalts and associated silicic volcanic rocks 17 to 14 m.y. old widens northward and becomes diffuse as it approaches the Nevada-Idaho-Oregon border area. The axial belt crosses the Owyhee Plateau region (Figs. 13-1, 13-4) as a zone

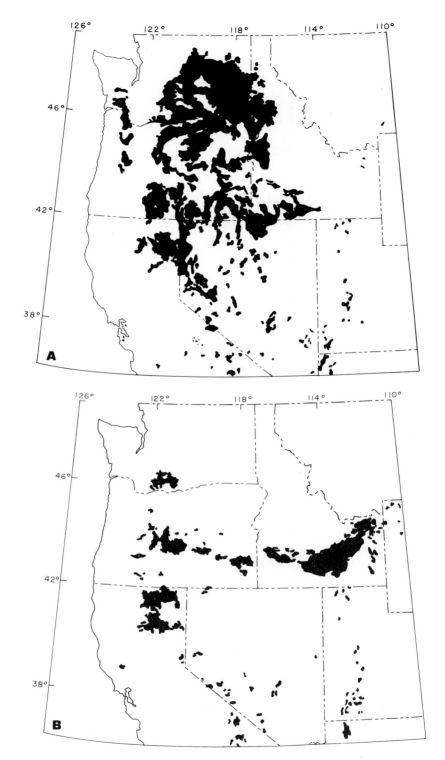

Figure 13-4. Maps showing distribution of Tertiary and Quaternary basalt in the Great Basin and Columbia Intermontane regions. (A) Tertiary basalts less than 17 m.y. old. (B) Quaternary basalts. (In part after Stewart and Carlson, 1976.)

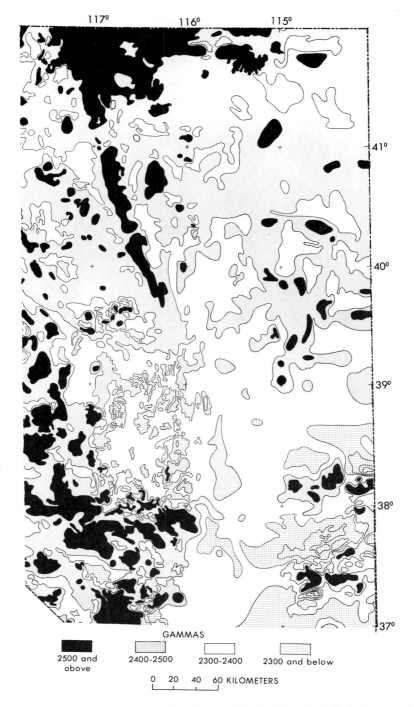

Figure 13-5. Aeromagnetic map of central and northeastern Nevada. Note the similarity between the outcrop pattern of basalts (Fig. 13-4) and many of the aeromagnetic highs, especially the north-trending belt in the central to northern part of the state. (After Stewart and others, 1977.)

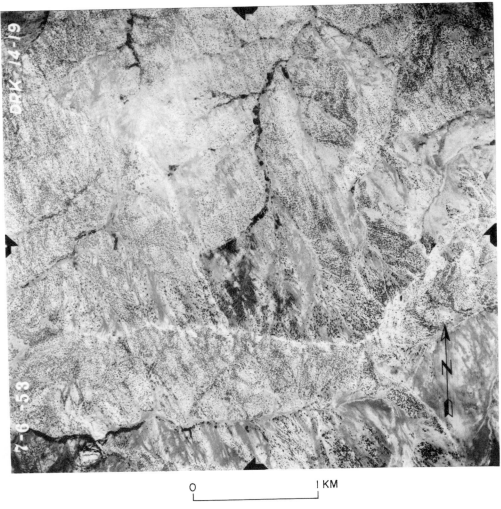

Figure 13-6. Aerial photograph of basaltic dike swarm in the western part of the Roberts Creek Mountains, near the southern end of the linear aeromagnetic high of central Nevada (Fig. 13-5). Lines trending north-northwest across topographic features are nearly vertical dikes.

of poorly defined north-northwest–trending lineaments and basaltic dikes. Basaltic flows of this age are widespread.

Successively younger basalts of the Great Basin region generally occur in successively narrower belts near the margins of the province (Fig. 13-4; also see Armstrong, 1970; Stewart and Carlson, 1976 and this volume). That is, although basaltic volcanism appears to have begun along the axis of the province, by at least about 12 to 14 m.y. ago such volcanism was active over most of the surface of the region all the way to its margins, just as were the basin-range faults. With further evolution of the province and progressive outward concentration of faulting, basaltic volcanism has tended to be excluded from a progressively larger central region and to be concentrated in successively narrower zones around the margins of the Great Basin.

Most basalts of the Great Basin region are alkali-rich (generally K-rich) types (Leeman and Rogers, 1970). Some tholeiitic basalts occur near the province boundaries.

Upper Miocene and younger silicic rocks are less abundant than basaltic rocks in most of the Great Basin, but the volcanic rocks as a whole form a bimodal rhyolite-basalt suite. Rhyolitic volcanic rocks younger than 17 m.y. are particularly abundant in a belt of caldera-related source areas along the southern margin of the region (Noble, 1972), including the Kane Wash caldera (Noble, 1968), the Timber Mountain caldera complex (Christiansen and others, 1977), the Black Mountain caldera (Christiansen and Noble, 1965, 1968; Noble and Christiansen, 1968, 1974), the Long Valley caldera (Bailey and others, 1976), and the Little Walker volcanic center (Noble and others, 1974). Abundant silicic volcanic rocks also are associated with the basaltic rocks of the northern Great Basin (Willden, 1964; McKee and Silberman, 1970; Noble and Parker, 1975; McKee and others, 1976; Stewart and Carlson, 1976 and this volume). Minor rhyolites are associated with basalts in other localities such as the Mineral Range and Cove Fort areas of Utah (Lipman and others, 1975). The ages of these silicic rocks span the same ranges as the basalts with which they are areally associated.

Snake River Plain–Yellowstone Plateau

The Snake River Plain is a predominantly basalt-covered belt across Idaho, but along the margins of the plain are both basaltic and rhyolitic volcanic rocks that mainly predate the rocks of the plain itself. The plain comprises two segments. The northwest-trending western segment consists of younger basalts superimposed across a zone of middle and upper Miocene rhyolitic and basaltic volcanic rocks (Malde and Powers, 1962; McIntyre, 1972; Mabey, 1976). By contrast, stratigraphic and geochronologic relations among volcanic rocks marginal to the eastern segment show that major volcanism along it has propagated from southwest to northeast (Armstrong and others, 1975). This volcanism began at least 14 m.y. ago in the region of southwestern Idaho and has migrated northeastward at a rate of several centimetres per year to form a linear zone (Christiansen and Blank, 1969; Armstrong and others, 1975). The youngest part of this propagating zone is the Yellowstone Plateau, which overlies a very large active rhyolitic magma chamber (Christiansen, 1974; Eaton and others, 1975).

Armstrong and others (1975) interpreted geochronologic data for the eastern Snake River Plain and its margins as showing episodes of contemporaneous volcanism along the plain characterized by a facies relationship, with rhyolitic rocks in the northeast, basaltic rocks to the southwest, and an area of both rhyolitic and basaltic rocks between; as the northeastern facies has migrated northeastward, the southwestern facies has followed and overprinted the older rhyolitic rocks. Christiansen and Blank (1969), Christiansen and Lipman (1972), and Christiansen (ms. in prep.) have emphasized a somewhat different view of this relationship; they have stressed that the fundamental character of the volcanism is basaltic, but that cycles of rhyolitic volcanism mark the eastward propagation of the system. A sequence is repeated in each new area that becomes the head of the volcanic zone as it propagates northeastward along the axis of the eastern Snake River Plain. This cyclic sequence begins with basaltic or contemporaneous basaltic and rhyolitic volcanism of small to moderate volume. This is followed by the evolution of a very large rhyolitic magma chamber that sustains the eruption of rhyolites. The sequence climaxes in voluminous ash-flow eruptions and associated caldera-collapse of the source area and culminates in postcollapse rhyolitic volcanism. The Yellowstone Plateau is now in this stage (Christiansen, 1974). One rhyolitic cycle in each successive area has a duration on the order of 1 to 2 m.y., during which time basaltic volcanism continues around the margins but not within the principal rhyolitic source area. A rhyolitic cycle ends with solidification and fracturing of the large rhyolitic magma chamber; this allows basalts to erupt through the former rhyolitic source area, accompanied only occasionally by small volumes of rhyolite or mixed basalt-rhyolite complexes. Island Park, just west of Yellowstone, is now in this stage (Hamilton, 1965; Christiansen, 1975). Ultimately, as the head of the volcanic zone propagates eastward past an area, the activity in that area reverts to continued basaltic volcanism while the axis of the volcanic zone subsides, as on most of the Snake River Plain.

The basalts of the Snake River Plain–Yellowstone region are predominantly olivine tholeiites (Stone, 1967; Hamilton, 1963, 1965; Tilley and Thompson, 1970; Christiansen, ms in prep.). The younger tholeiites that flood the subsided axis of the Snake River Plain are generally somewhat more potassic and iron-rich than the earlier tholeiites associated more closely with rhyolitic volcanism, now exposed on the plains margins and around Yellowstone.

Oregon High Lava Plains

Basaltic flows are widespread in southeastern Oregon, blanketing large parts of the region and locally, as at Steens Mountain, forming sequences 1,000 m or more thick. More typically, one or a few flows form mesas of 10 km^2 or less. The basaltic vents are not commonly seen, but where observed, they generally form north- or northwest-trending dike swarms. In most of the region, basalts are the youngest volcanic rocks, but they occur throughout the upper Cenozoic sequence as well. The oldest basalts are about 17 to 16 m.y. old and are known from the Steens Mountain area eastward and southeastward toward Idaho and northern Nevada. The Quaternary basalts of central and eastern Oregon (Fig. 13-4B) are restricted to a zone slightly oblique to the trend of the Brothers fault zone, extending from the Newberry volcano to near the Oregon-Idaho border.

Rhyolitic domes, lava flows, and ash flows crop out widely in southeastern Oregon (Fig. 13-7). More than 100 dome complexes have been recognized between Newberry volcano and the Owyhee Plateau in western Idaho (MacLeod and others, 1976). These domes are distributed more or less equally across the region with local concentrations such as near Newberry. Rhyolitic ash-flow tuffs are associated with some of the dome complexes, especially near the Harney basin, about midway between Newberry and the Owyhee Plateau (Greene and others, 1972; Walker, 1973). The combined volume of rhyolite in these ash-flow sheets probably is greater than 300 km^3.

The oldest rhyolitic rocks lie mainly in eastern Oregon, western Idaho, and north-central Nevada and have ages in the 17- to 14-m.y. range. A few occur in central Oregon as well. The younger rhyolitic domes, flows, and tuffs record progressively younger periods of volcanism from southeast to northwest across Oregon, ranging from about 11 m.y. east of the Harney basin to less than 1 m.y. at Newberry volcano (Fig. 13-7; also see MacLeod and others, 1976). The succession of rhyolitic domes defines a broad belt more than 250 km long trending N75°W across southeastern Oregon in which the age progression is well defined and uniform. The rate of propagation calculated by MacLeod and others (1976) is about 1 cm/yr in the western part of the belt, where the domes are <1 to 5 m.y. old, and about 3 cm/yr along the central and eastern parts of the belt, where the domes are 5 to more than 11 m.y. old.

This west-northwest–trending belt of progressively younger rhyolitic rocks virtually mirrors the northeastward rhyolitic progression of the eastern Snake River Plain and Yellowstone, although the eruptive volumes are less in Oregon. The lengths of the belts, rates of propagation of rhyolitic volcanism, and general age spans are similar in the two belts, and the starting area for both is in the Owyhee region near the Idaho-Oregon-Nevada boundary. Geochronologic evidence indicates similar successions of events in the two propagating volcanic systems. In Oregon as in the eastern Snake River Plain and Yellowstone, both basaltic and rhyolitic volcanism occurred at an early stage. They are closely associated now in the very young Newberry volcano. Rhyolitic volcanism occurs in a given area for only about 1 m.y. or less after the initial volcanism, but basaltic activity has continued all along the principal volcanic axes of both systems through Quaternary time.

Columbia Plateau

The Columbia River Group forms by far the largest Cenozoic basalt field in North America. It represents the northern part of the fundamentally basaltic volcanic association of the Great Basin and Columbia Intermontane region. The basaltic flows commonly are relatively thick (20 m or more) and very widespread, some occurring tens or even hundreds of kilometres from their linear fissure vents (Wright and others, 1973; Swanson and others, 1975). That individual flows can be traced so far attests to the subdued topography of the region at the time of eruption. Many flows ponded in the Pasco basin, in the center of the Columbia Plateau; this indicates continued subsidence of the basin during the volcanism. Within a region about 300 km across there is an estimated 200,000 km^3 of erupted basalts. A variety of compositional types is present, but all are tholeiitic (Wright and others, 1973; McDougall, 1976).

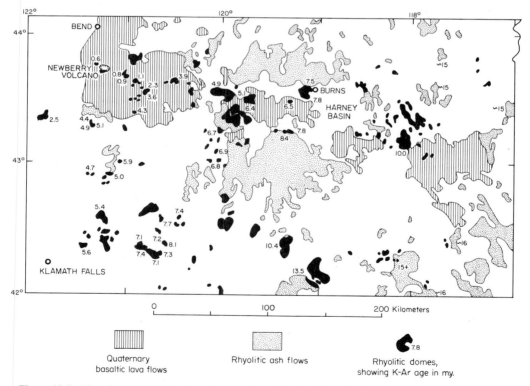

Figure 13-7. Map showing distribution of upper Cenozoic volcanic rocks in south-central Oregon. Rhyolite domes and ash-flow tuffs decrease in age from about 17 m.y. in the southeast to less than 1 m.y. in the northwest. (After MacLeod and others, 1976.)

Known feeder dikes are found only in the central and eastern part of the Columbia Plateau. Most of these dikes occur in the Monument dike swarm in north-central Oregon and the Chief Joseph dike swarm of northeastern Oregon and adjacent parts of Washington and Idaho. Within these large swarms, vents for specific flows are not easily recognized because of subsequent burial by younger flows. However, several vent systems for regionally extensive flows or groups of flows have been recognized; they form linear systems that are tens of kilometres long and a few kilometres wide (for example, see Swanson and others, 1975). These include the Roza vent system near the east edge of the Columbia Plateau and the Ice Harbor system near the center of the plateau. Other linear vents are inferred from flow-outcrop patterns, alignments of cinder and spatter cones, and local accumulations of thin pahoehoe flows. All the known vents and vent systems are parallel, trend roughly north to northwest, and are located east of the center of the Columbia Plateau.

Most of the Columbia River Group has been determined by K-Ar dating to be between about 16 and 14 m.y. old (Watkins and Baksi, 1974). Successively erupted volumes generally became successively smaller although with some exceptions. A relatively small amount of basalt that forms intracanyon flows along the Snake River floods the center of the Pasco basin (Wright and others, 1973; Swanson and others, 1975) and is as young as about 6 m.y. (McKee and others, 1977), but these final basaltic eruptions marked a protracted waning stage of volcanism.

Regional Volcanic Pattern

Cenozoic volcanic rocks less than about 17 m.y. old throughout the Great Basin and Columbia Intermontane region define a marked pattern in time and space. A predominantly basaltic or bimodal rhyolite-basalt suite is the main volcanic assemblage. The association of regional tectonic extension, especially the normal faulting of the Great Basin, with basaltic or bimodal volcanism is comparable to extensional and rift environments elsewhere in the world.

The following summarizes the regional volcanic pattern: (1) The general pattern of late Cenozoic volcanism across the region begins with basalts about 17 to 14 m.y. old, erupted from elongate north- or northwest-trending vent and feeder systems that lie near the central axis of the region; in the Great Basin region, this system is now west of the symmetry axis, whereas it is east of the center of the Columbia Intermontane region. (2) These 17- to 14-m.y.-old basalts occur in increasing amounts from the central Great Basin northward into the Columbia Plateau. The more voluminous basalts, erupted farther north, are more tholeiitic and commonly quartz-normative; basalts of the High Lava Plains are less voluminous olivine tholeiites and high-alumina types; the basalts of the Great Basin are generally more alkalic. (3) By about 14 to 12 m.y. ago, basaltic and bimodal volcanism was occurring across most of the Great Basin region and had begun to decline in the Columbia Plateau region; successively younger basalt and associated rhyolites in the Great Basin were generally restricted to progressively narrower zones around the margins of the earlier, broad, volcanically active region. (4) In a general way, the age of basalts probably correlates locally with the times of most intense basin-range faulting; both began regionally about 17 m.y. ago, and outward restriction of volcanism in the Great Basin appears to have followed outward concentration of extensional faulting. The youngest basalts occur mainly within the seismic belts close to the Sierran and Wasatch fronts. (5) The volcanic systems of the eastern Snake River Plain and southeastern to west-central Oregon define symmetrical but oppositely directed propagating systems; at an early stage as each of these systems propagated into a new area, the system evolved into a major rhyolitic center, represented now by the active Yellowstone Plateau volcanic field at the eastern end and the basaltic and rhyolitic Newberry volcano at the western end. (6) The northeastward-propagating system has produced greater volumes of eruptive rocks than the northwestward-propagating system; the major rhyolitic magmatic systems along the Snake River Plain–Yellowstone axis have been of batholithic size. (7) Since about 14 m.y. ago, the two systems have propagated from their common area of origin at about the same rate that the inner margins of the principal zones of Great Basin basaltic volcanism have retreated outward from the center. Basalts, however, continue to erupt along the axes of both propagating systems.

A MODEL FOR VOLCANO-TECTONIC EVOLUTION

Background

We propose here a qualitative conceptual model for late Cenozoic volcanic and tectonic evolution of the Great Basin region, the Snake River Plain–Yellowstone system, the plateaus of southern Oregon, and the Columbia Plateau. Such an ambitious objective can only be approached in the most tentative manner, yet such models can be worthwhile in helping to focus studies toward an understanding and integration of tectonic and magmatic histories of the region. The principal relations that our model seeks to represent are as follows:

1. The position of the whole region relative to the East Pacific and Juan de Fuca ridge systems, the Cascade arc, the San Andreas fault system, and regional uplift of the Western United States.

2. Generally bilateral symmetry of the Great Basin and Columbia Intermontane regions, the former having an axis of low magnetic intensity and dampened high-frequency aeromagnetic anomalies that may represent especially high temperatures in the crust.

3. Generally high regional heat flow, thin crust, and low upper-mantle seismic velocities, particularly in the Great Basin.

4. A regional tectonic pattern of generally east-west to northwest-southeast extension.

5. A basaltic and bimodal rhyolite-basalt volcanic suite associated in time and space, both regionally and locally, with extensional tectonism.

6. Superposition since about 17 m.y. ago of the extensional tectonic pattern and fundamentally basaltic volcanism across a diverse orogenic and postorogenic predominantly andesitic terrane of late Mesozoic to middle Tertiary age, although pre-existing crustal features as old as Precambrian and as young as Tertiary strongly influence the late Cenozoic structures.

7. The lowest elevations, greatest cumulative extension, and most active seismicity of the Great Basin along its western and eastern margins; the southern margin too is seismically active.

8. Abrupt northward decline in tectonic extension at the High Lava Plains of Oregon and Idaho but a general lack of seismicity along this transitional northern boundary.

9. Earliest basaltic volcanism 17 to 14 m.y. ago along a north-trending belt generally near the regional axis but now displaced westward from the axis of bilateral symmetry in the Great Basin and eastward from the center of the Columbia Plateau.

10. Northward increase in volume rate of eruption 17 to 14 m.y. ago along the north-trending axis, and asymmetry of chemical types along this axis—most tholeiitic in the voluminous basalts of the Columbia Plateau, most alkalic in the less voluminous basalts of the Great Basin.

11. General decline in basaltic volcanism in the Columbia Plateau after about 14 m.y. ago and nearly total lack of such volcanism after about 8 m.y. ago.

12. Since about 14 m.y. ago, progressive restriction of Great Basin basaltic volcanism to successively narrower marginal zones with time apparently coincided with a outward concentration of major extensional faulting.

13. Propagation of rhyolitic volcanism both northwestward and northeastward after about 14 m.y. ago along the High Lava Plains from a starting area across the north-trending axis of earlier volcanism, apparently at about the same rate as the outward concentration of faulting and volcanism toward the margins of the Great Basin.

14. Occurrence of major rhyolitic volcanism in the propagating systems of the High Lava Plains within only about the first 2 m.y. of the initial volcanic pulse in each new area but continuation of basaltic volcanism along both belts of the High Lava Plains in the wake of volcanic propagation.

15. Much greater volume productivity of the Snake River Plain–Yellowstone propagating system than of the Oregon system.

16. Parallelism of the Snake River Plain–Yellowstone axis with two directions of possible tectonic significance—(a) a regional structural feature of the Precambrian basement along this axis that continues beyond it in both directions (Eaton and others, 1975) and (b) the probable direction of motion of the North American plate relative to the asthenosphere if the Hawaiian melting anomaly is assumed to mark the motion of the Pacific plate relative to the asthenosphere (compare with Smith and Sbar, 1974; Shaw and Jackson, 1973; Suppe and others, 1975).

Previous Concepts

Models applied recently to all or parts of the late Cenozoic volcanic and tectonic evolution of the region are considered in some detail in this section. They can be considered in two general categories:

1. Models relating this evolution to contemporary interactions of the North American, Pacific, and Farallon (or its remnants) plates; these have sometimes been called "passive" models.

2. Models based upon an active upwelling beneath the North American lithospheric plate; these models can be further subdivided by the primary mechanism envisioned for upwelling—(a) overriding by the North American plate over the East Pacific Rise, which is pictured as a convective source of energy and spreading motion; (b) buoyant upward displacement of previously subducted lithosphere of the Farallon and Pacific plates, perhaps after cessation of subduction (active "back-arc spreading" or "ensialic marginal basin"); and (c) uplift and rifting over a rising central thermal anomaly ("hot spot," "convection plume," or "chemical plume") or in the wake of one or more "hot spots" that are laterally stable relative to the lower mantle but in motion relative to the North American plate.

In this paper we adopt a plate-interaction model, but first we consider some arguments concerning models that feature primary upwelling beneath the lithosphere. The concept of an overridden East Pacific Rise, once widely popular (Menard, 1964; Cook, 1969; McKee, 1971), is no longer tenable from a plate-tectonics view of the role of spreading mid-oceanic ridges. The ridges are now recognized as results rather than causes of plate divergence; they represent merely thinning of the lithospheric plate and accretion of new lithospheric material at the divergent plate boundary (for example, Le Pichon and others, 1973). The East Pacific Rise has not been subducted and overridden by the North American plate. Rather, the geometrical and physical constraints of relative motion when the Pacific and American plates came in direct contact caused cessation of the processes that were active at both the rise and the subduction zone. Both ceased to exist as functional entities and were replaced by two triple junctions and a transform fault system (McKenzie and Morgan, 1969; Atwater, 1970).

The concept of the Great Basin as an ensialic marginal basin is currently popular, but several arguments persuade us that it is less attractive than a plate-interaction model. One of the original suggestions for such an active "back-arc spreading" model came from Scholtz and others (1971) on the basis of a concept advanced by Armstrong and others (1969) that Cenozoic volcanism in the Great Basin began in a core area about 40 m.y. ago and has since migrated outward toward the margins. That history, however, especially before 17 m.y. ago, has been shown to be incorrect (see McKee and others, 1976; Stewart, this volume; Stewart and others, 1977). Scholtz and others (1971) further suggested that block faulting moved outward in a similar manner, but we have cited evidence earlier in this paper that major faulting near the present margins of the Great Basin is at least 10 to 14 m.y. old. Also, the region of the proposed ensialic marginal basin, as several papers in this symposium volume emphasize, is a region mainly of uplift, not of subsidence like the supposedly analogous features of the western Pacific (Karig, 1971). By comparison, subduction along the margin of South America has continued beneath a continental plate longer than it did beneath North America but has not resulted in a major episode of back-arc spreading; thus, such a process is not, as sometimes proposed, an inevitable consequence of longevity in continental-margin subduction systems. Triggering of back-arc spreading by the cessation of compressive subduction (as suggested by Scholtz and others, 1971) seems problematical because of the relative timing of these events. Regional extension and basaltic and bimodal volcanism did begin in the southern part of the Basin and Range province and Southern Rocky Mountains at the time of cessation of subduction, but this tectonic and volcanic activity migrated with time as subduction ceased over a widening region (Christiansen and Lipman, 1972). However, similar processes affected the entire Great Basin and Columbia Intermontane regions simultaneously while subduction continued in a shrinking zone to the west (Christiansen and Lipman, 1972; Snyder and others, 1976).

If there were one part of this entire volcano-tectonic system that was in some ways analogous to an oceanic marginal basin, it would be the Columbia Plateau, which was a region of extension, voluminous tholeiitic volcanism, and subsidence for a period of several million years. Elsewhere the concept does not fit well.

"Hot spot" models are too diverse to discuss succinctly in general. However, the most commonly held forms of those models argue that a melting anomaly at Yellowstone (or at Yellowstone and at Raton in northeastern New Mexico) represents a convection plume from the deep mantle (Morgan, 1972; Matthews and Anderson, 1973; Smith and Sbar, 1974; Suppe and others, 1975). One of us (Christiansen, 1973, and ms. in prep.) has shown that such a model does not explain well the geologic, petrologic, or regional tectonic data for the Yellowstone–Snake River Plain region. In particular, (1) the mantle-plume model does not account for spontaneous origin of the melting anomaly between 17 and 14 m.y. ago at the same time as the inception of regional extension; (2) it does not explain the occurrence of volcanism not merely as a propagating "hot spot" but also as continuing basaltic activity for at least 14 m.y. along the trace; (3) it does not account for the simultaneous and symmetrical northeastward and northwestward propagation of rhyolitic volcanism along the eastern Snake River Plain and across central Oregon; and (4) although the eastern Snake River Plain correlates in orientation with the kinematic prediction of the mantle-plume model, the orientation also correlates with the prediction of the shear-melting concept and with an ancient structural feature of the crust.

Another variant model places a broader "hot spot" at the junction between the two propagating volcanic axes of the High Lava Plains and the central axis of the Great Basin; this model derives the two axes of volcanic propagation, the Great Basin, and the Columbia Plateau as tensional rifts across a resulting uplift (Prostka and others, 1976). In our view, such a model fails to relate closely either to (1) the Great Basin as a wide region of extension rather than as a linear rift or (2) the decline of the voluminous early basaltic volcanism in the Columbia Plateau just when the rhyolitic systems of the High Lava Plains began to propagate. Furthermore, since such a system could not reflect a deep-mantle convection plume unless that plume were migrating synchronously with the North American plate, Prostka and others (1976) proposed that an anomalous concentration of radiogenic heat-producing elements in the lithosphere drives the upwelling. That concentration presumably was the result of a "chemical plume" from the deep mantle before or during formation of the lithospheric plate. Two points argue against this concept, in our minds. First, there is no record of this chemical anomaly before 17 m.y. ago in any pattern that is coherent with the effects ascribed to it since. Second, this feature, presumably derived from some ancient deep upwelling, is centered in an area where regional geology and geophysics show that the lithosphere has grown in several accretionary events. East of the central zone of the proposed chemical plume—in the Yellowstone region and its surroundings—is an area of very old Precambrian crust and upper mantle; younger Precambrian basement underlies areas farther northwest in Montana and eastern Washington; farther west and southwest the lithosphere formed during Mesozoic arc-type magmatism and tectonism; the Columbia Plateau on the north side of the proposed chemical anomaly probably was oceanic lithosphere in Mesozoic time.

Basis of a Plate-Interaction Model

The foregoing arguments help to persuade us against any of the specific models proposed for dynamic upwelling from the asthenosphere or the deep mantle; our underlying reason for preferring a plate-interaction model, however, is that it can be viewed as a relatively simple, even inevitable consequence of the tectonic interactions between moving lithospheric plates. In this vein, we accept in their most general form the concepts of Atwater (1970) that the region of late Cenozoic tectonic extension is a diffuse zone of transform displacement between the Pacific plate and the tectonically stable part of the North American plate. Our model builds on Atwater's view that a region of oblique extension, having components both of rifting and of lateral shear, has resulted as a migrating triple junction caused the previously intervening Farallon plate to cease interacting with the Pacific and North American plates in a progressively widening zone. We also follow the general concepts of Christiansen and Lipman (1972) that "fundamentally basaltic" magma generation in this region of tectonic extension is a consequence of that extension and that the Mesozoic and Cenozoic thermal and stress histories of each part of the region determine just how that part responds to superposition of the extensional stress system.

We wish to emphasize that regional extension in such a model does not require that shear stresses generated at the San Andreas fault be transmitted across the entire region, as some critics of plate-interaction models have supposed. In most plate-tectonics models, regardless of the motive forces called upon to drive the plates, it is presumed that each plate is in some state of stress that characterizes it throughout. Although the stress field within a plate may vary in space and time and although there are pre-existing anisotropies in the continental lithosphere, each part of each plate reflects in some way the stresses that are imposed upon it by the motions of the plates and by their mutual interactions. The stresses are resolved at the plate boundaries in ways that are consistent with geometric and physical constraints of the relative plate motions. When the Pacific and North American plates began to interact, probably about 29 m.y. ago (Atwater and Molnar, 1973), the stresses along their mutual boundary began to reflect both the narrow zone of contact and the continuing interactions with the two large remnants of the Farallon plate, which was being subducted. At present, there is a wide zone of Pacific–North American plate interaction and a very narrow zone of weak interaction between the North American plate and the Juan de Fuca plate (the northern remnant of the Farallon plate). The temporal change between these conditions resulted in an altered stress field within the North American plate. The new stress field causes the regional extension shown by the course of late Cenozoic tectonic and magmatic evolution of the Cordilleran region.

The direct result of the extensional tectonic regime in the past 25 m.y. or so has been a chain of heating events. A few million years probably were required for thermal equilibration of the plates after initial contact of the Pacific and North American plates and local cessation of plate-margin spreading and subduction (Atwater, 1970). With the change from subduction to tectonic extension, stress at the base of the North American lithosphere is reduced; this facilitates partial melting of upper-mantle peridotites that were initially part of the lithosphere, and basaltic magmas are produced. In this model, a relatively small percentage of partial melting at considerable depths produces the alkali-rich basalts that characterize much of the region. Generally only at the boundaries between tectonic provinces has there been greater extension, locally enough to produce higher percentages of partial melting, perhaps at shallower depths in the upper mantle, to generate tholeiitic magmas. The basaltic magmas tend to rise buoyantly into and through the crust where favorable structures or particularly large deviatoric stresses accommodate this rise. Thus, the magmas erupted mainly in areas where extensional faulting was active simultaneously. Elsewhere these magmas would have tended to intrude the crust and to reside there for long periods. The presence of basaltic melts in the upper mantle and crust—both passing locally upward through the crust and residing in the crust over much of the region—has increased the regional heat flow, lowered the rigidity of the lithosphere, caused further extension, thinned the crust, produced thermal expansion and regional uplift, caused isostatically compensating mantle flowage, and lowered seismic velocities in the upper mantle.

The actual course of volcanic and tectonic evolution of each part of the Cordilleran region has varied, depending partly on the tectonic response of that region as conditioned by its preceding stress and thermal history. The zone of extension in the North American plate started at the continental margin of Mexico and southern California about 29 m.y. ago (Atwater and Molnar, 1973) and produced its first notable effects within the basin-range region (Christiansen and Lipman, 1972; Armstrong and Higgins, 1973; Snyder and others, 1976). During the period of 25 to 17 m.y. ago, only those areas that were directly inland from the coastal transform zone and that had undergone previous Laramide compressive deformation and Laramide or early to mid-Tertiary intermediate to silicic volcanism subsequently underwent extensional normal faulting and basaltic volcanism. These were the Southern Rocky Mountains and the southern part of the present Basin and Range province in New Mexico, Arizona, Sonora, and southeastern California; the Colorado Plateau remained unaffected. A pre-existing structural grain from Paleozoic and Mesozoic deformations has forced a pattern of oblique extension in the Great Basin during the past 17 m.y.

The approximate northern boundary of the region of maximum cumulative extension coincides in its eastern portion with an ancient structural alignment. With increasing extension in the Great Basin after about 14 m.y. ago, this alignment and a symmetrical tear to the west appear to have acted as a boundary for that extension. Whereas cumulative extension across the central and northern Great Basin probably is more than 100 km, comparable extension in areas north of the High Lava Plains is generally a few kilometres and, at most, less than a few tens of kilometres. Thus, the High Lava Plains are a transitional transform boundary zone of the Great Basin. In the same way that Atwater's basin-range model represents a "soft" plate margin, our model emphasizes the axis of the High Lava Plains as a "soft" edge to the zone of oblique extension that lies mainly to the south (Fig. 13-8). This transitional boundary is essentially aseismic and is characterized along its entire length, especially in the Snake River Plain, by Quaternary volcanic fields, presumably indicative of high crustal temperatures. The transitional zone continues northward from the eastern Snake River Plain to the Idaho seismic belt.

Interpretation of Late Cenozoic Volcano-Tectonic Evolution

By 17 m.y. ago the remnant volcanic arc along the continental margin was restricted to a narrow belt north of the latitude of southern or south-central Nevada (Fig. 13-8). For a long time before 17 m.y. ago, the eastern Pacific spreading ridge probably had intersected the North American continental margin near Vancouver Island (Atwater, 1970). Therefore, by about 17 m.y. ago, the remnant arc and subduction system was no more than about 2,200 km long and was bracketed at both ends by transform boundaries between the Pacific and North American plates. The stresses that resulted within the North American plate from this changing configuration of relative plate motions and interactions caused the western part of the North American plate to be dominated by the stress field resulting from the growing

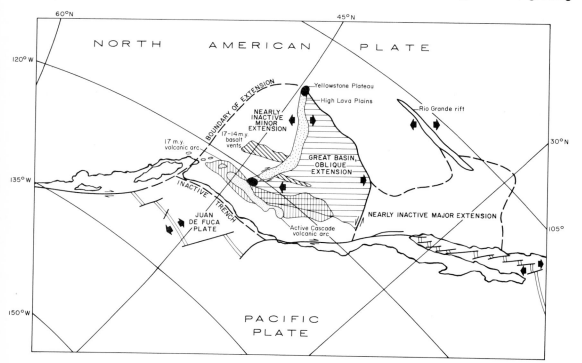

Figure 13-8. Map of the Western United States outlining the late Cenozoic tectonic features and upper Cenozoic volcanic rocks in relation to Pacific–North American plate interaction. Shown on the oblique Mercator projection of Atwater (1970); horizontal lines of the projection are directions of pure transform displacement; vertical lines are directions of pure dilation or compression.

Pacific–North American interaction instead of the declining stress field resulting from the Farallon–North American interaction. Under these conditions, in our model, the previously heated and deformed part of the North American plate did not behave in the ideally rigid manner of plate tectonics but allowed partial coupling between the two separated transform zones. This forced extension across an axis roughly parallel to the continental-margin arc and about 300 to 400 km inland from it (Fig. 13-8). Extension roughly perpendicular to this axis opened a linear rift that was continuous from central Nevada, through the western Snake River Plain, to the eastern Columbia Plateau. Subsequently, this rift axis has been offset by transform displacement at the northern boundary of the Great Basin. Stress relief at the base of the lithosphere along this rift produced the initial basaltic magma generation in the upper mantle. Under that part of the lithosphere having a Precambrian crystalline basement or a tectonically thickened Paleozoic geosynclinal accumulation, slight partial melting at considerable depth produced alkali-rich basalts. Farther north, where the 17- to 14-m.y.-old volcanic axis crossed the Mesozoic plutonic-orogenic belt, larger volumes of melt were produced at shallower levels and less-alkalic basalts, commonly of high-alumina character, were erupted. In the Columbia basin, according to the model, basalts which formed voluminous tholeiitic floods were generated by high percentages of partial melting under a thin crust that had been oceanic in Mesozoic time.

Within a few million years, continued tectonic extension was expressed across the entire region previously heated by early and mid-Tertiary volcanism; regional melting at the base of the lithosphere decreased lithospheric rigidity so that much subsequent deformation took place at considerable depth by thinning of the crust and flowage of the mantle. Under these conditions, much of the basaltic magma accumulated in the crust; this further increased regional heat flow, caused regional uplift, and reduced lithospheric rigidity. Continuously higher temperatures and decreasing rigidity of the lithosphere along the axis of the region resulted in progressive restriction of brittle deformation at upper-crustal levels away from the axial zone. Basaltic eruption occurred mainly in areas that were broken at shallow levels by normal faults. This accounts for the progressive concentration of both faulting (seismicity) and basaltic volcanism toward the Great Basin margins, even though the entire region is under extension and the axial zone, with its low seismicity and aeromagnetic "quiet zone," may even be hotter at depth than the margins.

The approximate northern boundary of the region of maximum cumulative extension (an ancient structural boundary in the east and a symmetrical tear in the west) is a diffuse transform boundary to the region of oblique extension (Fig. 13-8) and localizes the zone of stress relief. The eastern part (the eastern Snake River Plain) is a linear zone of subsidence flanked by uplifts. The western part (the Brothers fault zone and parallel zones) is essentially a right-lateral en echelon system (Lawrence, 1976). Basaltic volcanism along the full boundary zone has tended to narrow the transition with time so that tectonic extension has become nearly inactive in most of the region farther north. Progressive transform displacement along the transitional boundary has tended to displace the 17- to 14-m.y.-old rift axis westward in the Great Basin relative to the vent systems of the Columbia Plateau.

The most voluminous and diverse volcanism has tended to concentrate where the two zones of successively more localized lithospheric extension and stress relief marginal to the Great Basin intersect its northern transform boundary. As the inner edges of these zones have retreated toward the margins, their intersections with the boundary have propagated away from the initial axis of regional extension. At these intersections there has been intensely localized lithospheric extension and a relatively high percentage of melting in the upper mantle to produce significant volumes of tholeiitic magma. Intrusion of this magma into and through the lower crust has caused localized melting of the lower crust. The rhyolitic magmas thus produced in the lower crust are emplaced to shallow levels and sustain cycles of rhyolitic volcanism. Continued tectonic extension to the south and transform displacement along the High Lava Plains result in continued basaltic magmatism, heating of the crust, and an aseismic shadow within the northern part of the border zone of earthquakes around the Great Basin.

By far the most productive system of magma generation marginal to the Great Basin has been the eastern Snake River Plain–Yellowstone system, along which very large volumes of basaltic and rhyolitic magma have erupted. Christiansen (ms. in prep.) suggests that the generation of large volumes of rhyolitic and basaltic magma by the Yellowstone melting anomaly reflects a system of magma generation that has become self-sustaining in much the same way as the Hawaiian melting anomaly. The Yellowstone melting anomaly is not a "hot spot" in the sense of a fundamental thermal anomaly of the mantle; the geologic history of the Yellowstone melting anomaly cannot readily be reconciled with a deep-mantle convection plume. We thus favor a concept like that proposed by Shaw and Jackson (1973) for Hawaii, plus the additional effect of lower-crustal melting in a continental plate. In our view, the ancestor of the Yellowstone and Newberry systems was initiated where the original axis of extension and basaltic magma generation from the Great Basin through the Columbia Plateau intersected the structurally controlled transform northern boundary zone of the Basin and Range province. The melting anomalies have been augmented by concentration of extension and stress relief at this intersection as the central axis was replaced by two zones of progressively more concentrated faulting toward the margins of the Great Basin. The Yellowstone melting anomaly probably has become a self-sustaining system by the accumulated effects, guided by the fortuitously oriented old structural boundary, of (1) shear melting at the base of the lithosphere where relative motion parallels this structure, (2) the onset of a thermal feedback cycle, (3) partial melting in the lower crust as well as the mantle, and (4) development of a deep root, in which a zone of inward flow compensates for downward displacement of a dense unmelted mantle residuum (the "gravitational anchor" of Shaw and Jackson, 1973). This flow replenishes the supply of undepleted mantle for continued basaltic magma generation. The concept of such a root accounts for the seismic observations of Iyer (1975; also 1977, written commun.) that indicate a velocity anomaly to a depth of about 300 km in the mantle beneath Yellowstone. The high-velocity core of the mantle structure that Hadley and others (1976) interpreted beneath Yellowstone would be at least as consistent with the concept of a dense refractory residuum as it is with their hypothesized "chemical plume" for Yellowstone.

A model for the late Cenozoic volcano-tectonic evolution of the Great Basin and Columbia Intermontane regions may be summarized as follows: (1) The late Cenozoic relative motions of three mutually interacting lithospheric plates formed two separate zones of transform displacement between the Pacific and North American plates. (2) By 17 m.y. ago, the remnant subduction zone and continental-margin arc became so short between these transform zones that the two became partially coupled through a previously stressed and heated region behind the arc. This coupling produced an east-west to northwest-southeast minimum compressive stress within part of the North American plate, and the resulting extension caused a rift to open along an axis 300 to 400 km behind the arc. (3) Because of its complex recent tectonic and magmatic history, the response of the continental plate to the lithospheric stress field evolved by 14 m.y. ago to a pattern of oblique extension; an old structural boundary guided development of the transform northern boundary of this oblique extension. (4) Stress relief at the base of the lithosphere in the extensional region causes partial melting in the upper mantle and regional basaltic volcanism. (5) Continued basaltic magma generation and intrusion of the crust by the basaltic magma, in addition to continued extension, cause increased regional heat flow, progressive thinning of the crust, flowage in the upper mantle, decreasing rigidity of the crust, and thermal expansion with regional uplift; brittle deformation tends to be progressively excluded from the central, hottest part of the region. (6) The consequent progressive concentration of normal faulting toward the Great Basin margins has localized maximum cumulative extension toward the margins and has guided a similar outward narrowing of the zones of most active basaltic and related volcanism. (7) Intersection of these marginal zones with the transform northern boundary of the Great Basin has produced localized zones of intense lithospheric extension, basaltic magmatism, and lower-crustal partial melting to sustain brief cycles of rhyolitic volcanism; the parallelism of part of this transform boundary with the direction of plate motion relative to the asthenosphere has caused the eastern Snake River Plain–Yellowstone system to become a self-sustaining melting anomaly.

ACKNOWLEDGMENTS

We thank the organizers and participants of the 1975 Penrose Conference on Regional Geophysics and Tectonics of the Intermountain West for much stimulating discussion of the main concepts presented here. Helpful reviews of this paper were given J. H. Stewart, L.J.P. Muffler, and R. L. Armstrong. We also appreciate the suggestions, discussions, criticisms, and dissents of R. E. Anderson, M. G. Best, W. Hamilton, E. D. Jackson, P. W. Lipman, N. S. MacLeod, D. C. Noble, and D. A. Swanson.

REFERENCES CITED

Armstrong, R. L., 1970, Geochronology of Tertiary igneous rocks, eastern Basin and Range province, western Utah, eastern Nevada, and vicinity, U.S.A.: Geochim. et Cosmochim. Acta, v. 34, p. 203–232.

Armstrong, R. L., and Higgins, R. E., 1973, K-Ar dating of the beginning of Tertiary volcanism in the Mojave Desert, California: Geol. Soc. America Bull., v. 84, p. 1095–1100.

Armstrong, R. L., Ekren, E. B., McKee, E. H., and Noble, D. C., 1969, Space-time relations of Cenozoic silicic volcanism in the Great Basin of the Western United States: Am. Jour. Sci., v. 267, p. 478–490.

Armstrong, R. L., Leeman, W. P., and Malde, H. E., 1975, K-Ar dating, Quaternary and Neogene volcanic rocks of the Snake River Plain, Idaho: Am. Jour. Sci., v. 275, p. 225–251.

Atwater, Tanya, 1970, Implications of plate tectonics for the Cenozoic tectonic evolution of western North America: Geol. Soc. America Bull., v. 81, p. 3513–3536.

Atwater, Tanya, and Molnar, P. H., 1973, Relative motion of the Pacific and North American plates deduced from sea-floor spreading in the Atlantic, Indian and South Pacific Oceans, in Kovach, R. L., and Nur, Amos, eds., Tectonic prob-

lems of the San Andreas fault system, Conf., Proc.: Stanford Univ. Pubs. Geol. Sci., v. 13, p. 136–149.

Bailey, R. A., Lanphere, M. A., and Dalrymple, G. B., 1976, Volcanism, structure, and geochronology of Long Valley caldera, Mono County, California: Jour. Geophys. Research, v. 81, p. 725–744.

Barazangi, M., and Dorman, J., 1969, World seismicity map of ESSA Coast and Geodetic epicenter data for 1961–1967: Seismol. Soc. America Bull., v. 59, p. 369–380.

Blackwell, D. D., 1969, Heat-flow determinations in the northwestern United States: Jour. Geophys. Research, v. 74, p. 992–1007.

Christiansen, R. L., 1973, Rhyolitic ash flows of the Yellowstone Plateau volcanic field—Wyoming, Idaho, Montana—and generation of the rhyolite-basalt association: Geol. Soc. America Abs. with Programs, v. 5, p. 21–22.

——1974, Quaternary volcanism of the Yellowstone rhyolite plateau region, Wyoming-Idaho-Montana [abs]: EOS (Am. Geophys. Union Trans.), v. 56, p. 1189.

——1975, Origin and geothermal potential of Island Park, eastern Idaho: Geol. Soc. America Abs. with Programs, v. 7, p. 595–596.

Christiansen, R. L., and Blank, H. R., Jr., 1969, Volcanic evolution of the Yellowstone rhyolite plateau and eastern Snake River Plain, U.S.A. [abs.], in Symposium on volcanoes and their roots, volume of abstracts: Oxford, England, Internat. Assoc. Volcanology and Chemistry Earth's Interior, p. 220–221.

Christiansen, R. L., and Lipman, P. W., 1972, Cenozoic volcanism and plate tectonic evolution of the western United States—Pt. II, late Cenozoic: Royal Soc. London Philos. Trans., ser. A, v. 271, p. 249–284.

Christiansen, R. L., and Noble, D. C., 1965, Black Mountain volcanism of southern Nevada [abs.]: Geol. Soc. America Spec. Paper 82, p. 246.

——1968, Geologic map of the Trail Ridge quadrangle, Nye County, Nevada: U.S. Geol. Survey Geol. Quad. Map GQ-774.

Christiansen, R. L., Lipman, P. W., Carr, W. J., Byers, F. M., Jr., Orkild, P. P., and Sargent, K. A., 1977, The Timber Mountain–Oasis Valley caldera complex of southern Nevada: Geol. Soc. America Bull., v. 88, p. 943–959.

Cook, K. L., 1969, Active rift system in the Basin and Range province: Tectonophysics, v. 8, p. 469–511.

Eaton, G. P., Christiansen, R. L., Iyer, H. M., Pitt, A. M., Mabey, D. R., Blank, H. R., Jr., and Gettings, M. E., 1975, Magma beneath Yellowstone National Park: Science, v. 188, p. 787–796.

Eaton, G. P., Wahl, R. R., Prostka, H. J., Mabey, D. R., and Kleinkopf, M. D., 1978, Regional gravity and tectonic patterns: Their relation to late Cenozoic epeirogeny and lateral spreading in the western Cordillera, in Smith, R. B., and Eaton, G. P., eds., Cenozoic tectonics and regional geophysics of the western Cordillera: Geol. Soc. America Mem. 152 (this volume).

Greene, R. C., Walker, G. W., and Corcoran, R. E., 1972, Geologic map of the Burns quadrangle, Oregon: U.S. Geol. Survey Misc. Geol. Inv. Map I-680, scale 1:250,000.

Hadley, D. M., Stewart, G. S., and Ebel, J. E., 1976, Yellowstone: Seismic evidence for a chemical mantle plume: Science, v. 193, p. 1237–1239.

Hamilton, Warren, 1963, Petrology of rhyolite and basalt, northwestern Yellowstone Plateau: U.S. Geol. Survey Prof. Paper 475-C, p. C78–C81.

——1965, Geology and petrogenesis of the Island Park caldera of rhyolite and basalt, eastern Idaho: U.S. Geol. Survey Prof. Paper 504-C, 37 p.

Hamilton, Warren, and Myers, W. B., 1966, Cenozoic tectonics of the Western United States: Rev. Geophysics, v. 4, p. 509–550.

Hill, D. P., 1978, Seismic evidence for the structure and Cenozoic tectonics of the Pacific Coast States, in Smith, R. B., and Eaton, G. P., eds., Cenozoic tectonics and regional geophysics of the western Cordillera: Geol. Soc. America Mem. 152 (this volume).

Iyer, H. M., 1975, Anomalous delays of teleseismic P-waves in Yellowstone National Park: Nature, v. 253, p. 425–427.

Karig, D. E., 1971, Origin and development of marginal basins in the western Pacific: Jour. Geophys. Research, v. 76, p. 2542–2561.

Lachenbruch, A. H., and Sass, J. H., 1977, Heat flow in the United States and the thermal regime of the crust: Am. Geophys. Union Geophys. Mon. 20, p. 626–675.

Lawrence, R. D., 1976, Strike-slip faulting terminates the Basin and Range province in Oregon: Geol. Soc. America Bull., v. 87, p. 846–850.

Leeman, W. P., and Rogers, J.J.W., 1970, Late Cenozoic alkali-olivine basalts of the Basin-Range province, U.S.A.: Contr. Mineralogy and Petrology, v. 25, p. 1–24.

Le Pichon, X., Francheteau, J., and Bonin, J., 1973, Plate tectonics: Amsterdam, Elsevier, 300 p.

Lipman, P. W., Prostka, H. J., and Christiansen, R. L., 1972, Cenozoic volcanism and plate-tectonic evolution of the Western United States—Pt. I, early and middle Cenozoic: Royal Soc. London Philos. Trans., ser. A., v. 271, p. 217–248.

Lipman, P. W., Rowley, P. D., and Pallister, J. S., 1975, Pleistocene rhyolite of the Mineral Range, Utah—Geothermal and archeological significance: Geol. Soc. America Abs. with Programs, v. 7, p. 1173.

Longwell, C. R., Pampayen, E. H., Bowyer, B., and Roberts, R. J., 1965, Geology and mineral deposits of Clark County, Nevada: Nevada Bur. Mines Bull. 62, 218 p.

Love, J. D., 1956, Summary of geologic history of Teton County, Wyoming, during late Cretaceous, Tertiary and Quaternary times: Wyoming Geol. Assoc., Guidebook 11, p. 140–150.

Love, J. D., Reed, J. C., Jr., Christiansen, R. L., 1972, Geologic block diagram and tectonic history of the Teton region, Wyoming-Idaho: U.S. Geol. Survey Misc. Geol. Inv. Map I-730.

Mabey, D. R., 1966, Regional gravity and magnetic anomalies in part of Eureka County, Nevada: Soc. Exploration Geophysicists Mining Geophysics Case Histories, v. 1, p. 77–83.

——1976, Interpretation of a gravity profile across the western Snake River Plain, Idaho: Geology, v. 4, p. 53–55.

Mabey, D. R., Zeitz, I., Eaton, G. P., and Kleinkopf, M. D., 1978, Regional magnetic patterns in part of the Cordillera in the Western United States, *in* Smith, R. B., and Eaton, G. P., eds., Cenozoic tectonics and regional geophysics of the western Cordillera: Geol. Soc. America Mem. 152 (this volume).

MacLeod, N. S., Walker, G. W., and McKee, E. H., 1976, Geothermal significance of eastward increase in age of upper Cenozoic rhyolitic domes in southeastern Oregon, *in* Proc, Second United Nations symposium on the development and use of geothermal resources, Vol. 1; Washington, D.C., U.S. Govt. Printing Office, p. 465–474.

MacNeil, F. S., 1939, Fresh-water invertebrates and land plants of Cretaceous age from Eureka, Nevada: Jour. Paleontology, v. 13, p. 355–360.

Malde, H. E., and Powers, H. A., 1962, Upper Cenozoic stratigraphy of western Snake River Plain, Idaho: Geol. Soc. America Bull., v. 73, p. 1197–1220.

Marvin, R. F., Byers, F. M., Jr., Mehnert, H. H., and Orkild, P. P., 1970, Radiometric ages and stratigraphic sequence of volcanic and plutonic rocks, southern Nye and western Lincoln Counties, Nevada: Geol. Soc. America Bull., v. 81, p. 2657–2676.

Matthews, V., III, and Anderson, C. E., 1973, Yellowstone convection plume and break-up of the Western United States: Nature, v. 243, p. 158–159.

McDougall, I., 1976, Geochemistry and origin of basalt of the Columbia River Group, Oregon and Washington: Geol. Soc. America Bull., v. 87, p. 777–792.

McIntyre, D. H., 1972, Cenozoic geology of the Reynolds Creek experimental watershed, Owyhee County, Idaho: Idaho Bur. Mines and Geology Pamph. 151, 115 p.

McKee, E. H., 1971, Tertiary igneous chronology of the Great Basin of the Western United States—Implications for tectonic models: Geol. Soc. America Bull., v. 82, p. 3497–3502.

McKee, E. H., and Noble, D. C., 1974, Timing of late Cenozoic crustal extension in the Western United States: Geol. Soc. America Abs. with Programs, v. 6, p. 218.

McKee, E. H., and Silberman, M. L., 1970, Geochronology of Tertiary igneous rocks in central Nevada: Geol. Soc. America Bull., v. 81, p. 2317–2328.

McKee, E. H., Noble, D. C., and Silberman, M. L., 1970, Middle Miocene hiatus in volcanic activity in the Great Basin area of the western United States: Earth and Planetary Sci. Letters, v. 8, p. 93–96.

McKee, E. H., Tarshis, A. L., and Marvin, R. K., 1976, Summary of radiometric ages of Tertiary

volcanic and selected plutonic rocks in Nevada—Pt. V: Northeastern Nevada: Isochron/West, no. 16, p. 15–27.

McKee, E. H., Swanson, D. A., and Wright, T. L., 1977, Duration and volume of Columbia River Basalt volcanism, Washington, Oregon, and Idaho: Geol. Soc. America Abs. with Programs, v. 9, p. 463–464.

McKenzie, D. P., and Morgan, W. J., 1969, Evolution of triple junctions: Nature, v. 244, p. 125–133.

Menard, H. W., 1964, Marine geology of the Pacific: New York, McGraw-Hill, 271 p.

Morgan, W. J., 1972, Plate motions and deep mantle convection: Geol. Soc. America Mem. 132, p. 7–22.

Noble, D. C., 1968, Kane Springs Wash volcanic center, Lincoln County, Nevada, *in* Eckel, E. D., ed., Nevada Test Site: Geol. Soc. America Mem. 110, p. 109–116.

——1972, Some observations on the Cenozoic volcano-tectonic evolution of the Great Basin, western United States: Earth and Planetary Sci. Letters, v. 17, p. 142–150.

Noble, D. C., and Christiansen, R. L., 1968, Geologic map of the southwest quarter of the Black Mountain quadrangle, Nye County, Nevada: U.S. Geol. Survey Misc. Geol. Inv. Map I-562.

——1974, Black Mountain volcanic center, *in* Guidebook to the geology of four Tertiary volcanic centers in central Nevada: Nevada Bur. Mines and Geology Rept. 19, p. 27–54.

Noble, D. C., and Parker, D. F., 1975, Peralkaline silicic volcanic rocks of the Western United States: Bull. Volcanol., v. 38, p. 803–827.

Noble, D. C., and Slemmons, D. B., 1975, Timing of Miocene faulting and intermediate volcanism in the central Sierra Nevada and adjacent Great Basin: California Geology, v. 28, p. 105.

Noble, D. C., Slemmons, D. B., Korringa, M. K., Dickinson, W. R., Al-Rawi, Y., and McKee, E. H., 1974, Eureka Valley Tuff, east-central California and adjacent Nevada: Geology, v. 2, p. 139–142.

Nolan, T. B., Merriam, C. W., and Williams, J. S., 1956, The stratigraphic section in the vicinity of Eureka, Nevada: U.S. Geol. Survey Prof. Paper 276, 77 p.

Pardee, J. T., 1950, Late Cenozoic block faulting in western Montana: Geol. Soc. America Bull., v. 61, p. 359–406.

Proffett, J. M., Jr., 1977, Cenozoic geology of the Yerington District, Nevada, and implications for the nature and origin of Basin and Range faulting: Geol. Soc. America Bull., v. 88, p. 247–266.

Prostka, H. J., Eaton, G. P., and Oriel, S. S., 1976, Cordilleran thermotectonic anomaly: II. Interaction of an intraplate chemical plume mass and related mantle diapir: Geol. Soc. America Abs. with Programs, v. 8, p. 1054–1055.

Reynolds, M. W., 1969, Stratigraphy and structural geology of the Titus and Titanothere Canyons area, Death Valley, California [Ph.D. dissert.]: Berkeley, Univ. California, 310 p.

Robinson, E. S., 1970, Relations between geological structure and aeromagnetic anomalies in central Nevada: Geol. Soc. America Bull., v. 81, p. 2045–2060.

Robinson, G. D., 1960, Middle Tertiary unconformity in southwestern Montana: U.S. Geol. Survey Prof. Paper 400-B, p. B227–B228.

——1961, Origin and development of the Three Forks basin, Montana: Geol. Soc. America Bull., v. 72, p. 1003–1014.

Robinson, P. T., McKee, E. H., and Moiola, R. J., 1968, Cenozoic volcanism and sedimentation, Silver Peak region, western Nevada and adjacent California, in Coats, R. R., and others, eds., Studies in Volcanology (Williams volume): Geol. Soc. America Mem. 116, p. 577–612.

Sass, J. H., Lachenbruch, A. H., Munroe, R. J., Greene, G. W., and Moses, T. H., Jr., 1971, Heat flow in the Western United States: Jour. Geophys. Research, v. 76, p. 6376–6413.

Scholz, C. H., Barazangi, M., and Sbar, M. L., 1971, Late Cenozoic evolution of the Great Basin, western United States, as an ensialic interarc basin: Geol. Soc. America Bull., v. 82, p. 2979–2990.

Scott, R. B., Nesbitt, R. W., Dasch, E. J., and Armstrong, R. L., 1971, A strontium isotope evolution model for Cenozoic magma genesis, eastern Great Basin, U.S.A.: Bull. Volcanol., v. 35, p. 1–26.

Shaw, H. R., and Jackson, E. D., 1973, Linear island chains in the Pacific: Result of thermal plumes or gravitational anchors?: Jour. Geophys. Research, v. 78, p. 8634–8652.

Slemmons, D. B., 1967, Pliocene and Quaternary crustal movements of the Basin and Range province, U.S.A.: Jour. Geoscience, v. 10, p. 91–103.

Smith, R. B., 1978, Seismicity, crustal structure, and intraplate tectonics of the interior of the western Cordillera, in Smith, R. B., and Eaton, G. P., eds., Cenozoic tectonics and regional geophysics of the western Cordillera: Geol. Soc. America Mem. 152 (this volume).

Smith, R. B., and Sbar, M. L., 1974, Contemporary tectonics and seismicity of the western United States with emphasis on the Intermountain seismic belt: Geol. Soc. America Bull., v. 85, p. 1205–1218.

Snyder, W. S., Dickinson, W. R., and Silberman,

M. L., 1976, Tectonic implications of space-time patterns of Cenozoic magmatism in the Western United States: Earth and Planetary Sci. Letters, v. 32, p. 91–106.

Stewart, J. H., 1971, Basin and Range structure: A system of horsts and grabens produced by deep-seated extension: Geol. Soc. America Bull., v. 82, p. 1019–1044.

——1978, Basin-range structure in western North America—A review, in Smith, R. B., and Eaton, G. P., eds., Cenozoic tectonics and regional geophysics of the western Cordillera: Geol. Soc. America Mem. 152 (this volume).

Stewart, J. H., and Carlson, J. E., 1976, Cenozoic rocks of Nevada—Four maps and a brief description of distribution, lithology, age, and centers of volcanism: Nevada Bur. Mines and Geology Map Sheet 52.

——1978, Generalized maps showing distribution, lithology, and age of Cenozoic igneous rocks in the Western United States, in Smith, R. B., and Eaton, G. P., eds., Cenozoic tectonics and regional geophysics of the western Cordillera: Geol. Soc. America Mem. 152 (this volume).

Stewart, J. H., Walker, G. W., and Kleinhampl, F. J., 1975, Oregon-Nevada lineament: Geology, v. 3, p. 265–268.

Stewart, J. H., Moore, W., and Zietz, I., 1977, East-west patterns of Cenozoic igneous rocks, aeromagnetic anomalies, and mineral deposits, Nevada and Utah: Geol. Soc. America Bull., v. 88, p. 67–77.

Stock, Chester, and Bode, F. D., 1935, Occurrence of lower Oligocene mammal-bearing beds near Death Valley, California: Natl. Acad. Sci. Proc., v. 21, p. 571–579.

Stone, G. T., 1967, Petrology of upper Cenozoic basalts of the western Snake River Plain, Idaho [Ph.D. dissert.]: Boulder, Univ. Colorado.

Suppe, J., Powell, C., and Berry, R., 1975, Regional topography, seismicity, Quaternary volcanism and the present-day tectonics of the Western United States: Am. Jour. Sci., v. 275-A.

Swanson, D. A., Wright, T. L., and Helz, R. T., 1975, Linear vent systems and estimated rates of magma production and eruption for the Yakima Basalt on the Columbia Plateau: Am. Jour. Sci., v. 275, p. 877–905.

Taubeneck, W. H., 1970, Dikes of Columbia River basalt in northeastern Oregon, western Idaho, and southeastern Washington, in Gilmour, E. H., and Stradling, Dale, eds., Proc. 2nd Columbia River Basalt symposium: Cheney, Eastern Washington State College Press, p. 73–95.

Thompson, G. A., and Burke, D. B., 1974, Regional geophysics of the Basin and Range province: Ann. Rev. Earth and Planetary Sci., v. 2,

p. 213–237.

Thornbury, W. D., 1965, Regional geomorphology of the United States: New York, Wiley, 609 p.

Tilley, C. E., and Thompson, R. N., 1970, Melting and crystallization relations of the Snake River basalts of southern Idaho, U.S.A.: Earth and Planetary Sci. Letters, v. 8, p. 79–92.

Tschanz, C. M., 1960, Regional significance of some lacustrine limestones in Lincoln County, Nevada, recently dated as Miocene: U.S. Geol. Survey Prof. Paper 400-B, p. 293–295.

Walker, G. W., 1973, Preliminary geologic and tectonic maps of Oregon east of the 121st meridian: U.S. Geol. Survey Misc. Field Studies Map MF-495.

——1974, Some implications of late Cenozoic volcanism to geothermal potential of south-central Oregon: Ore Bin, v. 36, p. 109–119.

Watkins, N. D., and Baksi, A. K., 1974, Magnetostratigraphy and oroclinal folding of the Columbia River, Steens, and Owyhee basalts in Oregon, Washington, and Idaho: Am. Jour. Sci., v. 274, p. 148–189.

Willden, R., 1964, Geology and mineral deposits of Humboldt County, Nevada: Nevada Bur. Mines Bull. 59, 154 p.

Winfrey, W. M., Jr., 1960, Stratigraphy, correlation and oil potential of the Sheep Pass Formation, east-central Nevada, in Guidebook to the geology of east-central Nevada: Salt Lake City, Utah, Intermtn. Assoc. Petroleum Geologists, p. 126–133.

Wright, L. A., 1976, Late Cenozoic fault patterns and stress fields in the Great Basin and westward displacement of the Sierra Nevada block: Geology, v. 4, p. 489–494.

Wright, T. L., Grolier, M. J., and Swanson, D. A., 1973, Chemical variation related to the stratigraphy of the Columbia River Basalt: Geol. Soc. America Bull., v. 84, p.371–386.

Manuscript Received by the Society August 15, 1977

Manuscript Accepted September 2, 1977

Printed in U.S.A.

Geological Society of America
Memoir 152

14

Origin of the northern Basin and Range province: Implications from the geology of its eastern boundary

Myron G. Best
W. Kenneth Hamblin
Geology Department
Brigham Young University
Provo, Utah 84602

ABSTRACT

Dynamic crustal processes in the Basin and Range province during the late Cenozoic are characterized by regional uplift, extensional block faulting, and fundamentally basaltic volcanism. The products of these processes are well expressed along the boundary with the western Colorado Plateaus where they provide special insight into the evolution of these two provinces.

In Utah, the uplift and associated displacement on major faults along the western Colorado Plateaus has been proceeding at a vigorous rate (as much as 400 m/m.y.) over the past few million years. In Arizona, on the other hand, tectonism seems to have stagnated since a period of major uplift in the Pliocene, or earlier. The inception of fundamentally basaltic volcanism has migrated from central Arizona northward along the boundary at a rate of about 3 cm/yr since middle Miocene time, and eastward into the Colorado Plateaus at a slower rate, about 1 cm/yr.

Regional features of the Basin and Range province important to any interpretive model include (1) an east-west geologic and geophysical discontinuity at approximately lat 37°N and (2) bilateral symmetry about a north-south axis in topography, gravity, basaltic volcanism, and regional upwarp of flanking Sierra Nevada and Colorado Plateaus.

We believe that the region from the Sierra Nevada to the Colorado Plateaus is a broad crustal upwarp whose central part—the Basin and Range province—has been collapsing by block faulting. Upwarping, lateral east-west extension, faulting, and basaltic volcanism have operated concurrently as a by-product of upwelling asthenospheric mantle. There is evidence that after inception of block faulting in the middle Miocene throughout the Basin and Range province, a resurgent northward-migrating wave of dynamic processes developed. This produced a zone of broad upwarp and extensional faulting, followed by a zone of faulting and basaltic volcanism, and, in the southern wake, a zone of reduced tectonism where erosion and sedimentation are the dominant geologic processes.

The cause of widespread upwelling of the asthenosphere to initiate block faulting is unresolved, whereas the resurgent northward-migrating wave of Basin and Range dynamics may be related to thermal instabilities in the asthenosphere that are associated with a northward-moving hole in the Farallon plate which is descending beneath the southwestern United States.

INTRODUCTION

The origin of the Basin and Range province and adjacent Colorado Plateaus has intrigued geologists for several decades, but many important questions remain unanswered. Even with the recent development of the plate-tectonics concept, it is still difficult to explain many features of these regions and how they are related to the basic elements of global dynamics. Is the Basin and Range province simply a projection of a mid-oceanic rift system beneath the Western United States, or does it overlie an independent isolated mantle upwelling? If the province is a rift system, why is it so wide? How is it related to major plate-tectonics elements along the coast of western North America? These and many other questions are not easily answered, but over the past decade or so geophysical investigations of the Western United States have proliferated, so that now a considerable body of new factual and interpretive data are available to integrate with classical geologic observations.

We believe that in any tectonic synthesis of the Basin and Range province and the Colorado Plateaus it is essential to focus on their major characterizing and defining properties, developed during late Cenozoic time, that set them apart from other segments of the continent, namely, (1) regional uplift, (2) extensional block faulting in the Basin and Range province, and (3) basaltic volcanism concentrated along the margins of the province. Earlier geologic processes may have been prejudicial, but the complexity of the evolution of these two regions throughout recorded geologic history is so great that we can only concentrate on the Cenozoic interval of time here.

In focusing attention on the principal late Cenozoic geologic characteristics of the Basin and Range province and Colorado Plateaus, their mutual boundary in Utah and northwestern Arizona (Fig. 14-1) is especially significant because such features are well developed there. In northwestern Arizona and southern Utah, faults are well expressed by long continuous scarps, and in many areas details of fault-plane features are clearly exposed. A scattered sequence of late Cenozoic basaltic flows extruded over much of the area is particularly important. The flows can be dated radiometrically and thus provide data on absolute strains and strain rates where they are uplifted, tilted, and faulted. In addition, they provide insight into dynamic and thermal processes in the underlying upper mantle from which magma flows were derived as partial-melt extracts.

The exact position of the boundary between the Colorado Plateaus and the Basin and Range province depends upon the particular criteria employed in the definition of the two provinces. The classical physiographic boundary (Fenneman, 1931) is located along the Wasatch-Hurricane fault zone, although block faulting extends eastward some 100 km (Fig. 14-2). Moreover, magnetic properties and seismic refraction studies suggest that a boundary based on characteristics of the deep crust lies approximately 50 km east of the Fenneman line (Shuey and others, 1973).

In our geologic investigations the physiographic boundary defined by Fenneman seems most useful, although we recognize the structural boundary may be transitional.

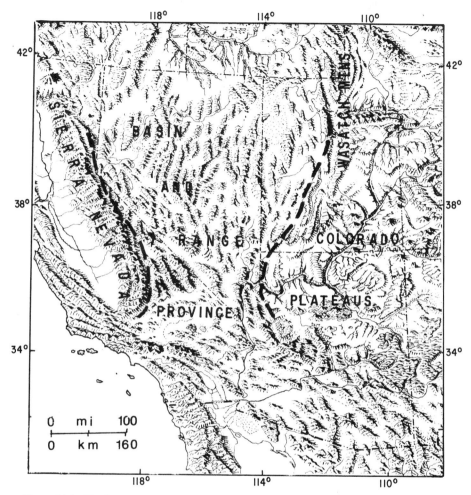

Figure 14-1. Physiographic provinces of the southwestern United States (base map of E. Raisz).

In this paper we summarize pertinent geologic data obtained to date in our continuing study of the eastern boundary of the Basin and Range province. These data, together with information that we perceive as important from other areas, will be used to formulate a tentative tectonic model for the part of the Basin and Range province lying mostly between lat 35° to 40°N. (Much of the area will be discussed in this paper corresponds to the Great Basin. However, as we will consider geophysical and geologic features independent of the surface drainage on which basis the Great Basin was originally defined, the term Basin and Range province is the preferred term.)

GEOLOGIC CHARACTERISTICS OF THE EASTERN BOUNDARY OF THE BASIN AND RANGE PROVINCE

From the vicinity of the Colorado River in Arizona to central Utah, the boundary between the Colorado Plateaus and the Basin and Range province, hereafter referred to simply as

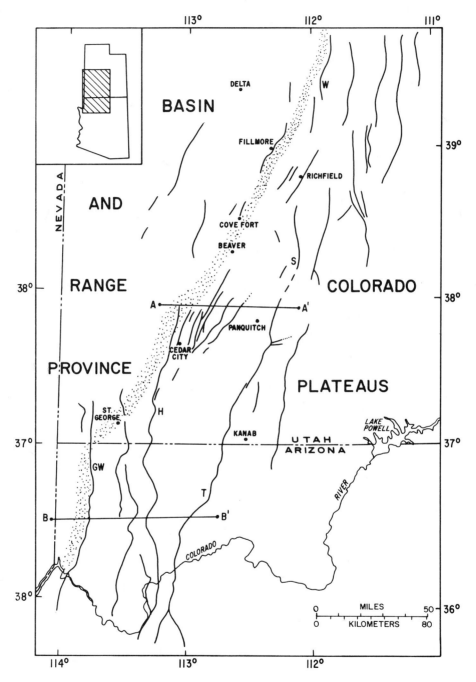

Figure 14-2. Major extensional block faults along the western Colorado Plateaus–Basin and Range boundary showing sinuous trends and their occurrence east of conventional physiographic boundary (stippled). For faults in the Basin and Range province refer to Stewart (this volume). AA′ and BB′ are positions of cross sections in Figure 3. W, Wasatch fault; GW, Grand Wash fault; T,S, Toroweap-Sevier fault; H, Hurricane fault. Data from state geologic maps and our own work.

"the boundary," is characterized by three major late Cenozoic processes. These are (1) block faulting, (2) a regional uplift and tilt of the strata to the northeast, and (3) basaltic volcanism.

Faulting

Our observations support other studies (for example, Stewart, this volume) that view Basin and Range faulting as a complex system of variably tilted horsts and grabens formed in response to roughly east-west crustal extension.

Along the boundary north of the Colorado River the major faults (Fig. 14-2) display properties similar to faults to the west within the Basin and Range province. Their general strike is slightly east of north and only slightly oblique to the more northeastward-trending physiographic boundary of the Colorado Plateaus. On almost all scales of observation from a small-scale regional map down to the outcrop, the faults have sinuous or arcuate traces or they may be en echelon in which the dominant offset migrates to east or west. This aspect produces ramp structures with scissorlike displacement shown in a block diagram in Stewart (1971, Fig. 6). Dips on faults between the Colorado River and the Utah State line are generally steeply westward with the west block down (Fig. 14-3B). There is slight eastward tilting of strata, especially very near the fault plane where accentuated by reverse drag (Hamblin, 1965). Northward into Utah the pattern of faulting along the western margin of the Colorado Plateaus is not a simple stair-step down to the west but consists of alternating horsts and grabens more like the situation in the Basin and Range province (Fig. 14-3A). Wherever actual fault surfaces are observable, slickensides indicate movements within 5° to 10° of the true dip direction. In contrast to faulting along the western boundary of the Basin and Range province, we have found no unequivocal evidence of substantial strike-slip movement.

A significant number of faults are manifest within broad alluvium-covered grabens along the boundary. Although small faults of a few metres displacement are commonly discernible in alluvium, offsets in resistant lava flows are better expressed and show properties similar to major faults with several hundred metres of displacement. In the Cove Fort and southern Black Rock Desert volcanic fields that lie within a broad graben, Hoover (1974) and Clark (1976) found more than 300 faults in lava flows that are mostly less than 1 m.y. old, with displacements up to 100 m (Fig. 14-4). Ninety-eight percent of the faults are normal and have no demonstrable stirke-slip component. A thrust component is only locally evident on some ramp faults. Clark observed actual fault planes with dips from 55° to 70° and found indirect evidence for a similar range of dips for most faults cutting lava flows. Along a typical east-west path, there is slightly less than 1% of horizontal extensional strain, assuming a uniform 60° dip angle on all the faults and using measured dip separations of the offset lava surface for faults along the path line.

Recurrent movement along most of the major fault systems in the boundary area is documented by displaced basalt flows that have crossed the fault line at various times. Fault scarps in alluvial fans produced by recurrent movement on these major normal faults substantiate the continued tectonic activity of the area. We are currently determining ages of lava flows offset by faults in an effort to elucidate strain rates and recurrence intervals in a spatial context.

The time of inception of faulting along the boundary is not well known but can be approximated in two ways. In southwestern Utah stratigraphic relations of Miocene ash-flow tuffs indicate that major boundary faults along the west side of the Colorado Plateaus might have been in existence approximately 20 m.y. ago (Anderson and others, 1975). Age dates on the middle Miocene Peach Spring Tuff in the vicinity of the Grand Wash fault just south of the Colorado River suggest initial movement about 18 m.y. B.P. (Young and Brennen, 1974).

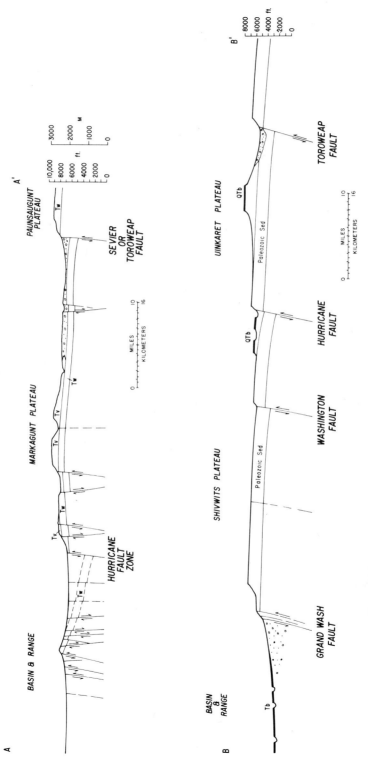

Figure 14-3. East-west structural sections across the western margin of the Colorado Plateaus in southern Utah (AA') and northwestern Arizona (BB'). Refer to Figure 2 for locations. QTb, Tb, basaltic flows; Tv, undifferentiated Oligocene andesitic volcanic fields; Tw, Eocene Wasatch Formation.

We have not studied the faulting along the boundary south of the Colorado River in central Arizona, but available data indicate that the Colorado Plateaus margin there is quite different from that north of the river. Both north and northwest-striking faults occur (Stewart, this volume), but their control on the evolution of the northwest-trending Mogollon Rim, which defines the southwestern margin of the Colorado Plateaus, is uncertain. The Mogollon Rim may owe its origin, at least in part, to northeastward migration of an erosional scarp.

Crustal Upwarp

Upwarp and extension of the crust in the Basin and Range province is implied from a number of geologic phenomena (Stewart, this volume). Additional information concerning the nature of this upwarp is evident along the western margin of the Colorado Plateaus in regionally uplifted and tilted Paleozoic through middle Tertiary strata and in uplifted and dissected lava-capped mesas, buttes, and alluvial fans. A regional tilt to the northeast is expressed physiographically in the vicinity of the Utah-Arizona State line by a series of cuestas known as the Grand Staircase and structurally in a contour map on the top of Permian strata in the area between the Colorado River and the Utah State line (Fig. 14-5). Here, a regional structural slope of approximately 15 m/km northeastward is apparent with more than 1 km of total structural relief. The total structural relief between lat 35° and 38°N along the western boundary of the Colorado Plateaus is in excess of 3 km.

We believe that the tilted western margin of the Colorado Plateaus is a flanking remnant of the upwarped and collapsed Basin and Range province, an idea supported by paleodrainage from the area of the present province toward the northeast. Remnants of this drainage are preserved in numerous localities along the plateaus margin and in the western Grand Canyon (McKee and McKee, 1972; Lucchitta, 1972; Young and Brennan, 1974).

Age of the upwarp that deformed the western margin of the Colorado Plateaus is important in constructing a tectonic model for the region. As Eocene Wasatch (Claron) strata and Oligocene tuffs of the Needles Range Formation are involved in the regional tilting in Utah, it is clear that there the deformation is at least in part post-Oligocene.

Crucial insight into the timing of uplift is provided by the late Cenozoic basaltic lava flows that were extruded in scattered fields throughout much of the plateaus margin during the past 14 m.y. and that have subsequently been uplifted, locally faulted, and eroded. McKee and McKee (1972) obtained K-Ar dates on basaltic cobbles in Tertiary gravels deposited by northeastward-flowing streams and on basaltic flows within the present south-flowing Verde drainage system near Flagstaff, Arizona. Their brackets on the time of major relative uplift of the southwestern margin of the plateaus lie between 10 and 5 m.y. ago.

The basaltic lava flows can be utilized as time markers for uplift in another way. It has been demonstrated that perennial streams in the region have established a profile of quasi-equilibrium and that when uplift or faulting occurs to upset the profile, adjustment to restore a profile of equilibrium occurs within a matter of a few thousands of years (Hamblin and others, 1975). The amount and rate of erosional downcutting of a stream is governed chiefly by the amount and rate of tectonic uplift. Areas of greater uplift suffer greater stream incision. Application of this principle to the inverted and dissected topography associated with the late Cenozoic lava flows of the western Colorado Plateaus margin shows that uplift has been active during the past 14 m.y. In detail, the uplift of the western margin of the Colorado Plateaus has occurred along different boundary faults in different places at different times. In the vicinity of Lake Mead most of the displacement (over 5 km) on the Grand Wash fault occurred 18 to 11 m.y. B. P. (Lucchitta, 1972). North of the Colorado River, displacement

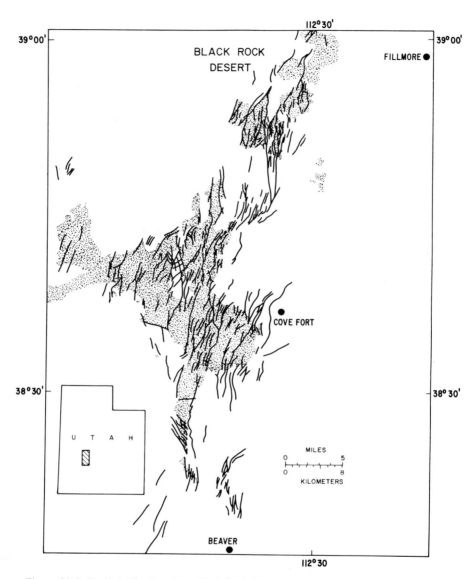

Figure 14-4. Faults in the Cove Fort–Black Rock Desert, graben after Clark (1976) and Hoover (1974). Lava flows, all <1 m.y. old, are stippled.

on the Grand Wash fault diminishes, whereas it increases on the Hurricane fault that lies to the east. Cumulative vertical offset of about 3 km is evident on the Hurricane fault near the town of Hurricane, Utah, 16 km north of the Utah-Arizona line. Several radiometrically dated lava flows (P. D. Damon, E. H. McKee, and M. G. Best, unpub. data) in this vicinity all show about 400 m of displacement, and corresponding uplift of the Colorado Plateaus margin, in the past million years or so. Assuming a steady rate of displacement, the 3 km of total vertical displacement on the Hurricane fault could have accumulated over the past 8 m.y.

These preliminary chronologic data suggest that displacement on the Grand Wash fault

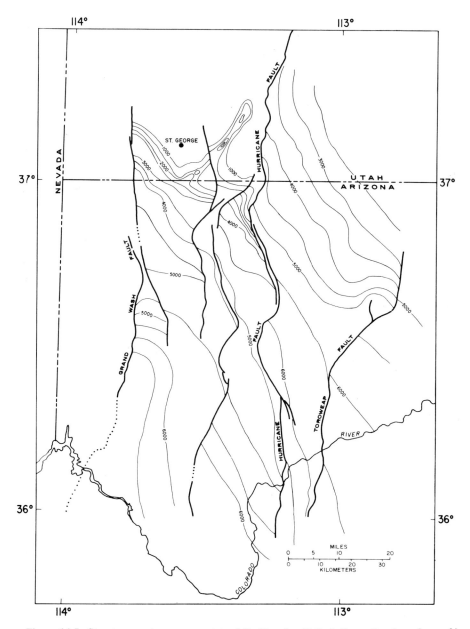

Figure 14-5. Structure contour map on top of the Permian Kaibab Formation (unpub. work) showing regional northeast dip. Contours are in feet above sea level.

and uplift of the plateaus margin south of the Colorado River essentially ceased several million years ago but that faulting and uplift along the margin to the north, mainly along the Hurricane fault, has been very vigorous over the past few million years. This recent activity extends far to the north into the Wasatch fault system where alluvial fans, moraines, and Pleistocene Lake Bonneville sediments show significant offset. Thus, there appears to be northward migration of a *termination* of fault activity and associated uplift.

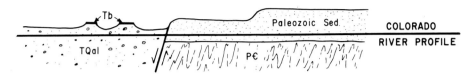

Figure 14-6. Schematic enlargement of west end of section BB' (Fig. 3B) across the Grand Wash graben showing inverted valley development on 7-m.y.-old lava flows (Tb).

Uplift of the basins, as well as the ranges, within the Basin and Range province is evident in dissected pediments, alluvial fans, and lake beds (Hunt and others, 1953, p. 39). Near Mormon Mesa, immediately north of Lake Mead, 200 m of uplift is indicated. In the major graben west of the Grand Wash fault, basalt flows 6.7 m.y. old (P. D. Damon, unpub. data) stand 160 m above the present level of the Colorado River (Fig. 14-6). At Sandy Point on the north shore of Lake Mead, a 3.3-m.y.-old lava flow lies 50 to 100 m above the river grade (Lucchitta, 1972). Although downcutting of these streams could reflect lowering of base level or changes in climatic factors, sediment load, and so forth, the local geologic relations suggest that none of these have exerted a controlling influence on the downcutting and that tectonic uplift is the major process involved (Hamblin and others, 1975).

Volcanism

Late Cenozoic volcanism of a fundamentally basaltic character (Christiansen and Lipman, 1972) is concentrated around the margins of the Basin and Range province (Fig. 14-7) and has been broadly contemporaneous with the uplift and extensional block faulting just described. North of lat 36°N along the boundary, volcanic rocks contemporaneous with extensional faulting occur in relatively isolated fields of generally less than a few tens of flows with associated cinder spatter cones. Large central volcanic complexes as, for example, the San Francisco Peaks complex near Flagstaff, Arizona, do not occur.

The volcanic rocks are chiefly basaltic, with only minor volumes of more silicic rock types represented in Utah (Fig. 14-8). The basaltic rocks themselves define a broad, continuous compositional spectrum including ankaramite, analcime basanite, alkali olivine basalt, and olivine tholeiite. By far the most voluminous type, however, is one with just under 50% SiO_2, 4% ($Na_2O + K_2O$), small concentrations of either hypersthene or nepheline in the norm, and normative plagioclase between 40% to 55% An (called hawaiite by Best and Brimhall, 1974). Less abundant rock types along the boundary are more siliceous andesites, latites (some with $K_2O/Na_2O \sim 1$), and high-silica, high-alkali rhyolites that are separated from the other rocks by a 10% gap in SiO_2.

Volcanic fields within the eastern Basin and Range province tend to be mainly (or wholly) tholeiitic, whereas fields on the Colorado Plateaus, particularly the Uinkaret Plateau (location shown in Figs. 14-2 and 14-3) are more alkalic and undersaturated (Fig. 14-8). However, essentially the same hawaiite, as defined above, is found in all fields. This geographic-compositional pattern is reminiscent of that found by Lipman (for example, Lipman and Mehnert, 1975) in the area of the Rio Grande depression on the east side of the Colorado Plateaus in Colorado and New Mexico. There, tholeiitic rocks are largely confined to the depression, whereas undersaturated alkalic flows occur only on the bordering Colorado Plateaus and High Plains. Saturated to undersaturated hawaiite (called silicic alkalic basalt by Lipman) occurs throughout the region. These patterns possibly reflect differing regimes of magma generation in the upper mantle associated with the contrasting tectonic provinces.

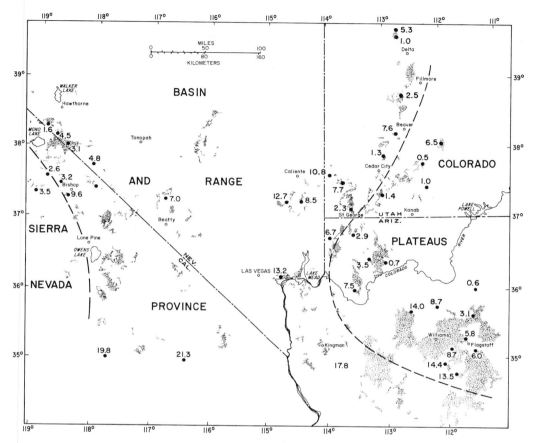

Figure 14-7. Late Cenozoic fundamentally basaltic volcanic fields in the Basin and Range province and adjoining regions. Mafic lava flows of late Cenozoic age around Lake Mead and between Hawthorne and Tonopah, Nevada, are excluded because of calc-alkalic affinities. Some of the volcanic fields stippled south of the Colorado River in Arizona and in California may not be fundamentally basaltic as defined by Christiansen and Lipman (1972) but for lack of data are included. Numbers are available representative K-Ar dates (in million years) showing that the inception of fundamentally basaltic volcanism has migrated northward along the margins of the Basin and Range province and eastward into the Colorado Plateaus. Source of dates as follows: 8.5, Armstrong (1970); 1.3, 5.3, 1.0, R. E. Anderson (unpub. data); 0.7, 2.3, 7.6, 2.5, 6.5, 10.8, 1.0, 0.5, 7.7, 1.4, 12.7, Best and E. H. McKee (unpub. data); 6.7, 2.9, 3.5, P. D. Damon (unpub. data); 7.5, Lucchitta and McKee (1975); 13.2, Anderson and others (1972); 14.0, 14.4, McKee and McKee (1972); 5.8, 6.0, 8.7, 3.1, 8.7, 0.6, Damon and others (1974); 13.5, McKee and Anderson (1971); 17.8, Young and others (1975); 4.8, 4.5, 3.1, 1.6, Silberman and McKee (1972); 2.6, 3.2, 3.5, 9.6, Dalrymple (1973); 19.8, 21.3, Armstrong and Higgins (1973).

For the basaltic rocks in Utah and adjacent Arizona, processes of magmatic differentiation can account for little of the observed spectrum of compositions (Best and Brimhall, 1974) and instead appeal must be made to diverse melting conditions in the region of magma generation in the upper mantle, or to heterogeneities in mantle composition.

Along the western margin of the Colorado Plateaus, relatively highly alkalic undersaturated basaltic rocks, which locally carry mantle-derived peridotitic inclusions (Best, 1974), have been erupted from north-trending swarms of vents on the Uinkaret Plateau and its northward extension just into Utah. In contrast to these undoubted tappings of deep magma sources in the upper mantle, less undersaturated rocks—locally with phenocrysts of plagioclase,

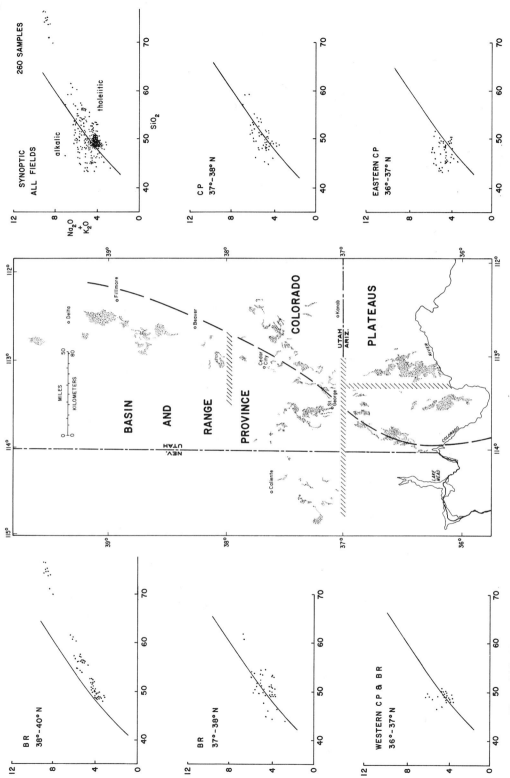

Figure 14-8. Alkali-silica variation diagrams for fundamentally basaltic upper Cenozoic lava flows. Data from Best and Brimhall (1974) and Best (unpub. data). Curved line in each diagram separates the alkalic field from the tholeiitic field (Irvine and Baragar, 1971). On facing page, diagrams on left are for eastern margin of the Basin and Range province; two lower diagrams on right are for Colorado Plateaus. Subdivisions shown on map by rows of slashes.

pyroxene, and olivine which, therefore, had some residence time in the crust—have been erupted from vents trending northeast and northwest. A substantial crustal residence is implied for the late Cenozoic lava flows erupted from mostly randomly scattered vents in the San Francisco volcanic field farther into the Colorado Plateaus about 200 km southwest of the Uinkaret Plateau (Moore and others, 1976). There, the volcanic rocks, though chiefly basaltic, range widely through intermediate to widespread rhyolitic compositions. Abundant xenoliths in basaltic extrusions are preponderantly, if not wholly, of crustal origin.

The time of inception of the fundamentally basaltic volcanism along the boundary of the western Colorado Plateaus and the Basin and Range provinces is important to the tectonic evolution of these two regions. North of about lat 36°N the inception is clearly separated from mid-Tertiary, predominantly intermediate calc-alkalic volcanism, by a hiatus of several million years. Southward in Arizona and Nevada in the Basin and Range province, however, fundamentally basaltic volcanism apparently merges with the intermediate episode, frustrating efforts to pick a wholly reliable specific date of inception (R. E. Anderson, in prep.; Christiansen and Lipman, 1972). Figure 14-7 shows representative K-Ar dates on the inception of fundamentally basaltic volcanism along the boundary (compare Armstrong and Higgins, 1973). The inception of basaltic activity has migrated northward along the boundary from north-central Arizona into Utah at a rate of about 3 cm/yr. At any particular point along the boundary there is an eastward component of migration into the plateaus at a lesser rate, about 1 cm/yr. Once started, basaltic volcanism has continued behind the frontal wave. Where volcanism started earliest, in middle to late Miocene time in central Arizona, the width of cumulative activity perpendicular to the plateaus margin is now broadest. Northward, the band of activity thins somewhat. North of lat 38°N activity has been confined to a narrow north-south corridor along the margin of the Basin and Range province during the Pliocene and Quaternary. An isolated hawaiite field well within the Colorado Plateaus 6.5 to 5.0 m.y. in age is older than would be expected from the age pattern to the south.

REGIONAL FEATURES OF THE NORTHERN BASIN AND RANGE PROVINCE

The East-West Discontinuity in Southern Nevada

Several properties of the Basin and Range province north of lat 37°N set it apart from the area to the south (Eaton, 1975; Stewart and others, 1977), although superficially the two regions appear similar. Eardley (1962, p. 425) summarized the geologic contrasts as follows:

A glance at the Geologic Map of the United States will show that the ranges of southern California and Arizona and southwestern New Mexico are smaller, more irregular in shape, less linear and parallel, and separated by relatively wider basins than those of western Utah and Nevada. Hence, the inclusion of the Sonoran Desert of Arizona in the Basin and Range province from a structural point of view must be made with reservations. The crisp boundaries imparted to ranges by block faulting are generally absent, and if the region is one of extensive block faulting, then the faults are older than those in Utah and Nevada, and erosion has beaten the fault scarps back considerable distances to form broad flanking pediments.

The topographic characteristics referred to by Eardley are easily recognized on a map of regional topography (such as Fig. 3 of Suppe and others, 1975) and in Figure 14-9. Strengthening the contrasts north and south of the lat 37°N discontinuity are the observations that Precambrian rocks older than 800 m.y. are absent to the north (King and Beikman, 1974) and that the character of basin-range faulting differs (Wright, 1976). The east-west discontinuity at lat

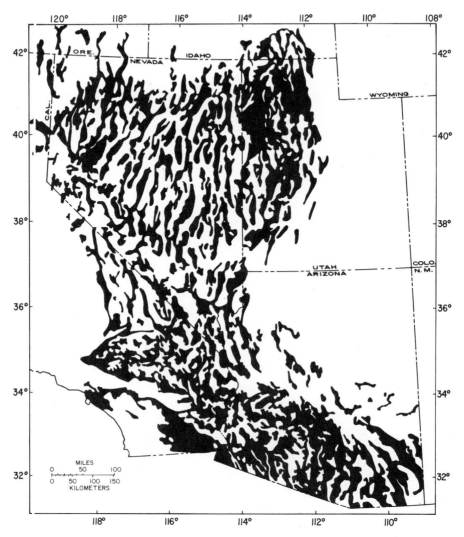

Figure 14-9. Quaternary and Pliocene basin deposits (black) in the Basin and Range province and Colorado Plateaus (King and Beikman, 1974).

37°N is further defined by a belt of seismicity swinging westward from the south end of the Intermountain seismic belt (Smith and Sbar, 1974).

Bilateral Symmetry North of Lat 35°N

In addition to the features just discussed, the northern Basin and Range province also has a strong bilateral symmetry about a north-south axis just east of long 116°W (see maps in Eaton and others, this volume). This symmetry (compare Proffett, 1977), which seems to extend south of the east-west discontinuity at lat 37°N to about lat 35°N, is expressed most dramatically in regional topography and the associated Bouguer gravity. It also appears, to a degree, in the thickness of the lithosphere, in the timing and geometry of bordering

uplifts, and in the pattern of inception of fundamentally basaltic volcanism.

The topographic symmetry of the northern Basin and Range province and flanking regions is a striking feature seen on regional topographic maps (see, for example, Fig. 3 of Suppe and others, 1975). The central area in Nevada is a topographic high in which several of the basins lie in excess of 1,900 m above sea level. In contrast, two large alluvial-lacustrine depressions—the Lahontan and the Bonneville basins—symmetrically located near the northwest and northeast corners of the province are topographically its lowest basins, lying only 1,300 m above sea level. Stewart (this volume) suggests that these two depressions experienced a greater extension, and Eaton and others (this volume) indicate that they have a thinner crust.

Striking bilateral symmetry is also to be expected (if isostatic equilibrium prevails) and is indeed found in the Bouguer gravity map of the northern Basin and Range province and flanking Colorado Plateaus and Sierra Nevada (Eaton, 1976; Eaton and others, Fig. 4, this volume). A complete explanation for all details of the gravity map is not yet possible; nonetheless, the long wave length of the gravity anomalies suggests that lateral variations in density of the lower crust and underlying asthenosphere are involved.

Scholz and others (1971, Fig. 3; compare Thompson and Burke, 1974, Fig. 10) portrayed a greatly thickened asthenosphere and a comparable thinned lithosphere beneath the northern Basin and Range province in Nevada. Keels of thickened lithosphere appear along the eastern and western margins, a configuration that corresponds to other symmetry elements just described.

Basaltic volcanism is concentrated along the east and west margins of the Basin and Range province with a union of the two marginal concentrations in southern Nevada (Fig. 14-7; also Stewart, this volume). Compositional and chronologic relationships along the western margin are complicated by widespread intermediate volcanism, which produced minor amounts of basalt, immediately preceding and merging with the fundamentally basaltic activity (Snyder and others, 1976). Nonetheless, reconnaissance sampling and analyses by Leeman and Rogers (1970) and by us (M. G. Best and W. K. Hamblin, unpub. data) disclose basalt types strikingly similar to those along the eastern margin of the province. Any possible pattern in the inception of fundamentally basaltic volcanism on the west side of the province is obscured by paucity of age dates and lack of compositional data on many lava flows. Our compilation (Fig. 14-7), however, reveals a gross pattern essentially compatible with northward migration of inception similar to that observed along the east margin of the province in Utah and Arizona.

Further symmetry is apparent in the uplift and tilting of flanking abutments of the Basin and Range province. The uplift of the Colorado Plateaus since about the middle Miocene is mirrored in the major uplift and westward tilting of the Sierra Nevada during the same time period (Hay, 1976; Christensen, 1966). The amount of tilt and magnitude of uplift, deduced from paleostream channels, decreases northward from the central Sierra, where it is 2,000 ± 300 m. In the southern Sierra, part of the range was elevated by vertical translation as well as by rotational tilting, and the western margin has been warped and faulted like the eastern margin. The more mature, dissected topography and the greater structural relief on basement rocks in the southern Sierra indicate substantially greater uplift than in the central and northern parts of the range (see Christensen, 1966, Fig. 1; Hay, 1976).

Additional elements of symmetry in the northern Basin and Range have been recognized by Christiansen and McKee (this volume) and Mabey and others (this volume) who point out that two significant late Cenozoic features coincide approximately with the axis of symmetry. The first, in northern Nevada, is a swarm of basaltic andesite dikes dated at approximately 16 m.y. B. P., and the second, in southern Nevada, is a magnetically "quiet zone" in which variations in field strength are much subdued.

Chronology of Late Cenozoic Tectonism in the Northern Basin and Range Province

Block faulting and regional uplift are significant characteristics of the Basin and Range province and certainly provide the key to a synthesis of the tectonics of the region. Two questions are especially important in considering the origin of the province and merit special attention. They are, When did initiation of block faulting and uplift occur? and What has been the subsequent nature and history of displacements? These questions are far from being completely answered, but certain observations seem pertinent.

The inception of major block faulting is perhaps best documented by the development of local basins, internal drainage systems, and resulting sedimentary deposits. Although the early sedimentary record in the basins is largely covered and obscure, there appears to have been widespread proliferation of fluvial-lacustrine sedimentary sequences after 17 m.y. B.P. (Stewart and Carlson, 1976). On the basis of integrated stratigraphic, structural, and radiometric data from many localities, McKee and Noble (1974) concluded that extensional tectonics began throughout most of the Basin and Range province between 21 and 16 m.y. B.P. There is direct evidence for inception of block faulting in several locales throughout the northern Basin and Range province 15 to 17 m.y. B.P. (Noble, 1972). Although we conclude that the major episode of extensional block faulting in the northern Basin and Range province began about 17 m.y. B.P., we recognize that many workers (for example, Loring, 1976) have found evidence of local normal faulting as early as the late Mesozoic.

It is not necessarily true that block faulting since inception has everywhere been at a steady rate or uniform in geometry. Two investigations in separate parts of the northern Basin and Range province, in fact, show just the opposite. Between Mono Lake and Yerington, Nevada, in the western part of the province, Gilbert and Reynolds (1973) documented a period of faulting along northeast and northwest directions before 7.5 m.y. B.P. that was followed by an interval of tectonic quiescence during which a well-developed erosion surface was created. This interval was then succeeded by an episode of northward-striking block faulting and tilting since 4 m.y. B.P. Ekren and others (1968) documented block faulting in southern Nevada along northwest and northeast directions in rocks older than 17 m.y. and north-trending block faulting in volcanic rocks less than 14 m.y. old. They further indicated that much of the present topography likely developed between 11 to 7 m.y. ago. In the Death Valley area, Hunt and Mabey (1966, p. A119) noted that the trend of faulting since late Pliocene time has changed from northwest to north. Dates on lava flows faulted along the eastern escarpment of the Sierra Nevada indicate as much as 1,900 m of displacement in the past 3 m.y. (Christensen, 1966), a rate which, if extrapolated uniformly backward in time, would lead to an implausible amount of displacement since about 17 m.y. B.P.

Although block faulting, uplift, and basaltic volcanism have been concurrently active along the east and west margins of the Basin and Range province for the past few million years, this triad of activity has not always existed there. At the time of inception of block faulting about 17 m.y. B.P., fundamentally basaltic volcanism was occurring only in northern Nevada, Oregon, and southern Idaho (possibly related to activity in the Columbia River Plateau and the Snake River Plain) and in southern Arizona and California. Since then, basaltic volcanism has migrated as a wave from west-central Arizona and southeastern California northward into eastern California, southern Nevada, and Utah (Fig. 14-7).

There are no data to show whether regional uplift occurred throughout the Basin and Range province north of lat 36°N immediately after inception of block faulting about 17 m.y. B.P.

In the central Sierra Nevada, two episodes of post–middle Miocene uplift with an intervening period of relative tectonic quiescence are suggested by geomorphic relationships (Hay, 1976;

Christensen, 1966). The second of the two uplifts occurred after 9 m.y. B.P. (Dalrymple, 1963).

SUMMARY OF OBSERVATIONS

Before proceeding onto interpretive models for development of the northern Basin and Range province, it is advantageous to summarize the late Cenozoic features on which the models are based.

1. Tectonic uplift and northeastward tilting of the western margin of the Colorado Plateaus; uplift of the eastern margin of the Basin and Range province with the basins also rising, but lagging behind the ranges.

2. Extensional block faulting and fundamentally basaltic volcanism *broadly* contemporaneous with uplift along the boundary of these two provinces.

3. Uplift and block faulting appear to have been active generally before Pliocene time (5 to 10 m.y. B.P.) south of about lat 36°N and to have stagnated since then. In contrast, tectonism northward into Utah has been vigorously active up to the present time. Inception of basaltic volcanism has migrated northward and eastward along the western margin of the Colorado Plateaus.

4. A fundamental bilateral symmetry in the Basin and Range province and flanking Sierra Nevada and Colorado Plateaus is evident in regional topography, Bouguer gravity, character and timing of uplift of the Sierra Nevada and Colorado Plateaus, and pattern of migration of basaltic volcanism.

5. A major east-west discontinuity lies near lat 37°N. To the north, ranges are more abundant (Fig. 14-9), more regularly oriented (north-south), have a sharper definition indicative of more recent uplift, regional topography is higher, Bouguer gravity is lower, and seismicity is greater.

6. Since inception of block faulting, apparently simultaneously throughout the Basin and Range province about 17 m.y. B.P., further tectonic development appears to have been resurgent rather than at a steady state (Fig. 14-10).

INTERPRETIVE MODEL FOR THE ORIGIN OF THE BASIN AND RANGE PROVINCE

Upwelling Mantle and Secondary Crustal Upwarp and Extension

The striking bilaterally symmetric properties of the northern Basin and Range province imply a bilaterally symmetric mechanism operative during late Cenozoic time. This important concept, in combination with the observed geologic and geophysical properties (Thompson and Burke, 1974), such as (1) concurrent extensional faulting and regional uplift, (2) high heat flow, (3) thin crust, (4) low-mantle seismic velocity, (5) thickened asthenosphere, (6) high electrical conductivity, and (7) diffuse shallow seismicity, all together strongly support upwelling of anomalously hot asthenospheric mantle as the primary cause of the Basin and Range dynamics. We envision a broad crustal upwarp whose central region—the Basin and Range province—is fragmented and collapsed by block faulting and whose flanking tilted abutments—the Colorado Plateaus and Sierra Nevada—are remnants of the upwarp (Fig. 14-11; Proffett, 1977, Fig. 18).

Atwater (1970) hypothesised that the Basin and Range province is a region of secondary oblique megashear in a "soft" segment of the continent behind the San Andreas transform

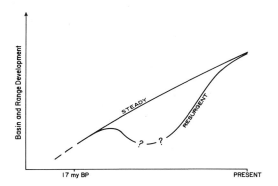

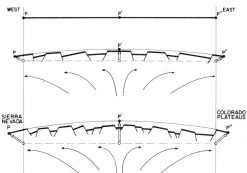

Figure 14-10. Highly schematic contrast in the concepts of steady-state versus resurgent development of the Basin and Range province since about 17 m.y. ago. The time of resurgence of activity is uncertain but may have begun earlier in the southern part of the province.

Figure 14-11. Schematic east-west section across the Basin and Range province showing our concept of progressive regional tumescence with contemporaneous upwarping and horizontal extension. P, P′, and P″ are reference points to show relative amounts of upwarp and extension. Slight curvature of Earth is ignored in the initial configuration at top of diagram.

fault. This mechanism accommodates some problematical excess motion along the continental margin and is kinematically plausible but fails to account for the bilateral symmetry. In addition, the mechanism evades the point that the "soft" property cannot be primary but instead must be a symptom of a more fundamental dynamic process. Late Cenozoic strike-slip faulting does indeed exist within the western and southern parts of the province, but we would interpret it in either of two ways. First, right-lateral faults along the western margin of the province could be associated in some manner with the San Andreas system but superposed upon the more fundamental phenomena of mantle upwelling and consequent crustal uplift and extension affecting the entire province and flanking Colorado Plateaus and Sierra Nevada. Second, these right-lateral faults, together with left-lateral faults widespread in eastern California and western and southern Nevada, might constitute conjugate shears formed in the same extensional setting as the normal faults (Wright, 1976).

The concept of the Basin and Range province as a fault-fragmented upwarp is not new and has been expresssed in one form or another for nearly 90 yr (Wallace, 1975). LeConte (1889, p. 262) wrote: "At the end of the Tertiary the whole region from the Wasatch to the Sierra, inclusive, was lifted by intumescent lava into a great arc, the abutments of which were the Sierra on the one side and the Wasatch on the other. . . . The arch broke down and the broken parts readjusted themselves by gravity into the ridges and valleys of the Basin regions, leaving the raw faces of the abutments overlooking the Basin toward one another." (The term "asthenosphere" should be substituted in lieu of LeConte's "lava"!) A two-stage evolution involving first uparching and then collapse lucidly described by LeConte is, however, inadequate in the face of modern information that supports contemporaneous upwarp and collapse during east-west extension. Most recent workers concerned with the structure of the province (for example, Stewart, 1971, this volume; Thompson, 1972) claimed a *minimum* of 10% east-west crustal extension, which is more than an order of magnitude greater than can occur in a simple, single-stage upwarp of realistic dimensions. Obviously, an additional component besides vertical tumescence alone must be involved and this, we believe, is the drag exerted on the crust by outward flow of the asthenosphere. If a crustal segment is extensionally strained, however great, by vertical uparching alone and then is allowed to collapse, the overall strain is reduced back to zero. Within the Basin and Range province, however, uparching, collapse, and lateral extension must all operate concurrently to produce the minimum of 10% horizontal extension (see the quantitative modeling of Stewart, this volume).

Thompson (1960) argued for concurrent uplift and extensional faulting in his synthesis of the western margin of the province and adjoining Sierra Nevada as a broad topographic swell with superposed ranges and basins. Thompson and Burke (1974) noted that substantial extension of the crust should cause a large isostatic gravity anomaly, but the anomaly in the Basin and Range province is only −10 mgal. They, therefore, postulate considerable "back-flow" in the mantle to compensate.

Relationships in central Nevada near Fish Lake Valley (Ekren and others, 1974, p. 116) imply that, "the block faulting and east and west tilting of the ranges developed as a single process without an initial central uplift or large magnitude . . . the central part of the Great Basin was constantly buoyed up during rifting of the province. . . ."

Northward-Migrating Wave of Basin and Range Dynamics

In additon to the bilaterally symmetric mantle upwelling and consequent basaltic volcanism and crustal upwarping and extensional faulting, we believe there is tantalizing evidence for a north-south component in these Basin and Range dynamics. This component is viewed as a northward-migrating wave of dynamic processes whose chief kinematic basis is the northward-migrating basaltic activity (Figs. 14-7, 14-12). The same hypothesis of northward-migrating Basin and Range dynamics has been advanced by Proffett (1977). The migration in time and space of other dynamic aspects of the wave, namely, upwarp and faulting, may be open to question, although the unified coherence of the whole concept is attractive to us and certain data appear compatible with it. The concept embodies the idea of resurgent Basin and Range development proposed above (Fig. 14-10) in that inception of block faulting throughout the province about 17 m.y. B.P. was subsequently followed by a northward-sweeping triplet of Basin and Range dynamic activity—basaltic volcanism, uplift, and extensional faulting.

We perceive at least the following three zones in the resurgent wave (Fig. 14-12):

Zone I is a region of broad crustal upwarp and extensional block faulting within the northern Basin and Range province. Throughout most of this region faulting is basically along a north-northeast trend, with sinuousity apparent at various scales. A northwest-trending set of faults appears in northwestern Nevada and adjoining parts of California and Oregon (see Stewart, this volume, Fig. 1).

Zone II is also a region of uplift and extensional faulting but with additional fundamentally basaltic volcanism that has been migrating northward in two major prongs along the east and west margins of the Basin and Range province. A secondary component of migration runs eastward into the Colorado Plateaus. Lack of data on basalts in south-central Nevada hinder conception of any pattern of migration there.

Zone III is a region of much reduced tectonism lying in the wake of the more dynamic zones I and II. Both extensional faulting and uplift are much older in this zone, and now the dominant geologic processes appear to be erosion and sedimentation, although fundamentally basaltic volcanism does locally persist.

Our preliminary data and understanding of patterns of uplift and faulting along the western margin of the Colorado Plateaus imply vigorous activity for the past several million years up to the present time in the area north of about lat 37°N. To the south, most of the activity seems to have occurred more than 10 m.y. B.P. and since then at a much reduced tempo. Mountain ranges expose large areas of Precambrian rocks and have been greatly subdued

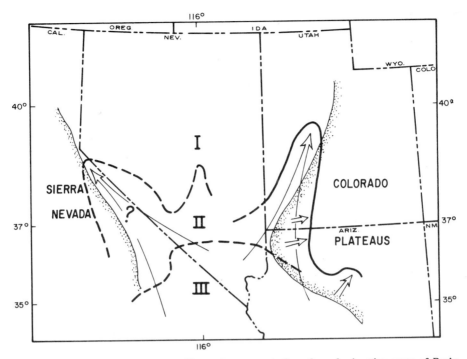

Figure 14-12. Schematic diagram illustrating concept of northward-migrating wave of Basin and Range dynamic processes. Arrows show components of migration of inception of fundamentally basaltic volcanism concentrated along the east and west margins of the province. Such volcanism in south-central Nevada is relatively less voluminous. Dashed lines indicate sparse data control. Zone I, upwarp and extensional block faulting; zone II, upwarp and extensional block faulting and basaltic volcanism; zone III, diminished tectonism, continued volcanism, erosion dominant.

by erosion. Complementary sedimentation in the basins has created expansive bajadas between ranges. Active faulting and uplift ceased long ago.

The east-west discontinuity near lat 37°N previously discussed can be at least partially integrated into the northward-migrating dynamic wave concept. Current seismicity (Smith and Sbar, 1974) follows a U-shaped pattern near the leading edge of the wave of basaltic volcanism (zone II). The break in regional topography, in the erosional development of ranges and basins, and the character of bedrock in the ranges falls between zones II and III.

Wright (1976, Fig. 2) has independently perceived a pattern of late Cenozoic faulting in the Great Basin strikingly similar to and supportive of our tectonic zone model. His field I of modest east-west extension along north-northeast normal faults corresponds closely to our zone I; his field II of greater east-west extension along normal and conjugate strike-slip shear faults corresponds to our zone II.

Marginal Concentration of Late Cenozoic Basaltic Volcanism

The near absence of fundamentally basaltic volcanism less then 17 m.y. in age in the core of the northern Basin and Range province and its concentration along peripheral areas to the west and east is one of the more stubborn problems of Basin and Range dynamics. Following a suggestion of Thompson (1966), Lachenbruch and others (1976) argued that passive basaltic dikes fill tensile fractures in the lower crust and that the anomalous surface heat flux of the Basin and Range province, as well as the local extensional strain rate, is proportional to the upward flux of magma from the immediately underlying partially melted asthenosphere. Although historic seismicity appears to be concentrated along the California-Nevada border and along the Wasatch Front in central Utah, there is clear evidence of late Cenozoic faulting all the way across the province in Nevada. Thus, if basaltic melts invaded tensile fractures in the lower crust, why, during the late Cenozoic, did they reach the surface mostly along the boundaries of the province and not throughout?

The concept of a northward-migrating wave of Basin and Range dynamics (Fig. 14-12) seems to offer no special insight into the problem but does cast it into a different perspective. The marginal concentrations of volcanism thus assume the form of two northward-extending prongs. A third prong is weakly expressed in south-central Nevada. The prongs could reflect zones of major tensile fractures opening in the crust so that mantle-derived magmas can be extruded at the surface. The central third prong and associated zone of tensile fracturing might be localized along the symmetry axis because of the divergent lateral flow of the underlying asthenosphere. But why should open tensile fractures be better developed along margins of the province? Could they represent a more brittle failure of the crust, in contrast to ductile extension of the lower crust in the center of the province directly over the hotter upwelling mantle?

The marginal concentration of volcanism, as well as the eastward component of migration of its inception into the Colorado Plateaus, could reflect some sort of interaction between upwelling, eastward-flowing asthenosphere beneath the Basin and Range province (Fig. 14-11) and a deepened lithospheric keel along the boundary (Best and Brimhall, 1974). A perturbed flow regime in the asthenosphere near the keel would cause additional melting in a thermal "runaway" related to viscous heating (Anderson and Perkins, 1974). This thermal instability might not only create sufficient additional basaltic liquid that could be buoyed to the surface but also, by raising the temperature of the peridotitic keel slightly above its solidus, it would be "eroded" by conversion from rigid solid lithosphere to partially melted asthenosphere capable of flow with the mainstream upwelling. Subcrustal erosion of the keel would displace it eastward and move the zone of enhanced melting and consequent volcanic activity with it.

Lateral, eastward-directed, outflow of asthenosphere against the keel of the southwestward-moving North American plate (Atwater and Molnar, 1973) might lead to an enhanced rate of local extensional strain in the overlying crust along the western Colorado Plateaus boundary. This effect could allow a greater upward flux of basaltic magma through tensile fractures and possibly a greater degree of volcanism at the surface. Such a mechanism might operate independently of the viscous melting phenomena, or more likely in conjunction with it. In any case, it is interesting to note that not only does the upper mantle everywhere beneath the Basin and Range province have an anomalously high electrical conductivity but along its eastern boundary the conductivity is either still greater, or the high conductivity layer is thicker, or both (Gough, 1974). By way of support of the conductivity models, Keller and others (1975) delineated by seismic methods a ridge of low-velocity (7.5 km/s) mantle extending to depths as shallow as 25 km along the boundary. These interpretive geophysical models allow for enhanced melting along the keel or enhanced emplacement of basaltic magma into the lower crust, or both.

In view of the bilateral symmetry of certain properties of the Basin and Range province, it might be anticipated that localization of basaltic volcanism along the west boundary could be similarly related to a lithospheric keel. But any throttling effect owing to westward outflow of asthenosphere against such a keel beneath the eastern Sierra might be less than on the opposite side of the province because of the southwestward movement of the North American plate. Insufficient chronologic data are available on the inception of basaltic volcanism in the western part of the province to test for a westward component of migration. Lachenbruch and others (1976), however, speculated that one possible reason for the very abrupt thermal transition (as manifest in surface heat flux) between the Sierra Nevada and Basin and Range province is a westward encroachment of the Basin and Range thermal regime into the Sierra in the past several million years.

PLATE-TECTONICS RELATIONSHIPS

Two basic first-order problems that have not been answered in our synthesis of Basin and Range dynamics are (1) the cause of upwelling of the asthenosphere and (2) the reason for apparent resurgence of activity in which an early episode of block faulting throughout the province was succeeded by a northward-migrating wave of upwarping, extensional faulting, and fundamentally basaltic volcanism. These problems are now examined in the context of Cenozoic plate systems along the western North American continental border (Fig. 14-13).

As the juncture between the Pacific and Farallon plates impinged upon the continent (North American plate) shortly after 29 m.y. ago at about lat 31°N, differential plate motion was accommodated chiefly along the San Andreas transform fault lying between two triple plate junctions that migrated northward and southward along the continental margin as the North American plate moved relatively southwestward. Reference to Figure 14-13 shows that Basin and Range evolution cannot be linked wholly to drift of the American plate over the East Pacific Rise, as orginally suggested by Menard (1960), because, among other reasons, that would imply initiation of spreading-type activity near lat 31°N and subsequent northeastward encroachment into the continent, whereas extensional block faulting apparently began more or less simultaneously throughout the province about 17 m.y. B.P. Scholz and others (1971), Thompson and Burke (1974), Uyeda and Miyashiro (1974), and Stewart (this volume) have invoked back-arc spreading such as occurs in the marginal basins behind west Pacific volcanic arcs to initiate Basin and Range dynamics. But exactly how this spreading was abruptly and simultaneously triggered throughout the province is quite uncertain. The last compressional tectonism to affect the Basin and Range region was the Cretaceous Sevier orogeny, which terminated about 60 m.y. ago. Some 40 m.y. then elapsed before advent of Basin and Range development. Near the end of this period of apparent tectonic quiescence 35 to 20 m.y. ago, voluminous ash-flow sheets were erupted onto a surface of apparently little topographic relief. Scholz and others (1971) postulated that the inception of the Basin and Range development occurred when the compressive stress system exerted along the continental margin was relaxed as the San Andreas transform accommodated differential movement of the North American and Pacific plates. But as this relaxation would have been initiated in the southern part of the province near lat 31°N about 29 m.y. B.P., the model offers no particular advantage over that of the eclipsing of the East Pacific Rise.

Although an adequate explanation for the apparent widespread and more or less simultaneous inception of extensional block faulting throughout the Basin and Range province about 17 m.y. B.P. has not been found, the foregoing discussion provides a fruitful starting point to explore the reason for a resurgent, northward-migrating wave of uplift, faulting, and basaltic volcanism. As the American plate moved relatively southwestward no new plate material was generated and subducted on the Farallon plate opposite the San Andreas transform (Fig. 14-12). This triangular "window" or "hole" (Dickinson and Snyder, 1975; Stewart, this volume)

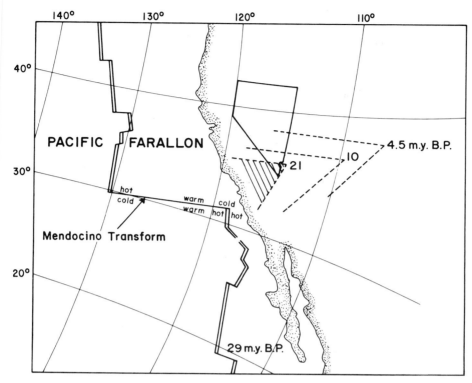

Figure 14-13. Late Cenozoic plate systems off the west coast of North America (after Atwater and Molnar, 1973). Position of the boundary between the Pacific and Farallon plates is shown for 29 m.y. B.P. The triangular areas labeled 21, 10, and 4.5 m.y. B.P. are the conjectured position of the "hole" in the Farallon plate descending beneath the North American continent, but uncorrected for dip. The outline of the state of Nevada is for geographic reference. Relative surface temperatures for parts of the Farallon and Pacific plates are shown.

in the descending plate became progressively larger as its northern and eastern edges moved farther toward the north and east beneath the continent. The exact positions of the northern and especially the eastern edges at any point in time are uncertain, not only because of the unknown dip of the subducting plate but also because the degree of preservation of the downgoing window and the behavior of the descending plate as it passes through the soft asthenosphere and encounters the more rigid underlying mesosphere are unknown. In spite of this uncertainty, the presence of a hole in the descending plate would perturb existing patterns of flow in the asthenosphere, especially in the vicinity of its apex. This perturbation might trigger thermal instability in the vicinity of the hole because of viscous heating (Anderson and Perkins, 1974) and would augment any asthenospheric upwelling already occurring by virtue of some sort of earlier back-arc phenomena. Because of the northward component of motion of the hole, a corresponding direction of migration of resurgent Basin and Range dynamics might occur. In reviewing an earlier version of our manuscript, G. P. Eaton (1977, written commun.) pointed out that temperature differentials in the descending Farallon plate (Fig. 14-12) could also affect asthenospheric flow in such a manner as to lead to an augmented or resurgent northward- migrating wave of Basin and Range dynamics.

Whatever the cause, the apparent lag in time of a few million years of the inception of fundamentally basaltic volcanism in the Basin and Range province (Fig. 14-7) behind the northward-advancing position of the Mendocino edge of the hole (Fig. 14-13) can be explained by dip of the plate and/or by the inertia of the asthenospheric system to respond to the thermal-flow perturbation.

The reality of a northward-migrating hole in the descending Farallon plate is supported in complementary fashion along the western boundary of the Basin and Range province by progressive switch-off of subduction-related volcanism (Snyder and others, 1976).

The eastward component of migration of the hole constitutes a viable alternative cause of the eastward component of migration of basaltic volcanism into the Colorado Plateaus. The influence of a lithospheric keel on localization of volcanism along the margins of the province remains, however, to be an integral facet of our overall tectonic model.

We recognize the problems of integrating our northern Basin and Range model with the broader Cenozoic evolution of western North America. For example, how does mantle upwelling and crustal upwarp relate to the epeirogenic uplift of the entire Western United States since the Cretaceous (Gilluly, 1973)? East-west extension in the Basin and Range province (Fig. 14-11) implies compression (compare Sbar and Sykes, 1973) in the Colorado Plateaus. But how intense was that compression and how was it related to southwestward motion of the North American plate? Further pursuit of these and allied important questions is beyond the scope of our effort here. Recently developed concepts (Anderson and Perkins, 1974; McKenzie and Weiss, 1975) of two scales of convective flow in the mantle may provide fresh insight, however.

SUMMARY AND CONCLUSIONS

Along the eastern boundary of the Basin and Range province between lat 36° and 40°N, the geologic record speaks strongly for broadly concurrent tectonic uplift, fundamentally basaltic volcanism, and extensional block faulting in late Cenozoic time. Uplift of the plateaus margin is in the form of a northeastward tilting; both basins and ranges within the province have been uplifted but the basins lag behind. Major displacement along boundary faults associated with relative uplift of the plateaus margin occurred in late Miocene time south of about lat 36°N. Since then tectonic activity has abated there. In contrast, the vigor of uplift and faulting to the north mainly along the Hurricane-Wasatch fault system has continued to the present time.

The inception of fundamentally basaltic volcanism along the boundary has migrated northward at a rate of about 3 cm/yr. An eastward component of migration into the plateaus is about 1 cm/yr. Basaltic rocks are chiefly undersaturated types on the plateaus but are more saturated tholeiitic basalts and andesites within the Basin and Range province.

A bilateral symmetry about a north-south axis near long 116°W north of about lat 35°N is clearly expressed in the Bouguer gravity of the Basin and Range province and flanking Sierra Nevada and Colorado Plateaus. The symmetry is strengthened by an apparent northward migration of inception of basaltic volcanism along the western margin of the Basin and Range province, which matches the pattern along the eastern margin, and by a similar pattern and chronology of uplift and tilt of the Sierra Nevada as for the Colorado Plateaus.

We postulate that the bilaterally symmetric northern Basin and Range province is the fault-fragmented core of a crustal upwarp whose tilted, uncollapsed abutments are the Sierra Nevada and Colorado Plateaus. Upwarp, east-west extension, and fundamentally basaltic volcanism have progressed concurrently as secondary responses to upwelling asthenospheric mantle whose horizontal axis of divergent lateral flow lies beneath the symmetry axis.

In addition to the bilateral symmetry associated with upwelling mantle, there is evidence for a north-south component in Basin and Range dynamic processes. We envision a northward-migrating wave of activity that has produced three zones of distinct geologic character (Fig. 14-12). The first zone is characterized by concurrent upwarp and extensional block faulting. The second, behind it to the south, is a zone of upwarp, extension, and fundamentally basaltic volcanism. The northern leading edge of this zone coincides roughly with a U-shaped belt of current seismicity and other geophysical-geologic discontinuities. The third zone—which lies south of lat 37°N in the wake of the other two zones—is a region of much reduced tectonism and limited basaltic volcanism, where the dominant geologic processes are erosion of ranges and sedimentation in basins.

The reason for concentration of basaltic volcanism along the eastern and western margins of the Basin and Range province is unresolved but may be caused by thermal instabilities resulting from viscous heating where the laterally flowing asthenosphere meets deepened lithospheric keels beneath the flanking Sierra Nevada and Colorado Plateaus.

Basic unresolved problems include (1) the cause of upwelling mantle, and (2) why, after apparent widespread inception of block faulting throughout the province about 17 m.y. B.P., there should be an apparent resurgent northward-migrating wave of extensional faulting, upwarping, and basaltic volcanism. Some insight into these questions is provided by an examination of Cenozoic plate systems along the west coast of North America. Back-arc spreading holds promise as a mechanism to initiate asthenospheric upwelling and consequent Basin and Range extensional faulting. A northeastward-migrating window in the descending Farallon plate beneath western North America may have caused augmented upwelling responsible for the later resurgent, northward-migrating wave of Basin and Range dynamics.

Further detailed investigations of the chronology and character of uplift, block faulting, and volcanism are urgently needed, as are quantitative geophysical modeling of asthenospheric systems, to critically evaluate these preliminary conclusions. Such studies would be vital in testing the fundamentally very significant but, at present, quite speculative concept of resurgent Basin and Range dynamics and would help to clarify how plate systems along the continental margin are involved.

ACKNOWLEDGMENTS

This research was supported by National Science Foundation Grant DES74-18384 and by Brigham Young University. Many of the ideas expressed in this paper were germinated at the Penrose Conference on Regional Geophysics and Tectonics of the Intermountain West convened by G. P. Eaton, D. R. Mabey, and R. B. Smith. We have benefited from unpublished K-ar dates supplied by R. E. Anderson and P. D. Damon, whose work at the Laboratory of Isotope Geochemistry, University of Arizona, has been supported by the National Science Foundation. Additional K-Ar dating has been done by E. H. McKee and M. G. Best at the U.S. Geological Survey Isotope Geology Laboratory in Menlo Park, and I (M.G.B.) express appreciation to G. B. Dalrymple and staff for their considerable assistance. We are indebted to many individuals for correspondence, preprints, and reviews of preliminary drafts of our manuscript; they include E. H. McKee, M. L. Silberman, G. P. Eaton, W. R. Dickinson, G. A. Thompson, L. F. Hintze, H. J. Bissell, I. Lucchitta, W. S. Snyder, J. H. Stewart, J. M. Proffett, Jr., and R. E. Anderson. Jack Stewart drew our attention to the concept of a hole in the descending Farallon plate and to the unpublished work of John Proffett, whose interpretations of Basin and Range dynamics parallels our own but which was perceived earlier on the basis of less data.

REFERENCES CITED

Anderson, J. J., Rowley, P.D., Fleck, R. J., and Nairn, A.E.M., 1975, Cenozoic geology of southwestern High Plateaus of Utah: Geol. Soc. America Spec. Paper 160, 88 p.

Anderson, O.L., and Perkins, P.C., 1974, Runaway temperatures in the asthenosphere resulting from viscous heating: Jour. Geophys. Research, v. 79, p. 2136–2138.

Anderson, R. E., Longwell, C. R., Armstrong, R. L., and Marwin, R. F., 1972, Significance of K-Ar ages of Tertiary rocks from the Lake Mead region, Nevada-Arizona: Geol. Soc. America Bull., v. 83, p. 273–288.

Armstrong, R. L., 1970, Geochronology of Tertiary igneous rocks, eastern Basin and Range province, western Utah, eastern Nevada, and vicinity, U.S.A.: Geochim. et Cosmochim. Acta, v. 34, p. 203–232.

Armstrong, R. L., and Higgins, R. E., 1973, K-Ar dating of the beginning of Tertiary volcanism in the Mojave Desert, California: Geol. Soc. America Bull., v. 84, p. 1095–1100.

Atwater, T., 1970, Implications of plate tectonics for the Cenozoic tectonic evolution of western North America: Geol. Soc. America Bull., v. 81, p. 3513–3536.

Atwater, T., and Molnar, P., 1973, Relative motion of the Pacific and North American plates deduced from sea-floor spreading in the Atlantic, Indian and South Pacific oceans, in Kovach, K. L., and Nur, A., eds., Tectonic problems of the San Andreas fault system Conf. Proc.: Stanford Univ. Pubs. Geol. Sci., v. 13, p. 136–148.

Best, M. G., 1974, Contrasting types of chromium-spinel peridotite xenoliths in basanitic lavas, western Grand Canyon, Arizona: Earth and Planetary Sci. Letters, v. 23, p. 229–237.

Best, M. G., and Brimhall, W. H., 1974, Late Cenozoic alkalic basaltic magmas in the western Colorado Plateaus and the Basin and Range transition zone, U.S.A., and their bearing on mantle dynamics: Geol. Soc. America Bull., v. 85, p. 1677–1690.

Christensen, M. N., 1966, Late Cenozoic crustal movements in the Sierra Nevada of California: Geol. Soc. America Bull., v. 77, p. 163–182.

Christiansen, R. L., and Lipman, P. W., 1972, Cenozoic volcanism and plate tectonic evolution of the Western United States; II, Late Cenozoic: Royal Soc. London Philos. Trans., ser. A, v. 271, p. 249–284.

Christiansen, R. L., and McKee, E. H., 1978, Late

Cenozoic volcanic and tectonic evolution of the Great Basin and Columbia Intermontane regions, in Smith, R. B., and Eaton, G. P., eds., Cenozoic tectonics and regional geophysics of the western Cordillera: Geol. Soc. America Mem. 152 (this volume).

Clark, E., 1976, Late Cenozoic volcanic and tectonic activity along the eastern margin of the Great Basin in the proximity of Cove Fort, Utah: Brigham Young Univ. Geology Studies, v. 24, p. 87–114.

Dalrymple, G. B. 1963, Potassium-argon dates of some Cenozoic volcanic rocks of the Sierra Nevada, California: Geol. Soc. America Bull., v. 74, p. 379–390.

Damon, P. E., Shafiqullah, M., and Leventhal, J. S., 1974, K-Ar chronology for the San Francisco volcanic field and rate of erosion of the Little Colorado River, in Karlstrom, T.N.V., Swann, G. A., and Eastwood, R. L., eds., Geology of northern Arizona. Pt. 1—Regional studies: Flagstaff, Northern Arizona Univ. and U.S. Geol. Survey Center of Astrogeology, p. 221–235.

Dickinson, W. R., and Snyder, W. S., 1975, Geometry of triple junctions and subducted lithosphere related to San Andreas transform activity: EOS (Am. Geophys. Union Trans.), v. 56, p. 1066.

Eardley, A. J., 1962, Structural geology of North America (2nd ed.): New York, Harper & Row, 743 p.

Eaton, G. P., 1975, Characteristics of a transverse crustal boundary in the Basin and Range province of southern Nevada: Geol. Soc. America Abs. With Programs, v. 7, p. 1062–1063.

——1976, Fundamental bilateral symmetry of the western Basin and Range province: Geol. Soc. America Abs. with Programs, v. 8, p. 583–584.

Eaton, G. P., Wahl, R. R., Prostka, H. J., Mabey, D. R., and Kleinkopf, M. D., 1978, Regional gravity and tectonic patterns: Their relation to late Cenozoic epeirogeny and lateral spreading of the western Cordillera, in Smith, R. B., and Eaton, G. P., eds., Cenozoic tectonics and geophysics of the western Cordillera: Geol. Soc. America Mem. 152 (this volume).

Ekren, E. B., Rogers, C. L., Anderson, R. E., and Orkild, P. P., 1968, Age of Basin and Range normal faults in Nevada Test Site and Nellis Air Force Range, Nevada, in Eckel, E. B., ed., Nevada Test Site: Geol. Soc. America Mem. 110, p. 247–250.

Ekren, E. B., Bath, G. D., Dixon, G. L., and

Quinlivan, W. D., 1974, Tertiary history of Little Fish Lake Valley, Nye County, Nevada, and implications as to the origin of the Great Basin: U.S. Geol. Survey Jour. Research, v. 2, p. 105–118.

Fenneman, N. M., 1931, Physiography of Western United States: New York, McGraw-Hill Book Co., 534 p.

Gilbert, C. M., and Reynolds, M. W., 1973, Character and chronology of basin development, western margin of the Basin and Range province: Geol. Soc. America Bull., v. 84, p. 2489–2510.

Gilluly, J., 1973, Steady plate motion and episodic orogeny and magmatism: Geol. Soc. America Bull., v. 84, p. 499–514.

Gough, D. I., 1974, Electrical conductivity under western North America in relation to heat flow, seismology and structure: Jour. Geomagnetism and Geoelectricity, v. 26, p. 105–123.

Hamblin, W. K., 1965, Origin of "reverse drag" on the downthrown side of normal faults: Geol. Soc. America Bull., v. 76, p. 1145–1164.

Hamblin, W. K., Damon, P. D., and Shafiqullah, M., 1975, Rates of erosion in the Virgin River drainage in southern Utah and northern Arizona: Geol. Soc. America Abs. with Programs, v. 7, p. 1097–1098.

Hay, E. A., 1976, Cenozoic uplifting of the Sierra Nevada in isostatic response to North American and Pacific plate interactions: Geology, v. 4, p. 763–766.

Hoover, J. D., 1974, Periodic Quaternary volcanism in the Black Rock Desert, Utah: Brigham Young Univ. Geology Studies, v. 21, p. 3–72.

Hunt, C. B., and Mabey, D. R., 1966, Stratigraphy and structure, Death Valley, California: U.S. Geol. Survey Prof. Paper 494-A, 162 p.

Hunt, C. B., Varnes, H. D., and Thomas, H. E., 1953, Lake Bonneville: Geology of northern Utah Valley, Utah: U.S. Geol. Survey Prof. Paper 257-A, 99 p.

Irvine, T. N., and Baragar, W.R.A., 1971, A guide to the chemical classification of the common volcanic rocks: Canadian Jour. Earth Sci., v. 8, p. 523–548.

Keller, G. R., Smith, R. B., and Braile, L. W., 1975, Crustal structure along the Great Basin–Colorado Plateau transition from seismic refraction studies: Jour. Geophys. Research, v. 80, p. 1093–1098.

King, P. B., and Beikman, H. M., 1974, Geologic map of the United States (exclusive of Alaska and Hawaii): U.S. Geol. Survey, scale 1:2,500,000.

Lachenbruch, A. H., Sass, J. H., Munroe, R. J., and Moses, T. H., Jr., 1976, Geothermal setting and simple heat conduction models for the Long Valley caldera: Jour. Geophys. Research, v. 81, p. 769–784.

LeConte, J., 1889, On the origin of normal faults and of the structure of the Basin region: Am. Jour. Sci., Third Ser., v. 38, p. 257–263.

Leeman, W. P., and Rogers, J.J.W., 1970, Late Cenozoic alkali-olivine basalts of the Basin-Range province, USA: Contr. Mineralogy and Petrology, v. 25, p. 1–24.

Lipman, P. W., and Mehnert, H. H., 1975, Late Cenozoic basaltic volcanism and development of the Rio Grande depression in the southern Rocky Mountains: Geol. Soc. America Mem. 144, p. 119–154.

Loring, A. K., 1976, Distribution in time and space of late Phanerozoic normal faulting in Nevada and Utah: Utah Geology, v. 3, p. 97–109.

Lucchitta, I., 1972, Early history of the Colorado River in the Basin and Range province: Geol. Soc. America Bull., v. 83, p. 1933–1948.

Lucchitta, I., and McKee, E. H., 1975, New geochronologic constraints on the history of the Colorado River and its Grand Canyon: Geol. Soc. America Abs. with Programs, v. 7, p. 342.

Mabey, D. R., Zietz, I., Eaton, G. P., and Kleinkopf, M. D., 1978, Regional magnetic patterns in part of the Cordillera in the Western United States, in Smith, R. B., and Eaton, G. P., eds., Cenozoic tectonics and regional geophysics of the western Cordillera: Geol. Soc. America Mem. 152 (this volume).

McKee, E. D., and McKee, E. H., 1972, Pliocene uplift of the Grand Canyon region—Time of drainage adjustment: Geol. Soc. America Bull., v. 83, p. 1923–1932.

McKee, E. H., 1971, Tertiary igneous chronology of the Great Basin of Western United States—Implications for tectonic models: Geol. Soc. America Bull., v. 82, p. 3497–3502.

McKee, E. H., and Anderson, C. A., 1971, Age and chemistry of Tertiary volcanic rocks in north-central Arizona and relation of the rocks to the Colorado Plateaus: Geol. Soc. America Bull., v. 82, p. 2767–2782.

McKee, E. H., and Noble, D. C., 1974, Timing of late Cenozoic crustal extension in the Western United States: Geol. Soc. America Abs. with Programs, v. 6, p. 218.

McKenzie, D., and Weiss, N., 1975, Speculations on the thermal and tectonic history of the earth: Royal Astron. Soc. Geophys. Jour., v. 42, p. 131–174.

Menard, H. W., Jr., 1960, The East Pacific Rise:

Science, v. 132, p. 1737–1746.

Moore, R. B., Wolfe, E. W., and Ulrich, G. E., 1976, Volcanic rocks of the eastern and northern parts of the San Francisco volcanic field, Arizona: U.S. Geol. Survey Jour. Research, v. 4, p. 549–560.

Noble, D. C., 1972, Some observations on the volcano-tectonic evolution of the Great Basin, Western United States: Earth and Planetary Sci. Letters, v. 17, p. 142–150.

Proffett, J. M., 1977, Cenozoic geology of the Yerington district, Nevada, and implications for the nature and origin of Basin and Range faulting: Geol. Soc. American Bull., v. 88, p. 247–266.

Sbar, M. L., and Sykes, L. R., 1973, Contemporary compressive stress and seismicity in eastern North America: An example of intra-plate tectonics: Geol. Soc. America Bull., v. 84, p. 1861–1882.

Scholz, C. H., Barazangi, M., and Sbar, M. L., 1971, Late Cenozoic evolution of the Great Basin, Western United States, as an ensialic interarc basin: Geol. Soc. America Bull., v. 82, p. 2979–2990.

Shuey, R. T., Schellinger, D. K., Johnson, E. H., and Alley, L. B., 1973, Aeromagnetics and the transition between the Colorado Plateau and Basin Range provinces: Geology, v. 1, p. 107–110.

Silberman, M. L., and McKee, E. H., 1972, A summary of radiometric age determinations of Tertiary volcanic rocks from Nevada and eastern California: Pt. II, Western Nevada: Isochron /West, no. 4, p. 7–28.

Smith, R. B., and Sbar, M. L., 1974, Contemporary tectonics and seismicity of the Western United States with emphasis on the Intermountain seismic belt: Geol. Soc. America Bull., v. 85, p. 1205–1218.

Snyder, W. S., Dickinson, W. R., and Silberman, M. L., 1976, Tectonic implications of space-time patterns of Cenozoic magmatism in the Western United States: Earth and Planetary Sci. Letters, v. 32, p. 91–106.

Stewart, J. H., 1971, Basin and Range structure: A system of horsts and grabens produced by deep-seated extension: Geol. Soc. America Bull., v. 82, p. 1019–1044.

——1978, Basin and Range structure in western North America, in Smith, R. B., and Eaton, G. P., eds., Cenozoic tectonics and regional geophysics of the western Cordillera: Geol. Soc. America Mem. 152 (this volume).

Stewart, J. H., and Carlson, J. E., 1976, Cenozoic rocks of Nevada: Nevada Bur. Mines Map 52, scale 1:1,000,000.

Stewart, J. H., Moore, W. J., and Zietz, I., 1977, East-west patterns of Cenozoic igneous rocks, aeromagnetic anomalies, and mineral deposits, Nevada and Utah: Geol. Soc. America Bull., v. 88, p. 67–77.

Suppe, J., Powell, C., and Berry, R., 1975, Regional topography, seismicity, Quaternary volcanism and the present-day tectonics of the Western United States: Am. Jour. Sci., v. 275-A, p. 397–436.

Thompson, G. A., 1960, Problem of late Cenozoic structure of the Basin Ranges: Internat. Geol. Cong., 21st, Copenhagen 1960, pt. XVIII, p. 62–68.

——1966, The rift system of the Western United States, in Irvine, T. N., ed., The world rift system: Canada Geol. Survey Paper 66-14, p. 280–290.

——1972, Cenozoic Basin Range tectonism in relation to deep structure: Internat. Geol. Cong., 24th, Montreal 1972, Proc., no. 24, p. 84–90.

Thompson, G. A., and Burke, D. B., 1974, Regional geophysics of the Basin and Range province: Earth and Planetary Sci. Ann. Rev, v. 2, p. 213–238.

Uyeda, S., and Miyashiro, A., 1974, Plate tectonics and the Japanese islands: A synthesis: Geol. Soc. America Bull., v. 85, p. 1159–1170.

Wallace, R. E., 1975, United States—Basin and Range province, in Fairbridge, R. W., ed., The encyclopedia of world regional geology, Pt. I, Western Hemisphere (including Antarctica and Australia): Stroudsburg, Penn., Dowden, Hutchinson and Ross, Encyclopedia of Earth science, v. VII, p. 541–548.

Wright, L., 1976, Late Cenozoic fault patterns and stress fields in the Great Basin and westward displacement of the Sierra Nevada block: Geology, v. 4, p. 489–494.

Young, R. A., and Brennan, W. J., 1974, Peach Springs Tuff: Its bearing on structural evolution of the Colorado Plateau and development of Cenozoic drainage in Mohave County, Arizona: Geol. Soc. America Bull., v. 85, p. 83–90.

Young, R. A., Brennan, W. J., Lucchitta, I., and McKee, E. H., 1975, Pre-Basin and Range uplift and Miocene volcanism along the west edge of the Colorado Plateau, Mohave County, Arizona: Geol. Soc. America Abs. with Programs, v. 7, p. 1324.

Manuscript Received by the Society August 15, 1977

Manuscript Accepted September 2, 1977

Printed in U.S.A.

Geological Society of America
Memoir 152

15

Bright Angel and Mesa Butte fault systems of northern Arizona

E. M. SHOEMAKER
Division of Geological and Planetary Sciences
California Institute of Technology, Pasadena, California 91125
and
U.S. Geological Survey, Flagstaff, Arizona 86001

R. L. SQUIRES
Department of Geoscience, California State University
Northridge, California 91324

M. J. ABRAMS
Jet Propulsion Laboratory
California Institute of Technology, Pasadena, California 91103

ABSTRACT

Regional geologic mapping using pictures from the first Earth Resources Technology Satellite (ERTS-1) has led to the recognition of two parallel northeast-trending systems of normal faults, each of which can be traced more than 100 km. Many eruptive centers appear to be localized along these fault systems or along their extensions. The faults are chiefly observed in Phanerozoic rocks and have minor displacement but are interpreted by us to reflect fault zones of major displacement in the crystalline Precambrian basement.

The Bright Angel fault system extends as a continuous zone of normal faults from Cataract Creek on the southwest to the Echo Cliffs on the northeast. Beyond the Echo Cliffs, the system continues northeastward to the vicinity of Monument Valley as a more diffuse, discontinuous zone of normal faults. The Bright Angel fault, Vishnu fault, and Eminence Break graben are among the larger individual members of the total system. The Navajo mountain intrusive center lies along the discontinuous part of the system. Three major eruptive centers of the Mount Floyd volcanic field lie on the southwestern projection of the Bright Angel fault system. If the eruptive centers are included as part of the recognizable structural system, the Bright Angel system has a total known length of slightly more than 300 km.

The Mesa Butte fault system, as now recognized, extends from Chino Valley on the southwest to Shadow Mountain on the northeast. Bill Williams Mountain, Sitgreaves Peak, and Kendrick Peak are principal silicic to intermediate eruptive centers of the San Francisco volcanic field that appear to be localized along the fault system. Red Mountain, Mesa Butte, and Shadow Mountain are prominent basaltic eruptive centers along the system; monchiquite diatremes at Tuba Butte and Wildcat Peak lie on the northeast projection of the fault system. The total distance from Chino Valley to Wildcat Peak is more than 200 km.

Comparison of the Bright Angel and Mesa Butte fault systems with a residual aeromagnetic map of Arizona reveals a close correspondence between the positions of the observed relatively minor normal faults and the margins of a series of large northeast-trending magnetic anomalies. Perhaps the most noteworthy feature of the aeromagnetic map is a 400-km-long northeast-trending belt of large positive aeromagnetic anomalies that extends from the vicinity of Congress to the northern border of Arizona. The Mesa Butte fault system lies along the southeast margin of this anomaly belt. Another large positive anomaly, bounded on the southeast by the Bright Angel fault, corresponds in the Grand Canyon to a belt of Precambrian amphibolite and schist. Most of the large positive aeromagnetic anomalies along the Bright Angel and Mesa Butte fault systems may correspond to similar bodies of mafic metavolcanic rocks, which have been offset along two major and perhaps several minor faults of Precambrian age. The normal faults that displace the overlying Phanerozoic rocks have been formed by renewed movement along these ancient fault zones, in response to dilation of the crust from late Tertiary time to the present.

The ancient fault zones inferred to be present along the Bright Angel and Mesa Butte fault systems may be related in origin to the Shylock and Chaparral fault zones in central Arizona described by Anderson (1967). Both the Shylock fault zone and the Chaparral fault have right-lateral transcurrent displacement. As shown by Anderson, the Shylock zone has a probable minimum horizontal displacement of 8 km. A large contrast in the magnetic properties of the rocks on opposite sides of the fault zone, indicated by the aeromagnetic map, suggests that the displacement may be several tens of kilometres or more. Comparably large right-lateral displacements may have occurred along the ancestral Bright Angel and Mesa Butte fault zones.

The location of epicenters of recent earthquakes and reports of earthquakes by residents in the region indicate that the Bright Angel and Mesa Butte fault systems are currently active.

INTRODUCTION

Among the structural features of northern Arizona, the Bright Angel fault has become perhaps one of the best known. It certainly must rank as the fault most often visited; thousands of tourists hike along part of the Bright Angel fault each year as they descend from the south rim of the Grand Canyon into the inner gorge along the Bright Angel trail, or from the north rim along the Kaibab trail. How many of these visitors are aware that a fault has controlled the route of easiest descent and, indeed, the course of Bright Angel Canyon, through which the Kaibab trail passes, is not known. It is clear, however, that the easy access into the Grand Canyon provided by these trails has led to close inspection of the fault by many geologists.

The Bright Angel fault was first described by Ransome (1908), who recognized that displacement on the fault had occurred during at least two widely separated periods of time, one Precambrian and one post-Paleozoic. The first map to portray the fault was a reconnaissance

Figure 15-1. ERTS-1 picture of Coconino Plateau showing Bright Angel and Mesa Butte fault systems. NASA picture ERTS E-1104-17382-5, November 4, 1972.

map of northern Arizona and New Mexico by Darton (1910), who also briefly described the fault and noted the two episodes of displacement. In 1914, a part of the fault lying within the Bright Angel quadrangle south of the Colorado River was mapped in detail by Noble; although he contributed some important observations about the fault (Noble and Hunter, 1916, p. 101; Noble, 1918), the map was never published. Later, McKee described the Bright Angel fault in a short paper in 1929.

The problem of multiple episodes of faulting on the Bright Angel and related faults attracted the attention of Maxson and Campbell (1934), as they worked on the ancient crystalline rocks of the Grand Canyon. Much later, Maxson published a detailed map of the Bright Angel quadrangle and described a complex history of displacement on the Bright Angel fault and on intersecting faults (Maxson, 1961). Most recently, the Bright Angel fault was remapped, where it cuts exposed Precambrian rocks, by Sears (1973), who reinterpreted the history of displacement.

We became interested in the Bright Angel fault in the course of a regional geologic investigation of the hydrographic basin of Cataract Creek, the principal drainage of the Coconino Plateau, which forms the south rim of the Grand Canyon. Starting from the southern boundary of the Bright Angel quadrangle, we traced the Bright Angel fault 30 km to the southwest. At this point the fault dies out, but other parallel faults continue to the southwest at least another 20 km. In this investigation we employed pictures taken from an altitude of 907 km by the first Earth Resources Technology Satellite (ERTS-1). While studying these pictures, we noted a remarkably long zone or system of faults of which the Bright Angel fault is a member. From the center of the Cataract Creek region, the fault system can be traced northeastward for at least 270 km. A parallel system of faults that lies about 50 km southeast of the Bright Angel system can also be recognized in the ERTS-1 pictures (Fig. 1). Some of the principal Cenozoic eruptive centers on the Coconino Plateau appear to be localized along these two fault systems.

At this stage in our investigation the evidence suggested that the two fault systems, which are chiefly observed in Phanerozoic rocks, might be related to major ancient structures in the underlying Precambrian crystalline complex. To test this hypothesis we compiled a map of the faults in northwestern Arizona (Fig. 2) and compared this map with a residual aeromagnetic map of Arizona published by Sauck and Sumner (1971). About 60% of the fault map is based on previous published and unpublished maps. The remainder is based on new detailed and reconnaissance field mapping by us and by Ivo Lucchitta, using ERTS-1 pictures and aerial photographs. Comparison of the pattern of faults with the aeromagnetic map revealed many features of interest. In this paper we will discuss mainly the Bright Angel fault system and the parallel system of faults to the southeast, referred to here as the Mesa Butte fault system. Other systems of faults are briefly described.

BRIGHT ANGEL FAULT SYSTEM

The Bright Angel fault system extends as a continuous zone of northeast-trending normal faults from Cataract Creek on the southwest to the Echo Cliffs on the northeast. Beyond the Echo Cliffs, the system may be traced northeastward to the Monument upwarp in Utah as a more diffuse discontinuous zone of normal faults.

South of the Grand Canyon, the Bright Angel system comprises several distinct faults, each of which is some tens of kilometres long. The Bright Angel and Vishnu faults are the principal members of the system (Figs. 3 and 4). Here the faults cross the Coconino Plateau, which is capped by Kaibab Limestone of Permian age. Maximum displacement of the exposed Permian rocks on each of the larger faults is on the order of 100 m. On most faults, displacement is down to the southeast. The downthrown side commonly exhibits reverse drag; narrow prisms of beds belonging to the Moenkopi Formation of Triassic age are preserved in a few places on the hanging wall. Stripping away of the easily eroded Moenkopi from much of the rest of the plateau has left relatively prominent fault-line scarps formed by the upthrown resistant beds of Kaibab Limestone.

A curious feature of the Bright Angel fault system, which is well illustrated on the Coconino Plateau, is the tendency of individual faults to die out as they approach northwest-trending monoclines (Fig. 4). Two small monoclines lie athwart the Bright Angel fault system on the plateau. None of the individual faults in the system crosses either of these monoclines.

The Bright Angel fault is one of the longest members of the entire fault system. It can be traced for a total distance of 65 km. At its north end, the fault swings around from a northeasterly to a northerly strike and then dies out as it approaches the East Kaibab monocline (Fig. 4). A series of en echelon faults continues farther to the northeast along the main trend of the Bright Angel fault.

Displacement of Permian beds on the Bright Angel fault is down to the southeast, both on the Coconino Plateau and on the Kaibab Plateau. Where it crosses the Grand Canyon, however, a complex set of relationships may be observed along the fault in older beds, particularly in Precambrian rocks. The net displacement of beds belonging to the Grand Canyon Supergroup, of late Precambrian age, is up on the southeast. Their displacement is due primarily to one or more episodes of reverse faulting in late Precambrian time (Maxson, 1961; Sears, 1973).

Northeast of the East Kaibab monocline, along the projected trend of the main part of the Bright Angel fault, the narrow Eminence Break graben can be traced across the Marble Platform (Fig. 4), a stripped surface capped by the Kaibab Limestone. Beds southeast of the graben lie at a higher elevation than the beds northwest of the graben. Thus the displacement is greater on the fault that bounds the graben on the southeast side. A prominent northwest-facing fault-line scarp, the Eminence Break, is developed along the bounding fault on the southeast side of the graben. This fault is 40 km long and has a maximum displacement of about 100 m. It dies out to the northeast at the Echo Cliffs monocline. The fault bounding the northwest side of the graben is only 30 km long and dies out before reaching the Echo Cliffs monocline. To the southwest, toward the East Kaibab monocline, the graben bends to the south and dies out precisely at the lower axis of the monoclinal flexure. A few shorter faults that are parallel or subparallel to the main part of the Eminence Break graben occur on nearby parts of the Marble Platform.

A zone of en echelon grabens continues for a distance of 12 km northeast of the Echo Cliffs along the projected trend of the Eminence Break graben. Beyond this point, the Bright Angel fault system becomes a broad diffuse zone of relatively short faults (Fig. 2). In general, faults become more and more widely spaced to the northeast, and the northeastern limit of the system is not well defined. The system extends at least as far as the Monument upwarp, Utah. One branch of the system may be represented by a set of faults that extends to the Comb monocline, just north of the San Juan River in Utah. The easternmost fault of this set was described by Sears (1956). Between the Echo Cliffs and Comb Ridge, the displacement on faults in the Bright Angel system is generally less than 100 m, and none exceeds 20 km in length. At the surface, the faults displace Triassic and Jurassic rocks and, in the vicinity of Monument Valley, Permian rocks.

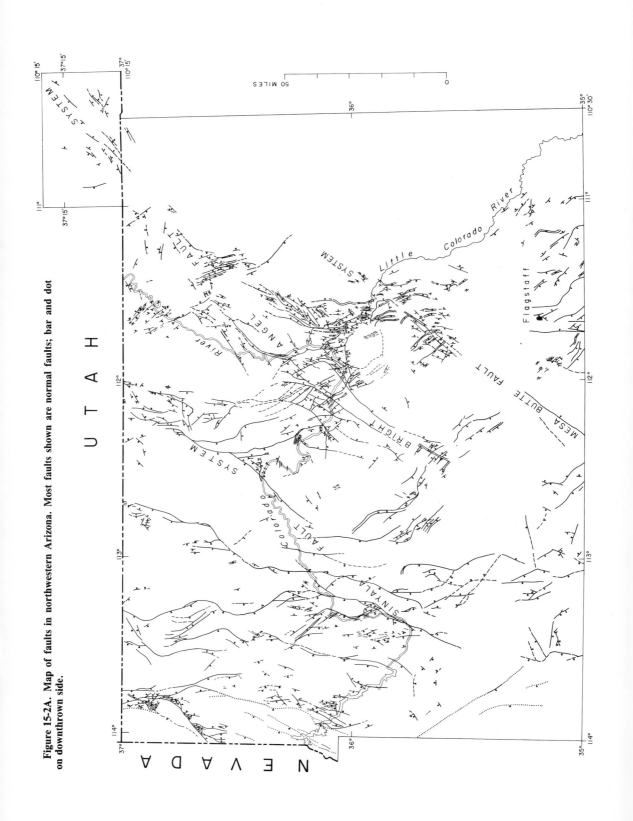

Figure 15-2A. Map of faults in northwestern Arizona. Most faults shown are normal faults; bar and dot on downthrown side.

Figure 15-2B. Index to sources of information shown in Figure 2A.

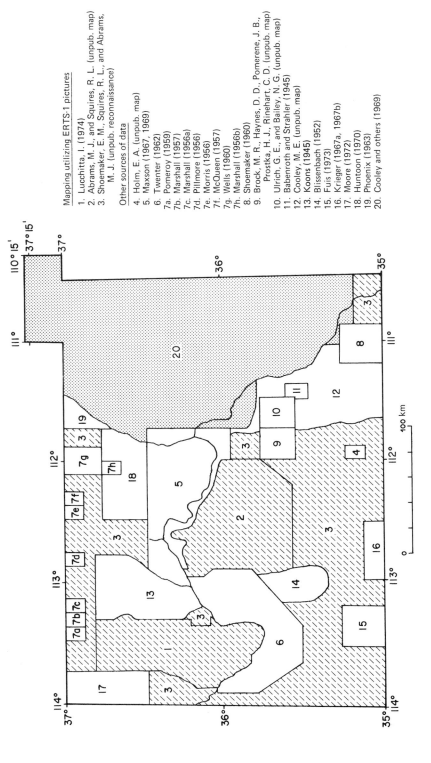

Mapping utilizing ERTS-1 pictures

1. Lucchitta, I. (1974)
2. Abrams, M. J., and Squires, R. L. (unpub. map)
3. Shoemaker, E. M., Squires, R. L., and Abrams,
   M. J. (unpub. reconnaissance)

Other sources of data

4. Holm, E. A. (unpub. map)
5. Maxson (1967, 1969)
6. Twenter (1962)
7a. Pomeroy (1959)
7b. Marshall (1957)
7c. Marshall (1956a)
7d. Pillmore (1956)
7e. Morris (1956)
7f. McQueen (1957)
7g. Wells (1960)
7h. Marshall (1956b)
8. Shoemaker (1960)
9. Brock, M. R., Haynes, D. D., Pomerene, J. B.,
   Prostka, H. J., Rinehart, C. D. (unpub. map)
10. Ulrich, G. E., and Bailey, N. G. (unpub. map)
11. Babenroth and Strahler (1945)
12. Cooley, M. E. (unpub. map)
13. Koons (1945)
14. Blissenbach (1952)
15. Fuis (1973)
16. Krieger (1967a, 1967b)
17. Moore (1972)
18. Huntoon (1970)
19. Phoenix (1963)
20. Cooley and others (1969)

MESA BUTTE FAULT SYSTEM

The Mesa Butte fault system, as now recognized, extends from Chino Valley, near Paulden, Arizona, on the southwest, to Shadow Mountain on the northeast. The known length of the fault system is about 150 km, but detailed mapping of Precambrian terrane west and southwest of Prescott, Arizona, may extend the system many tens of kilometres farther to the southwest. About midway along its known length, the fault system is concealed beneath Quaternary

Figure 15-3A. Oblique high-altitude aerial photograph of Mesa Butte fault system, looking southwest across Coconino Plateau; (1) Mesa Butte, (2) Cedar Ranch fault, (3) Mesa Butte fault, (4) Red Mountain, (5) Slate Mountain, (6) source of Tappan Wash flow, (7) Kendrick Peak, (8) Sitgreaves Mountain, (9) Bill Williams Mountain.

lava flows of the San Francisco volcanic field. The recognized faults in the system are thus grouped into a northeastern segment and a southwestern segment.

The Mesa Butte fault is the principal member of the northeastern segment of the fault system. It was first noted by Johnson (1909) and was mapped and named by Babenroth and Strahler (1945). As may be seen in Figure 3A, Mesa Butte, an elongate basaltic cinder cone, was formed by fissure eruption along the fault. Southwest of Mesa Butte, the fault is covered by Pleistocene lava flows that lap against the base of the fault scarp. The position of the

Figure 15-3B. Oblique high-altitude aerial photograph of Bright Angel fault system, looking southwest across Coconino Plateau; (1) Howard Hill, (2) Mount Floyd.

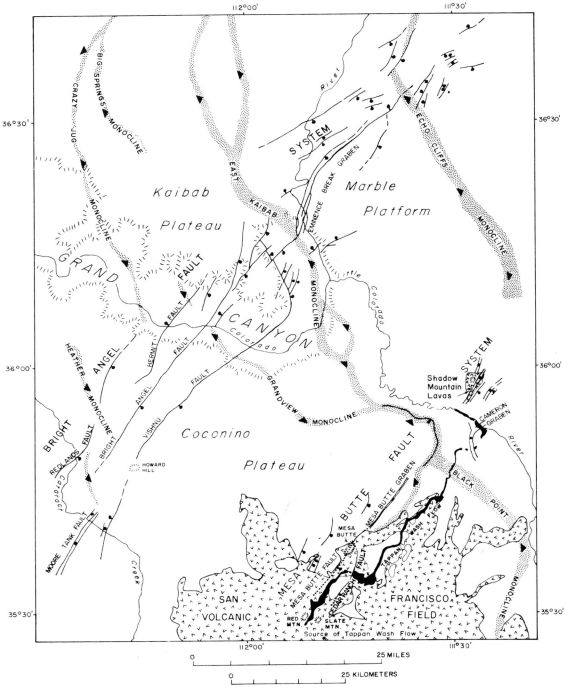

Figure 15-4. Map of part of Bright Angel and Mesa Butte fault systems, showing relation of faults to monoclines and Cenozoic volcanic rocks. Bar and dot are on downthrown side of faults; triangles indicate direction of dip of steep limb of monocline.

fault and its displacement can be estimated with some precision, however, for a distance of about 10 km southwest of Mesa Butte. Farther to the southwest the fault scarp is completely concealed beneath younger lava flows and basaltic cinder cones (Fig. 4). Northeast of Mesa Butte, the fault emerges from beneath the lavas and becomes the bounding fault on the northwest side of the spectacular, long, narrow Mesa Butte graben (Fig. 3). The graben is 300 to 400 m wide, 15 km long, and forms a trench in the surface of the Coconino Plateau 100 to 200 m deep. Formations exposed in the walls of the trench are the Coconino Sandstone, Toroweap Formation, and Kaibab Limestone, all of Permian age. The Mesa Butte fault can be recognized at the surface for a distance of about 35 km; maximum observed displacement is about 100 m.

A branch fault, the Cedar Ranch fault, joins the Mesa Butte fault near Mesa Butte. It can be traced southwestward about 15 km, where it disappears beneath younger lavas. The Cedar Ranch fault is covered by lava that laps against the fault scarp, but locally the lava is offset by faulting. As along the Mesa Butte fault, displacement is down to the southeast.

Northeast of the Mesa Butte graben and directly in line with it is the sharply flexed, locally faulted southern salient of the Grandview monocline (Fig. 4). Farther to the northeast the monocline swings around to the west and veers away from the Mesa Butte fault system. About 30 km beyond the Mesa Butte graben, a cluster of small northeast-trending faults near Shadow Mountain marks the northeastern limit of exposed faults in the Mesa Butte system. However, a series of monoclines along the northwest edge of Black Mesa and a segment of the Comb monocline along the southern margin of the Monument upwarp, near Kayenta, Arizona, may be controlled by the northeast extension of the ancestral Mesa Butte fault in the Precambrian rocks.

Faults belonging to the southwestern segment of the Mesa Butte fault system were mapped along Hell Canyon in the Paulden quadrangle and were described by Krieger (1965). The principal fault in this segment was traced by Krieger 8 km northeastward from the Paulden quadrangle across the Tonto Rim, where it cuts Pennsylvanian and Permian rocks, and is shown on the geologic map of Arizona (Wilson and others, 1969). The continuation of this fault through the southern part of the San Francisco volcanic field can be recognized in the ERTS-1 pictures (Fig. 1) and traced an additional 22 km. The total recognized length of the fault is 38 km. Displacement is down to the southeast; maximum displacement of the exposed rocks is about 150 m. The fault disappears northeastward in a field of lava flows of Pleistocene age. Whether the displacement simply dies out or the fault is covered by younger lavas is not known, as the relations have not been studied in detail in the field.

Directly in line with the 38-km-long fault, 12 km southwest of its last recognized exposure in Hell Canyon, rocks correlated by Krieger (1965) with the Texas Gulch Formation of Precambrian age are in contact with the Mazatzal Quartzite of Precambrian age, the Tapeats Sandstone of Cambrian age, and the Martin Limestone of Devonian age, along closely spaced northeast-trending faults. As mapped by Krieger (1965), the displacement that brought Mazatzal Quartzite in contact with the Texas Gulch Formation(?) at this locality is Precambrian (pre-Tapeats). The later displacement of Cambrian and Devonian beds may have been controlled by a Precambrian fault zone. It appears possible that the ancestral Mesa Butte fault is partly exposed in the Precambrian rocks at this place.

Still farther to the southwest, on the trend of the Mesa Butte fault system, rocks mapped as the Yavapai Series of Precambrian age are in contact with Precambrian granitoid rocks northwest of Granite Mountain (Wilson and others, 1969). This contact may be the southwestward continuation of the ancestral Mesa Butte fault. Work in the field is needed to test this hypothesis.

OTHER FAULT SYSTEMS

In addition to the Bright Angel and Mesa Butte fault systems, other swarms or systems of faults can be recognized from the pattern of faulting shown in Figure 2. One of these, the Sinyala system, which comprises a swarm of northeastward-trending faults, is roughly parallel with the Bright Angel and Mesa Butte systems. Other systems trend northwest and approximately north (Fig. 5). Each system is a relatively broad lane or zone of faults in which the individual faults tend to be parallel or subparallel with the overall trend of the lane. Where the lanes intersect, faults belonging to two or more systems are present. At these intersections, some individual faults follow the direction of one system for part of their length and then turn abruptly and follow the direction of another system.

The Sinyala fault system consists, in large part, of a set of en echelon faults. It is named for a fault 50 km long that crosses the Colorado River midway along the length of the system (Fig. 2) and trends parallel with the total system. Although this fault is very long, its displacement generally is less than a few metres. Some other faults in the Sinyala system with much larger displacement are segments of long faults that follow one system for a distance and then turn and follow another. The faults are observed in rocks ranging in age from Precambrian to Permian.

Four northwest-trending systems of faults are recognized in northern Arizona (Figs. 2 and 5): the Chino Valley system, the Cataract Creek system, the Kaibab system, and the Mormon Ridges system (Fig. 5). The Chino Valley system of faults is observed mainly in Paleozoic rocks, but a set of fault scarps 20 km long is developed in alluvium in the northwestern Chino Valley along the Big Chino fault of Krieger (1965, 1967a).

The Cataract Creek system consists chiefly of a newly mapped swarm of faults that cut Permian, Triassic, and Tertiary rocks in the southern and western part of the Cataract Creek basin. Faults belonging to this system extend into the Shivwits Plateau on the northwest and across the southern part of the San Francisco volcanic field on the southeast.

The Kaibab system is a relatively broad lane of northwest-trending faults that extends from the southern Kaibab Plateau across the Grand Canyon onto the northern Coconino Plateau. Rocks ranging in age from Precambrian to Permian are cut by this system. As in the case of the northeast-trending faults of the Bright Angel system, many faults in the Kaibab system can be shown to have had a long history of displacement that began in Precambrian time.

The Mormon Ridges system is observed principally in Mesozoic rocks. It is named for ridges on the Kaibito Plateau formed along a closely spaced set of faults in the Navajo Sandstone of Triassic(?) and Jurassic age (Cooley and others, 1969).

Two roughly north-trending swarms of faults are present in the fault pattern of northern Arizona. They are referred to here as the Toroweap system and the Oak Creek Canyon system. Both of these systems cut rocks ranging in age from Cambrian to Tertiary. The Toroweap system includes a north-trending segment of the Toroweap fault (Koons, 1945; Twenter, 1962), a parallel segment of the Hurricane fault (Twenter, 1962), and a roughly north-trending segment of the Aubrey fault (Blissenbach, 1952), each of which has a displacement of several hundred metres. A number of smaller faults are also present in the system.

The Oak Creek Canyon system extends north from Oak Creek Canyon, through Flagstaff, Arizona, to the Kaibab Plateau. Just north of Flagstaff, the Oak Creek Canyon system is concealed by Quaternary and Pliocene(?) volcanic rocks of the San Francisco volcanic field; the large stratovolcano of San Francisco Mountain lies astride the system.

Northwest of the Sinyala system, faults are somewhat more uniformly distributed than in the region to the east and south (Fig. 2). Major faults are present near this corner of the

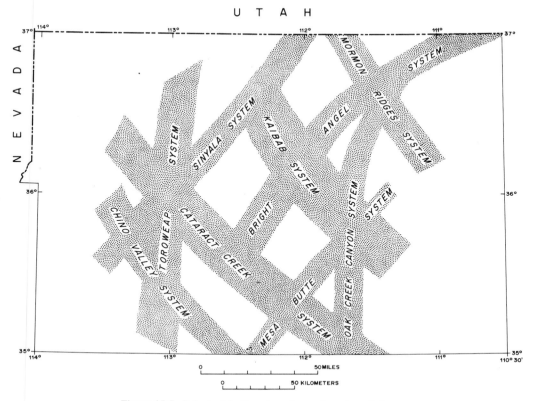

Figure 15-5. Principal fault systems in northwestern Arizona.

state, and northeast-, north-, and northwest-trending sets of faults are represented in this area, but the faults are not as clearly grouped into discrete lanes as they are in the adjacent region.

RELATION OF ERUPTIVE CENTERS TO FAULTS

Inspection of the distribution of volcanoes in the San Francisco volcanic field reveals that about half of the silicic to intermediate volcanoes, described by Robinson (1913), lie along the Mesa Butte fault system. This relationship is easily recognized on the ERTS-1 pictures (Fig. 1); these volcanoes form the most prominent peaks in the region. Closer study shows that Sitgreaves Mountain lies precisely on the projected trend of the Mesa Butte fault (Fig. 7). Bill Williams Mountain lies a few kilometres northwest of this line, and Kendrick Peak lies a few kilometres southeast of the line, close to the projected trend of the Cedar Ranch fault. Slate Mountain, a relatively small silicic volcanic center, lies on the trend of the Cedar Ranch fault (Figs. 3B and 4). The remaining silicic volcanoes in the San Francisco field occur either on the Oak Creek Canyon system of faults or on the Cataract Creek system.

Several hundred basaltic cinder cones are present in the San Francisco volcanic field, but only a few of these are related in an obvious way to the major fault systems. Red Mountain, one of the largest cinder cones, lies on the line of the Mesa Butte fault (Figs. 3A and 4). Mesa Butte, formed by fissure eruption along the fault, is one of the northernmost eruptive centers in the San Francisco field. Well separated from the rest of the volcanic field, Shadow Mountain lies 45 km farther to the northeast along the Mesa Butte fault system (Figs. 4 and 7). It is, however, a basaltic eruptive center of the San Francisco type (Condit, 1974).

Beyond Shadow Mountain lie the monchiquite diatremes and dikes at Tuba Butte and Wildcat Peak (Fig. 7). These isolated volcanic centers are far removed from most other diatremes and alkalic basalts of the Navajo country (Gregory, 1917; Williams, 1936) but are close to the projected trend of the Mesa Butte fault system.

Eruptive centers are more widely spaced along the Bright Angel fault system. In the vicinity of Monument Valley, Utah, a broad swarm of minette dikes and two kimberlite pipes occur near the extreme northeastern end of the Bright Angel fault system. Navajo Mountain, a prominent structural dome in Utah, just north of the Arizona state line, occurs on the margin of the relatively diffuse part of the fault system. For many years the Navajo Mountain dome was thought to have been formed over a laccolith or stock (Gilbert, 1877; Gregory, 1917; Baker, 1936; Hunt, 1942) but without any definite evidence. The dome is distant from the known laccolithic mountain groups of the Colorado Plateau. A small syenite porphyry intrusion near the summit of the dome was reported by Condie (1964).

Near the southwestern end of the Bright Angel fault system, Howard Hill, a small structural dome that resembles the Navajo Mountain dome in shape but not in size, lies just beyond the end of the Vishnu fault (Figs. 3B and 4). This dome may be located over a small stock. Still farther to the southwest, Trinity Mountain, Round Mountain, and Mount Floyd, the principal eruptive centers of the Mount Floyd volcanic field, occur on the trend of the Bright Angel fault system. Mount Floyd, itself, lies almost precisely on the projection of the Bright Angel fault (Figs. 3B and 7).

PRECAMBRIAN ORIGIN OF FAULTS

The distribution of eruptive centers along the Bright Angel and Mesa Butte fault systems and the relatively great length of these systems suggest that the faults observed in Phanerozoic rocks may be related to more profound, deep-seated structures in the crust. Where the Bright Angel and Kaibab fault systems cross the Grand Canyon, the relationship of the faults to the deeper-lying Precambrian rocks is exposed. Here the principal displacement of Paleozoic beds along both systems of faults was controlled by more ancient faults in the underlying Precambrian rocks.

Walcott (1889) was the first to demonstrate that displacement of Phanerozoic rocks in the Grand Canyon had occurred along a Precambrian fault. He found that the northwest-trending Butte fault, which occurs along the East Kaibab monocline and which had reverse movement in Tertiary time, had much larger displacement, but of the opposite sense, in Precambrian time. Ford and Breed (1973) estimated as much as 1.5 km of normal displacement on the Butte fault after deposition of the Grand Canyon Supergroup and before deposition of the Tapeats Sandstone of Cambrian age.

Noble (1914) recognized a number of other post-Paleozoic faults that occur along Precambrian lines of displacement. He found that both northwest-trending and northeast-trending faults are controlled by Precambrian structure. On these faults, Phanerozoic displacement generally is smaller and in the opposite sense from that which occurred in late Precambrian time. The northwest-trending Muav fault (Fig. 6) was shown by Noble to have had a minimum normal displacement of nearly 2 km in Precambrian time. A faulted monocline with much smaller throw is developed in the Paleozoic rocks over the Muav fault.

Maxson and Campbell (1934) found that the crystalline rocks of the Grand Canyon had been displaced by faults prior to deposition of the Grand Canyon Supergroup. These faults trend in two different directions: (1) N15° to 30°E, parallel with the schistosity of the crystalline metamorphic complex (Vishnu Schist); and (2) N20° to 30°W, parallel with a direction of master joints, which also were formed prior to deposition of the Grand Canyon Supergroup. Some of these faults are overlain by unbroken Precambrian strata; other faults of the same network that displace overlying strata presumably were active in post–Vishnu Schist pre–Grand Canyon Supergroup time (Maxson, 1961).

The Bright Angel fault belongs to the latter category. Maxson (1961) inferred six episodes of displacement on the Bright Angel fault:

1. Displacement of crystalline metamorphic rocks of early Precambrian age prior to intrusion of the Zoroaster Granite. Foliated migmatites in metamorphic complex have been dated at 1,695 ± 15 m.y. B.P. and the Zoroaster Granite, at 1,725 ± 15 m.y. B.P.; these dates indicate a major episode of Precambrian deformation at about 1,700 m.y. B.P. (Pasteels and Silver, 1966).

2. Post-Zoroaster displacement of the crystalline complex prior to deposition of the Grand Canyon Supergroup.

3. Displacement after deposition of the Dox Sandstone of Precambrian age and prior to intrusion of diabase. Diabase intrusions in the Grand Canyon have been dated at 1,150 to 1,200 m.y. B.P. by L. T. Silver (1974, oral commun.).

4. Reverse faulting after deposition of the Grand Canyon Supergroup.

5. A second episode of reverse displacement prior to deposition of the Tapeats Sandstone of Cambrian age.

6. Normal and strike-slip displacement during the Laramide orogeny.

Sears (1973) inferred a minimum of seven episodes of displacement on the Bright Angel fault:

1. Displacement of 60 m, up on the southeast, during deposition of the Shinumo Quartzite of Precambrian age.

2. Post–Dox Sandstone, pre-diabase displacement.

3. Local displacement during intrusion of diabase.

4. Post-diabase reverse displacement of 200 m.

5. Local post–Grand Canyon Supergroup scissors displacement.

6. Reverse displacement after deposition of Redwall Limestone of Mississipian age and prior to deposition of the lowermost beds of the Supai Formation, which are of Pennsylvanian age.

7. Post-Paleozoic normal displacement.

It appears that at least seven and possibly as many as nine distinct episodes of displacement can be documented at various places along the Bright Angel fault. Most of this displacement occurred during Precambrian time. Maxson indicated that there had been a significant strike-slip component of displacement in post–Paleozoic time, but Sears has strongly challenged this interpretation.

The decipherable network of faults cutting the Grand Canyon Supergroup, which were active in Precambrian time, has been carefully re-examined by Sears (1973). His synthesis of this network is illustrated in Figure 6. The more ancient northwest and northeast directions of faulting in the crystalline rocks are closely reflected in the late Precambrian pattern of displacement. This pattern is reflected, in turn, by Phanerozoic displacement along the Bright

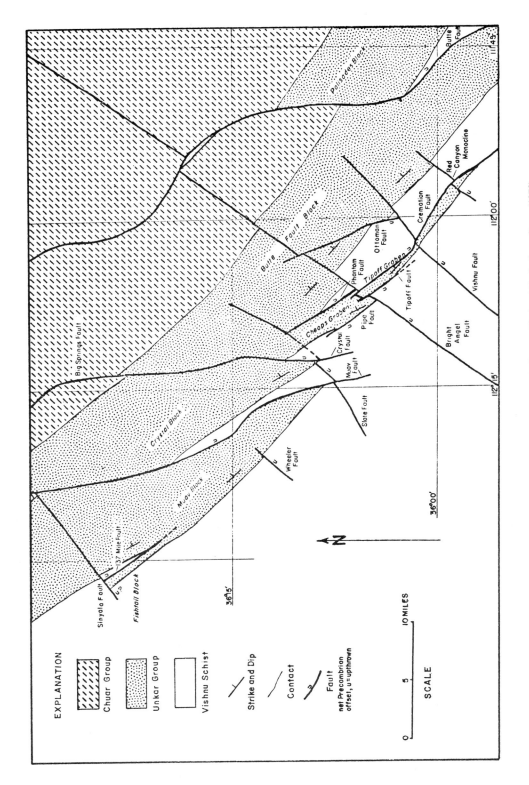

Figure 15-6. Sub–Paleozoic geologic map of Grand Canyon National Park, from Sears (1973).

Angel and Kaibab fault systems. Most of the faults that were active in late Precambrian time were reactivated in Phanerozoic time.

The early history of displacement on the Mesa Butte fault system is much less well known. An episode of Precambrian displacement is documented in the Paulden quadrangle (Krieger, 1965). Northeast of the Mesa Butte graben, minor displacement on the Grandview monocline in Permian time may be indicated by slump blocks in the Toroweap Formation of Permian age (Marshall, 1972).

AEROMAGNETIC AND GRAVITY ANOMALIES

Exposures of Precambrian rocks in northern Arizona are relatively limited, and it would be desirable to determine the relationship of Phanerozoic fault displacements to Precambrian structure over a broader area than the Grand Canyon. Aeromagnetic and gravity data provide powerful tools to examine Precambrian structure, especially in northwestern Arizona, where Phanerozoic rocks are generally less than 2 km thick and, except for Cenozoic volcanic rocks, very weakly magnetic. Local magnetic anomalies of limited extent and with steep gradients, associated chiefly with Cenozoic volcanic rocks in the San Francisco volcanic field, are readily distinguished from relatively broad anomalies with shallow gradients related to the Precambrian rocks. A residual aeromagnetic map of Arizona by Sauck and Sumner (1971) and a Bouguer gravity anomaly map of the state by West and Sumner (1973) reveal prominent structural trends in the Precambrian that are parallel with the observed systems of faults.

Comparison of the principal faults in the Bright Angel, Mesa Butte, and Sinyala fault systems with the residual aeromagnetic map reveals a close correspondence between the positions of these faults and the margins of a series of large northeast-trending magnetic anomalies. This correspondence has emboldened us to infer the traces of the concealed ancestral Bright Angel, Mesa Butte, and Sinyala faults shown in Figure 7. Large displacement of the crystalline Precambrian rocks is postulated to have occurred on these faults in order to account for aligned linear margins of the anomalies. Significant displacement of the crystalline rocks along the Bright Angel fault is suggested by the observations of Noble (Noble and Hunter, 1916, p. 101).

The ancestral Bright Angel fault is exposed in the Grand Canyon for a distance of 14 km, but the ancestral Mesa Butte fault is not exposed in the area shown in Figure 7. Precambrian displacement on the Sinyala fault system has not been demonstrated on the outcrop; the ancestral Sinyala fault is inferred on the basis of the fault pattern in observed Phanerozoic rocks and on the relationship of the observed faults to the magnetic anomalies.

Perhaps the most noteworthy feature of the aeromagnetic map of Arizona is a northeast-trending belt of large positive magnetic anomalies 400 km long that extends from the vicinity of Congress to the northern border of the state. The amplitude of these anomalies ranges from about 300 to 700 γ (1 γ = 1 nt). Over most of its length in northern Arizona the magnetic-anomaly belt corresponds to a belt of positive gravity anomalies with amplitudes of 10 to 30 mgal (1 mgal = 10^{-5} m/s^2); the gravity-anomaly belt, however, is less well defined. The Mesa Butte fault system lies along the southeastern margin of the magnetic anomaly belt, and the Mesa Butte fault follows some of the steepest anomaly gradients on this margin. The monoclines along the northeast projection of the Mesa Butte fault system also follow the margin of the anomaly belt. This relationship suggests that the ancestral Mesa Butte fault continues northeastward beneath these monoclines (Fig. 7).

Another large northeast-trending positive magnetic anomaly is bounded on the southeast by the Bright Angel fault. This anomaly, about 75 km long and 15 km wide, has a maximum amplitude of about 700 γ. Where the anomaly is most pronounced, it corresponds to a well-defined positive gravity anomaly with an amplitude of 5 to 10 mgal. Where it crosses the Grand Canyon, the magnetic anomaly coincides approximately with a belt of amphibolite, migmatite, and schist referred to by Maxson (1961) as the Brahma Schist.

Maxson interpreted the Brahma Schist to be a sequence of metavolcanic and metasedimentary rocks overlying dominantly metasedimentary rocks of the Vishnu Schist. According to Maxson, the Brahma Schist is folded down into the Vishnu terrane in a large isoclinal syncline. Subsequent work has not supported this interpretation, however. The rocks called Brahma Schist by Maxson appear to be part of the Vishnu Schist sequence (Ragan and Sheridan, 1970); therefore, the name Brahma Schist has been abandoned. The positive magnetic and gravity anomalies along the prevailing strike of Maxson's Brahma Schist, on the other hand, indicate that the rocks in this part of the Vishnu terrane have higher mean magnetic susceptibility and higher mean density than adjacent parts of the Vishnu Schist, at least northwest of the Bright Angel fault. Probably these physical properties can be attributed to a greater-than-average abundance of amphibolite in this block of Vishnu terrane. By analogy, we suggest that the large positive magnetic anomalies associated with positive gravity anomalies adjacent to the Mesa Butte fault system are also related to belts of mafic metavolcanic rocks in the crystalline Precambrian complex.

Other large positive magnetic anomalies are bounded by inferred extensions of the ancestral Bright Angel fault (Fig. 7), but these anomalies are more widely spaced than those along the Mesa Butte fault system. Some, but not all, of the positive magnetic anomalies along the Bright Angel system are associated with positive gravity anomalies.

The ancestral Sinyala fault has been drawn along the margins of a series of positive magnetic anomalies of relatively limited extent that are grouped into two broad aeromagnetic highs (Fig. 7). The northeast trend of individual anomalies in these groups is not as pronounced as along the Bright Angel and Mesa Butte fault systems. The positive magnetic anomalies along the Sinyala system correspond only very roughly with positive gravity anomalies.

Faults belonging to the northwest- and north-trending fault systems also follow linear margins of aeromagnetic anomalies, although these trends are not as obvious as the northeast trend on the aeromagnetic map. A good example is a north-trending magnetic-anomaly boundary that coincides with the north-striking Oak Creek Canyon fault. We conclude that the northwest- and north-trending fault systems are also controlled by faults with large displacement in the Precambrian crystalline basement.

SHYLOCK AND CHAPARRAL FAULT ZONES

The ancestral Bright Angel and Mesa Butte faults may be related in origin to the Shylock and Chaparral fault zones in the central Arizona mountain belt, which were described by Anderson (1967). Here Precambrian rocks are exposed over a wide area, and the detailed structure of the Precambrian has been worked out by Anderson and his colleagues (Anderson, 1959, 1967; Anderson and Creasey, 1958, 1967; Anderson and others, 1971; Blacet, 1966; Krieger, 1965). The Shylock and Chaparral fault zones are easily recognized in ERTS-1 pictures (Fig. 8A), and the Shylock has been traced south of the area mapped by Anderson on the basis of the ERTS-1 data (extension shown by dashed lines in (Fig. 8B).

The Shylock fault zone is a north–trending, 1- to 3-km-wide belt of roughly parallel, interlacing faults cutting Precambrian rocks. As interpreted by Anderson (1967), the zone represents

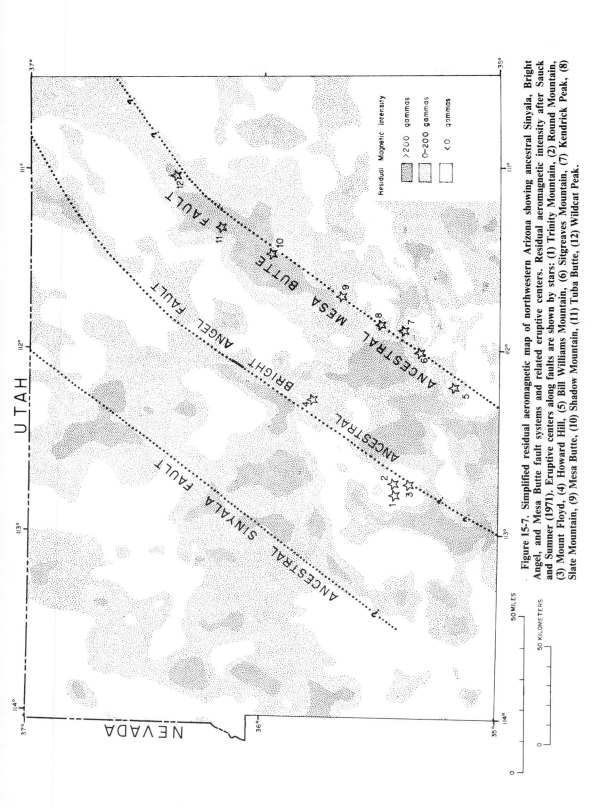

Figure 15-7. Simplified residual aeromagnetic map of northwestern Arizona showing ancestral Sinyala, Bright Angel, and Mesa Butte fault systems and related eruptive centers. Residual aeromagnetic intensity after Sauck and Sumner (1971). Eruptive centers along faults are shown by stars: (1) Trinity Mountain, (2) Round Mountain, (3) Mount Floyd, (4) Howard Hill, (5) Bill Williams Mountain, (6) Sitgreaves Mountain, (7) Kendrick Peak, (8) Slate Mountain, (9) Mesa Butte, (10) Shadow Mountain, (11) Tuba Butte, (12) Wildcat Peak.

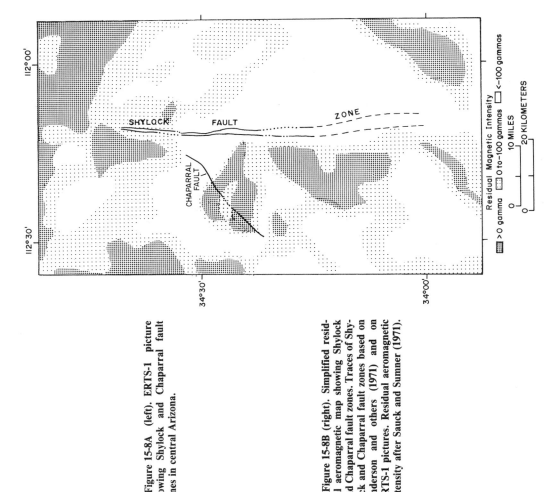

Figure 15-8A (left). ERTS-1 picture showing Shylock and Chaparral fault zones in central Arizona.

Figure 15-8B (right). Simplified residual aeromagnetic map showing Shylock and Chaparral fault zones. Traces of Shylock and Chaparral fault zones based on Anderson and others (1971) and on ERTS-1 pictures. Residual aeromagnetic intensity after Sauck and Sumner (1971).

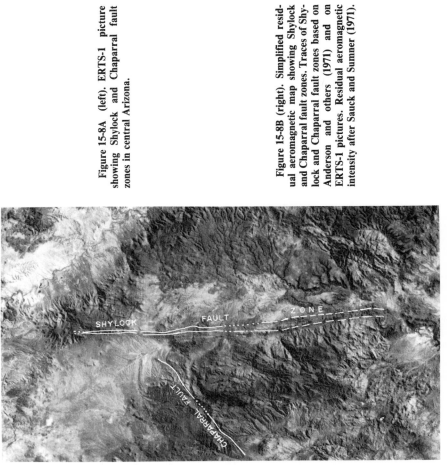

a major transcurrent fault with a minimum right-lateral displacement of 8 km. Estimates of this displacement are based on the offset of slices of quartz diorite in the fault zone. The total horizontal displacement may be much greater than 8 km.

A large contrast in the magnetic properties of the Precambrian rocks on opposite sides of the Shylock fault zone, indicated by aeromagnetic data (Fig. 8B), suggests the displacement may be as great as several tens of kilometres. Positive magnetic anomalies on the west side of the fault are related, in a general way, to metavolcanic rocks of the Big Bug Group of Precambrian age. The anomalies do not correspond closely to mapped geologic units, however. A broad magnetic low on the east side of the fault corresponds, at least in part, to a cluster of plutonic rocks. In contrast to the relations observed along the Bright Angel and Mesa Butte fault systems, the positive magnetic anomalies along the Shylock zone are gravity lows, and the magnetic low coincides with a broad positive gravity anomaly with an amplitude of 25 mgal. The plutonic rocks east of the Shylock fault zone are evidently less strongly magnetized but have higher density than the metavolcanic rocks of the Big Bug Group.

In the Mingus Mountain quadrangle, along the southern border of the Black Hills, the Shylock fault zone is locally overlain by unbroken Tapeats Sandstone of Cambrian age (Anderson and Creasey, 1958). The large transcurrent displacement occurred before the Cambrian and after emplacement of the quartz diorite pluton that is offset by the fault. The quartz diorite has been dated at 1,760 ± 15 m.y. B.P. (Anderson and others, 1971). It is possible that the transcurrent displacement along the Shylock zone is related in time to faulting of the Vishnu terrane that occurred before deposition of the Grand Canyon Supergroup.

Near the Black Hills, the Coyote fault, which branches north-northwest from the Shylock fault zone, displaces Paleozoic and Tertiary rocks (Anderson and Creasy, 1958). North of the Black Hills, directly in line with the main Shylock zone, the Orchard fault also displaces Paleozoic and Tertiary rocks. At least two episodes of normal displacement have occurred on both faults (Lehner, 1958). The old transcurrent fault zone has clearly controlled the pattern of Phanerozoic normal displacement.

The Chaparral fault zone is a northeast-trending zone of distributive shear, as much as 1 km wide, that cuts Precambrian rocks (Anderson and Creasey, 1958; Krieger, 1965). Detailed structural features within the zone indicate right-lateral slip (Krieger, 1965), and Anderson (1967) suggested that appreciable right-lateral displacement is required to account for separation of intrusive rocks common to both sides of the fault zone. The Chaparral fault zone trends toward the Shylock zone but does not offset it. It appears likely that the Chaparral joins the Shylock tangentially, but the critical area is concealed by Phanerozoic deposits.

We suggest that the northeast- and north-trending fault systems of northern Arizona are controlled by ancestral transcurrent faults in the Precambrian basement similar to the Shylock and Chaparral fault zones. The 1,700- to 1,800-m.y.-old Precambrian terrane, represented by the Vishnu Schist in northern Arizona and the Yavapai Series (Anderson and others, 1971) in central Arizona and by associated plutonic rocks, probably was riven by right-lateral faults prior to the deposition of the Grand Canyon Supergroup. Displacement on the ancestral Bright Angel and Mesa Butte faults may have been comparable to that on the Shylock fault zone. Major displacement probably occurred on a few main faults, which divide the crust into blocks tens of kilometres across. Minor shearing occurred within these blocks, particularly along their margins. Both the major and minor faults controlled later, dominantly vertical displacement in late Precambrian and Phanerozoic time.

Whether the ancestral northwest-trending faults are related tectonically to the northeast- and north-trending ancestral transcurrent faults is a problem that awaits further investigation. Evidence should be sought in the central Arizona mountain belt.

CENOZOIC HISTORY OF DISPLACEMENT

Cenozoic deformation in northern Arizona occurred in two widely separated periods. Folding on a monocline parallel with the northern end of the East Kaibab monocline took place after deposition of Upper Cretaceous strata and before deposition of beds now assigned to the Paleocene Series (Bowers, 1972). The principal folding of some monoclines on the eastern side of the Colorado Plateau occurred near the end of the Eocene (Shoemaker, 1956). Presumably, the monoclinal folding and broad regional warping of strata in northern Arizona took place during this episode of compressive deformation that lasted from latest Cretaceous to late Eocene time, which commonly has been referred to as the Laramide orogeny. Some reverse faulting along monoclines probably occurred during this early period of deformation. Subsequently, Paleozoic and Mesozoic strata were deeply eroded and, in central Arizona, stripped away entirely. By mid-Tertiary time a widespread, relatively mature erosion surface had developed (McKee and others, 1967), which is locally preserved beneath mid-Tertiary sediments and Oligocene and Miocene volcanic rocks (Krieger and others, 1971). Where the history of displacement on faults can be documented by stratigraphic evidence, most or all Cenozoic normal displacement has occurred after emplacement of these mid-Tertiary sedimentary and volcanic rocks.

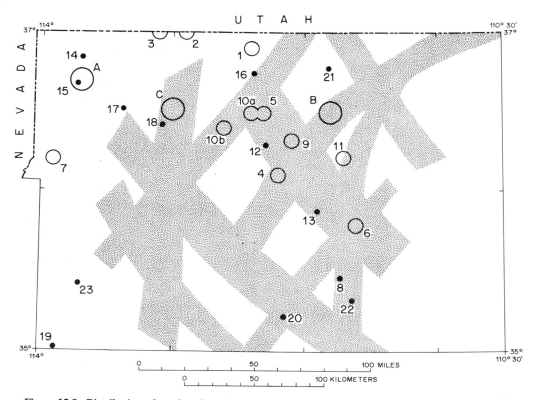

Figure 15-9. Distribution of earthquake epicenters in northwestern Arizona for the period 1938 to 1973. Epicenters shown with large circles generally are least accurately known; those shown with solid dots generally are most accurately known. Principal fault systems are shown with stipple. Numbers and letters correspond to list in Table 15-1.

In northern Arizona, large displacement took place on normal faults in Miocene time and has continued to the present. Displacement on the Grand Wash fault occurred mainly after emplacement of the 17- to 18-m.y.-old Peach Springs Tuff (Young and Brennan, 1974) but before deposition of the Muddy Creek Formation of late Miocene and Pliocene(?) age (Lucchitta, 1972, 1974). Some normal displacement along the Cottonwood Cliffs occurred prior to emplacement of the Peach Springs Tuff (Fuis, 1973). On faults farther to the east, most displacement postdates middle to upper Miocene lavas but predates Pleistocene volcanic rocks. Faulting has continued into Quaternary time, however, as shown by relatively fresh fault scarps in alluvium in Chino Valley and near Cameron, Arizona (Akers and others, 1962), by minor offset of Pleistocene lava flows, and by the present seismicity of the region.

The Cenozoic history of displacement of the Mesa Butte fault system is fairly well known. On the southwestern segment of the system, most displacement postdates old basaltic lavas in the southern San Francisco volcanic field and predates lavas and sedimentary rocks now assigned by Krieger and others (1971) to the Perkinsville Formation of late Miocene to Pliocene age. McKee and Anderson (1971) reported K-Ar ages of 4.5 to 6.0 m.y. on basalts in the Perkinsville. In some places older units of the Perkinsville are displaced and younger units are not. Old basaltic lavas in the southern part of the San Francisco volcanic field have been dated at 11.1 ± 0.5 and 14.4 ± 0.6 m.y. B.P. from a locality near Sycamore Canyon by McKee and McKee (1972). Damon and others (1974) have dated a basalt from another locality near Sycamore Canyon at 8.68 ± 0.98 m.y. B.P. These localities are 15 to 25 km east of the prominent fault scarp where basalts (probably of comparable age) are displaced.

On the northeastern segment of the Mesa Butte fault system, most displacement occurred before the extrusion of lavas that follow the base of fault scarps and conceal the faults. Mesa Butte and the lava flows from this vent are clearly later than most of the displacement on the Mesa Butte fault system. Locally these flows are offset, however, both on the Mesa Butte fault and on the Cedar Ranch fault. On the basis of many dated basaltic lavas in nearby parts of the San Francisco volcanic field, normal displacement on the fault systems near Mesa Butte appears to have occurred between about 4.0 and 0.5 m.y. ago (Damon and others, 1974).

The youngest lavas displaced by faults of the Mesa Butte system are a flow from the

TABLE 15-1. LIST OF EARTHQUAKES PLOTTED ON FIGURE 9

Map No.	Date	Time (GMT)	Magnitude	Map No.	Date	Time (GMT)	Magnitude
A	August 17, 1938	09:08:06	. .	11	September 4, 1967	23:27:45	. .
B	January 7, 1945	22:25:32	5.1	12	November 24, 1970	16:47:56	3.0
C	November 2, 1949	02:29:36	4.7	13	December 3, 1970	03:47:25	2.8
1	February 15, 1962	07:12:43	4.5	14	December 16, 1970	13:44:19	2.6
2	February 15, 1962	09:06:45	. .	15	December 16, 1970	13:46:47	2.2
3	August 28, 1964	06:50:47	. .	16	March 27, 1971	04:39:12	2.6
4	June 7, 1965	14:28:01	. .	17	May 1, 1971	03:11:20	2.9
5	September 3, 1966	07:53:20	. .	18	May 6, 1971	16:57:18	2.2
6	October 3, 1966	16:03:51	. .	19	May 23, 1971	21:31:52	3.0
7	December 1, 1966	09:20:41	3.7	20	November 4, 1971	02:18:59	3.7
8	March 28, 1967	03:48:59	. .	21	December 15, 1971	12:58:15	3.0
9	July 20, 1967	13:51:10	. .	22	April 20, 1972	13:28:16	3.7
10a	August 7, 1967	16:24:49	. .	23	April 12, 1973	10:57:48	. .
10b	August 7, 1967	16:40:32	. .				

Note: Leaders (. .) indicate no data.

Shadow Mountain vent, dated at 0.62 ± 0.23 m.y. B.P. (Damon and others, 1974), and the Tappan Wash flow, dated at 0.510 ± 0.079 m.y. B.P. (Damon and others, 1974). The Shadow Mountain lava is displaced about 13 m by a graben that cuts the flow (Condit, 1974). The Tappan Wash flow has been traced by M. Malin, E. M. Shoemaker, and R. B. Moore from a cinder cone on the east flank of Kendrick Peak (Fig. 4) to previously recognized outcrops in Tappan Wash and along the Little Colorado River. Where it crosses the Cedar Ranch fault, it appears to be displaced several metres. Along the Little Colorado River (Fig. 4) the Tappan Wash flow is displaced about 20 m by the Cameron graben (Reiche, 1937). According to Reiche, the graben is entirely younger than the flow. A channel was cut into the flow by the Little Colorado River, filled with alluvium, and then abandoned, all before development of the graben (Reiche, 1937).

Evidence of the time of displacement on the Bright Angel fault system is indirect. The principal displacement on some faults of the Cataract Creek system, which intersects the Bright Angel system near Cataract Creek, occurred after deposition of mid-Tertiary sediments and extrusion of basalt flows in the nearby Mount Floyd volcanic field. Basalt flows at Long Point, on the northern edge of this field, have been dated by McKee and McKee (1972) at 7.4 ± 0.4 and 14.0 ± 0.6 m.y. B.P. A well-integrated dendritic drainage system in the Cataract Creek basin appears to be antecedent to the Bright Angel fault (Fig. 3B). However, the prisms of Moenkopi Formation preserved along several faults in the system show that displacement began before the Moenkopi Formation was stripped from most of the Coconino Plateau. Several faults in the system have controlled the development of side canyons in the Grand Canyon (see, for example, Fig. 3B). Probably the drainage pattern is anteposed (Hunt, 1956) to the fault system.

Several lines of evidence (for example, Hamblin, 1970) indicate that normal faulting in northern Arizona has continued into Holocene time. Indeed, the fault systems are seismically active. Three moderately strong earthquakes (intensity VI to VIII on the modified Mercalli scale) occurred in 1906, 1910, and 1912 near or north of Flagstaff (Townley and Allen, 1939; Sturgul and Irwin, 1971; Coffman and von Hake, 1973). Newspaper accounts suggest the epicenters of the 1910 and 1912 earthquakes may have been near the intersection of the Mesa Butte, Oak Creek Canyon, and Kaibab fault systems. Sturgul and Irwin (1971) reported 17 earthquakes of intensity IV or greater in northwestern Arizona during the period from 1850 to 1966. The epicenters of almost all these earthquakes, however, are not known with sufficient precision for comparison with the fault systems.

As of January 11, 1974, 27 earthquakes in northwestern Arizona (Fig. 9), for which the epicenters have been estimated to 0.1° latitude and longitude or better, were on record in the hypocenter data file of the National Oceanic and Atmospheric Administration. Only locations determined with five or more seisomographic stations are included. The actual error in location of several of the epicenters plotted in Figure 9 probably is greater than 0.1°. All focal depths are shallow, probably in the crust. The epicenter locations of relatively minor earthquakes that occurred after 1969 probably are the most reliable.

All but one of the epicenters shown in Figure 9 lie either within the previously defined fault systems, on the projection of these systems, or along a major fault. The earthquake plotted as no. 2 followed earthquake no. 1 by less than 2 h (see Table 1). It is likely that the distance between the epicenters of these two earthquakes is less than indicated in Figure 1; one or both of the epicenters may be significantly in error. Another pair of events (10a and 10b) is separated by 10 min in time and by about 20 km on the map; a third pair (14 and 15), along the Grand Wash fault, is separated by 2.5 min and 20 km.

The Toroweap, Sinyala, Bright Angel, Mesa Butte, Kaibab, and Oak Creek Canyon fault systems are demonstrably active. A magnitude 5.1 earthquake appears to have occurred along or near the Eminence Break graben in 1945. Five epicenters shown near the western border of the map are along or near the surface trace of the Grand Wash fault.

ACKNOWLEDGMENTS

We wish to thank D. P. Elston, V. L. Freeman, Ivo Lucchitta, P. W. Huntoon, and R. B. Moore for helpful comments and suggestions. C. D. Condit was particularly helpful in bringing to our attention unpublished work and in tracking down information. J. W. Sears kindly gave permission to reproduce a map from his unpublished thesis. M. Brock, E. A. Holm, Ivo Lucchitta, and G. E. Ulrich gave us copies of their manuscript maps, and Ulrich secured for us the high-altitude oblique aerial photographs used in Figure 3. C. A. von Hake provided the data on earthquakes shown in Figure 9. Our colleague, A. F. Goetz provided many computer-processed ERTS-1 pictures used in the course of this study. P. E. Damon graciously sent isotopic age determinations in advance of publication. We especially wish to thank J. D. Strobell and C. S. Shoemaker for extensive help in the course of this study and in bringing the final manuscript together.

This investigation has been performed under NASA Contract NAS 7-100.

REFERENCES CITED

Akers, J. P., Irwin, J. H., Stevens, P. R., McClymonds, N. E., and Chenoweth, W. L., 1962, Geology of the Cameron quadrangle, Arizona: U.S. Geol. Survey Geol. Quad. Map GQ-162.

Anderson, C. A., 1959, Preliminary geologic map of the NW¼ Mayer quadrangle, Yavapai County, Arizona: U.S. Geol. Survey Map MF-228.

——1967, Precambrian wrench fault in central Arizona, in Geological Survey research 1967: U.S. Geol. Survey Prof. Paper 575-C, p. C60–C65.

Anderson, C. A., and Creasey, S. C., 1958, Geology and ore deposits of the Jerome area, Yavapai County, Arizona: U.S. Geol. Survey Prof. Paper 308, 184 p.

——1967, Geologic map of the Mingus Mountain quadrangle, Yavapai County, Arizona: U.S. Geol. Survey Geol. Quad. Map GQ-715.

Anderson, C. A., Blacet, P. M., Silver, L. T., and Stern, T. W., 1971, Revision of Precambrian stratigraphy in the Prescott–Jerome area, Yavapai County, Arizona: U.S. Geol. Survey Bull. 1324-C, p. C1–C16.

Babenroth, D. L., and Strahler, A. N., 1945, Geomorphology and structure of the East Kaibab monocline, Arizona and Utah: Geol. Soc. America Bull., v. 56, p. 107–150.

Baker, A. A., 1936, Geology of the Monument Valley–Navajo Mountain region, San Juan County, Utah: U.S. Geol. Survey Bull. 865, 106 p.

Blacet, P. M., 1966, Unconformity between gneissic granodiorite and overlying Yavapai Series (older Precambrian), central Arizona, in Geological Survey research 1966: U.S. Geol. Survey Prof. Paper 550-B, p. B1–B5.

Blissenbach, Erich, 1952, Geology of Aubrey Valley: Plateau, v. 24, p. 119–127.

Bowers, W. E., 1972, The Canaan Peak, Pine Hollow, and Wasatch Formations in the Table Cliff region, Garfield County, Utah: U.S. Geol. Survey Bull. 1331-B, p. B1–B39.

Coffman, J. L., and von Hake, C. A., 1973, Earthquake history of the United States: U.S. Natl. Oceanic and Atmospheric Admin. Pub. 41-1, rev. ed. (through 1970), 208 p.

Condie, K. C., 1964, Crystallization P_{O_2} of syenite porphyry from Navajo Mountain, southern Utah: Geol. Soc. America Bull., v. 75, p. 359–362.

Condit, C. D., 1974, Geology of Shadow Mountain, Coconino County, Arizona: in Geology of northern Arizona with notes on archaeology and paleoclimate, Pt. II—Area studies and field guides: Flagstaff, Northern Arizona Univ., p. 454–463.

Cooley, M. E., Harshbarger, J. W., Akers, J. P., and Hardt, W. F., 1969, Regional hydrogeology of the Navajo and Hopi Indian Reservations: U.S. Geol. Survey Prof. Paper 521-A, 61 p.

Damon, P. E., Shafiqullah, M., and Leventhal, J. S., 1974, K-Ar chronology for the San Francisco volcanic field and rate of erosion of the Little Colorado River, *in* Geology of northern Arizona with notes on archaeology and paleoclimate, Pt. I—Regional studies: Flagstaff, Northern Arizona Univ., p. 221–235.

Darton, N. H., 1910, A reconnaissance of parts of northwestern New Mexico and northern Arizona: U.S. Geol. Survey Bull. 435, 88 p.

Ford, T. D., and Breed, W. J., 1973, Late Precambrian Chuar Group, Grand Canyon, Arizona: Geol. Soc. America Bull., v. 84, p. 1243–1260.

Fuis, G. S., 1973, The geology and mechanics of formation of the Fort Rock dome, Yavapai County, Arizona [Ph.D. dissert.]: Pasadena, California Inst. Technology, 278 p.

Gilbert, G. K., 1877, Report on the geology of the Henry Mountains: U.S. Geog. and Geol. Survey, Rocky Mtn. Region, 160 p.

Gregory, H. E., 1917, Geology of the Navajo country; a reconnaissance of parts of Arizona, New Mexico, and Utah: U.S. Geol. Survey Prof. Paper 93, 161 p.

Hamblin, W. K., 1970, Structure of the western Grand Canyon region: Utah Geol. Soc. Guidebook 23, p. 3–19.

Hunt, C. B., 1942, New interpretation of some laccolithic mountains and its possible bearing on structural traps for oil and gas: Am. Assoc. Petroleum Geologists Bull., v. 26, p. 197–203.

——1956, Cenozoic geology of the Colorado Plateau: U.S. Geol. Survey Prof. Paper 279, 99 p.

Huntoon, P. W., 1970, The hydro-mechanics of the ground water system in the southern portion of the Kaibab Plateau, Arizona [Ph.D. dissert.]: Tucson, Univ. Arizona, 381 p.

Johnson, Douglas, 1909, A geological excursion in the Grand Canyon region: Boston Soc. Natl. History Proc., v. 34, p. 135–161.

Koons, E. D., 1945, Geology of the Uinkaret Plateau, northern Arizona: Geol. Soc. America Bull., v. 56, p. 151–180.

Krieger, M. H., 1965, Geology of the Prescott and Paulden quadrangles, Arizona: U.S. Geol. Survey Prof. Paper 467, 127 p.

——1967a, Reconnaissance geologic map of the Picacho Butte quadrangle, Yavapai and Coconino Counties, Arizona: U.S. Geol. Survey Misc. Geol. Inv. Map I-500.

——1967b, Reconnaissance geologic map of the Turkey Canyon quadrangle, Yavapai County, Arizona: U.S. Geol. Survey Misc. Geol. Inv. Map I-501.

Krieger, M. H., Creasey, S. C., and Marvin, R. F., 1971, Ages of some Tertiary andesitic and latitic volcanic rocks in the Prescott-Jerome area, north-central Arizona: U.S. Geol. Survey Prof. Paper 750-B, p. B157–B160.

Lehner, R. E., 1958, Geology of the Clarkdale quadrangle, Arizona: U.S. Geol. Survey Bull. 1021-N, p. 511–592.

Lucchitta, I., 1972, Early history of the Colorado River in the Basin and Range province: Geol. Soc. America Bull., v. 83, p. 1933–1948.

——1974, Structural evolution of northwest Arizona and its relation to adjacent Basin and Range province structures, *in* Geology of northern Arizona with notes on archaeology and paleoclimate, Pt. I—Regional studies: Flagstaff, Northern Arizona Univ., p. 336–354.

Marshall, C. H., 1956a, Photogeologic map of the Lost Spring Mountain NW quadrangle, Mohave County, Arizona: U.S. Geol. Survey Misc. Geol. Inv. Map I-146.

——1956b, Photogeologic map of the Jacob Lake quadrangle, Coconino County, Arizona: U.S. Geol. Survey Misc. Geol. Inv. Map I-194.

——1957, Photogeologic map of the Hurricane Cliffs—Two NE quadrangle, Mohave County, Arizona: U.S. Geol. Survey Misc. Geol. Inv. Map I-252.

Marshall, D. R., 1972, Gravity gliding at Gray Mountain, Coconino County, Arizona [M.S. dissert.]: Flagstaff, Northern Arizona Univ., 83 p.

Maxson, J. H., 1961, Geologic map of the Bright Angel quadrangle, Grand Canyon National Park, Arizona: Grand Canyon Natl. History Assoc.

——1967, Preliminary geologic map of the Grand Canyon and vicinity, Arizona, eastern section: Grand Canyon Natl. History Assoc.

——1969, Preliminary geologic map of the Grand Canyon and vicinity, Arizona, western and central sections: Grand Canyon Natl. History Assoc.

Maxson, J. H., and Campbell, Ian, 1934, Faulting in the Bright Angel quadrangle, Arizona [abs.]: Geol. Soc. America Proc., p. 301.

McKee, E. D., 1929, The Bright Angel fault: Grand Canyon Nature Notes, v. 4, p. 21–22.

McKee, E. D., and McKee, E. H., 1972, Pliocene uplift of the Grand Canyon region: Time of drainage adjustment: Geol. Soc. America Bull., v. 83, p. 1923–1932.

McKee, E. D., Wilson, R. F., Breed, W. J., and Breed, C. S., 1967, Evolution of the Colorado River in Arizona—A hypothesis developed at the symposium on Cenozoic geology of the Colorado Plateau in Arizona, August 1964: Mus. Northern Arizona Bull. 44, 67 p.

McKee, E. H., and Anderson, C. A., 1971, Age and chemistry of Tertiary volcanic rocks in north-central Arizona and relation of the rocks to the Colorado Plateau: Geol. Soc. America Bull., v. 82, p. 2767–2782.

McQueen, Kathleen, 1957, Photogeologic map of the Shinarump NE quadrangle, Coconino County, Arizona: U.S. Geol. Survey Misc. Geol. Inv. Map I-255.

Moore, R. T., 1972, Geology of the Virgin and Beaverdam Mountains, Arizona: Arizona Bur. Mines Bull. 186, 65 p.

Morris, R. H., 1956, Photogeologic map of the Shinarump NW quadrangle, Coconino County, Arizona: U.S. Geol. Survey Misc. Geol. Inv. Map I-139.

Noble, L. F., 1914, The Shinumo quadrangle, Grand Canyon district, Arizona: U.S. Geol. Survey Bull. 549, 96 p.

——1918, Geologic history of the Bright Angel quadrangle, Arizona: U.S. Geol. Survey, text on back of topographic sheet, Bright Angel quadrangle, Coconino County, Arizona.

Noble, L. F., and Hunter, J. F., 1916, A reconnaissance of the Archean complex of the Granite Gorge, Grand Canyon, Arizona: U.S. Geol. Survey Prof. Paper 98-I, p. 95–113.

Pasteels, Paul, and Silver, L. T., 1966, Geochronologic investigations in the crystalline rocks of the Grand Canyon, Arizona [abs.]: Geol. Soc. America Spec. Paper 87, p. 124.

Phoenix, D. A., 1963, Geology of the Lees Ferry area, Coconino County, Arizona: U.S. Geol. Survey Bull. 1137, 86 p.

Pillmore, C. L., 1956, Photogeologic map of the Short Creek NE quadrangle, Mohave County, Arizona: U.S. Geol. Survey Misc. Geol. Inv. Map I-142.

Pomeroy, J. S., 1959, Photogeologic map of the Hurricane Cliffs—2 NW quadrangle, Mohave County, Arizona: U.S. Geol. Survey Misc. Geol. Inv. Map I-293, scale 1:24,000.

Ragan, D. M., and Sheridan, M.F., 1970, The Archean rocks of the Grand Canyon, Arizona: Geol. Soc. America Abs. with Programs, v. 2, p. 132–133.

Ransome, F. L., 1908, Pre–Cambrian sediments and faults in the Grand Canyon of the Colorado: Science, v. 27, p. 667–669.

Reiche, Parry, 1937, Quaternary deformation in the Cameron district of the Plateau province: Am. Jour. Sci., 5th ser., v. 34, p. 123–138.

Robinson, H. H., 1913, The San Franciscan volcanic field, Arizona: U.S. Geol. Survey Prof. Paper 76, 213 p.

Sauck, W. A. and Sumner, J. S., 1971, Residual aero-magnetic map of Arizona: Tucson, Univ. Arizona.

Sears, J. D., 1956, Geology of Comb Ridge and vicinity north of San Juan River, San Juan County, Utah: U.S. Geol. Survey Bull. 1021-E, p. 176–207.

Sears, J. W., 1973, Structural geology of the Precambrian Grand Canyon series, Arizona [M.S. dissert.]: Laramie, Univ. Wyoming, 100 p.

Shoemaker, E. M., 1956, Structural features of the central Colorado Plateau and their relation to uranium deposits, in Page, L. R., ed., Contributions to the geology of uranium and thorium: U.S. Geol. Survey Prof. Paper 300, p. 155–170.

——1960, Penetration mechanics of high velocity meteorites: Internat. Geol. Cong., 21st, Copenhagen 1960, Rept., pt. 18, p. 418–434.

Sturgul, J. R., and Irwin, T. D., 1971, Earthquake history of Arizona and New Mexico: Arizona Geol. Soc. Digest, v. 9, p. 1–37.

Townley, S. D., and Allen, M. W., 1939, Descriptive catalog of earthquakes of the Pacific coast of the United States, 1769–1928: Seismol. Soc. America Bull., v. 29, p. 1–297.

Twenter, F. R., 1962, Geology and promising areas for groundwater development in the Hualapai Indian Reservation: U.S. Geol. Survey Water–Supply Paper 1576-A, p. A1–A37.

Walcott, C. D., 1889, Study of a line of displacement in the Grand Cañon of the Colorado, in northern Arizona: Geol. Soc. America Bull., v. 1, p. 49–64.

Wells, J. D., 1960, Stratigraphy and structure of the House Rock Valley area, Coconino County, Arizona: U.S. Geol. Survey Bull. 1081-D, p. 117–158.

West, R. E., and Sumner, J. S., 1973, Bouguer gravity anomaly map of Arizona: Tucson, Univ. Ariz.

Williams, Howell, 1936, Pliocene volcanoes of the Navajo-Hopi country: Geol. Soc. America Bull., v. 47, p. 111–172.

Wilson, E. D., Moore, R. T., and Cooper, J. R., 1969, Geologic map of Arizona: Tucson, Arizona Bur. Mines.

Young, R. A., and Brennan, W. J., 1974, Peach Springs Tuff: Its bearing on structural evolution of the Colorado Plateau and development of Cenozoic drainage in Mohave County, Arizona: Geol. Soc. America Bull., v. 85, p. 83–90.

MANUSCRIPT RECEIVED BY THE SOCIETY AUGUST 15, 1977

MANUSCRIPT ACCEPTED SEPTEMBER 2, 1977

Printed in U.S.A.

Geological Society of America
Memoir 152

16

Some quantitative aspects of orogenic volcanism in the Oregon Cascades

CRAIG M. WHITE
ALEXANDER R. MCBIRNEY
Center for Volcanology
University of Oregon
Eugene, Oregon 97403

ABSTRACT

Quantitative data on the rates of production and compositional variations of igneous rocks are needed before any of the various models for magmatic activity at convergent plate boundaries can be realistically evaluated. The Cascade Range provides an excellent opportunity to obtain this information, because it contains a record of igneous activity in various structural settings and through most of late Cenozoic time.

When analytical data are combined with volumetric estimates and are grouped according to the measured age relations of the principal eruptive episodes, they show significant compositional trends, both in space and time. The most notable changes since late Oligocene time have been a general decline in the volumes and average SiO_2 contents of erupted rocks. These changes have been accompanied by a marked drop in the degree of Fe enrichment, a steady increase in Na, Sr, and K/Rb and a simultaneous decrease in the Rb contents of rocks at the same stage of differentiation. Sr-isotope ratios remained essentially constant with time and show no relation to SiO_2 or Rb contents. The abundance of Rb in Quaternary rocks is inversely related to the volume of Tertiary rocks erupted the same region.

No single mechanism can account for all the observed variations. At least three distinct stages of crystal-liquid equilibration seem to be required, one in the mantle source region, another near the base of a steadily thickening lithosphere, and a third in the shallow reservoirs of mature volcanoes. There is no evidence that the magma is generated in a subducted plate.

INTRODUCTION

It is impossible to place quantitative constraints on the variety of competing models for magmatic activity at convergent plate boundaries without a better knowledge of the character, distribution, and rates of production of the principal rock units as a function of time and space. In an effort to alleviate the deficiency of such data, we have undertaken a study of the Cenozoic volcanic history of a section of the Cascade Range of central Oregon where an extensive series of Tertiary and Quaternary rocks can be mapped, sampled, and compared with similar rocks elsewhere.

We selected this area because it has an unusually complete Neogene section and has relatively clear stratigraphic and structural relations. Earlier workers, such as Thayer (1937), Williams (1957), and Peck and others (1964), established the basic geologic framework of the units, and more recent studies by the staff and students of the Center for Volcanology have provided a wealth of petrographic and geochemical data on individual groups of rocks and eruptive centers. Radiometric dating by John F. Sutter (Sutter, 1978) has provided essential chronologic control for stratigraphic correlations and estimates of the duration of magmatic events.

Using the results of these studies as a basis for additional sampling, we have attempted to fill the most conspicuous gaps in the stratigraphic and areal distribution of analyzed samples, so that better estimates can be made of the volumes and spatial relations of the principal rock types. A preliminary compilation of these data (McBirney and others, 1974) showed that late Cenozoic volcanism has been dominantly basaltic and that activity has been concentrated in four or five brief episodes that appear to have been synchronous with volcanism elsewhere in and around the Pacific Ocean (Kennett and others, 1977). Our purpose here is to present additional data on compositional trends that may reflect the mechanisms responsible for generation and differentiation of calc-alkalic magmas.

METHODS

Published analyses of Cascade rocks are strongly biased in favor of youthful differentiated lavas. The impression that andesites are the dominant rocks of the region is the result of the tendency for most workers to concentrate on large composite volcanoes and to pay less attention to flat-lying basaltic lavas. Although they are not topographically imposing, some individual basaltic flows have volumes greater than that of a large andesitic volcano. Our sampling has been designed to redress this imbalance by obtaining more data from flows beneath or between the major composite cones. In addition, we have attempted to carry out a more systematic sampling of stratigraphic sections through the Quaternary volcanoes and their underlying Tertiary basement.

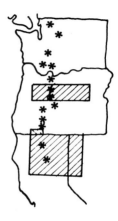

Figure 16-1. Areas for which proportions of upper Cenozoic rocks have been calculated (McBirney and others, 1974). The small area in Oregon near the California border has been studied by Naslund (1977) in order to provide a direct correlation between recent studies in central Oregon and older work in northern California. Asterisks show main Cascade axis.

Analyses have been compiled in a computer file from which the data can be selectively retrieved and statistically evaluated. All analyses from published sources, theses, and unpublished results from the Center for Volcanology are coded according to composition, age, location, and rock name. The file currently contains more than 1,000 analyses from the entire Cascade province; nearly half of these analyses are from the area of central Oregon where we have concentrated our main attention.

Rocks have been divided into three broad categories—basalt, andesite, and dacite-rhyolite—on the basis of SiO_2 content. Divisions have been placed at 53.5% and 63% SiO_2. A further subdivision has been made of Quaternary rocks by grouping rocks between 53.5% and 57% SiO_2 in a separate group of basaltic andesites and by dividing dacite from rhyolite at 68% SiO_2.

Averages for each compositional and age group have been calculated from available analyses of appropriate age and location. Overall averages of all rocks of a given area or magmatic episode have been weighted according to the volumetric proportions of the three main rock types. Volumetric proportions were calculated from the areal distribution, thickness, and probable original extent of each unit, as shown on geologic maps (McBirney and others, 1974).

Volumetric estimates have been made for two parts of the Cascade system, one in central Oregon and another in northern California (Fig. 16-1). Only three volcanic episodes are represented by enough data in northern California to justify volumetric calculations, but in central Oregon, where the section is deeper, four full units and the upper part of a fifth are exposed. One of these, the Elk Lake formation (about 9 to 10 m.y. old) is much less important than the others, and too few rocks, especially in the range of dacites and rhyolites, have been analyzed to provide a valid comparison with other groups. For this reason, the rocks have been included with those of the mid-Miocene Sardine Formation to which they have a close resemblance.

The numbers of major-element analyses from the four principal units in central Oregon are Oligocene–lower Miocene Little Butte Formation, 41; middle Miocene Sardine Formation and upper Miocene Elk Lake formation, 136; Pliocene Outerson Formation, 99; and Quaternary High Cascade rocks, 176.

RESULTS

Quaternary Rocks

Pleistocene and Holocene rocks of the High Cascade volcanoes have been examined in terms of their compositional relations in the chain as a whole and in the development of a single cone.

The largest Quaternary volcano in the central Oregon section that has been studied in detail is Mount Jefferson (Thayer, 1937; Condie and Swensen, 1974; Sutton, 1974; C. M. White, in progress). The rocks forming the base and main cone of Mount Jefferson were erupted during three stages, all of which took place within the present period of normal magnetic polarity (that is, less than 690,000 yr ago). The earliest-erupted lavas make up the glaciated Minto sequence, which formed scattered cones and shields covering an area nearly twice as broad as the basal diameter of Mount Jefferson. The composite cone of Mount Jefferson developed during two main stages, which were followed by minor postglacial flank eruptions (Forked Butte lavas). Sutton (1974) has shown that the volumes of these rocks tended to diminish with time as follows: Pleistocene shield lavas, 98 km^3; first stage of main cone,

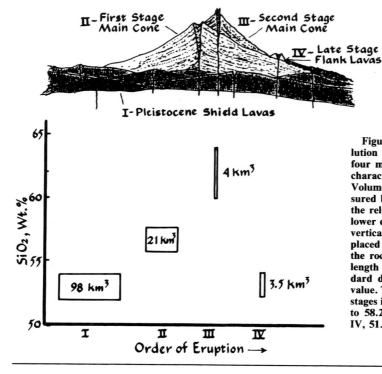

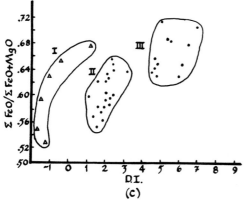

I-Pleistocene Shield Lavas

Figure 16-2. The Quaternary evolution of Mount Jefferson included four main stages, each of which was characterized by distinctive rocks. Volumes of rocks in each stage (measured by Sutton, 1974) are shown by the relative areas of rectangles in the lower diagram. The mid-point on the vertical dimension of the rectangles is placed at the mean value of SiO_2 for the rocks of that stage, and vertical length of the edge indicates one standard deviation from the mean SiO_2 value. Total range of SiO_2 for the four stages is I, 51.6% to 54.3%; II, 54.5% to 58.2%; III, 60.1% to 64.3%; and IV, 51.5% to 54.3% by weight.

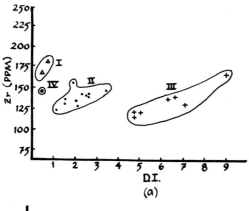

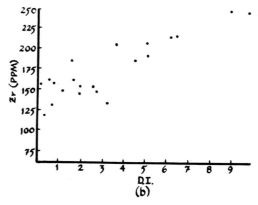

Figure 16-3. (a) Concentration of Zr in Quaternary volcanic rocks of Mount Jefferson plotted against a modified Larsen differentiation index (D.I. = $\frac{1}{3}Si - Mg - Ca + K$). Zr values determined by Sutton (1974). Stages of activity are the same as indicated in Figure 16-2. (b) Concentration of Zr in Miocene volcanic rocks of the central Oregon Cascade Range plotted against the same index as in Figure 16-3a. (c) $\Sigma FeO/(\Sigma FeO + MgO)$ of Quaternary volcanic rocks of the Mount Jefferson area plotted against the same index as in Figures 16-3a and 16-3b.

21 km^3; second stage of main cone, 4 km^3; and Forked Butte lavas, 3.5 km^3. If the entire Quaternary sequence for the central Oregon section of Figure 16-1 is considered, the three age groups show a similar decline of volumes with time: glaciated shield lavas, 1,282 km^3; main Cascade cones, 189 km^3; and Holocene cinder cones and lava, 55 km^3.

Each of the Quaternary age groups (stages) of Mount Jefferson is made up of rocks that appear to have been products of separate batches of magma with distinctive major- and trace-element compositions. There is, however, a general progression toward more-differentiated rocks through most of the evolution of the complex as a whole (Fig. 16-2). As the volume of eruptive rocks declined, the degree of differentiation increased, and although each group contains rocks that may not have appeared in any regular order of SiO$_2$ content, there is no overlap of the SiO$_2$ contents between the prinicipal groups.

In contrast to the behavior of SiO$_2$, factors such as the FeO/(ΣFeO + MgO) ratio and Zr concentration do not increase linearly in successive suites (Fig. 16-3). The earliest lavas were slightly tholeiitic, whereas later rocks are strongly calc-alkalic. The initial Zr content of each sequence of lavas is lower than that of the previous group. The "resetting" of trace-element concentrations without a corresponding return to low SiO$_2$ suggests that a magma reservoir was replenished with magma that was strongly depleted in Zr but produced lavas that were differentiated to a pre-established level, which may have been controlled by some structural or thermal condition beneath the volcano. The Zr content of the Quaternary rocks as a whole does not differ significantly from that of the Miocene series in the same region (Fig. 16-3).

The very latest lavas are a conspicuous exception to the trend of increasing differentiation with time; they revert to more primitive compositions similar to those of the Pleistocene shield lavas. This feature of late-stage rocks is characteristic of most mature volcanoes of the Cascade Range of Oregon and California. Sequences of this type have been called "divergent" as distinguished from "coherent" suites in which the order of differentiation is continuous and unidirectional with time (McBirney, 1968). Divergent suites among the Quaternary rocks of the Oregon Cascades are characterized by late-stage eruptions of essentially contemporaneous basalt and dacite or rhyolite, mainly from satellite vents on the flanks of mature andesitic cones.

Average compositions of basalt, basaltic andesite, andesite, dacite, and rhyolite for the central Oregon Cascades are given in Table 16-1, and a volumetrically weighted average of these rocks is given in Table 16-2. Data on the section in northern California, though less complete, serve as a useful comparison to the central Oregon Cascades. The two areas differ markedly in crustal structure and prior tectonic history. The Quaternary volcanoes of northern California stand on thick continental crust, whereas those of central Oregon have been built on relatively thin crust consisting mainly of mafic lavas and volcanic sediments. The relative proportions of rocks in these two regions are shown in Figure 16-4, together with those of central Washington where the thickness of continental crust is also great, but the total volume of Cenozoic volcanic rocks is small. The proportion of andesite is greatest in the Washington Cascades, less in northern California, and least in central Oregon, and the relative abundance of andesite in the three areas varies inversely with the total volume of volcanic rocks. Where the volume is large, basalt is by far the dominant rock type; the absolute volumes of andesite in the three areas are essentially equal.

Too few analyses are available to estimate the overall compositions of rocks in central Washington, but average compositions weighted for the relative abundances of each rock type in northern California can be compared with the corresponding average for central Oregon (Table 16-2). When this is done, the resulting averages are more similar than the differences in relative volumes would indicate. The reason for this apparent inconsistency lies in the tendency for basalts of northern California to have lower SiO_2 and alkali contents and higher MgO and iron contents than those of central Oregon. These differences balance the effect of differing volumetric proportions.

Andesites in the northern and southern High Cascade volcanoes are slightly more potassic than those of the central part of the range. The average K_2O contents (normalized to 60% SiO_2) of andesites from volcanoes in the three regions are Rainier (10 analyses), 1.66% K_2O; Hood and Jefferson (41 analyses), 1.43% K_2O; and Lassen and Shasta (10 analyses), 1.59% K_2O. K_2O content appears to vary directly with the proportion of andesite in these three parts of the chain. There is an even greater variation in Rb contents, which, as we shall

TABLE 16-1. AVERAGE COMPOSITION OF QUATERNARY ROCKS, CENTRAL OREGON CASCADES

	Basalt 43 analyses (<53.5% SiO_2)		Basaltic andesite 57 analyses (53.5%–57% SiO_2)		Andesite 56 analyses (57%–63% SiO_2)		Dacite 16 analyses (63%–68% SiO_2)		Rhyolite 15 analyses (>68% SiO_2)	
	(wt %)	(S.D.)	(wt %)	(S.D.)	(wt %)	(S.D.)	(wt %)	(S.D.)	(wt %)	(S.D.)
SiO_2	51.1	1.84	55.4	0.91	60.0	1.72	64.9	1.78	71.6	2.47
TiO_2	1.4	0.33	1.0	0.17	0.9	0.16	0.7	0.22	0.3	0.13
Al_2O_3	17.3	0.99	17.9	0.72	17.4	0.64	16.2	0.42	13.9	0.65
ΣFeO	9.4	2.40	7.6	0.79	6.4	0.79	4.7	0.83	2.4	0.57
MnO	0.2	0.02	0.1	0.1	0.1	0.03	0.1	0.03	0.1	0.02
MgO	6.1	2.03	4.6	0.92	2.8	0.73	1.7	0.56	0.5	0.34
CaO	8.9	0.78	7.5	0.78	6.1	0.75	4.4	0.99	1.7	0.36
Na_2O	3.6	0.56	3.9	0.29	4.3	0.34	4.5	0.56	4.5	0.43
K_2O	0.8	0.24	0.9	0.22	1.2	0.33	1.6	0.33	3.0	0.35
P_2O_5	0.3	0.12	0.2	0.07	0.2	0.06	0.2	0.08	0.1	0.05

TABLE 16-2. WEIGHTED AVERAGE COMPOSITIONS OF CASCADE VOLCANIC ROCKS

	Quaternary rocks Northern California	Quaternary rocks	Pliocene rocks	Middle and upper Miocene rocks	Oligocene-Miocene rocks
		Central Oregon			
	Average composition (in wt %)				
SiO_2	52.2	52.7	52.7	57.3	62.4
TiO_2	1.2	1.4	1.3	1.1	0.9
Al_2O_3	18.0	17.4	17.3	16.6	15.7
ΣFeO	9.6	9.1	8.6	7.4	6.1
MnO	0.1	0.1	0.1	0.1	0.1
MgO	5.5	5.3	5.6	3.7	2.3
CaO	9.1	8.4	8.7	6.7	5.3
Na_2O	3.2	3.8	3.5	3.5	3.4
K_2O	0.9	0.9	0.9	1.3	1.7
P_2O_5	0.2	0.3	0.2	0.3	0.2
Total	100.0	99.4	98.9	98.0	98.1
	No. of analyses and percentage of rock type by volume (in parentheses)				
Basalt	7 (69)	33 (85)	17 (90)	20 (39)	6 (10)
Andesite	22 (29)	112 (13)	76 (9)	99 (41)	25 (45)
Dacite-rhyolite	35 (2)	31 (2)	6 (1)	17 (20)	10 (45)
	Total volume of rocks (in km^3)				
	3,095	4,600	2,150	24,850	10,000

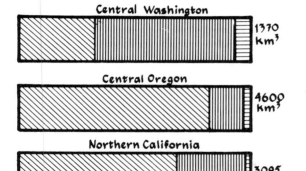

Central Washington

1370 km³

Central Oregon

4600 km³

Northern California

3095 km³

Figure 16-4. Proportions of Quaternary volcanic rocks in central Washington, central Oregon, and northern California. North-south dimensions of the three areas are approximately equal. Diagonal pattern indicates basalt; vertical, andesite; horizontal, dacite and rhyolite.

Figure 16-5. Distribution of analyzed samples of Pliocene basalts. K_2O and K/Rb values are shown as a function of longitude and distance behind the main Cascade axis (shown by asterisks). Some points on the location map may have more than one analyzed sample, and some samples have not been analyzed for Rb.

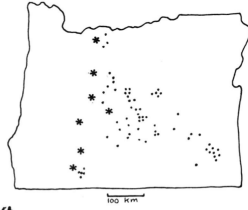

100 km

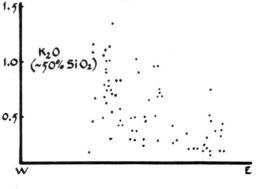

1.5

K_2O
(~50% SiO_2)

1.0

0.5

W   E

TABLE 16-3. AVERAGE COMPOSITION OF 33 PLIOCENE BASALTS FROM CENTRAL OREGON

	Wt %	S.D.
SiO_2	48.5	1.92
TiO_2	1.4	0.46
Al_2O_3	17.1	1.03
Fe_2O_3	2.9	1.98
FeO	7.4	
MnO	0.2	0.03
MgO	8.0	1.36
CaO	9.9	1.12
Na_2O	3.0	0.46
K_2O	0.5	0.33
P_2O_5	0.3	0.20

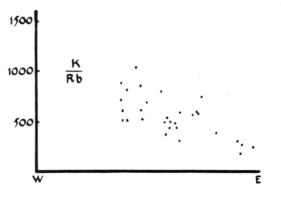

1500

1000

$\dfrac{K}{Rb}$

500

W   E

note, appear to be related to the volume of Tertiary volcanic rocks erupted in the same parts of the chain.

Pliocene Rocks

Pliocene rocks have been considered in two groups, one near the Cascade axis and another extending eastward across central Oregon.

The Pliocene rocks of central Oregon consist mainly of a distinctive group of thin but widespread basaltic lava flows and much smaller volumes of rhyolitic pumice and obsidian. Most basaltic eruptions occurred during a period between 3 and 6 m.y. ago and were scattered in seemingly random fashion over a broad region between the Cascade axis and eastern Oregon. Rhyolites are restricted to two linear belts in which eruptions appear to have migrated westward over a period of at least 10 m.y. (MacLeod and others, 1977).

Because the basalts are restricted to a narrow time range but have a wide areal distribution (Fig. 16-5), they are well suited for comparing compositional features at differing distances from the Cascade axis. Although the range of their SiO_2 contents is rather narrow and their average composition is that of high-alumina basalt (Table 16-3), they cover a spectrum between nepheline-normative and quartz-normative compositions. The compositional range is greatest in the west and becomes narrower toward the east. The K contents, for example, have the highest average values and greatest spread in the west and diminish eastward (Fig. 16-5).

Too few Pliocene rocks have been analyzed within the Cascade province to say whether they have regional variations along the axis. Our data are most complete for the rocks of central Oregon and somewhat less so for northern California. The total volumes and proportions of Pliocene rocks are similar to those of the Quaternary group in both areas, but the proportions of differentiated rocks are slightly lower in the Pliocene group. The weighted average composition for central Oregon (Table 16-2) reflects this difference in its lower Na_2O and higher MgO and CaO content.

Most of the Pliocene rocks of the Cascade region were erupted from small central-vent volcanoes scattered over a zone about twice as wide as that of the Quaternary High Cascade chain. We have found no large composite volcanoes in this age group in the area we have studied.

Middle and Upper Miocene Rocks

Two series of rocks have been combined in this age group, one a minor sequence (the Elk Lake formation) produced during a short episode between 9 and 10 m.y. ago and a much more voluminous series (the Sardine Formation) erupted between about 13.5 and 16 m.y. ago. The present status of mapping and dating is inadequate to separate these rocks in most areas. Our data on volumes and proportions of rock types are confined to the section in central Oregon.

The combined volumes of volcanic rocks in this age group exceed that of all younger calc-alkalic rocks combined. The locus of this episode of intense volcanism was concentrated in a chain of large composite volcanoes similar to but probably larger than those of the Quaternary High Cascade Range and displaced a few tens of kilometres to the west. The axis of Miocene centers trended slightly east of north at an angle of about 15° to the younger Cascade chain, which the Miocene axis intersects near the Oregon-Washington border. Despite its great thickness on the west side of the High Cascades, the Sardine Formation has not been found on the east side of the range in central Oregon, where minor amounts of coarse sedimentary debris are the only materials derived from the huge western Cascades volcanic pile. For this reason the volume of mid-Miocene rocks beneath the High Cascades is uncertain, and the calculations are subject to a large possible error.

Most mid-Miocene volcanic centers coincide crudely with shallow stocks and associated aureoles of low-grade metamorphism. The intrusions are only slightly younger than the lavas they intrude. Their compositions are similar to those of their volcanic counterparts, but too few rocks have been analyzed to date to permit an accurate comparison.

The proportions of rock types in the middle and upper Miocene groups differ from those of the younger episodes (Table 16-2). Andesite makes up a larger fraction and is approximately equal in volume to basalt. Much of the andesite was erupted as lahars and other types of clastic rocks. Tephra and shallow-marine volcanic sediments make up a major fraction of the Miocene section, especially around the lower flanks of the large cones.

Upper Oligocene and Lower Miocene Rocks

An extensive sequence of siliceous pyroclastic rocks and subordinate andesitic and basaltic lavas underlies the Sardine Formation in most of the central and southern parts of the western Cascades Range (Peck and others, 1964). The rocks of this group belong to the Little Butte Formation and include minor members, such as the Breitenbush Tuff of Thayer (1937). The boundary between the Little Butte and Sardine Formations has previously been taken to coincide with a thin but widespread basaltic unit that is part of the Columbia River Group, but radiometric dating has shown that the major time break is somewhat lower in the section (Sutter, 1978), and the proportion of Miocene rocks we have included with the Sardine Formation is greater than that of earlier workers, whereas the proportion in the Little Butte is correspondingly less.

The main feature of the Little Butte Formation is its high proportion of differentiated rocks. Tephra beds, including air-fall tuffs, ignimbrites, and water-laid pyroclastic deposits of dacitic and rhyolitic compositions, are by far the most abundant rocks. Elsewhere in the western Cascades, however, basaltic and andesitic lavas are common, especially in the lower part of the formation. It is difficult to determine the eruptive centers that were active during this period. The few that have been identified tend to follow close to the line of mid-Miocene centers, and there seems to have been no major shift in the axis of volcanism between the early and middle Miocene episodes.

Geochemical Variations

In addition to these changes in the absolute and relative volumes of rock types, the various age groups show systematic spatial and temporal variations of their major- and trace-element compositions, most notably in Fe, Na, and Rb. Comparisons of these elements are given in Table 16-4 and Figure 16-7. The differences are best seen in rocks of basaltic composition in which the effects of differentiation have not been superimposed on the more primitive character of the mafic magmas.

Fe versus SiO_2. Oligocene and lower Miocene basalts in the Little Butte Formation near the base of the exposed section in central Oregon differ markedly from younger basalts of the Cascades in that they are distinctly more Fe-rich and tholeiitic. Because their compositions resemble those of quartz-normative tholeiites of the Columbia River Group, they have been correlated with that group of basalts and have been said to have originated outside the main Cascades system. Work by Nathan and Fruchter (1974) has shown that the rocks differ from

Columbia River basalts in their trace-element compositions, and our own radiometric dating indicates that the Cascade tholeiites predate the Columbia River basalts by as much as 10 m.y. A few Columbia River basalts have been identified within the Sardine Formation well above the Little Butte tholeiites, but the relationship, if any, between the Cascade basalts and those of the Columbia River Group remains unclear.

The Fe enrichment of Little Butte rocks (Fig. 16-6) is greater than that of any younger lavas in the Cascade series. The general trend with time is from a distinctly tholeiitic suite to one that is strongly calc-alkalic. Baker (1968), Jakeš and Gill (1970), and others have recognized the tholeiitic nature of early volcanic series in island arcs, where tholeiites are common near the base of the volcanic sequence and tend to give way to more calc-alkalic rocks as the crust thickens with time (Miyashiro, 1974). It may be significant that we have only found strongly tholeiitic rocks in the central part of the Cascade province where there is no evidence for thick pre-Tertiary continental crust. The abrupt change from tholeiitic to calc-alkalic rocks in the same region after the Oligocene–early Miocene episode is one of the most conspicuous features of the temporal evolution of the Cenozoic Cascade system.

In other respects, however, the change in the compositions of rocks with time differs from the pattern that Miyashiro (1974) considered typical of the evolution of island arcs and continental margins. Miyashiro indicated that the change to more calc-alkalic compositions is accompanied by an increase in average SiO_2 content, but our data show the reverse; SiO_2-rich rocks are most abundantly formed in the older episodes, and the average SiO_2 content declines sharply thereafter (Table 16-2). In these respects, it appears that the trend of SiO_2 and Fe contents differs from what is thought to be the rule elsewhere, but we suspect that the difference is due to the fact that plots of analytical data in variation diagrams that ignore relative volumes can be misleading, at least in terms of the overall character of igneous rock series.

Osborn and Watson (1977) have pointed out that a change from tholeiitic to calc-alkalic compositions can result from an expansion of the spinel stability field with an increase of either f_{O_2} (O_2 fugacity) or dry load pressure. Eggler and Burnham (1973) have found that magnetite is not a liquidus phase in a melted Mount Hood andesite at elevated water pressures and any geologically reasonable f_{O_2}, but experimental studies of basalt and basaltic andesite from the Oregon Cascades by Osborn and Watson (1977) indicate that dry pressures of 10 kb and modest values of f_{O_2} result in an Fe-rich spinel phase at near-liquidus temperatures.

TABLE 16-4. MEAN Rb, Sr, Na_2O, AND K_2O VALUES FOR ROCKS OF CENTRAL OREGON CASCADES

Age group	Rock type	Rb (ppm)	Sr (ppm)	Na_2O (wt %)	K_2O (wt %)	Rb/Sr	K/Rb
	Basalt (14)	10	761	3.6	0.8	0.013	808
Quaternary	Andesite (14)	13	621	4.1	1.0	0.022	765
	Rhyolite (3)	23	558	4.5	1.8	0.037	775
	Basalt (3)	12	538	3.5	1.0	0.020	807
Pliocene	Andesite (1)	18	539	3.9	1.1	0.033	594
	Rhyolite (1)	25	349	4.7	1.9	0.072	756
Late and	Basalt (5)	17	611	3.1	1.0	0.028	602
Middle	Andesite (7)	31	397	3.7	1.5	0.081	491
Miocene	Rhyolite (6)	64	225	3.9	3.1	0.373	480
Early	Basalt (6)	25	365	2.9	0.9	0.068	369
Miocene	Andesite (3)	52	400	3.5	2.1	0.133	396
	Rhyolite (5)	66	264	3.4	2.3	0.287	353

Note: Numbers in parentheses after rock names are numbers of analyses used in calculating averages for Rb and Sr. Rock-type divisions have been placed at 53.5% and 63.0% SiO_2.

Their results, though preliminary, indicate that basic tholeiitic liquids would result from equilibration of magmas under thin lithosphere and that the liquids would change to a more calc-alkalic trend with an increasing depth of equilibration under a thickening lithosphere.

Subsequent differentiation at higher levels in the crust should follow trends that are essentially similar, regardless of the degree of Fe enrichment of the mafic parental magma. The trends of andesites, dacites, and rhyolites in both tholeiitic and calc-alkalic series should converge on end members that are very similar, as they do in the Cascade rocks, in which there is little difference in the rhyolitic differentiates of the tholeiitic and calc-alkalic series.

Na and K. If only rocks at the same stage of differentiation are considered, we observe a slight decrease in the K_2O contents of rocks from the oldest to the youngest series (Table 16-4), but taken as a whole, the weighted average K_2O content of the rocks declines markedly with time (Table 16-2). Miyashiro (1974) indicated that K_2O should increase with time as the rocks become more calc-alkalic, but we find no evidence that this is so. Miyashiro (1974) and Sugisaki (1976) thought that alkalic rocks are abundant where rates of plate convergence are low, but the subduction rate off the coast of the Pacific Northwest is one of the lowest known, and we have found no alkalic rocks in the main Cascade system. And finally, it is often said that the average K_2O contents at a constant SiO_2 value increase toward the continental side of a volcanic front, but our data on Pliocene rocks behind the Cascade axis show the opposite to be true (Fig. 16-5).

Na, unlike K, shows an almost perfect linear relationship to the mean age of the units we have sampled (Fig. 16-7). The increase of sodium with time is less pronounced in more differentiated rocks that have probably evolved through high-level fractionation of plagioclase. The tendency for andesites of various ages to be more uniform than basalts may indicate that they have equilibrated at shallow depths and in so doing, have lost the contrasting Na contents that characterized their parental basaltic liquids. The temporal increase of Na in basalts is consistent with two possible relations, one resulting from increasing depths of melting and the other from increasing depths of differentiation.

The increase is the opposite of what one would expect from progressive melting of a solid-solution series in which the sodic end member is enriched in early liquids. It is, however,

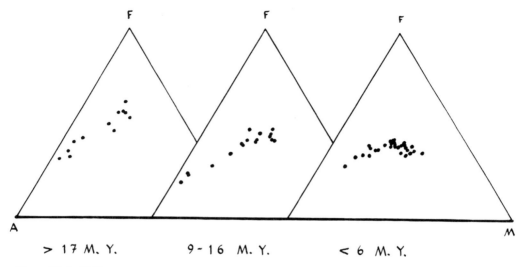

Figure 16-6. AMF [(Na_2O + K_2O):MgO:ΣFeO] relations for rocks in three major age groups in the central Oregon Cascade Range.

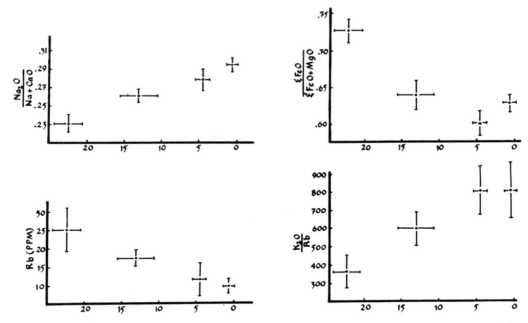

Figure 16-7. Variations of Fe, Na, Rb, and K/Rb with time (millions of years) in basaltic rocks of the central Oregon Cascade Range. Error bars show 1 standard deviation in chemical data and range in radiometric dates.

consistent with either of two melting relations at pressures where plagioclase is not a stable mineral. Kushiro (1973) observed that the Na content of partial melts of garnet lherzolite increased with pressure, while Mysen (1974) found that increasing f_{O_2} caused a similar increase in partial melts of a spinel lherzolite. The relations we have already noted in the Fe contents of the same rocks are consistent with both of these possibilities, but it seems less likely that the variations in basaltic liquids of the Cascade system are the result of increasing f_{O_2}; the geologic relations are consistent with thickening of the lithosphere with time but suggest no apparent mechanism by which f_{O_2} might increase concurrently.

The increase of Na with time could also be caused by an increase in the depth of differentiation. At shallow depths, plagioclase should have a large field of stability, and crystallization of this mineral in place of another aluminous phase such as spinel or garnet would inhibit the enrichment of Na. Such a relation is consistent with an increasing depth of segregation of liquids at the base of the lithosphere and could be independent of the original depth of melting.

Rb and Sr. With the exception of the Pliocene units for which we have little data, Sr contents of basalts have increased regularly with time in a manner similar to Na, but Rb varies inversely with Na and Sr and declines with time (Table 16-4, Fig. 16-8). The fact that Sr, which should be fractionated into plagioclase, behaves in a manner similar to Na lends support to the second of the two interpretations considered in the preceding section. As the thickness of the lithosphere increased with time, the pressure at which basalt last equilibrated in the mantle became higher, and the stability field of plagioclase must have been reduced.

The trend of Rb is emphasized in differentiated rocks (Fig. 16-8), owing, no doubt, to the incompatible nature of the large Rb ion. Because basalts of all ages have nearly constant K contents, their K/Rb ratios increase with time.

Rb contents of lavas from seven Quaternary Cascade volcanoes are shown in Figure 16-9. Despite a certain amount of scatter, at least part of which results from the varied sources of the analytical data, it is apparent that the Rb contents of lavas from the central part of the Cascade Range are lower than those of volcanoes at the northern and southern ends. A comparison of the Rb contents at the same stage of differentiation is given in Figure 16-10, together with measured thicknesses of the Tertiary volcanic sections in corresponding segments of the chain. There appears to be a crude inverse correlation between the amount of Tertiary volcanism in a given region and the Rb contents of Quaternary lavas in the same part of the chain. Where there was a large amount of Tertiary volcanism, Quaternary lavas are poor in Rb, and where there was little Tertiary volcanism, Rb contents are high.

Systematic variations in Rb contents have been cited from a variety of tectonic settings (Kesson, 1973; Baxter, 1976), and several explanations for these variations have been proposed: (1) continued melting of a given source rock would result in large amounts of Rb entering the first liquid. If melting continues without separation of this liquid, the Rb content should decline as it is diluted in a larger volume, but if the first liquid is removed and melting continues, a sharp decrease will be seen in successive melts from the same source. (2) If Rb is concentrated in phlogopite and the mineral persists through a wide range of melting, there could be a steady decline of both Rb and K at a rate that depends on the distribution coefficient between phlogopite and basaltic liquids rather than on the frequency of separation of liquids. (3) Rb may be introduced in varying amounts by subduction of sedimentary material

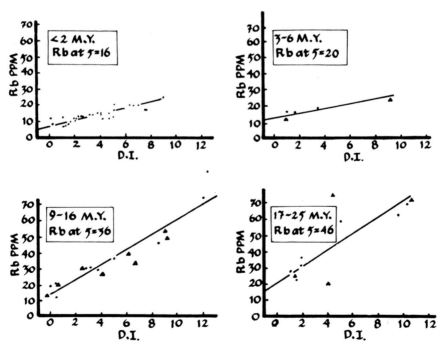

Figure 16-8. Concentrations of Rb in volcanic rocks of four age groups in the central Oregon Cascades plotted against a modified Larsen differention index (D.I. = $\frac{1}{3}$Si − Mg − Ca + K). Solid triangles represent chemically analyzed splits of radiometrically dated samples; dots represent rocks for which the age is inferred from stratigraphic relations. The Rb content at an intercept of the least-squares line at a D.I. of 5 (andesite) is given for each group.

and by transfer of Na and K along with H_2O and other mobile components into a melt generated in or above the zone of subduction. (4) Rising magma may scavenge Rb from the rocks through which it passes and gradually deplete them as successive magmas follow similar paths toward the surface. And (5), the mantle may be compositionally zoned with decreasing amounts of Rb with depth; melting may migrate downward through these zones with time. Below, we examine the Rb relations we observe in the light of these various possibilities.

1. Gast (1968) pointed out that differing degrees of partial melting of the same source

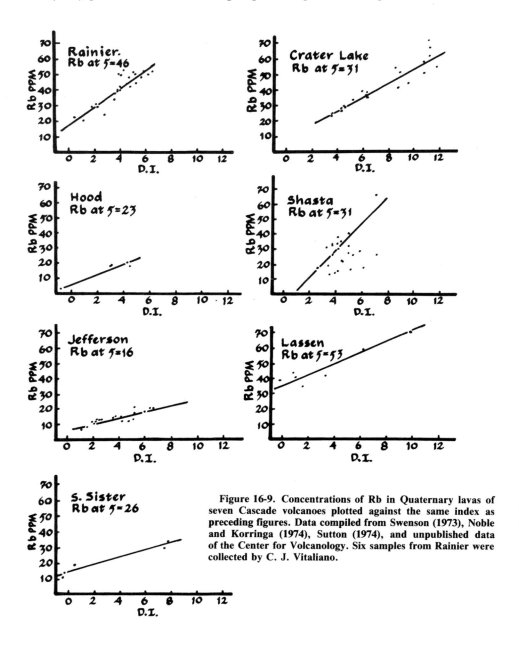

Figure 16-9. Concentrations of Rb in Quaternary lavas of seven Cascade volcanoes plotted against the same index as preceding figures. Data compiled from Swenson (1973), Noble and Korringa (1974), Sutton (1974), and unpublished data of the Center for Volcanology. Six samples from Rainier were collected by C. J. Vitaliano.

should result in either a steady or step-wise decline of the abundances of strongly excluded elements, depending on the frequency of extraction of successive melts. Our estimates of volumes indicate that post-Miocene rocks comprise only 15% of the total volume of the central Oregon Cascades and could represent only a small volume compared to earlier episodes. Such a relation argues against dilution as an explanation of the low Rb contents of young basalts, because the rate of decline is far from proportional to the volumes of erupted magma. Similarly, we can rule out different degrees of melting of different source rocks, because the volumetric relations are the reverse of what such a model predicts. These conclusions depend, of course, on the assumption that the volumes of eruptive rocks are roughly proportional to the amount of magma generated at depth.

2. The fact that the decline in Rb is steady and crudely proportional to the amount of volcanism may be consistent with progressive melting of a source in which phlogopite persists through a wide range of melting, but the amount of phlogopite necessary to satisfy this condition seems too large for the observed heat-flow from the mantle (Yoder and Kushiro, 1969). Beswick (1976) predicted that progressive melting of phlogopite should cause K/Rb ratios of basalts to decline with time; the fact that we find the opposite trend also argues against this mineral remaining through long melting intervals.

3. Fyfe and McBirney (1975) have considered the problem of the stability of hydrous minerals during subduction and their possible roles in magma generation. They concluded that, because phlogopite is the only hydrous mineral that is likely to persist to depths where melting can take place below a volcanic front, there should be a relation between the amount of H_2O, K, and other components released when this mineral breaks down and the volumes and compositions of magma produced by flux melting. If Rb was supplied from a descending slab of lithosphere, its abundance and ratio to K should be a function of the rate of subduction and composition of sedimentary material and independent of the earlier history of the region. Hence, the model is not supported by our data. It cannot be ruled out, however, because other factors may intervene between the locus of melting and final equilibration of the magma and may influence the compositional relations.

4. Best (1975) has proposed that alkalies may be scavenged from the lithosphere as magma rises from deeper sources. Presumably other components would be added to the magma at the same time. The decline of Rb with time and the fact that it is more depleted where

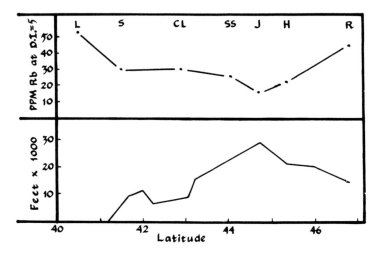

Figure 16-10. Relations of Rb contents of Quaternary volcanoes in Figure 16-9 to measured stratigraphic thicknesses of Tertiary volcanic rocks in the same area. Volcanoes are shown in order from south to north: L, Lassen: S, Shasta; CL, Crater Lake; SS, South Sister; J, Jefferson; H, Hood; R, Rainier (stratigraphic thicknesses from Fiske and others, 1963; Kays, 1970; Maynard, 1974; Wells, 1956; Wise, 1969, 1970).

TABLE 16-5. Sr-ISOTOPE DATA FOR ROCKS OF CENTRAL OREGON

No. in Figure 11	Field no.*	Rock type	K/Ar age (m.y.)	Rb (ppm)	Sr (ppm)	$^{87}Sr/^{86}Sr$
1	BX-99	Basalt	24.2	21	325	0.7029§
2	DMS-77	Granodiorite	15.9	97	161	0.7029
3	DMS-24	Basalt	15.4	18	516	0.7027
4	SJ-201	Basalt	<0.7†	7	658	0.7026#
5	SJ-70	Andesite	<0.7†	19	522	0.7031
6	DMS-23	Basalt	15.8	32	340	0.7054

Note: Rb and Sr concentration determined by X-ray fluorescence analysis. Samples 1 and 6 were analyzed in duplicate. All ratios have been corrected for isotope fractionation to a standard value of $^{86}Sr/^{88}Sr = 0.1194$. The average $^{87}Sr/^{86}Sr$ ratio for seven analyses of the Eimer and Amend Sr-isotope standard was 0.70807 ± 0.0002 (1σ). The initial Sr-isotope ratios of the Tertiary samples were calculated by correcting the measured values for the decay of ^{87}Rb since the rocks crystallized.

* Sample localities and analytical data are on file in the computer data bank of the Center for Volcanology, University of Oregon.
† Normally polarized flows from the main cone of Mount Jefferson, not dated by K/Ar method.
§ Split analyzed by R. L. Armstrong at the University of British Columbia yielded a value of 0.7035.
# Split analyzed by R. L. Armstrong at the University of British Columbia yielded a value of 0.7030.

there has been the greatest amount of previous volcanism is consistent with this interpretation. K shows a similar but less-pronounced relation, but we find no systematic decline in Zr (Fig. 16-3) or in other lithophile elements that should be affected more strongly than Rb.

In an effort to clarify this question, we have measured the Sr-isotope compositions of samples of various ages and compositions (Table 16-5 and Fig. 16-11). The values we obtained are remarkably uniform (0.70283 ± 0.00014 at one standard deviation) and unusually low. In fact, they are among the lowest yet reported from the Cascade Range and are considerably lower than that of a Columbia River basalt in the same section (no. 6, Table 16-5). The older lavas do not have higher ratios than younger ones, and rocks that are more differentiated and richer in Na and K are not more radiogenic than basalts. If Rb were scavenged from the crust, the process would have to have operated in such a way that the radiogenic Sr formed by decay of Rb was not assimilated, at least in measurable amounts. Scavenging from deeper levels in the lithosphere is not ruled out by the Sr-isotope data.

5. The uniformity of Sr-isotope ratios also argues strongly against derivation from a compositionally inhomogeneous mantle. If source regions differed in their original Rb contents, their Sr-isotope ratios must also have differed, and we have found no correlation between these two factors.

Rare-Earth Elements. All rare-earth elements (REE) for which we have data show a progressive decline in abundance from older to younger units (Fig. 16-12). This trend is especially apparent in the light REE and is analogous to the general trends in Rb discussed above. The similarity in the configurations of the chondrite-normalized REE plots suggests that the lavas of each age group were derived from mineralogically similar source regions.

SUMMARY AND CONCLUSIONS

The principal observations that can be made from these data are as follows: (1) The volumes of volcanic rocks erupted since late Oligocene time have declined with time in an irregular fashion. (2) At the same time, the proportions of basaltic rocks have increased. The greatest

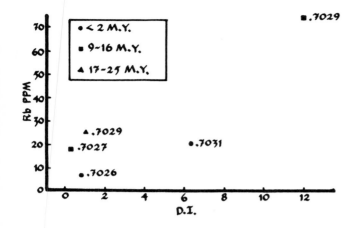

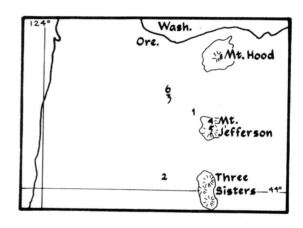

Figure 16-11. Distribution of Sr-isotope ratios listed in Table 16-5 according to their Rb content, age, and
differentiation index (above). Locations of samples are shown in the sketch
map to right.

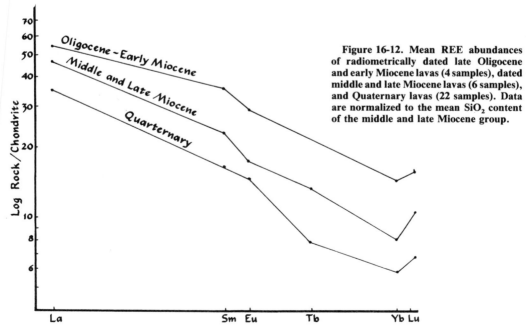

Figure 16-12. Mean REE abundances
of radiometrically dated late Oligocene
and early Miocene lavas (4 samples), dated
middle and late Miocene lavas (6 samples),
and Quaternary lavas (22 samples). Data
are normalized to the mean SiO$_2$ content
of the middle and late Miocene group.

increase followed the large Miocene event, and since that time, basalt has outweighed all other rocks in the central part of the Cascade Range. Andesite is relatively more important in the northern and southern parts of the chain where there was less Tertiary volcanism and the thickness of continental crustal rocks is greater. (3) When basaltic rocks of each episode are compared, they show a decrease in Fe, Rb, REE, and to a minor extent, K, with time. Simultaneously, Na, Sr, and K/Rb increase, and Sr-isotope ratios remain essentially constant. (4) The abundance of Rb in Quaternary rocks is highest at the extremities of the chain. Its abundance varies inversely with the volume of Tertiary volcanism in the same segment of the chain. (5) The variations of K in basalts erupted at differing distances behind the volcanic front differ from the conventionally accepted pattern. Both the range and mean abundances of K are highest near the main axis and become lower toward the east. (6) Quaternary volcanism responsible for construction of Mount Jefferson was fed by distinct batches of magma, which became more calc-alkalic and poorer in Zr with time, even though SiO_2 contents increased steadily throughout all but the last stage of activity.

No single mechanism of partial melting or differentiation is adequate to explain all these variations. The only process for which we find no supporting evidence is one that requires a steady input of components from a subducted slab; temporal variations we observe would require improbable changes in the rates of subduction or the compositions of subducted material on a short scale of Cenozoic time. A more complex multi-stage process seems to be required. The combination that best fits our present knowledge is a three-stage sequence that includes (1) progressive depletion of a source region in the mantle, (2) re-equilibration of each large batch of magma at a greater depth than the preceding one, possibly near the base of the lithosphere, and (3) subsequent differentiation at a high level in the crust.

The evidence for these three stages has been considered in the discussions of individual components and can be briefly summarized. Progressive melting of a single source in the mantle is consistent with the decline of volumes, Fe/Mg ratios, SiO_2, K, Rb, and REE with time, but these same variations are also consistent with successive batches of magma passing through the same rocks overlying the source and depleting them in components that would be fractionated into the liquid. We have no way of making a distinction between these two possibilities, but we can rule out significant contamination by old continental crust.

The change from a tholeiitic to a more calc-alkalic trend of differentiation with time can be explained as the result of equilibration of the rising liquid with its co-existing mineral phases and segregation of the liquid at a level that has become deeper with time, possibly as a result of accumulation of crustal rocks and underplating of the lithosphere with the dense residue of earlier batches of rising magma.

The increasing Na and Sr contents of basalts also lend support to this interpretation and indicate that plagioclase was fractionated from the earlier magmas but with time became less important than an Fe-rich spinel.

Most of the final differentiation from basaltic to more SiO_2-rich compositions seems to have taken place at shallow depths, possibly immediately below the volcanic superstructure. Crustal contamination does not appear to have played an important role in this process. At least three separate batches of magma rose beneath Mount Jefferson, but they were drawn from a source that was progressively depleted in Fe and Zr until the latest stage, which appears to have come from a small but separate pulse of magma similar to that of the earliest stage. Studies of subvolcanic stocks may clarify these processes.

We have not attempted to relate our observations to possible changes in the tectonic regime of the Pacific Northwest. Changes of the structure and stress distribution in the crust must affect the manner in which magmas rise through the lithosphere and could also influence the depth and degree of differentiation. If so, they may have altered the compositions and proportions of rock types we find at the surface. We consider the most important conclusion to be drawn from the data we have presented to be that volcanic rocks can only be seen in their true perspective when they are considered in relation to the compositions, volumes, and structural settings of the volcanic series as a whole.

ACKNOWLEDGMENTS

This research has been supported by National Science Foundation Grant No. GA-35129 to A. R. McBirney. Field work was supported in part by Geological Society of America Research Grants Nos. 1650–72, 1769–73, 1844–74 to C. M. White. We are grateful to Gunter Faure for allowing us the use of laboratory facilities to determine the Sr-isotope ratios cited in this report.

REFERENCES CITED

Baker, P. E., 1968, Comparative volcanology and petrology of the Atlantic island-arcs: Bull. Volcanol., v. 32, p. 189–206.

Baxter, A. N., 1976, Geochemistry and petrogenesis of primitive alkali basalt from Mauritius, Indian Ocean: Geol. Soc. America Bull., v. 87, p. 1028–1034.

Best, M. G., 1975, Migration of hydrous fluids in the upper mantle and potassium variation in calc-alkaline rocks: Geology, v. 3, p. 429–432.

Beswick, A. E., 1976, K and Rb relation in basalts and other mantle derived materials: Is phlogopite the key?: Geochim. et Cosmochim. Acta, v. 40, p. 1167–1183.

Condie, K. C., and Swenson, D. H., 1974, Compositional variation in three Cascade stratovolcanoes: Jefferson, Rainier, and Shasta: Bull. Volcanol., v. 37, p. 205.

Eggler, D. H., and Burnham, C. W., 1973, Crystallization and fractionation trends in the system andesite-H_2O-CO_2-O_2 at pressures to 10 kb: Geol. Soc. America Bull., v. 84, p. 2517–2532.

Fiske, R. S., Hopson, C., and Waters, A., 1963, Geology of Mount Rainier National Park, Washington: U.S. Geol. Survey Prof. Paper 444.

Fyfe, W. S., and McBirney, A. R., 1975, Subduction and the structure of andesitic volcanic belts: Am. Jour. Sci., v. 275-A, p. 285–297.

Gast, P. W., 1968, Trace element fractionation and the origin of tholeiitic and alkaline magma types: Geochim. et Cosmochim. Acta, v. 32, p. 1057–1086.

Jakeš, P., and Gill, J., 1970, Rare earth elements and the island arc tholeiitic series: Earth and Planetary Sci. Letters., v. 9, p. 17–28.

Kays, M. A., 1970, Western Cascades volcanic series, South Umpqua Falls region, Oregon: Ore Bin, v. 32, p. 81–96.

Kennett, J. P., McBirney, A. R., and Thunell, R. C., 1977, Episodes of Cenozoic volcanism in the Circum-Pacific region: Jour. Volcanology and Geothermal Research, v. 2, p. 145–163.

Kesson, S. E., 1973, The primary geochemistry of the Monaro alkaline volcanics, southeastern Australia, evidence for upper mantle heterogeneity: Contr. Mineralogy and Petrology, v. 42, p. 93–108.

Kushiro, I., 1973, Partial melting of garnet lherzolites from kimberlite at high pressures, in Nixon, P. H., ed., Lesotho kimberlites: Maseru, Lesotho, Lesotho Mat. Development Corp., p. 294–299.

MacLeod, N. S., Walker, G. W., and McKee, E. H., 1977, Geothermal significance of eastward increase in age of upper Cenozoic rhyolitic domes in southeastern Oregon, in Proc. 2nd United Nations Symposium on the development and use of geothermal resources, Vol. 1: Washington, D.C., U.S. Govt. Printing Office, p. 465–474.

Maynard, L., 1974, Geology of Mt. McLaughlin [M.S. thesis]: Eugene, Univ. Oregon.

McBirney, A. R., 1968, Petrochemistry of the Cascade andesitic volcanoes: Oregon Dept. Geology and Mineral Industries Bull. 62, p. 101–107.

McBirney, A. R., Sutter, J. F., Naslund, H. R., Sutton, K. G., and White, C. M., 1974, Episodic volcanism in the central Oregon Cascade Range: Geology, v. 2, p. 585–589.

Miyashiro, A., 1974, Volcanic rock rock series in

island arcs and active continental margins: Am. Jour. Sci., v. 274, p. 321–355.

Mysen, B. O., 1974, The oxygen fugacity (fO_2) as a variable during partial melting of peridotite in the upper mantle: Carnegie Inst. Washington Yearbook, v. 73, p. 237–240.

Naslund, H. R., 1977, The geology of the Hyatt Reservoir and Surveyor Mountain quadrangles, Oregon [M.S. thesis]: Eugene, Univ. Oregon.

Nathan, S., and Fruchter, J., 1974, Geochemical and paleomagnetic stratigraphy of the Picture Gorge and Yakima basalts (Columbia River Group) in central Oregon: Geol. Soc. America Bull., v. 85, p. 63–76.

Noble, D. C., and Korringa, M. K., 1974, Strontium, rubidium, potassium, and calcium variations in Quaternary lavas, Crater Lake, Oregon, and their residual glasses: Geology, v. 2, p. 187–190.

Osborn, E. F., and Watson, E. B., 1977, Studies of phase relations in subalkaline volcanic rock series: Carnegie Inst. Washington Year Book, v. 76 (in press).

Peck, D. C., Griggs, A. B., Schlicker, H. G., Wells, F. G., and Dole, H. M., 1964, Geology of the central and northern parts of the western Cascade Range in Oregon: U.S. Geol. Survey Prof. Paper 449, 56 p.

Sugisaki, R., 1976, Chemical characteristics of volcanic rocks: Relation to plate movements: Lithos, v. 9, p. 17–30.

Sutter, J. F., 1978, K/Ar ages of Cenozoic volcanic rocks from the Oregon Cascades west of 121°31': Isochron/West (in press).

Sutton, K. G., 1974, Geology of Mt. Jefferson [M.S. thesis]: Eugene, Univ. Oregon.

Swenson, D. H., 1973, Geochemistry of three Cascade volcanoes [M.S. thesis]: Socorro, New Mexico Inst. Mining and Technology.

Thayer, T. P., 1937, Petrology of later Tertiary and Quaternary rocks of the north-central Cascade Mountains in Oregon: Geol. Soc. America Bull., v. 48, p. 1611–1652.

Wells, F. G., 1956, Geology of the Medford quadrangle, Oregon-California: U.S. Geol. Survey Geol. Quad. Map GQ-89.

Williams, H., 1957, A geologic map of the Bend quadrangle, Oregon, and a reconnaissance geologic map of the central potion of the High Cascade Mountains: Oregon Dept. Geology and Mineral Industries Geol. Map Ser.

Wise, W. S. 1969, Geology and petrology of the Mt. Hood area: A study of High Cascade volcanism: Geol. Soc. America Bull., v. 80, p. 969–1006.

——1970, Cenozoic volcanism in the Cascade Mts. of southern Washington: Washington Div. Mines and Geology Bull., v. 50, 45 p.

Yoder, H. S., Jr., and Kushiro, I., 1969, Melting of a hydrous phase: Phlogopite: Carnegie Inst. Washington Year Book, v. 67, p. 161–167.

MANUSCRIPT RECEIVED BY THE SOCIETY AUGUST 15, 1977

MANUSCRIPT ACCEPTED SEPTEMBER 2, 1977

Printed in U.S.A.

Accompanying maps — 9

QE 690 .C44 1978

Cenozoic tectonics and
  regional geophysics of the